Remote Sensing for Climate Change

Remote Sensing for Climate Change

Editor

Xander Wang

MDPI • Basel • Beijing • Wuhan • Barcelona • Belgrade • Manchester • Tokyo • Cluj • Tianjin

Editor
Xander Wang
University of Prince Edward Island
Canada

Editorial Office
MDPI
St. Alban-Anlage 66
4052 Basel, Switzerland

This is a reprint of articles from the Special Issue published online in the open access journal *Remote Sensing* (ISSN 2072-4292) (available at: https://www.mdpi.com/journal/remotesensing/special_issues/RS_for_Climate_Change).

For citation purposes, cite each article independently as indicated on the article page online and as indicated below:

LastName, A.A.; LastName, B.B.; LastName, C.C. Article Title. *Journal Name* **Year**, *Volume Number*, Page Range.

ISBN 978-3-0365-6628-3 (Hbk)
ISBN 978-3-0365-6629-0 (PDF)

Contents

About the Editor

Xander Wang

Dr. Xander Wang is an Associate Professor in the School of Climate Change and Adaptation at the University of Prince Edward Island (UPEI). He is also the Director of Climate Smart Lab in the Canadian Centre for Climate Change and Adaptation. Dr. Wang's research is mainly focused on regional climate modeling, climate downscaling, hydrological modeling and flooding risk analysis, energy systems modeling under climate change, climate change impact assessment and adaptation studies, GIS and remote sensing applications, and big data analysis and visualization.

remote sensing

MDPI

Editorial

Remote Sensing Applications to Climate Change

Xander Wang [1,2]

[1] Canadian Centre for Climate Change and Adaptation, University of Prince Edward Island, St. Peter's Bay, PE C0A 2A0, Canada; xxwang@upei.ca

[2] School of Climate Change and Adaptation, University of Prince Edward Island, Charlottetown, PE C1A 4P3, Canada

1. Introduction

Climate change research remains a challenging task, as it requires vast quantities of long-term data to investigate the past, present, and future scenarios of Earth's climate system and other biophysical systems at global to local scales. Both traditional ground-based observation methods and remote-sensing technologies are available options for gathering data for climate change research. Observations from weather stations have been widely used to study climate change over long periods of time. However, due to the scarcity of point-based weather observations, our understanding of the Earth's changing climate is very limited. This impedes the advancement of our knowledge of the Earth's climate system and our ability to develop well-suited climate models to simulate future climate change, which further results in considerable uncertainties associated with future climate projections. Thus, the determination of a method for quantifying and minimizing these uncertainties is quickly becoming one of the most challenging issues yet to be addressed by climate change impact assessment and adaptation studies. Remote sensing offers a new method for observing the Earth's climate system with continuous and high-resolution spatial coverage through satellite-based, aircraft-based, or drone-based sensor technologies. This can significantly improve our understanding of climate change and its potential impacts at global, regional, and local scales. The data collected with remote-sensing technologies can also be used to validate our climate models, improve our knowledge of the physical and dynamic processes of the climate system, and help us project future climate change and its impacts with minimized uncertainties.

This Special Issue intends to capture the latest research advances regarding remote-sensing technologies and their applications in climate change research. Sixteen original research articles authored by one hundred and five researchers were published in this Special Issue, presenting recent advances in remote-sensing technologies (two articles) and the application of remote-sensing technologies to climate change modelling (five articles), monitoring (six articles), and impact assessment (three articles). While the articles span multidisciplinary perspectives and methodologies, they are clustered into four themes.

Citation: Wang, X. Remote Sensing Applications to Climate Change. *Remote Sens.* **2023**, *15*, 747. https://doi.org/10.3390/rs15030747

Received: 16 January 2023
Accepted: 27 January 2023
Published: 28 January 2023

2. The Advances in Remote-Sensing Technologies

The continuous development of remote-sensing technologies, including advanced satellites and new procedures for large scale data processing, is essential for increasing the accuracy and reliability of climate change research based on remote-sensing data. Meftah et al. [1] present the technological development of Ultraviolet and Infrared Sensors at high Quantum efficiency onboard a small Satellite (UVSQSAT). They also present the findings of the first in-orbit observations of the mapping of solar radiation reflected by the Earth and of the outgoing longwave radiation at the top of the atmosphere throughout February 2021. Their study shows the feasibility of using a miniaturized satellite to measure the Earth's energy imbalance (EEI) over a short period.

While satellite technologies are improving, the performance of remote-sensing data analysis techniques remains of great concern. Basheer et al. [2] evaluate the land-use land-cover (LULC) classification performance of two commonly used platforms (i.e., ArcGIS Pro and Google Earth Engine) with different satellite datasets (i.e., Landsat, Sentinel, and Planet) through a case study concerning the city of Charlottetown, Canada, over the period of 2017 to 2021. The study provides the scientific basis for the selection of remote-sensing classifiers and satellite imagery with which to develop accurate LULC maps.

3. Climate Change Modelling

The application of remote sensing to climate change modelling constitutes a crucial domain in climate research. This Special Issue includes five articles presenting regional-level climate change-modelling studies using different applications of remote-sensing data. Yan et al. [3] explore potential changes in future extreme precipitation events in China based on Coupled Model Intercomparison Project Phase 6 (CMIP6) (under SSP2-4.5 scenario), using a machine learning approach to integrate and fit multiple models. The study estimates the distribution and trends in the precipitation amount (PRCPTOT), very heavy precipitation days (R20mm), extreme precipitation intensity (SDII95), extreme precipitation amount (R95pTOT), maximum consecutive 5-day precipitation (Rx5day), and precipitation intensity (SDII) for the early 21st century (2023–2050), mid-21st century (2051–2075), and late 21st century (2076–2100). Gnitou et al. [4] adopt a two-way approach to CORDEX-CORE RegCM4-7 seasonal precipitation simulations' Added Value (AV) analysis over Africa with the aim of quantifying the potential improvements introduced by a downscaling approach at high- and low-resolution using satellite-based observational products. Zhou et al. [5] use Regional Climatic Model (RegCM) simulations to downscale the boundary conditions of Geophysical Fluid Dynamics Laboratory Earth System Model Version 2M (GFDL-ESM2M) over the Prairie Provinces, and extract the daily mean, maximum, and minimum temperatures from the historical and future climate simulations. The study investigates temperature variations in two future periods (i.e., 2036 to 2065 and 2065 to 2095) compared to the baseline period (i.e., 1985 to 2004). Lu et al. [6] explore the long-term spatial and seasonal variations of three dominant variables with respect to the water cycle (i.e., precipitation, evapotranspiration, and runoff) in China using the Regional Climate Model system (RegCM) developed by the International Centre for Theoretical Physics. The research compares the simulation results with remote-sensing data and gridded observations to validate the model's performance. Ahmad et al. [7] present a two-dimensional (2D) hydrodynamic model combined with remote sensing (RS) and a geographic information system (GIS)-based approach to generate additional flood characteristic maps (e.g., flood velocity, duration, arrival time, and recession time) for the transboundary river Deg Nullah in Pakistan. The study simulates the extreme flood event that occurred in the study area in 2014 to evaluate the performance of the approach.

4. Monitoring Climate Change

Remote-sensing technologies are now being extensively applied to climate change monitoring at the global, regional, and local scales at an unprecedented rate, especially where ground observation data are scarce. Accordingly, six articles published in this Special Issue demonstrate the remote-sensing-based, long-term monitoring of climate parameters (e.g., precipitation, temperature, etc.) and glaciers. Lu et al. [8] explore possible causes for the interdecadal shift in the interannual variability in summertime precipitation (IVSP) over South China (SC) following the mid-2000s. The study uses climate datasets such as the Precipitation Reconstruction dataset (from NOAA), monthly atmospheric circulation (from NCEP/NCAR), a daily outgoing longwave radiation (OLR) dataset (from NOAA), and the NOAA OI SST V2 High-Resolution Dataset to analyze summertime precipitation variability over South China. Ran et al. [9] present a deep-learning algorithm for fog detection at dawn and dusk under terrain restrictions and an enhanced channel domain attention mechanism (DDF-Net) using advanced Himawari-8 imager data (H8/AHI). Satellite- and ground-based observational climate data concerning northern China, including the Inner Mongolia Plateau and the Loess Plateau, during the winter months (November to December) of 2015 to 2017 were used in this study. Fan et al. [10] employ downscaled climate data using machine learning algorithms and develop a Batch Gradient Descent Linear Regression model to calculate the contributions of temperature and precipitation to runoff in data-scarce high mountains. A case study of six mountainous basins originating from the Tianshan Mountains in Northwest China are used to demonstrate the application of this novel approach. Li et al. [11] demonstrate the use of ERA5-Land, a reanalysis dataset with high spatial and temporal resolution, to quantify the trends and variations in Frost-Free Periods (FFP) and Frost Days (FD) across China from 1950 to 2020. Zhou et al. [12] investigate glacier velocity changes and glacier flow patterns in the Himalayas over the period from 1999 to 2018 using 220 scenes of Landsat-7 panchromatic images taken between 1999 and 2000 and Sentinel-2 panchromatic images taken between 2017 and 2018. Wang et al. [13] analyze the spatial and temporal variations in the glaciers in the Ebi Lake basin during the period from 1964 to 2019 based on the first and second Chinese Glacier Inventories (CGI) and remote-sensing data. The study also investigates the response of glaciers to the warming climate by analyzing digital elevation modeling and meteorology.

5. Climate Change Impact Assessment

The assessment of climate change's impacts on biophysical systems is a highly complex area of research wherein remote-sensing technologies are widely used to evaluate the changes in these systems over time. In this Special Issue, three articles present case studies demonstrating integrated impact assessments with the aid of remote-sensing data. Wang et al. [14] investigate the changes in vegetation across the Three-Rivers Headwaters Region (TRHR) (the Yangtze, the Yellow, and the Lancang (Mekong) rivers) by mapping the normalized difference vegetation index (NDVI) over the growing season from 1982 to 2015. Moreover, the study examines how the vegetation cover has changed with the changing temperature and precipitation at different altitudes of the TRHR region throughout the past few decades. Guild et al. [15] estimate forest loss within Kutai National Park (KNP) in Indonesia precipitated by illegal logging and wildfires over various time periods since 1997 using an extensive catalogue and the processing power of the Google Earth Engine. Comparing time-series estimates of precipitation, the ENSO index, burned area, and forest loss, the paper demonstrates that the risk of fire within KNP mainly depends on drought severity, and that the rates of non-fire- (gradual) and fire-related (extreme) forest loss threaten the remaining forests of this National Park. Chen et al. [16] investigate the relative importance of plant nutrient traits and soil nutrient availability with respect to mediating the climatic sensitivity of desert scrub biomes in the Tibetan Plateau. This study analyzes the vegetation sensitivity index (VSI) of desert scrub for the period from 2000 to 2015 and field measurements of the nutrients in the soil and plant leaves from research conducted in 2016 for seven types of desert scrub communities in the Qaidam Basin (NE Tibetan Plateau). Multiple linear and structural equation models were used to reveal how leaf and soil nutrient regimes affect desert scrubs' sensitivity to climate variability.

This collection of articles addresses some of the knowledge gaps in the field of 'remote sensing applied to climate change.' I hope this will encourage further investigation in this area and thus improve the performance of remote-sensing technologies and data analysis techniques as well as widen applications in research related to climate change modelling, monitoring, and impact assessment.

Funding: This research was funded by the Natural Science and Engineering Research Council of Canada, the New Frontiers in Research Fund, and the Government of Prince Edward Island.

Acknowledgments: I would like to thank all the authors for their contributions and all the reviewers for their valuable comments and feedback.

Conflicts of Interest: The author declares no conflict of interest.

References

1. Meftah, M.; Boutéraon, T.; Dufour, C.; Hauchecorne, A.; Keckhut, P.; Finance, A.; Bekki, S.; Abbaki, S.; Bertran, E.; Damé, L.; et al. The UVSQ-SAT/INSPIRESat-5 CubeSat Mission: First In-Orbit Measurements of the Earth's Outgoing Radiation. *Remote Sens.* **2021**, *13*, 1449. [CrossRef]
2. Basheer, S.; Wang, X.; Farooque, A.A.; Nawaz, R.A.; Liu, K.; Adekanmbi, T.; Liu, S. Comparison of Land Use Land Cover Classifiers Using Different Satellite Imagery and Machine Learning Techniques. *Remote Sens.* **2022**, *14*, 4978. [CrossRef]
3. Yan, Y.; Wang, H.; Li, G.; Xia, J.; Ge, F.; Zeng, Q.; Ren, X.; Tan, L. Projection of Future Extreme Precipitation in China Based on the CMIP6 from a Machine Learning Perspective. *Remote Sens.* **2022**, *14*, 4033. [CrossRef]
4. Gnitou, G.T.; Tan, G.; Hongming, Y.; Nooni, I.K.; Lim Kam Sian, K.T.C. Resolution-Sensitive Added Value Analysis of CORDEX-CORE RegCM4-7 Past Seasonal Precipitation Simulations over Africa Using Satellite-Based Observational Products. *Remote Sens.* **2022**, *14*, 2102. [CrossRef]
5. Zhou, X.; Huang, G.; Li, Y.; Lin, Q.; Yan, D.; He, X. Dynamical Downscaling of Temperature Variations over the Canadian Prairie Provinces under Climate Change. *Remote Sens.* **2021**, *13*, 4350. [CrossRef]
6. Lu, C.; Huang, G.; Wang, G.; Zhang, J.; Wang, X.; Song, T. Long-Term Projection of Water Cycle Changes over China Using RegCM. *Remote Sens.* **2021**, *13*, 3832. [CrossRef]
7. Ahmad, I.; Wang, X.; Waseem, M.; Zaman, M.; Aziz, F.; Khan, R.Z.N.; Ashraf, M. Flood Management, Characterization and Vulnerability Analysis Using an Integrated RS-GIS and 2D Hydrodynamic Modelling Approach: The Case of Deg Nullah, Pakistan. *Remote Sens.* **2022**, *14*, 2138. [CrossRef]
8. Lu, W.; Zhu, Y.; Zhong, Z.; Hu, Y.; Ha, Y. Weakened Impacts of the East Asia-Pacific Teleconnection on the Interannual Variability of Summertime Precipitation over South China since the Mid-2000s. *Remote Sens.* **2022**, *14*, 5098. [CrossRef]
9. Ran, Y.; Ma, H.; Liu, Z.; Wu, X.; Li, Y.; Feng, H. Satellite Fog Detection at Dawn and Dusk Based on the Deep Learning Algorithm under Terrain-Restriction. *Remote Sens.* **2022**, *14*, 4328. [CrossRef]
10. Fan, M.; Xu, J.; Chen, Y.; Fan, M.; Yu, W.; Li, W. Temperature Contributes More than Precipitation to Runoff in the High Mountains of Northwest China. *Remote Sens.* **2022**, *14*, 4015. [CrossRef]
11. Li, H.; Liu, G.; Han, C.; Yang, Y.; Chen, R. Quantifying the Trends and Variations in the Frost-Free Period and the Number of Frost Days across China under Climate Change Using ERA5-Land Reanalysis Dataset. *Remote Sens.* **2022**, *14*, 2400. [CrossRef]
12. Zhou, Y.; Chen, J.; Cheng, X. Glacier Velocity Changes in the Himalayas in Relation to Ice Mass Balance. *Remote Sens.* **2021**, *13*, 3825. [CrossRef]
13. Wang, L.; Bai, C.; Ming, J. Current Status and Variation since 1964 of the Glaciers around the Ebi Lake Basin in the Warming Climate. *Remote Sens.* **2021**, *13*, 497. [CrossRef]
14. Wang, K.; Zhou, Y.; Han, J.; Chen, C.; Li, T. Long-Term Tibetan Alpine Vegetation Responses to Elevation-Dependent Changes in Temperature and Precipitation in an Altered Regional Climate: A Case Study for the Three-Rivers Headwaters Region, China. *Remote Sens.* **2023**, *15*, 496. [CrossRef]
15. Guild, R.; Wang, X.; Russon, A.E. Tracking Deforestation, Drought, and Fire Occurrence in Kutai National Park, Indonesia. *Remote Sens.* **2022**, *14*, 5630. [CrossRef]
16. Chen, B.; Chen, H.; Li, M.; Fiedler, S.; Mărgărint, M.C.; Nowak, A.; Wesche, K.; Tietjen, B.; Wu, J. Climate Sensitivity of the Arid Scrublands on the Tibetan Plateau Mediated by Plant Nutrient Traits and Soil Nutrient Availability. *Remote Sens.* **2022**, *14*, 4601. [CrossRef]

remote sensing

MDPI

Article

Long-Term Tibetan Alpine Vegetation Responses to Elevation-Dependent Changes in Temperature and Precipitation in an Altered Regional Climate: A Case Study for the Three Rivers Headwaters Region, China

Keyi Wang [1,2,3], Yang Zhou [2,3,*], Jingcheng Han [2,3], Chen Chen [3,4] and Tiejian Li [3,4,5]

1 Shenzhen Institute of Information Technology, College of Transportation and Environment, Shenzhen 518109, China
2 Water Science and Environmental Engineering Research Center, College of Chemical and Environmental Engineering, Shenzhen University, Shenzhen 518060, China
3 State Key Laboratory of Hydroscience and Engineering, Tsinghua University, Beijing 100190, China
4 School of Environment and Civil Engineering, Dongguan University of Technology, Dongguan 523820, China
5 State Key Laboratory of Plateau Ecology and Agriculture, Qinghai University, Xining 810016, China
* Correspondence: yangzhou@uregina.ca

Abstract: Recent studies offer more evidence that the rate of warming is amplified with elevation, indicating thereby that high-elevation ecosystems tend to be exposed to more accelerated changes in temperature than ecosystems at lower elevations. The phenomenon of elevation-dependent warming (EDW), as one of the regional climate-change impacts, has been observed across the Tibetan Plateau. Studies have often found large-scale greening trends, but the drivers of vegetation dynamics are still not fully understood in this region, such that the local implications of vegetation change have been infrequently discussed. This study was designed to quantify and characterize the seasonal changes in vegetation across the Three Rivers Headwaters Region (TRHR), where the land cradles the headwaters of the Yangtze, the Yellow, and the Lancang (Mekong). By mapping the normalized difference vegetation index (NDVI) over the growing season from 1982 to 2015, we were able to evaluate seasonal changes in vegetation cover over time. The results show a slightly increased tendency in green vegetation cover, which could possibly be attributed to sustained warming in this region over the past three decades, whereas a decline in the green-up rate with elevation was found, indicating an inconsistent trend of vegetation greening with EDW. The cause of the green-up rate decline at high elevations could be linked to the reduced soil water availability induced by the fast increase in warming rates associated with EDW. The findings of this study have important implications for devising adaptation strategies for alpine ecosystems in a changing climate.

Keywords: elevation-dependent warming; climate change; vegetation greening; NDVI; Three Rivers Headwaters Region

Citation: Wang, K.; Zhou, Y.; Han, J.; Chen, C.; Li, T. Long-Term Tibetan Alpine Vegetation Responses to Elevation-Dependent Changes in Temperature and Precipitation in an Altered Regional Climate: A Case Study for the Three Rivers Headwaters Region, China. *Remote Sens.* **2023**, *15*, 496. https://doi.org/10.3390/rs15020496

Academic Editor: Inge Jonckheere

Received: 30 November 2022
Revised: 10 January 2023
Accepted: 10 January 2023
Published: 13 January 2023

1. Introduction

Climate change has become a recognized cause of major alterations to natural ecosystems, with many climate-driven changes already being observed in biotic community structure and plant species composition [1,2]. Growing evidence suggests that the rate of warming is amplified with elevation, such that high-elevation ecosystems tend to be exposed to more accelerated changes in temperature than ecosystems at lower elevations [3–5]. The phenomenon of elevation-dependent warming (EDW), as one of the regional climate-change impacts, has been observed and recognized as a contributing cause of shifts in species distribution, hydrological regimes, and biochemical cycles in high-elevation regions [6]. Alpine ecosystems are known to be particularly susceptible to climate change because of their unique thermodynamic properties [7]. In light of the vulnerability of ecosystem services in alpine regions, e.g., biodiversity, water cycling, and carbon sequestration, shifts in ecosystem composition, structure, and function are likely to have far-reaching social and economic implications for local and regional areas [8]. Vegetation is a vital part of healthy, functioning ecosystems, and serves as a foundation for terrestrial food webs and habitat for animals [9,10]. It helps to cycle energy and nutrients throughout an ecosystem as well as improving water quality and reducing soil erosion [11–13]. Therefore, monitoring vegetation change across remote regions and quantifying ecosystem feedbacks to climate change has become a topic of increasing interest among academics in recent years [14,15]. Rapid technological advancement has accelerated the growth of remote sensing and enabled a more comprehensive assessment of vegetation change across a variety of spatial and temporal scales [16–20]. The normalized difference vegetation index (NDVI) is a standardized metric that describes the difference between visible and near-infrared reflectance of vegetation cover, allowing for more detailed information that links vegetation greening patterns to drivers [21]. Moreover, because of its simple estimation, high availability, and noise elimination, the NDVI has been widely used for regional-to-global-scale vegetation monitoring and assessment [22–26].

It has been successfully used by many researchers to measure trends in greening at northern latitudes and high elevations [27–30]. In general, the interactions between different climatic variables, such as precipitation, temperature, and solar-radiation-driven evapotranspiration, could affect vegetation greenness [31]. It is reported that high warming rates are recognized as the essential driver of vegetation growth [32]. However, different results may be obtained for different terrestrial characteristics and regional climate environments. For instance, Angert et al. found drought-induced reduction in photosynthesis throughout the growing season at both middle and high latitudes, indicating that the relationship between temperature changes and vegetation growth may change with elevation [33]. Trujillo et al. demonstrated that vegetation greening trends varied with elevation and maximum greenness could be observed in mid-latitude mountain regions [34]. Piao et al. suggested that vegetation growth was generally positively correlated with high warming rates but the correlation could change over time following alternations in other environmental factors, particularly in areas above a latitude of 30 degrees north [35]. Kumari et al. indicated that vegetation growth was more related to evapotranspiration than precipitation in the UKR Basin in the Himalayas [25]. These findings indicate that the driving mechanism of the correlation between elevation-dependent warming (EDW) and vegetation greening is not yet fully understood and that there are large uncertainties regarding climate change impacts at higher elevations. Given the knowledge gaps in the current understanding of EDW and high-elevation vegetation change, expanded efforts to shed further light on this research topic could advance the understanding of the exact driving mechanism.

The Three Rivers Headwaters Region (TRHR) lies in the hinterland of the Tibetan Plateau, and its 363,000 km^2 of land cradles the headwaters of the Yangtze, the Yellow, and the Lancang (Mekong). This region is also known as China's water tower, providing runoff for these three rivers as well as for several hundred million people downstream. Specifically, The TRHR provides 25% of the annual runoff of the Yangtze, 49% of the Yellow, and 15% of the Lancang [36]. The alpine ecosystem of the TRHR is characterized by high biodiversity, including endemic species not found anywhere else, and a high-plateau climate, meaning large diurnal and monthly temperature variations across the region. The phenomenon of EDW, as one of the regional climate-change impacts, has been commonly observed here, possibly owing to the warming rate in this region, which is twice as high as the global average [37]. The increased rate of warming could have a profound effect on the biosphere processes in the TRHR, with many and diverse impacts on local biological resources, e.g., common vegetation and habitat types. In addition, the distinctive landform and geologic characteristics have also made the TRHR highly vulnerable to climate change; for example, soils in this region are thin and coarse-textured, making them more susceptible to soil erosion [38]. The TRHR is of significant value to nature conservation and ecological services within regional and national contexts, and has become a hotspot of regional climate concerns due to the escalating number of droughts, vegetation degradation, and extreme-climate-related conditions. Additionally, further research on this region can also improve the understanding of ecosystem dynamics and vegetation change at high altitudes [8,39–41]. However, some studies within this region also find positive ecological feedbacks associated with the local anthropogenic and climatic changes. For example, grassland productivity has been found to be significantly improved, possibly as a result of successful efforts to implement alpine conservation and ecological restoration programs as well as consistent increases in temperature [42,43] Given the spatial heterogeneity of vegetation change in high-elevation regions, researchers often focus their attention on ecological and environmental impacts with regard to specific factors, e.g., climate change, grazing activities, water-use patterns, and environmental conservation programs [43–45]. However, another factor in vegetation change, also driven by EDW, has not been studied adequately in the TRHR. As aforementioned, many studies have found that EDW plays a potentially important role in altering vegetation growth patterns at higher elevation. Therefore, there is a need to ensure adequate analysis and interpretation of EDW in this hotspot region. In addition, the TRHR is composed of a variety of vegetation cover types, including grasslands, barren lands, forests, croplands, wetlands, etc. In light of the diverse vegetation cover types, the vegetation patterns and dynamics within the TRHR may also be altered fundamentally by warmer temperatures, but few researchers have discussed the potential impacts of vegetation composition on vegetation change in this region. It is therefore imperative that researchers make new efforts to advance the understanding of the internal vegetation variability across vegetation types and the external drivers associated with spatial heterogeneity.

This study was designed to quantify and characterize the seasonal changes in vegetation using more than 30 years of remote sensing data across the TRHR so as to help further the understanding of the role of EDW in changing alpine vegetation growth patterns in this region. Specifically, our research steps included: (i) investigating the interannual variability in the growing-season NDVI ($NDVI_{gs}$) from May to September across the TRHR between 1982 and 2015; (ii) examining the correlation between $NDVI_{gs}$ and elevation as well as driving forces behind changes in $NDVI_{gs}$ using multiple statistical approaches; (iii) assessing the effects of temperature and precipitation changes on vegetation over time so as to identify the climate variables that most affect vegetation productivity; and (iv) exploring the linkages between vegetation composition and greening trends to help further the understanding of $NDVI_{gs}$ change across vegetation types. Results from this study will be beneficial for advancing the understanding of EDW in high-elevation regions, e.g., the TRHR, and of particular value to the local authorities in evaluating ecological and environmental conservation programs.

2. Materials and Methods

2.1. Study Area

The TRHR (31°39′–37°10′ N, 89°24′–102°27′ E) is located in the northeastern part of the Qinghai–Tibet Plateau, with a total area of about 363,000 km^2. The elevation in the TRHR ranges from about 2000 m in the northeastern corners to above 6800 m in the ridges of the western mountains, and the average elevation of the region is nearly 4500 m (Figure 1). This region has a typical continental climate with warm, humid summers and cold, arid winters [46]. Additionally, there is significant temporal variability in precipitation and temperature across the TRHR. For instance, about 60% of the region's annual rainfall occurs in the summer months, and the monthly average temperature is above 0 °C only during the growing season, from May to September [47]. Over the past decades, the TRHR has experienced a striking change in local climate and been confronted with a series of climate-related issues, such as grassland degradation, soil erosion, glacier retreat, and lake and wetland decline [48]. Therefore, central and local governments have long undertaken ecological interventions to reverse the degradation of local ecosystems and to help them regain their ecological functionality in a changing climate. The central government has also set aside protected land for the Three-Rivers Nature Reserve in this region to preserve the natural heritage of the Tibetan Plateau.

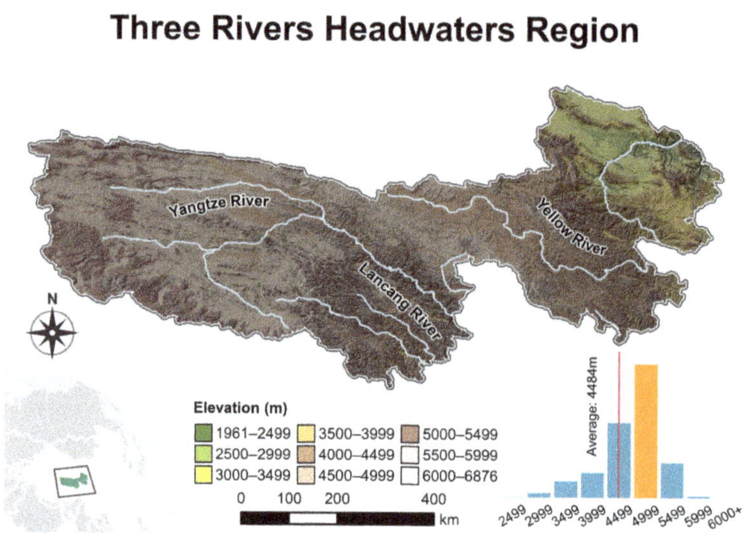

Figure 1. Location of study area in the Qinghai–Tibet Plateau, China. The TRHR cradles the headwaters of the Yangtze, the Yellow, and the Lancang, and is known as China's water tower. The elevation in the area ranges from 1961 m to 6876 m. The average elevation is 4484 m, but most of the area is above 4500 m.

2.2. Remote Sensing Data

2.2.1. Land Cover Data

The land cover data were derived from the Finer Resolution Observation and Monitoring of Global Land Cover dataset (FROM-GLC 2017v1, http://data.ess.tsinghua.edu.cn/ (accessed on 1 December 2021)). The FROM-GLC dataset is the first 30 m resolution global land cover product generated using Landsat Thematic Mapper (TM) and Enhanced Thematic Mapper Plus (ETM+) data, and is publicly accessible from Tsinghua University. As shown in Figure 2, the categories of grassland and barren land are the most prevalent in the TRHR and the most common types of land cover, accounting for about 94% coverage of the region (Figure 2b). The other categories account for the remaining 6%.

2.2.2. NDVI Data

We used the Global Inventory Monitoring and Modeling Studies (GIMMS) NDVI3g dataset in this study. This dataset was generated from several advanced very high resolution radiometer (AVHRR) sensors onboard the U.S. National Oceanic and Atmospheric Administration (NOAA) series satellites at 0.083° over the years 1982 through 2015. The GIMMS-NDVI3g dataset has been calibrated for sensor degradation, inter-sensor differences, cloud cover, solar zenith angle, etc., and has also been widely proved to be suitable for assessing temporal changes in vegetation [49,50]. The monthly NDVI data were calculated using the semi-monthly values and synthesized using the maximum-value composite method. This method can help greatly reduce the impact of clouds and aerosols on the vegetation index retrieval algorithm [51]. The mean growing-season NDVI (NDVIgs) was used to map the vegetation greenness changes across the TRHR (Figure 2a).

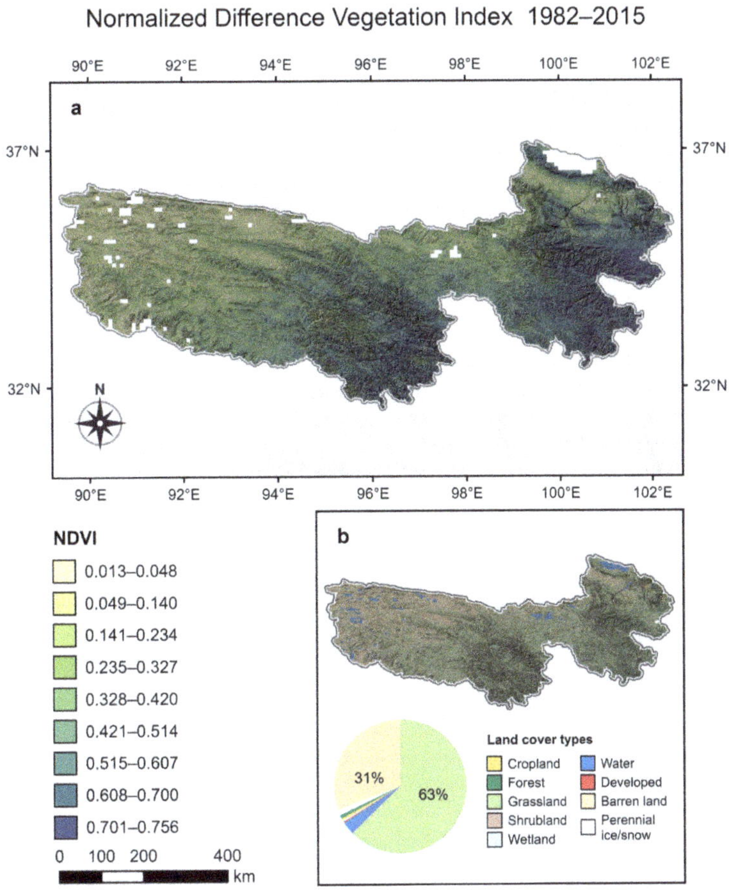

Figure 2. (**a**) The 34-year average growing-season NDVI greenness map for the TRHR (the pixels with $NDVI_{gs}$ values smaller than 0.01 were deemed to be non-vegetation areas and displayed as blank); (**b**) the land cover map for the TRHR, represented by nine land cover categories.

2.2.3. Climatic Data

The monthly precipitation data were obtained from the Climate Hazards Group InfraRed Precipitation with Stations (CHIRPS) dataset at 0.05° resolution for 1982 to 2015. The monthly temperature data were derived from the European Centre for Medium-Range Weather Forecasts (ECMWF) dataset at 0.125° resolution for 1982 to 2015. In order to allow for a finer resolution of data analysis, the climatic datasets were interpolated to a higher resolution, 0.0833°, to ensure the compliance of the climate data's resolution with that of the NDVI data.

2.2.4. Evapotranspiration Estimation from GLEAM

The monthly evapotranspiration (ET) and potential ET (PET) data were obtained from the Global Land Evaporation Amsterdam Model (GLEAM) dataset at 0.25° resolution for 1982 to 2015, which can reasonably capture the evaporation variability in all vegetation types and climate conditions [52,53]. GLEAM-ET is estimated by using reanalysis net radiation and air temperature, satellite-, reanalysis-, and gauge-based precipitation, satellite-based vegetation optical depth, and snow water equivalents. The ET modeling algorithm is defined by Martens et al. [52] as follows:

$$ET = E_p S + E_i \tag{1}$$

where E_p is the potential evapotranspiration, estimated using the Priestley and Taylor equation driven by observations of surface net radiation and near-surface air temperature; S is the evaporative stress factor, calculated from the observations of microwave vegetation optical depth (VOD) and estimates of root-zone soil moisture; E_i is the interception loss, calculated using a Gash analytical model.

The general information of data used in this study has been summarized in Table 1.

Table 1. Summary of data sources used for this study.

Data Types	Data Sources	Temporal Resolution	Spatial Resolution	Temporal Coverage
NDVI	GIMMS NDVI3g	15 days	0.083°	1982 to 2015
Precipitation	CHIRPS	Month	0.05°	1982 to 2015
Temperature	ERA-Interim	Month	0.125°	1982 to 2015
Evapotranspiration/Potential evapotranspiration	GLEAM	Month	0.25°	1982 to 2015

2.2.5. Methods

Multiple analysis methods were used in this study to reflect the characteristics of NDVI trends at different elevations. Firstly, the non-parametric Mann–Kendall (MK) test was used to evaluate long-term changes in vegetation cover and to identify greening patterns across the TRHR. Previous studies indicated that serial correlation in hydrological or climatic series may result in larger standard errors and lead to misleading trend estimates [54–56] Thus, the trend-free pre-whitening (TFPW) procedure was integrated with the MK test to improve the reliability of trend analysis [55]. After application of the TFPW-MK test, the trend slope was estimated as follows:

$$b = Median\left(\frac{x_j - x_i}{j - i}\right), \forall i < j \tag{2}$$

where b is the trend slope; x_i denotes the ith datum.

Next, by using both average-data and pixel-based methods, the trend changes in the NDVI across the region were linked to drivers such as elevation. This allows for a better understanding of the role of elevation in driving vegetation changes across elevational gradients.

The relationships between trends in the NDVI and climatic variables were investigated using the following linear least-squares regression model:

$$y_i = ax_i + b + \varepsilon_i,\ i = 1, \ldots, n \tag{3}$$

where y_i denotes the dependent variable; x denotes the independent variable; n is the sample size; a is the slope; b is the intercept; and ε_i is the random error.

The correlations between the trend changes in the NDVI and elevation as well as relevant climatic variables could be evaluated with Spearman rank correlation analysis:

$$r = \frac{\sum_i (x_i - \overline{x})(y_i - \overline{y})}{\sqrt{\sum_i (x_i - \overline{x})^2 \sum_i (y_i - \overline{y})^2}}, i = 1, \ldots, n \tag{4}$$

where x and y are the variables; n is the sample size. Student's t-test was applied to evaluate the statistical significance in a trend analysis.

With regard to the roles of climatic factors in vegetation growth, their effects on elevation-dependent changes in NDVI trends can be understood through partial correlation analysis. Partial correlation analysis is an effective way to investigate the net effect of one independent variable when multiple independent variables act on a dependent variable simultaneously [57,58]. The partial correlation coefficient is defined as follows:

$$R_{xy|z} = \frac{r_{xy} - r_{xz}r_{yz}}{\sqrt{(1 - {r_{xz}}^2)(1 - {r_{yz}}^2)}} \quad (5)$$

where x and y denote the independent variables; z denotes the dependent variable; r denotes the correlation coefficient between different variables; and R represents the degree of correlation. The null hypothesis assumed that there was no linear correlation between the two independent variables, and the significance level of the hypothesis test was set at $\alpha = 0.05$ in this study.

3. Results

3.1. Spatiotemporal Variations in $NDVI_{gs}$ and Climatic Variables

According to Figure 2a, the 34-year average growing-season NDVI greenness map demonstrated clear spatial delineations of vegetation changes in greenness. The southeast part of the TRHR was mostly covered in grassland with $NDVI_{gs}$ values spanning from 0.5 to 0.7, whereas barren land was prevalent in the northwest part with $NDVI_{gs}$ values of less than 0.2. To further investigate the spatiotemporal variations in vegetation and climatic variables across the TRHR, we applied trend analysis to the NDVI and related climatic variables in this study. Comparisons of trend patterns among long-term vegetation and climate records across the study region showed similar yet statistically divergent results with respect to the temporal changes in $NDVI_{gs}$ and climatic variables. First, there was a slight upward trend in $NDVI_{gs}$ at an average rate of 0.003 per decade during the period 1982–2015 (Figure 3a), although this overall trend ($p > 0.05$) was not sufficiently statistically significant to indicate that greening trends across the region became more prominent over time. However, expansive changes in $NDVI_{gs}$ over space showed that over 70% of the region experienced greening between 1982 and 2015, and the greening pattern was spatially fragmented (Figure 3b). Negative $NDVI_{gs}$ trends were found mainly in the southeastern corners of the region, whereas positive $NDVI_{gs}$ trends were more prevalent in the western and northeastern parts. Second, there was a clear uptrend in the growing-season temperature (T_{gs}) over time, as evident from the statistically significant p value ($p < 0.01$). The average increase rate was about 0.39 °C per decade during the period 1982–2015 (Figure 3c). Although the overall warming trend was evident across the TRHR, the northeastern corners of the region, near Qinghai Lake, had flat changes or declining trends in temperature; interestingly, the eastern half of the region also experienced the highest warming rate, whereas the most widespread warming was seen in the mostly higher-elevation western half (Figure 3d). The phenomenon of EDW appears to exist within the TRHR as well. Third, the growing-season precipitation (P_{gs}) also showed a statistically significant uptrend at an average rate of 31 mm per decade during the period 1982–2015 (Figure 3e). The westernmost parts of the TRHR had experienced greater increases in precipitation than others, while a few mountainous areas in the southern and southeastern corners had seen minimal and insignificant increases in precipitation (Figure 3f). We conducted correlation analysis to assess the association between $NDVI_{gs}$ and the two climatic variables (i.e., T_{gs} and P_{gs}). We found that there was a weak positive correlation between $NDVI_{gs}$ and T_{gs} ($r = 0.43$, $p < 0.05$), while no statistically significant correlation was found between $NDVI_{gs}$ and P_{gs} ($r = 0.18$, $p > 0.05$). These findings indicate that the vegetation change could be mainly attributed to the growing-season warming across the TRHR.

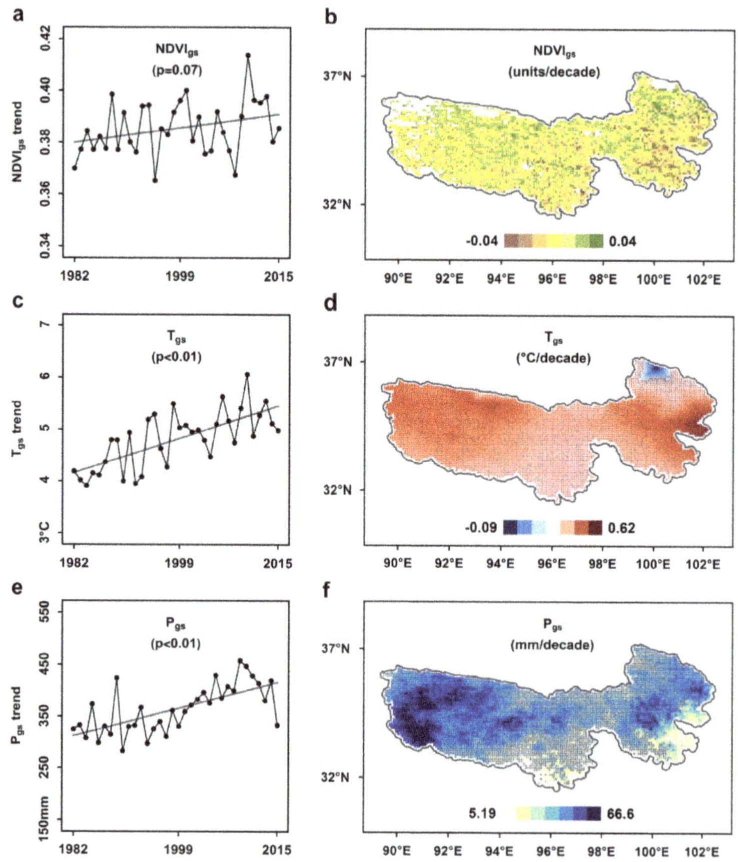

Figure 3. The average trends in (**a**) NDVI, (**c**) temperature, and (**e**) precipitation during the growing season from 1982 to 2015; the decadal changes in (**b**) NDVI, (**d**) temperature, and (**f**) precipitation across the TRHR. The pixels with significant changes at the 95% confidence level are marked by crosses.

3.2. *Elevation-Dependent Responses of Climatic Variables to $NDVI_{gs}$*

We used multi-year averages of the indicators to account for effects of the elevational gradient on spatial disparity in temperature and precipitation, and to investigate linkages between elevation-dependent changes in temperature and precipitation and trends in $NDVI_{gs}$. Spatially, the TRHR showed a gradual increase in the average temperature from 0.24 °C in the westernmost parts to 11.8 °C in the northeastern corners, indicating a strong correlation between temperature and elevation; temperature gradually increased as elevation decreased (Figure 4a). Mountainous areas in the southern and southeastern parts received much more annual rainfall than the flat-lying areas in the west and northeast, indicating that variations in precipitation are closely tied to landscapes rather than elevations (Figure 4b). The $NDVI_{gs}$ values show an inverted V-shape trend across elevational gradients, peaking at around 3900 m (Figure 4c). By comparing both $NDVI_{gs}$ and T_{gs} trends at different elevations, we noticed a negative correlation between $NDVI_{gs}$ and T_{gs} in low-elevation areas, which gradually turned positive with increases in elevation. In contrast, there was a stronger correlation between $NDVI_{gs}$ and P_{gs} over elevational gradients as a similar inverted V-shape trend was observed for the two indicators (Figure 4d).

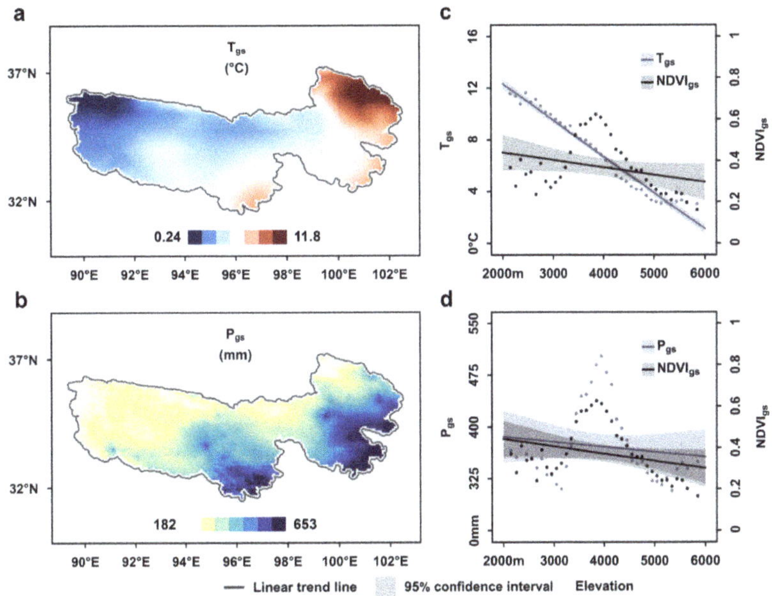

Figure 4. The growing-season (**a**) temperature and (**b**) precipitation patterns across the TRHR from 1982 to 2015; comparisons of elevational trends in (**c**) temperature and $NDVI_{gs}$, and (**d**) precipitation and $NDVI_{gs}$.

3.3. Elevation-Dependent Changes in Climatic and $NDVI_{gs}$ Trends

In order to better understand the rate trends in temperature, precipitation, and vegetation over elevation within the study region, we used both average-data and pixel-based methods to conduct a trend analysis. The average $NDVI_{gs}$ values showed a statistically significant declining trend with elevation, indicating that the green-up rate was slower at high elevations (Figure 5a). However, this type of trend analysis failed to factor in differences in the landscapes throughout the TRHR, so the statistical significance could be more or less over-stated. After re-analyzing the $NDVI_{gs}$ value across each pixel grid within the region, we found that the declining trend was confirmed but less statistically significant (Figure 5d). The phenomenon of EDW was highlighted in the results from both analysis methods, suggesting that the rate of warming was amplified at high elevations (Figure 5b,e). In fact, this regional manifestation of global warming had also been observed in other high-elevation regions across the Tibetan Plateau. In addition, the high-elevation areas within the TRHR experienced much greater yearly precipitation increments than the low-elevation areas between 1982 and 2015, and the widespread changes in precipitation exhibited a statistically significant dependency on elevation (Figure 5c,f). Overall, elevation-dependent changes in temperature and precipitation appeared to be happening faster at high elevations, whereas changes in the green-up rate were relatively insignificant.

3.4. Correlations between Changing Climatic Conditions and $NDVI_{gs}$ Trends

We also used both average-data and pixel-based methods to conduct a correlation analysis to improve our understanding of greening trends and drivers in an altered local climate. Both methods showed a statistically significant correlation between T_{gs} changes and $NDVI_{gs}$ trends (Figure 6a,c), and this correlation was overall negative (i.e., decreasing $NDVI_{gs}$ with warming). The average-data method found a weak positive correlation between P_{gs} changes and $NDVI_{gs}$ trends (i.e., increasing $NDVI_{gs}$ with higher precipitation increments), but this finding shows limited statistical significance (Figure 6b). After re-analyzing the multi-year average P_{gs} value across each pixel grid within the region, this weak positive correlation was confirmed statistically (Figure 6d). Despite the identified statistical significance of both climatic variables for the $NDVI_{gs}$ trends, temperature appeared to be a more explicit driver of vegetation change as warming resulted in more variability in $NDVI_{gs}$ across the study region.

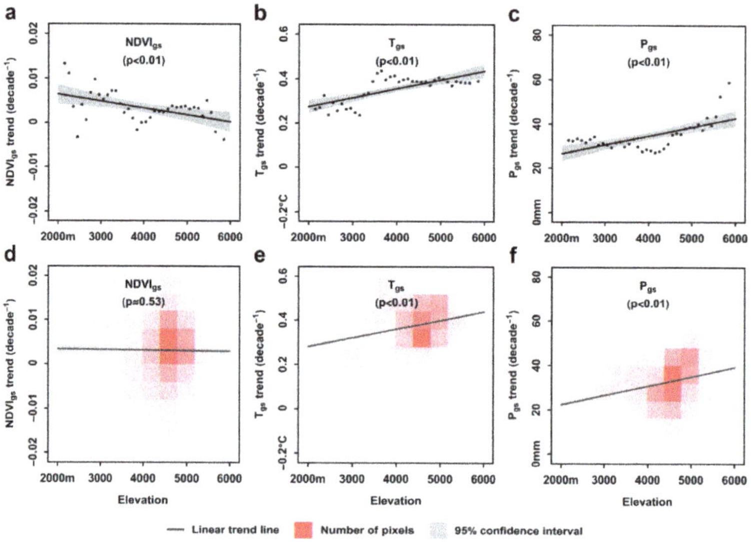

Figure 5. The elevational trends in (**a**) $NDVI_{gs}$, (**b**) temperature, and (**c**) precipitation using the average-data method; the elevational trends in (**d**) $NDVI_{gs}$, (**e**) temperature, and (**f**) precipitation using the pixel-based method.

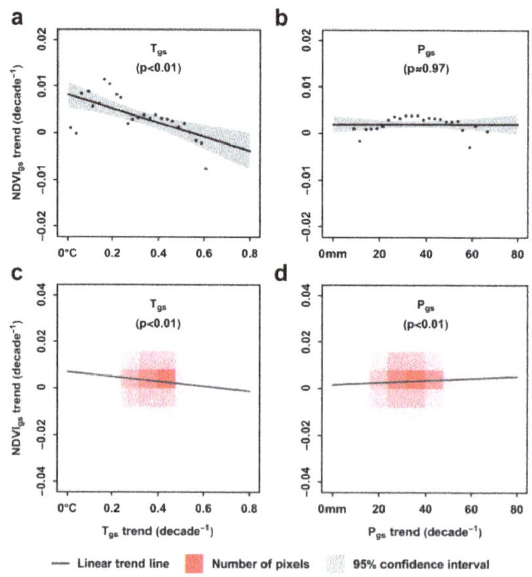

Figure 6. Correlation analysis between trends in temperature and precipitation and trend in $NDVI_{gs}$. (**a**,**b**) Correlations from the average-data method, and (**c**,**d**) correlations from the pixel-based method.

3.5. Spearman's Rank Correlation and Partial Correlation Analysis

Spearman's rank correlation and partial correlation coefficients were calculated to allow for more detailed statistical information that accounts for the elevation dependency of temperature, precipitation, and $NDVI_{gs}$ changes as well as the relationships between temperature and precipitation changes and $NDVI_{gs}$ trends (Figure 7). Both temperature and precipitation changes were positively correlated with elevation, and these strong correlations were of sufficient statistical significance. According to the Spearman correlation coefficient, $NDVI_{gs}$ trends were significantly and negatively correlated with elevation; however, the partial correlation coefficient suggested that this negative correlation was much less statistically significant if the confounding variables, i.e., temperature and precipitation, were factored out. This implied that elevation was not a direct driver of $NDVI_{gs}$ trends. In addition, $NDVI_{gs}$ trends were also negatively correlated with temperature changes, indicating warming is an identified driver of vegetation change. In contrast, precipitation changes exhibited no statistically significant correlation with $NDVI_{gs}$ trends.

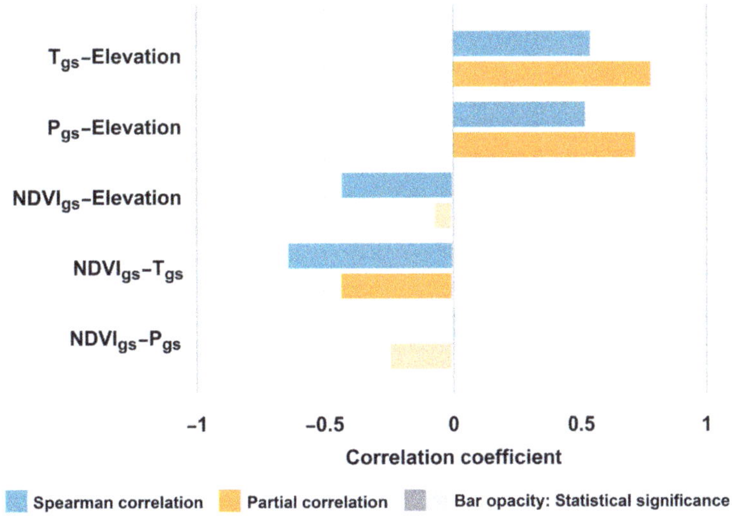

Figure 7. Summary of Spearman correlation and partial correlation analysis. Bar opacity represents statistical significance at the 95% confidence level.

3.6. Land-Cover-Based $NDVI_{gs}$ Quantitative Analysis

Land cover composition is an important explanatory factor when calculating $NDVI_{gs}$ values and evaluating their associated trends. In order to refine our understanding of the relationships between land coverage and $NDVI_{gs}$ values and trends, we used the pixel-based quantitative method to evaluate different land cover types, including grassland, barren land, cropland, developed land, forest and shrubland. The majority of the land within the TRHR was covered in grassland and barren land, accounting for 94% of the total land coverage. Thus, the two land cover types had more chances to have 100% coverage in a pixel grid. As the coverage of grassland increased, the $NDVI_{gs}$ values increased significantly (Figure 8a). In contrast, more coverage in barren land resulted in decreased $NDVI_{gs}$ values (Figure 8b). Furthermore, forest and shrubland, in addition to grassland, were the land cover types most positively related to the $NDVI_{gs}$ values (Figure 8e,f). The results from the pixel-based quantitative analysis also allowed for detection of the spatial patterns of land cover distributions. For example, cropland was mainly distributed in densely vegetated areas (Figure 8c), while developed land was evenly distributed between densely and sparsely vegetated areas (Figure 8d).

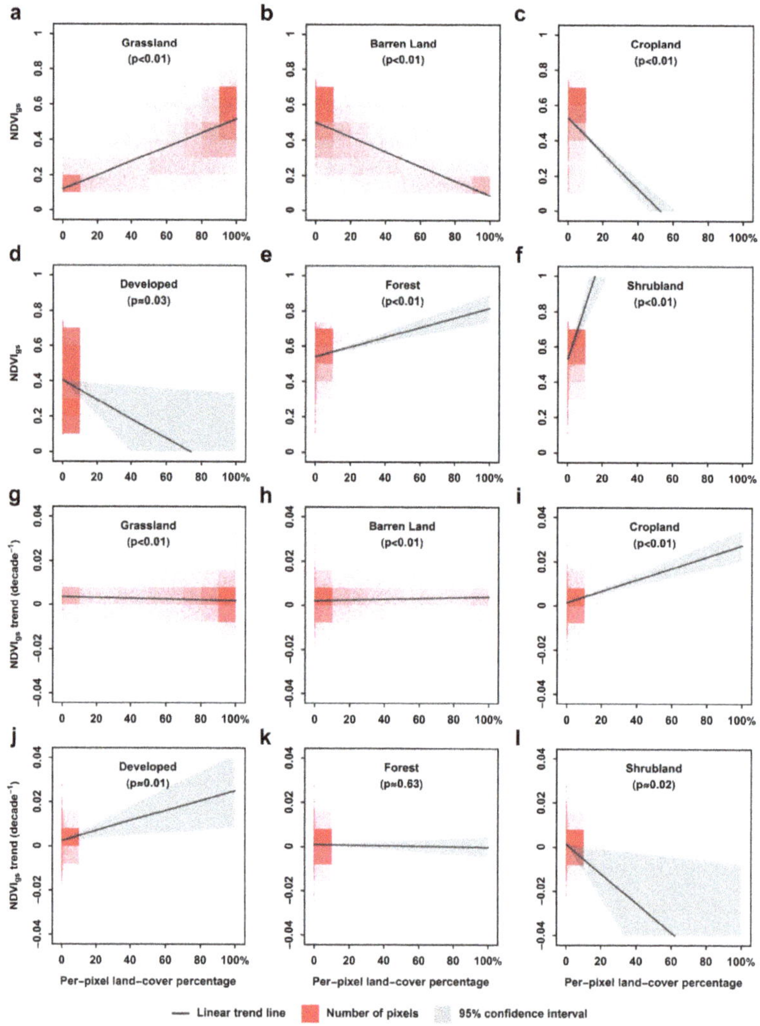

Figure 8. Correlations between $NDVI_{gs}$ and per-pixel land cover percentage (**a**–**f**); correlations between $NDVI_{gs}$ trend and per-pixel land cover percentage (**g**–**i**).

We also evaluated the decadal average $NDVI_{gs}$ changes for land cover types to improve the understanding of the trends of land cover changes within the TRHR between 1982 and 2015. There was a slight declining trend in the positive changes in $NDVI_{gs}$ with an increasing coverage rate of grassland within each pixel grid, indicating that the greening trends in grassland were likely to be slower than those in other land cover types between 1982 and 2015 (Figure 8g). The reason for this decline could be linked to the continued rise in livestock population and the adverse effects of grazing activities. In contrast, more areas covered in barren land appeared to experience slight positive changes in $NDVI_{gs}$, indicating a growing trend of barren land greening within the TRHR (Figure 8h). This implied that the significant changes in temperature and precipitation across the TRHR could probably have facilitated the process of vegetation rehabilitation in barren areas, while the warmer temperature and increased precipitation during the growing season could also have greatly slowed down the progress of grassland degradation. Both cropland and developed land had lower coverage rates within each pixel grid; however, more pronounced changes in greening were found in areas partially covered by these two land cover types (Figure 8i,j). This implies that the local government's long-term efforts to deliver targeted interventions for ecological restoration, e.g., the Grain for Green program, have brought positive results in areas where human activities are concentrated. Forest showed no specific trend that was of statistical significance (Figure 8k), while shrubland seemed to experience a negative decline in $NDVI_{gs}$ changes (Figure 8l).

4. Discussion

It is now accepted that continued climate warming could further boost vegetation growth at high elevations as warmer temperatures and increased rainfalls might potentially create more favorable growing environments in local areas. Our results show that since 1982, most parts of the TRHR have experienced a sustained warming trend, and are getting wetter during the growing season (Figure 3). This is consistent with findings from other studies in this region [37,59,60]. Moreover, the interannual variability in $NDVI_{gs}$ was found to be positively correlated with T_{gs} at a high significance level based on the correlation analysis results, suggesting a driving effect of temperature on vegetation greening.

Both our study and previous studies show that temperature appears to be a dominant driver of vegetation change within the TRHR; therefore, there do seem to be reasons to suppose that the change in the vegetation green-up rate would be consistent with the trend in warming. However, Figure 5a,d show that there was a slight downward trend in the vegetation green-up rate as elevation increased, which appears to be contradictory to the amplified effect of EDW at high elevations. This result indicates that, despite being a dominant driver of interannual vegetation greening, temperature may play a different role in the vegetation dynamics altering the green-up rate.

Water availability is critical for vegetation growth and is usually deemed to be closely related to precipitation. Our results show that high-elevation areas tend to experience greater increases in precipitation than other areas, and this tendency is statistically significant (Figure 5c,f). However, the decreased $NDVI_{gs}$ trend at high elevations indicates that precipitation plays a limited role in regulating the green-up rate. In fact, Figure 6b,d also presents a relatively weak correlation between P_{gs} and $NDVI_{gs}$ trends. Notwithstanding a few studies reporting a strong dependency of water availability on precipitation [30,35,48], the TRHR has seen no sign of an evident growth in water availability.

As we know, water availability on land is not only related to precipitation but may also be directly affected by increased evapotranspiration (ET). Previous studies suggest that the ratio of evapotranspiration to potential evapotranspiration (PET), i.e., ET/PET, is a useful indicator of soil moisture conditions, which can also be used to assess the water availability conditions for vegetation growth during the growing season [61]. To discern the roles of temperature and precipitation changes in altering the vegetation green-up rate, we further investigated the pattern of the growing-season ET/PET ratio (ET/PET_{gs}) with elevation. We observed that most areas had increased decadal ET_{gs}, particularly in the western TRHR, which is covered mostly by barren land (Figure 9a). The increased decadal ratios of ET/PET_{gs} were mainly found in the western and northeastern parts, while the decreased ET/PET_{gs} ratios were observed in the central and southeastern areas (Figure 9b). The ET_{gs} values showed a V-shape trend across elevational gradients, suggesting that ET tended to increase sharply at high elevations (Figure 9c). The ET/PET_{gs} ratios showed a statistically significant declining trend with elevation ($p < 0.01$), which was broadly consistent with the $NDVI_{gs}$ trend (Figure 9d). Moreover, we found that the vegetation green-up rate was positively correlated to the ET/PET_{gs} ratio at a higher significance level ($r = 0.52$, $p < 0.01$) and the correlation between ET and the green-up rate was less significant ($p > 0.05$). Therefore, the increased ET within the TRHR could mainly be attributed to the overall increase in temperature.

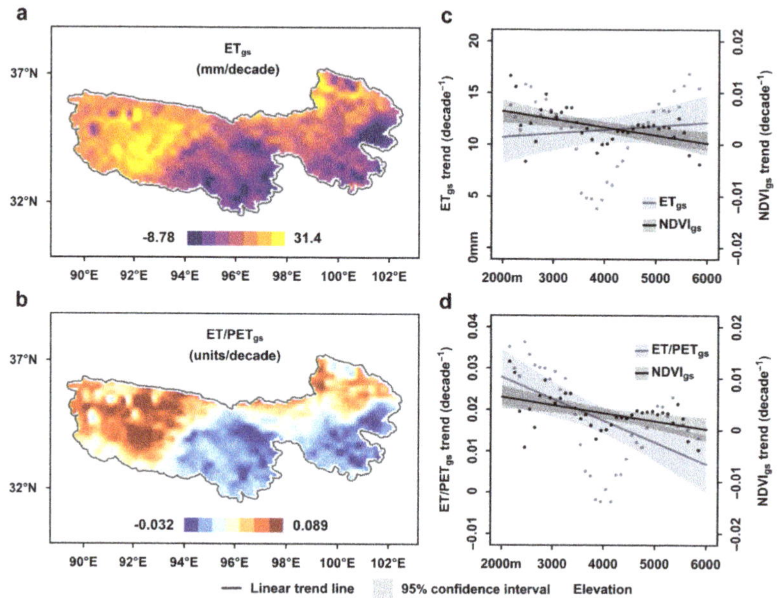

Figure 9. The growing-season (**a**) ET_{gs} and (**b**) ET/PET_{gs} patterns across the TRHR from 1982 to 2015; comparisons of elevational trends in (**c**) ET_{gs} and $NDVI_{gs}$, and (**d**) ET/PET_{gs} and $NDVI_{gs}$.

Overall, the reductions in water availability brought about by climate warming could greatly slow the vegetation green-up rate at high elevations, although EDW appears to create a warmer environment for vegetation growth. In addition, the negative correlation between T_{gs} and $NDVI_{gs}$ shown in Figure 6 also highlights the adverse effect of warming on vegetation greening within high-elevation areas. Thus, the decline in the vegetation green-up rate with elevation could be attributed to the sustained reductions in water availability for vegetation growth at high elevations.

5. Conclusions

In this study, we quantified and characterized the seasonal changes in vegetation across the Three Rivers Headwaters Region (TRHR) to advance the understanding of linkages between elevation-dependent changes in temperature and precipitation and vegetation greening trends over this region. By mapping the growing-season normalized difference vegetation index ($NDVI_{gs}$) from 1982 to 2015, we were able to evaluate vegetation changes across different years and detect greening trends over time. The results provide evidence that the rate and magnitude of elevation-dependent changes in temperature and precipitation may vary across elevation and vegetation types in alpine regions. The major findings of this study can be summarized as follows:

(i) For $NDVI_{gs}$ across the TRHR, a more or less gradual greening trend from west to east was observed, whereas an uneven yet consistent greening trend was noticed in decadal-scale $NDVI_{gs}$, indicating that about 70% of the land within the TRHR had undergone positive shifts in vegetation between 1982 and 2015.

(ii) The trends in vegetation greening were negatively correlated with the warming rates across the region, whereas the precipitation changes generally exhibited no strong correlation with greening trends.

(iii) The statistical results demonstrate a consistent declining trend in the growing-season green-up rate with increasing elevation within the TRHR. This trend was possibly attributed to the reduced soil water availability induced by the fast increase in warming rates associated with EDW.

(iv) The implementation of alpine conservation and ecological restoration programs within the TRHR appear to be effective in driving barren land greening. However, the presence of anthropogenic activities could also adversely affect local alpine ecosystems, and in some cases was also deemed the primary cause of environmental degradation, e.g., via loss of productivity in grasslands with overgrazing.

Author Contributions: Conceptualization, K.W., Y.Z., J.H., C.C. and T.L.; methodology, K.W.; formal analysis, K.W., Y.Z. and J.H.; data curation, K.W. and C.C.; writing—original draft preparation, K.W. and Y.Z.; writing—review and editing, Y.Z. and T.L.; visualization, Y.Z.; funding acquisition, K.W. and Y.Z. All authors have read and agreed to the published version of the manuscript.

Funding: This research was funded by the Open Research Fund Program of the State Key Laboratory of Hydroscience and Engineering (sklhse-2022-A-02; sklhse-2021-A-02), the Science and Technology Project of Shenzhen Institute of Information Technology (SZIIT2022KJ011), and the Key Research Project of Qinghai Province (2021-SF-A7-1).

Data Availability Statement: The NDVI data are available at https://data.tpdc.ac.cn/en/data/9775f2b4-7370-4e5e-a537-3482c9a83d88/ (accessed on 1 December 2020). The monthly precipitation data are available at https://data.chc.ucsb.edu/products/CHIRPS-2.0/ (accessed on 31 December 2020). The monthly temperature data are available at https://www.ecmwf.int/en/forecasts/dataset/ecmwf-reanalysis-v5 (accessed on 31 December 2020). The monthly ET and PET data are available at https://www.gleam.eu/ (accessed on 1 December 2020).

Conflicts of Interest: The authors declare no conflict of interest.

References

1. Blois, J.L.; Zarnetske, P.L.; Fitzpatrick, M.C.; Finnegan, S. Climate Change and the Past, Present, and Future of Biotic Interactions. *Science* **2013**, *341*, 499–504. [CrossRef] [PubMed]
2. Buermann, W.; Forkel, M.; O'Sullivan, M.; Sitch, S.; Friedlingstein, P.; Haverd, V.; Jain, A.K.; Kato, E.; Kautz, M.; Lienert, S.; et al. Widespread Seasonal Compensation Effects of Spring Warming on Northern Plant Productivity. *Nature* **2018**, *562*, 110–114. [CrossRef] [PubMed]
3. Palazzi, E.; Filippi, L.; von Hardenberg, J. Insights into Elevation-Dependent Warming in the Tibetan Plateau-Himalayas from CMIP5 Model Simulations. *Clim. Dyn.* **2017**, *48*, 3991–4008. [CrossRef]
4. Pepin, N.; Bradley, R.S.; Diaz, H.F.; Baraer, M.; Caceres, E.B.; Forsythe, N.; Fowler, H.; Greenwood, G.; Hashmi, M.Z.; Liu, X.D.; et al. Elevation-Dependent Warming in Mountain Regions of the World. *Nat. Clim. Chang.* **2015**, *5*, 424–430. [CrossRef]
5. Rangwala, I.; Miller, J.R. Climate Change in Mountains: A Review of Elevation-Dependent Warming and Its Possible Causes. *Clim. Chang.* **2012**, *114*, 527–547. [CrossRef]
6. Zhang, L.; Guo, H.D.; Wang, C.Z.; Ji, L.; Li, J.; Wang, K.; Dai, L. The Long-Term Trends (1982–2006) in Vegetation Greenness of the Alpine Ecosystem in the Qinghai-Tibetan Plateau. *Environ. Earth Sci.* **2014**, *72*, 1827–1841. [CrossRef]
7. Wang, G.; Wang, Y.; Li, Y.; Cheng, H. Influences of Alpine Ecosystem Responses to Climatic Change on Soil Properties on the Qinghai–Tibet Plateau, China. *Catena* **2007**, *70*, 506–514. [CrossRef]
8. Liu, J.; Xu, X.; Shao, Q. Grassland Degradation in the "Three-River Headwaters" Region, Qinghai Province. *J. Geogr. Sci.* **2008**, *18*, 259–273. [CrossRef]
9. Chen, H.; Zhu, Q.; Peng, C.; Wu, N.; Wang, Y.; Fang, X.; Gao, Y.; Zhu, D.; Yang, G.; Tian, J.; et al. The Impacts of Climate Change and Human Activities on Biogeochemical Cycles on the Qinghai-Tibetan Plateau. *Glob. Chang. Biol.* **2013**, *19*, 2940–2955. [CrossRef]
10. Cong, N.; Shen, M.; Yang, W.; Yang, Z.; Zhang, G.; Piao, S. Varying Responses of Vegetation Activity to Climate Changes on the Tibetan Plateau Grassland. *Int. J. Biometeorol.* **2017**, *61*, 1433–1444. [CrossRef]
11. Hamid, M.; Khuroo, A.A.; Malik, A.H.; Ahmad, R.; Haq, S.M. Early Evidence of Shifts in Alpine Summit Vegetation: A Case Study From Kashmir Himalaya. Front. *Plant Sci.* **2020**, *11*, 421. [CrossRef]
12. Sarkar, S.; Kafatos, M. Interannual Variability of Vegetation over the Indian Sub-Continent and Its Relation to the Different Meteorological Parameters. *Remote Sens. Environ.* **2004**, *90*, 268–280. [CrossRef]
13. Kumari, N.; Saco, P.M.; Rodriguez, J.F.; Johnstone, S.A.; Srivastava, A.; Chun, K.P.; Yetemen, O. The Grass Is Not Always Greener on the Other Side: Seasonal Reversal of Vegetation Greenness in Aspect-Driven Semiarid Ecosystems. *Geophys. Res. Lett.* **2020**, *47*, e2020GL088918. [CrossRef]
14. Cotto, O.; Wessely, J.; Georges, D.; Klonner, G.; Schmid, M.; Dullinger, S.; Thuiller, W.; Guillaume, F. A Dynamic Eco-Evolutionary Model Predicts Slow Response of Alpine Plants to Climate Warming. *Nat. Commun.* **2017**, *8*, 15399. [CrossRef]
15. Holzinger, B.; Hülber, K.; Camenisch, M.; Grabherr, G. Changes in Plant Species Richness over the Last Century in the Eastern Swiss Alps: Elevational Gradient, Bedrock Effects and Migration Rates. *Plant Ecol.* **2008**, *195*, 179–196. [CrossRef]

16. Srivastava, A.; Sahoo, B.; Raghuwanshi, N.S.; Singh, R. Evaluation of Variable-Infiltration Capacity Model and MODIS-Terra Satellite-Derived Grid-Scale Evapotranspiration Estimates in a River Basin with Tropical Monsoon-Type Climatology. *J. Irrig. Drain. Eng.* **2017**, *143*, 04017028. [CrossRef]
17. Pandey, R.; Kumar, P.; Archie, K.M.; Gupta, A.K.; Joshi, P.K.; Valente, D.; Petrosillo, I. Climate Change Adaptation in the Western-Himalayas: Household Level Perspectives on Impacts and Barriers. *Ecol. Indic.* **2018**, *84*, 27–37. [CrossRef]
18. Xie, Y.; Sha, Z.; Yu, M. Remote Sensing Imagery in Vegetation Mapping: A Review. *J. Plant Ecol.* **2008**, *1*, 9–23. [CrossRef]
19. Boegh, E.; Soegaard, H.; Hanan, N.; Kabat, P.; Lesch, L. A Remote Sensing Study of the NDVI-Ts Relationship and the Transpiration from Sparse Vegetation in the Sahel Based on High-Resolution Satellite Data. *Remote Sens. Environ.* **1999**, *69*, 224–240. [CrossRef]
20. Autovino, D.; Minacapilli, M.; Provenzano, G. Modelling Bulk Surface Resistance by MODIS Data and Assessment of MOD16A2 Evapotranspiration Product in an Irrigation District of Southern Italy. *Agric. Water Manag.* **2016**, *167*, 86–94. [CrossRef]
21. Tucker, C.J. Red and Photographic Infrared Linear Combinations for Monitoring Vegetation. *Remote Sens. Environ.* **1979**, *8*, 127–150. [CrossRef]
22. Bounoua, L.; Collatz, G.J.; Los, S.O.; Sellers, P.J.; Dazlich, D.A. Sensitivity of Climate to Changes in NDVI. *J. Clim.* **2000**, *13*, 2277. [CrossRef]
23. Chakraborty, A.; Seshasai, M.V.R.; Reddy, C.S.; Dadhwal, V.K. Persistent Negative Changes in Seasonal Greenness over Different Forest Types of India Using MODIS Time Series NDVI Data (2001–2014). *Ecol. Indic.* **2018**, *85*, 887–903. [CrossRef]
24. Wang, R.; Gamon, J.A.; Montgomery, R.A.; Townsend, P.A.; Zygielbaum, A.I.; Bitan, K.; Tilman, D.; Cavender-Bares, J. Seasonal Variation in the NDVI–Species Richness Relationship in a Prairie Grassland Experiment (Cedar Creek). *Remote Sens.* **2016**, *8*, 128. [CrossRef]
25. Kumari, N.; Srivastava, A.; Dumka, U.C. A Long-Term Spatiotemporal Analysis of Vegetation Greenness over the Himalayan Region Using Google Earth Engine. *Climate* **2021**, *9*, 109. [CrossRef]
26. Shahzaman, M.; Zhu, W.; Bilal, M.; Habtemicheal, B.A.; Mustafa, F.; Arshad, M.; Ullah, I.; Ishfaq, S.; Iqbal, R. Remote Sensing Indices for Spatial Monitoring of Agricultural Drought in South Asian Countries. *Remote Sens.* **2021**, *13*, 2059. [CrossRef]
27. Bastos, A.; Ciais, P.; Park, T.; Zscheischler, J.; Yue, C.; Barichivich, J.; Myneni, R.B.; Peng, S.; Piao, S.; Zhu, Z. Was the Extreme Northern Hemisphere Greening in 2015 Predictable? *Environ. Res. Lett.* **2017**, *12*, 44016. [CrossRef]
28. Ding, Z.; Peng, J.; Qiu, S.; Zhao, Y. Nearly Half of Global Vegetated Area Experienced Inconsistent Vegetation Growth in Terms of Greenness, Cover, and Productivity. *Earth's Futur.* **2020**, *8*, e2020EF001618. [CrossRef]
29. Lucht, W.; Prentice, I.C.; Myneni, R.B.; Sitch, S.; FriedLingstein, P.; Cramer, W.; Bousquet, P.; Buermann, W.; Smith, B. Climatic Control of the High-Latitude Vegetation Greening Trend and Pinatubo Effect. *Science* **2002**, *296*, 1687–1689. [CrossRef]
30. Mao, J.; Shi, X.; Thornton, P.E.; Piao, S.; Wang, X. Causes of Spring Vegetation Growth Trends in the Northern Mid–High Latitudes from 1982 to 2004. *Environ. Res. Lett.* **2012**, *7*, 14010. [CrossRef]
31. Srivastava, A.; Saco, P.M.; Rodriguez, J.F.; Kumari, N.; Chun, K.P.; Yetemen, O. The Role of Landscape Morphology on Soil Moisture Variability in Semi-arid Ecosystems. *Hydrol. Process.* **2021**, *35*, e13990. [CrossRef]
32. Myneni, R.B.; Keeling, C.D.; Tucker, C.J.; Asrar, G.; Nemani, R.R. Increased Plant Growth in the Northern High Latitudes from 1981 to 1991. *Nature* **1997**, *386*, 698–702. [CrossRef]
33. Geert, J.; Balmaseda, M.A.; Ferranti, L.; Stockdale, T.N.; Anderson, D. Evaluation of Atmospheric Fields from the ECMWF Seasonal Forecasts over a 15-Year Period. *J. Clim.* **2005**, *18*, 3250–3269. [CrossRef]
34. Trujillo, E.; Molotch, N.P.; Goulden, M.L.; Kelly, A.E.; Bales, R.C. Elevation-Dependent Influence of Snow Accumulation on Forest Greening. *Nat. Geosci.* **2012**, *5*, 705–709. [CrossRef]
35. Piao, S.; Nan, H.; Huntingford, C.; Ciais, P.; Friedlingstein, P.; Sitch, S.; Peng, S.; Ahlström, A.; Canadell, J.G.; Cong, N.; et al. Evidence for a Weakening Relationship between Interannual Temperature Variability and Northern Vegetation Activity. *Nat. Commun.* **2014**, *5*, 5018. [CrossRef]
36. Zhang, Y.; Zhang, S.; Zhai, X.; Xia, J. Runoff Variation and Its Response to Climate Change in the Three Rivers Source Region. *J. Geogr. Sci.* **2012**, *22*, 781–794. [CrossRef]
37. Wang, K.; Xu, Q.; Li, T. Does Recent Climate Warming Drive Spatiotemporal Shifts in Functioning of High-Elevation Hydrological Systems? *Sci. Total Environ.* **2020**, *719*, 137507. [CrossRef]
38. Xu, W.; Gu, S.; Zhao, X.Q.; Xiao, J.; Tang, Y.; Fang, J.; Zhang, J.; Jiang, S. High Positive Correlation between Soil Temperature and NDVI from 1982 to 2006 in Alpine Meadow of the Three-River Source Region on the Qinghai-Tibetan Plateau. *Int. J. Appl. Earth Obs. Geoinf.* **2011**, *13*, 528–535. [CrossRef]
39. Harris, R.B. Rangeland Degradation on the Qinghai-Tibetan Plateau: A Review of the Evidence of Its Magnitude and Causes. *J. Arid Environ.* **2010**, *74*, 1–12. [CrossRef]
40. Yang, J.; Ding, Y.; Chen, R. Spatial and Temporal of Variations of Alpine Vegetation Cover in the Source Regions of the Yangtze and Yellow Rivers of the Tibetan Plateau from 1982 to 2001. *Environ. Geol.* **2006**, *50*, 313–322. [CrossRef]
41. Zhang, J.; Yao, F.; Zheng, L.; Yang, L. Evaluation of Grassland Dynamics in the Northern-Tibet Plateau of China Using Remote Sensing and Climate Data. *Sensors* **2007**, *7*, 3312–3328. [CrossRef] [PubMed]
42. Pan, T.; Zou, X.; Liu, Y.; Wu, S.; He, G. Contributions of Climatic and Non-Climatic Drivers to Grassland Variations on the Tibetan Plateau. *Ecol. Eng.* **2017**, *108*, 307–317. [CrossRef]
43. Shen, X.; An, R.; Feng, L.; Ye, N.; Zhu, L.; Li, M. Vegetation Changes in the Three-River Headwaters Region of the Tibetan Plateau of China. *Ecol. Indic.* **2018**, *93*, 804–812. [CrossRef]

44. Chen, C.; Li, T.; Sivakumar, B.; Li, J.; Wang, G. Attribution of Growing Season Vegetation Activity to Climate Change and Human Activities in the Three-River Headwaters Region, China. *J. Hydroinformatics* **2020**, *22*, 186–204. [CrossRef]
45. Jiang, C.; Zhang, L. Ecosystem Change Assessment in the Three-River Headwater Region, China: Patterns, Causes, and Implications. *Ecol. Eng.* **2016**, *93*, 24–36. [CrossRef]
46. Zhang, S.; Hua, D.; Meng, X.; Zhang, Y. Climate Change and Its Driving Effect on the Runoff in the "Three-River Headwaters" Region. *Dili Xuebao/Acta Geogr. Sin.* **2011**, *66*, 13–24. [CrossRef]
47. Zheng, Y.; Han, J.; Huang, Y.; Fassnacht, S.R.; Xie, S.; Lv, E.; Chen, M. Vegetation Response to Climate Conditions Based on NDVI Simulations Using Stepwise Cluster Analysis for the Three-River Headwaters Region of China. *Ecol. Indic.* **2018**, *92*, 18–29. [CrossRef]
48. Liu, L.; Wang, Y.; Wang, Z.; Li, D.; Zhang, Y.; Qin, D.; Li, S. Elevation-Dependent Decline in Vegetation Greening Rate Driven by Increasing Dryness Based on Three Satellite NDVI Datasets on the Tibetan Plateau. *Ecol. Indic.* **2019**, *107*, 105569. [CrossRef]
49. Tucker, C.; Pinzon, J.; Brown, M.; Slayback, D.; Pak, E.; Mahoney, R.; Vermote, E.; El Saleous, N. An Extended AVHRR 8-Km NDVI Dataset Compatible with MODIS and SPOT Vegetation NDVI Data. *Int. J. Remote Sens.* **2005**, *26*, 4485–4498. [CrossRef]
50. Beck, H.E.; Mcvicar, T.R.; Dijk, A.; Schellekens, J.; Jeu, R.; Bruijnzeel, L.A. Global Evaluation of Four AVHRR–NDVI Data Sets: Intercomparison and Assessment against Landsat Imagery. *Remote Sens. Environ.* **2011**, *115*, 2547–2563. [CrossRef]
51. HOLBEN, B.N. Characteristics of Maximum-Value Composite Images from Temporal AVHRR Data. *Int. J. Remote Sens.* **1986**, *7*, 1417–1434. [CrossRef]
52. Martens, B.; Miralles, D.G.; Lievens, H.; van der Schalie, R.; de Jeu, R.A.M.; Fernández-Prieto, D.; Beck, H.E.; Dorigo, W.A.; Verhoest, N.E.C. GLEAM~v3: Satellite-Based Land Evaporation and Root-Zone Soil Moisture. *Geosci. Model Dev.* **2017**, *10*, 1903–1925. [CrossRef]
53. Miralles, D.G.; Holmes, T.R.H.; De Jeu, R.A.M.; Gash, J.H.; Meesters, A.G.C.A.; Dolman, A.J. Global Land-Surface Evaporation Estimated from Satellite-Based Observations. *Hydrol. Earth Syst. Sci.* **2011**, *15*, 453–469. [CrossRef]
54. Wolf, L.J.; Anselin, L.; Arribas-Bel, D. Stochastic Efficiency of Bayesian Markov Chain Monte Carlo in Spatial Econometric Models: An Empirical Comparison of Exact Sampling Methods. *Geogr. Anal.* **2018**, *50*, 97–119. [CrossRef]
55. Yue, S.; Wang, C.Y. Applicability of Prewhitening to Eliminate the Influence of Serial Correlation on the Mann-Kendall Test. *Water Resour. Res.* **2002**, *38*, 4-1–4-7. [CrossRef]
56. Zhang, X.; Vincent, L.A.; Hogg, W.D.; Niitsoo, A. Temperature and Precipitation Trends in Canada during the 20th Century. *Atmos.-Ocean* **2000**, *38*, 395–429. [CrossRef]
57. Yang, Q.; Zhang, H.; Peng, W.; Lan, Y.; Luo, S.; Shao, J.; Chen, D.; Wang, G. Assessing Climate Impact on Forest Cover in Areas Undergoing Substantial Land Cover Change Using Landsat Imagery. *Sci. Total Environ.* **2019**, *659*, 732–745. [CrossRef]
58. Liu, Y.; Li, Y.; Li, S.; Motesharrei, S. Spatial and Temporal Patterns of Global NDVI Trends: Correlations with Climate and Human Factors. *Remote Sens.* **2015**, *7*, 13233–13250. [CrossRef]
59. Kuang, X.; Jiao, J.J. Review on Climate Change on the Tibetan Plateau during the Last Half Century. *J. Geophys. Res. D. Atmos. JGR* **2016**, *121*, 3979–4007. [CrossRef]
60. Zhang, D.; Huang, J.; Guan, X.; Chen, B.; Zhang, L. Long-Term Trends of Precipitable Water and Precipitation over the Tibetan Plateau Derived from Satellite and Surface Measurements. *J. Quant. Spectrosc. Radiat. Transf.* **2013**, *122*, 64–71. [CrossRef]
61. Seddon, A.W.R.; Macias-Fauria, M.; Long, P.R.; Benz, D.; Willis, K.J. Sensitivity of Global Terrestrial Ecosystems to Climate Variability. *Nature* **2016**, *531*, 229–232. [CrossRef] [PubMed]

remote sensing MDPI

Article

Tracking Deforestation, Drought, and Fire Occurrence in Kutai National Park, Indonesia

Ryan Guild [1,2], Xiuquan Wang [1,2,*] and Anne E. Russon [3]

1 Canadian Centre for Climate Change and Adaptation, University of Prince Edward Island, St. Peters Bay, PE C0A 2A0, Canada
2 School of Climate Change and Adaptation, University of Prince Edward Island, Charlottetown, PE C1A 4P3, Canada
3 Psychology Department, Glendon College of York University, North York, ON M4N 3M6, Canada
* Correspondence: xxwang@upei.ca; Tel.:+1-902-739-2244

Abstract: The dry lowland and mangrove forests of Kutai National Park (KNP) in Indonesia provide invaluable ecosystem services to local human populations (>200,000 in number), serve as immense carbon sinks to recapture anthropogenic emissions, and safeguard habitats for thousands of wildlife species including the critically endangered Northeast Bornean orangutan (*Pongo pygmaeus morio*). With recent reports of ongoing illegal logging and large-scale wildfires within this National Park, we sought to leverage the extensive catalogue and processing power of Google Earth Engine to track the rates and influences of forest loss within KNP over various time periods since 1997. We present estimates of forest loss from the Hansen Global Forest Change v1.9 dataset (2000–2021) which detected a loss of 15% (272 km^2) of forest cover within KNP since 2000, half of which (137 km^2) coincided with the El Niño-induced wildfires of 2015–2016. Using the MCD64A1 C6.1 MODIS dataset, we found significant spatial overlap between burned area and forest loss detections during the 2015–2016 period but identified considerable omissions in the burned area dataset over smallholder farms within KNP. We discuss the implications of deforestation in areas of primary orangutan habitat and how patterns of forest loss have influenced drought and fire dynamics within KNP. Finally, we compare time-series estimates of precipitation, the ENSO index, burned area, and forest loss to demonstrate that fire risk within KNP depends largely—but not exclusively—on drought severity, and that rates of non-fire (gradual) and fire-related (extreme) forest loss threaten the remaining forests of this National Park.

Keywords: deforestation; wildfire; drought; satellite; Kutai National Park; Google Earth Engine

Citation: Guild, R.; Wang, X.; Russon, A.E. Tracking Deforestation, Drought, and Fire Occurrence in Kutai National Park, Indonesia. *Remote Sens.* **2022**, *14*, 5630. https://doi.org/10.3390/rs14225630

Academic Editor: Sandra Eckert

Received: 13 October 2022
Accepted: 5 November 2022
Published: 8 November 2022

1. Introduction

Kutai National Park (KNP) represents one of the last relatively intact areas of protected lowland and mangrove forest of Indonesian Borneo (Kalimantan) [1]. This protected area is home to over 300 species of birds, a significant population of the critically endangered Northeast Bornean orangutan (*Pongo pygmaeus morio*), and some of the last tracts of ecologically significant and commercially valuable Bornean ironwood trees in the region [1,2]. In addition to catastrophic wildfires reported during the extreme El Niño droughts of 1983–1984, 1997–1998 and 2015–2016 [3,4], KNP has continuously experienced illegal logging from nearby human populations for timber, expanding agricultural fields, and land speculation which the National Park authority struggles to address [5,6].

The use of high resolution (<30 m) multiband satellite imagery in such fields for ecosystem monitoring and biodiversity conservation has kept pace with gradual improvements in its access and availability [7]; today, it represents the primary resource by which researchers track changes in forest cover over time. Yet only two studies have attempted to track the extent of forest loss within KNP using satellite imagery. [5] first used visual inspection of Landsat imagery to detect areas of loss between 1990 and 2007 over KNP, while [8] used supervised classification of Landsat imagery to detect forest loss in the eastern (populated) region of the park between 2006 and 2009. The latter study estimated an annual deforestation rate of 2.15% in the park's eastern region and attributed much of it to conversion to settlements and agricultural land. At this rate, complete loss of forest in the eastern half of KNP would occur by the mid-century; however, plans to relocate the capital city of Indonesia from Java to East Kalimantan by 2024 are likely to intensify pressures on KNP's forests and accelerate this time frame. This would represent a substantial loss to both the *morio* orangutan habitat within Indonesia and to the ecosystem services upon which local residents—more than 200,000 people within and immediately surrounding KNP—rely for fresh water, forest resources, and tourism (see [9]). Thus, an updated analysis of the rates, extents, and causes of forest loss within KNP using modern sensing and processing technologies is imperative.

The objective of this study is to track forest loss, drought, and fires over the past two decades in KNP using historical collections of thermal and spectral satellite imagery mostly sourced from Google Earth Engine (GEE). Specifically, we utilize the Hansen Global Forest Change (v1.9) dataset to track annual forest loss (2000–2021); the MODIS MCD64A1 C6.1 product to track fire-affected areas (2002–2021); and the CHIRPS (Climate Hazard InfraRed Precipitation with Stations) Pentad dataset to track drought patterns within KNP (1997–2021). We attempt to validate each of these using the appropriate data and methods. The data presented here provide information to researchers and governments to demarcate vulnerable areas of protected forests and to devise strategies to avert future forest loss in this National Park.

2. Data and Methods

2.1. Study Area

KNP covers 1927 km^2 along the eastern coast of Borneo in the Indonesian province of East Kalimantan (Figure 1) [9]. The ecology of the region comprises inland areas of primary and secondary lowland forests (<250 m above sea level) dominated by Dipterocarpaceae and Lauraceae family trees and coastal areas featuring mangrove forests and brackish estuaries [9]. KNP is bounded by two cities at its southeastern (Bontang) and northeastern (Sangatta) corners (Figure 1) that have each seen considerable population expansion since the park's formal declaration in 1995, which has prompted the excision of several illegally settled areas from KNP's formal boundaries [10]. The Sangatta River forms part of KNP's eastern and northern boundaries, while the remaining northern, western, and southern perimeters abut a contiguous series of industrial concessions, including an open-pit coal mine, plus wood fibre, timber, and oil palm plantations (Figure 1). Several small villages of a few thousand people remain within the bounds of KNP in the eastern and coastal areas [9] where many active and abandoned smallholder farms dominate the landscape (pers. obs.). Oil palm monocultures represent a large portion of planted agriculture within KNP primarily because of the high returns on land and labour afforded by the crop [11]. Encouragement by the Indonesian government to cultivate oil palms as a poverty-reduction strategy and growing contracts offered by palm oil companies in the region [12,13] have further incentivized farmers to illegally establish and enlarge oil palm plantations throughout eastern KNP.

In an intact old-growth state prior to 1970, the forests of KNP rarely experienced wildfires due to low combinations of fire hazards (i.e., dryness and fuel loads) and fire risks (i.e., prevalence of ignition sources) [14]. While ignition sources including lightning, swidden agriculture, and burning coal seam have existed in this region for millenia, neither naturally nor human-induced fires are believed to have driven large-scale deforestation here prior to 1970 [14,15]. However, fire hazards in this region have since increased from more frequent extreme droughts and unprecedented amounts of logging-related fuel loads, as have fire risks from the proliferation of croplands that use fire for various purposes (see [6]). Over the past 40 years, three super El Niño droughts have facilitated the spread of wildfires throughout KNP's forests. In 1982/83, around half of the park's forests burned [16] and high mortality rates were reported for canopy trees, figs, and lianas [17], while in 1997/98, a much larger extent of the park burned and most of its forests were reset to a secondary successional stage [4,18]. The mixed-dipterocarp forests of this region can recover their structure when left undisturbed for a period of 10–20 years [19,20]. Most recently, during the super El Niño drought of 2015/16, island-wide studies revealed a significant number of fire detections over KNP [3,21], but the extent of—and damage caused by—burning during this event is yet unknown.

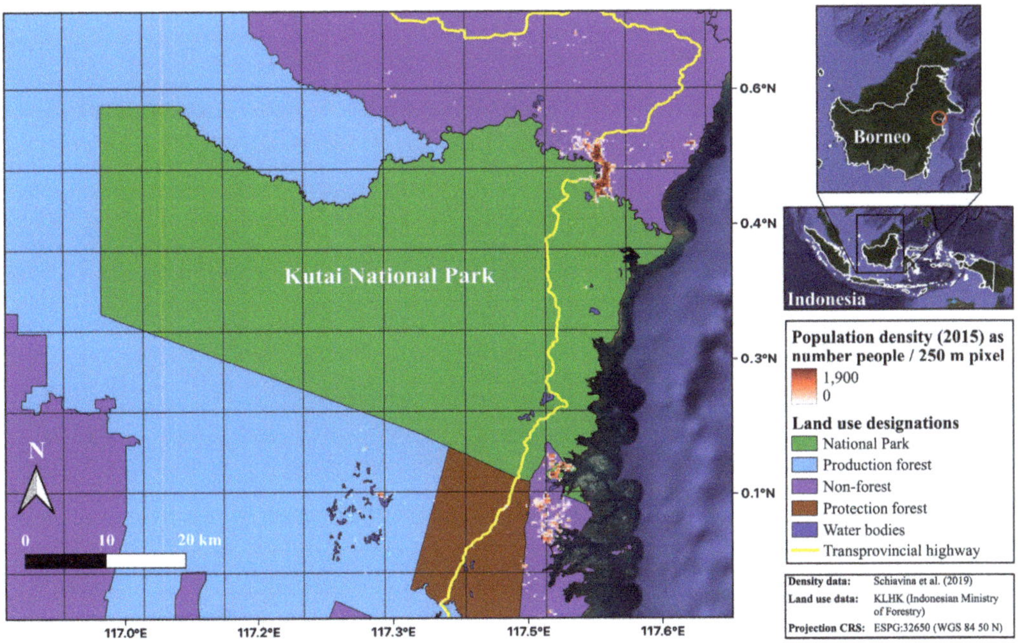

Figure 1. Land use designations and population density within and around Kutai National Park of East Kalimantan, Indonesia. Density displayed as estimated number of people per 250 m × 250 m grid cell [22] Southeastern and northeastern population centers represent the respective cities of Bontang and Sangatta.

2.2. Datasets

2.2.1. Hansen Global Forest Change (HGFC)

The HGFC v1.9 dataset is a global gridded (30 m) forest loss product available on GEE that uses a time-series change detection analysis of multispectral Landsat TM (L4 and L5), ETM+ (L7), and OLI/TIRS (L8) scenes to track annual changes in global forest cover between 2000 and 2021 [23,24]. The first release of the dataset (2000–2012) reported a producer's accuracy of 83.1% (overall accuracy 99.5%) over tropical domains [23]. The most recent version (v1.9) reports an improved detection accuracy since 2013 due to the inclusion of L8 data; thus, we only discretize post-2013 years annually in our analyses. Forest loss in HGFC is defined as a stand-replacement disturbance [of tree canopy cover] via removals or mortality and it uses a detection algorithm described by [24,25]. This dataset does not account for potential regrowth after a pixel has been assigned as 'lost', which prevents discernment of outcomes of canopy loss (i.e., regrowth, further degradation, conversion to other land uses) in our study area. Although high-resolution satellite imagery indicates that conversion to settlements and croplands predominate these outcomes over the frontline [loss] areas of eastern KNP, additional data and analysis is required to quantify the capacity for—and extents of—regrowth following canopy loss in our study area.

The 'treecover2000' and 'lossYear' bands of HGFC v1.9 contain classified pixels of (1) global tree canopy cover (vegetation >5 m in height) in the year 2000 (baseline) and (2) annual canopy loss between this baseline and 2021, respectively. We clipped these bands to the boundary of KNP, reduced them to area estimates using GEE functions, and exported them for colour rendering in QGIS. The full script (see Data Availability Statement) used to generate the raster images of annual forest loss in KNP was co-opted from a Google-based Earth Engine tutorial (https://developers.google.com/earth-engine/tutorials/tutorial_forest_02, accessed on 1 September 2022).

We employ two additional datasets to validate HGFC-detected areas of forest canopy loss over two sample regions within KNP between 2017–2020. First, we utilize the novel GEE-based Forest Canopy Disturbance Monitoring (FCDM) Tool using the developer-recommended parameter values to detect annual canopy loss within the sample regions. The FCDM Tool detects new canopy openings from L8 scenes across user-defined time periods and reports an overall accuracy of 77.8% within tropical evergreen rainforests [26]. Second, we visually inspected high-resolution (5 m) cloud-free composite images (June–November) via Planet/NICFI on GEE to manually delineate canopy loss over the sample regions for each year of the validation period. For this "reference" dataset (as referred to in this study), we first traced the extent of non-forest areas in the 2016 composite over both sample regions, then overlayed composites from each subsequent year to trace additional areas of canopy loss. Since each dataset uses a different temporal approach to detect canopy loss between years (see Figure A1), we focus our comparison on the total—rather than annual-extent(s) of detected forest loss between datasets.

2.2.2. MCD64A1 Burned Area

The MCD64A1 C6.1 burned area (BA) product (500 m) contains estimates of global monthly BA based on a time-series detection analysis of daily MODIS C6.1 observations between 07/2000 to 02/2022 [27]. Its most recent version (C6.1) reports an improved detection accuracy of BA (63% producer's; 97% overall), yet it still performs poorly over croplands due to spectral confusion between burnt and harvested land and a limited ability to detect small-burning fires [27–29]. We first generated annual mosaics of BA detections within KNP between 2002–2021 in GEE (see Data Availability and Source Code), and then colour-rendered and reduced each mosaic to area estimates (using the GRASS r.series algorithm) in QGIS. Using active fire (AF) observations from NASA's VIIRS (Variable Infrared Imaging Radiometer Suite) product (375 m) we graphed annual aggregates of daily fire counts within KNP (2012–2021) and compared the location of AF and BA detections during the wildfire years of 2015–2016. While the two datasets detect different signals of fire activity—thermal (AF) vs. post-fire burn scars (BA)—the AF (VIIRS) product can detect relatively smaller fires and is less susceptible to harvest-related confusion, making it valuable to compare against BA detections over cropland regions [30,31].

We attempted to validate the MCD64A1 dataset for the 2015 wildfires within KNP following the burn severity mapping approach developed by UN-SPIDER (United Nations Space-based Information for Disaster Management and Emergency Response). We first selected two sample areas of detected BA and determined the estimated date(s) of burning for each according to MCD64A1 and VIIRS datasets. We also selected one 'unburned' sample area (without BA detections) over SW KNP where numerous AF were detected in 2015. In GEE, we (1) generated cloud-free pre- and post-fire composite images over each sample region using L8 scenes, (2) calculated the normalized burn ratio (NBR) of each composite, and (3) produced a differenced normalized burn ratio (dNBR) image to classify burn severity for each of the three sample regions (see Data Availability Statement). We classified dNBR images using the burn severity thresholds and colour coding developed by the USGS (United States Geological Survey) and UN-SPIDER, respectively, and compared the classified images against the MCD64A1-detected burned areas over each sample region.

We chose to generate cloud-masked composite scenes of pre- and post-fire conditions to address the challenge of year-round high cloud cover that obscures most satellite scenes over KNP. This approach is not recommended given the potential for signal interference over the composited time-period (e.g., through changes in vegetation greenness, agricultural harvests, etc.). However, our approach detected signals of browned (i.e., burnt) areas that are uncommon in scenes or composites from non-wildfire years over this region, and thus this offers value to our analysis of BA in KNP.

2.2.3. Climate Hazard InfraRed Precipitation and Station (CHIRPS)

The CHIRPS Pentad reanalysis gridded rainfall dataset combines infrared measurements of Cold Cloud Duration taken every 5 days (from the Tropical Rainfall Measuring Mission Multisatellite Precipitation Analysis) with in situ weather station data sourced from the World Meteorological Organization's Global Telecommunication System [32]. This global dataset has a spatial resolution of 5.5 km and is available from January 1981 to present.

To generate a time series of monthly rainfall estimates, we first used a GEE mapping function to calculate total monthly precipitation (mm) per pixel across KNP for each year between 2002–2021 (date range of MCD64A1 data), then averaged the pixel estimates across KNP for each month using a reducer function (see Data Availability Statement). Monthly estimates were exported from GEE as csv files and transformed into rolling three-monthly total precipitation (e.g., DJF, JFM, FMA, etc.) to reduce monthly variation and to better reflect medium-term drought conditions (cf. [33]). To evaluate spatial variations in rainfall patterns, we also mapped the 25-year (1997–2021) average of total fire-season (Aug–Nov) rainfall for each pixel across KNP using the GRASS r.series algorithm in QGIS.

CHIRPS-based rainfall estimates were compared with daily observations from the only available rainfall gauge within KNP, located at the Bendili research station (est. July 2010; 0.56°N, 117.4°E; founded and operated by the Orangutan Kutai Project) in the NE corner of the park (Figure A2). It is important to note that this comparison serves to visualize conformity between datasets at a single location within KNP and that a standard validation analysis could not be performed without access to additional rainfall observations.

3. Results and Discussion

3.1. Forest Loss within KNP

Figure 2 depicts areas of baseline tree canopy cover (>30% coverage in 2000) and canopy loss throughout KNP between 2001–2021, as detected by the HGFC v1.9 dataset. Baseline canopy cover is shown in green and canopy loss is presented as an aggregate between 2001–2012 but annually thereafter (2013–2021) to reflect the greater detection accuracy of the post-2013 period (Figure 2). Much of the forest loss detected since 2000 is located along the trans-provincial highway that runs north-south through the eastern region of KNP (est. 1991) where illegal settlements and agricultural fields have proliferated over time. Outside of this highway region, notable patches of recent (dark blue) forest loss were also detected along the park's southern border (Figure 2A). The southwest patch borders an industrial tree plantation (oil palm and acacia) [34,35] while the easternmost patch borders an industrial open-pit coal mine [36].

To assess the validity of the HGFC estimates over KNP, we compared them with estimates from two other datasets for the years 2017–2020 over two sample areas of NE KNP (Figure 3). The three datasets conformed well in terms of the spatial pattern and area of canopy loss they detected over the two validation areas (Figure 3). Only two estimates by the reference dataset (in 2018 and 2020 over validation area B) were markedly different from those of the detection datasets (Figure 3)—perhaps an artifact of the different approaches that each use in the temporal detection of canopy loss (Figure A1). Overall, our validation supports the detection performance of the HGFC v1.9 dataset over this region, yet a standard validation analysis is warranted to quantify its accuracy over KNP.

The significant spike of canopy loss detected in 2016 (Figure 2C) coincides with the super El Niño drought of 2015–2016 that sparked major wildfires throughout the region according to Russon AE (pers. obs.) and [3,21]. Although this spike was detected in 2016, more than twice as many active fires and seven-times as much burned area were detected throughout KNP in 2015 than in 2016 (Figure 4). Since the HGFC dataset relies on clear-sky observations for accurate detection, it is plausible that wildfire smoke and high cloud cover obscured most Landsat scenes during the 2015 fire season (August–November) so that most of the fire-related forest loss from 2015 could not be detected until the following year. This is one known caveat of the HGFC dataset, making it more useful in detecting multi-year trends in forest canopy loss than deriving definitive area estimates per year. Overall, the HGFC dataset detected 272 km^2 of total forest canopy loss between 2001–2021 (15% of the 2000 baseline canopy cover), half of which (137 km^2) was detected between 2015–2016. On average, the annual rate of loss within KNP during non-wildfire years between 2013–2021 was 5.6 km^2 ($\pm$1.3) per year, or a loss of 0.3% of baseline forest cover each year.

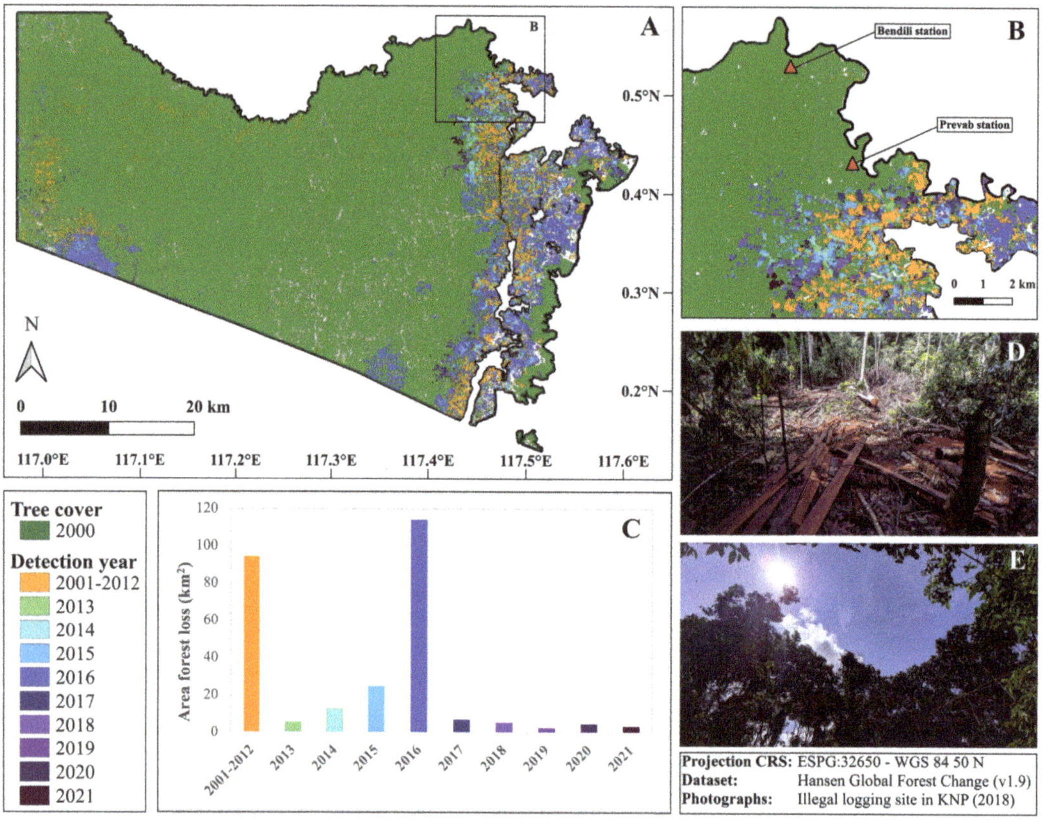

Figure 2. Classified areas of forest loss within Kutai National Park (**A**) and the Greater Mentoko Area [GMA] (**B**) between 2001–2021 and corresponding area estimates (**C**) as detected by the Hansen Global Forest Change (v1.9) dataset. Only years post-2013 are discretized annually due to higher detection accuracy of the 2013–2021 period. Tree cover with >70% canopy closure in year 2000 are depicted in green. Images of an illegal logging site (**D**) and the resultant canopy opening (**E**) in the GMA from 2018 by RG.

The only area of KNP for which accounts of pre-disturbance forest conditions are available is the Greater Mentoko Area (GMA) of NE KNP (Figure 2B). As the site of the first orangutan research in Indonesia in 1970 [37], its forests were described as "near-pristine", free of major logging activities, and only affected by natural disturbances such as wind-throws, droughts, and recurrent river flooding [38,39]. Since 2000, nearly half of its original forest cover has been lost (Figure 2B) to both fire- and non-fire disturbances, the latter of which include illegal logging, settlements, and agricultural expansion to support the local population (Sangatta) that has grown 500-fold over the past five decades [6]. Attributing precise contributions of fire- and non-fire disturbances to annual forest loss within KNP, however, requires further analysis beyond the scope of this study.

The forest loss detected in the GMA has replaced primary habitat for the resident orangutans in this area (est. 1000 individuals) [9] and placed their home ranges in closer proximity to humans and their livelihoods—exacerbating the record-setting rates of human-orangutan conflict reported within KNP [40,41]. Further loss in this region along a westward trajectory will also have negative implications for the local human population by weakening crucial forest-related ecosystem services (e.g., flood management, water quality, climate regulation, availability of non-timber forest products, etc.) and by compromising park tourism that relies in large part on the forested orangutan habitats around the Prevab research station (Figure 2B).

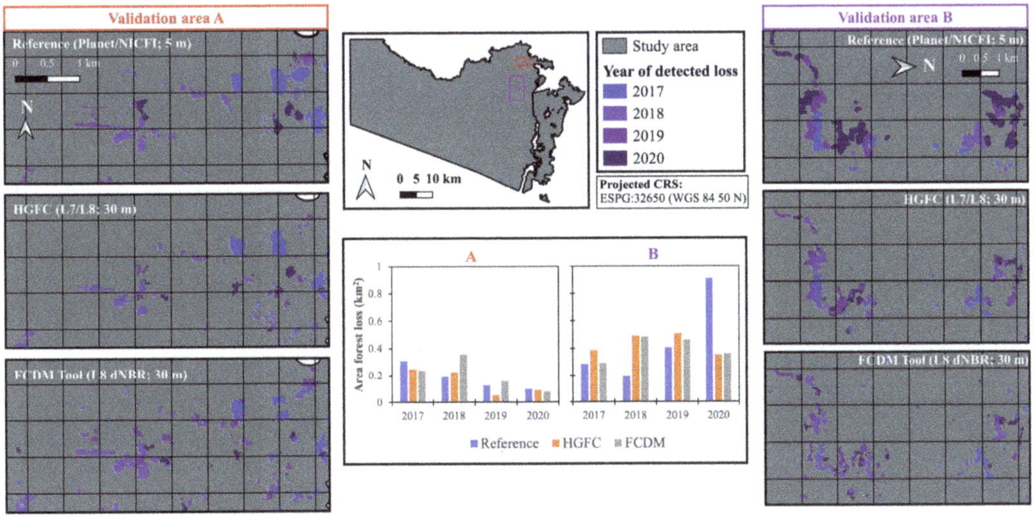

Figure 3. Validation of the Hansen Global Forest Change (HGFC; v1.9) product for the years 2017–2020 over two sample regions (**A**,**B**) in northeastern KNP using (1) high-resolution Planet/NICFI satellite imagery (Reference dataset) and (2) the Forest Canopy Disturbance Monitoring Tool (FCDM Tool) developed by [26].

3.2. Hotspot Mapping within KNP

Between 2002–2021, a significant amount of BA was detected throughout KNP in 2004, 2015, and 2016 by the MCD64A1 dataset (Figure 4). The BA from 2015 cover settlement areas and agricultural fields in NE KNP and coincides with recently disturbed forests in SW KNP that had lost canopy cover only a year prior (Figures 2A and 4A). Date-of-burning estimates suggest that fires (2015) in the latter region spread from the neighbouring industrial plantation into the recently disturbed area, resulting in a footprint of burning that dwarfs the patches of earlier (2014) forest loss (Figures 2A and 4A). The effect of recent canopy openings facilitating the spread and development of wildfires that cause further deforestation has been well documented in past studies in this region [42,43].

AF observations (VIIRS) revealed a significant footprint of fires in eastern KNP between 2015–2016 that coincide with detected forest loss over this period (Figure 2) but not with BA (Figure 4D). This is perhaps because relative to BA datasets, AF products are known to detect more small-scale fires [28], such as those originating from the individual smallholder farms that predominate eastern KNP. Lower detection accuracy over croplands and other areas of predominantly small fires is a known limitation of the MCD64A1 dataset, owing largely to its coarse resolution (500 m) [27,28,44].

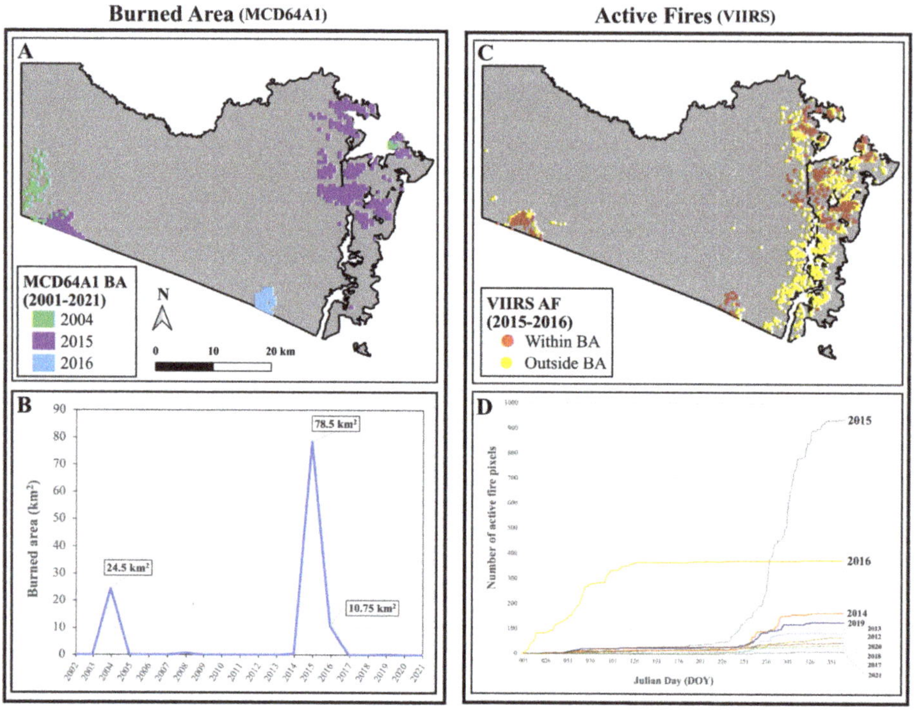

Figure 4. MCD64A1 (left): location (**A**) and annual estimates (**B**) of detected burned area (BA) within KNP between 2002–2021. VIIRS (right): active fire detections within and outside the perimeters of detected BA in KNP between 2015–2016 (**C**) and cumulative active fire detections across days of the year (DOY) between 2012–2021 within KNP (**D**).

To further assess the validity of MCD64A1-detected BA over KNP, we followed a validation approach defined by UN-SPIDER that maps burn severity using the difference of normalized burn ratios (dNBR) between pre- and post-fire satellite images (Figure 5). We chose to validate BA from the 2015 wildfires using L8 cloud-free composite images over three sample areas of KNP, covering both burned (regions A and C) and non-burned (region B) land according to MCD64A1 (Figure 5; see caveats in Section 2.2.2). The MCD64A1 dataset performed well over the SW region of KNP where croplands are absent, but poorly over the other sampled areas where croplands predominate; it missed all of the moderate- to severe-burn scars over SE KNP (Figure 5) despite a large number of active fire detections over this region in 2015 (Figure 4). Despite its poor performance over croplands, the MCD64A1 dataset still provides a good indicator of years of high BA within KNP (Figure 4), but alternative BA datasets and/or mapping methods are required to estimate the contributions of fire to forest loss in this region.

3.3. Climatic Influences of Fire and Forest Loss within KNP

According to estimates of total "fire season" (Aug–Nov) rainfall over the past 25 years, the deforested eastern region of the park receives the least amount of rainfall, as low as 5.5 mm/day (500 mm/3 mo.) on average around the bounds of Sangatta city (Figure 6A). Previous research over Borneo and other tropical regions has found patterns of lower rainfall and higher ambient temperatures over deforested areas (compared to unlogged), with the most pronounced effects observed during the dry season and El Niño-related droughts [45,46]. Deforested areas are also more accessible to humans, experience higher temperature extremes, and contain relatively higher fuel loads from logging debris and woody successional plants, all of which contribute to an elevated fire hazard/risk in such areas [43,47]. These implications are supported by studies that have connected rampant deforestation patterns throughout eastern Borneo with stronger El Niño droughts [45,47] and more severe and widespread wildfires [42].

During the 2015 fire season, rainfall estimates (<2.5 mm/day) and single-point observations (<0.75 mm/day) reveal extreme drought conditions over much of the burned area (Figure 6B), well below the reported 4–5 mm/day fire risk threshold for this region [3,45]. These severe drought conditions—driven by a 20-year peak of El Niño strength—combined with a prevalence of potential ignition sources (see [6]) and widespread degradation to facilitate large-scale wildfires within KNP (Figure 7). It is likely that such wildfires contributed both directly (i.e., drought- and fire-related mortality) and indirectly (i.e., facilitating clearing/conversion) to the spike in detected forest loss over this period (Figure 7), but further analysis is necessary to confirm this interpretation.

Rainfall estimates also suggest that most—but not all—of the six El Niño events over the past 20 years have induced drought conditions below the fire risk threshold for this region (Figure 7). Spikes in BA were only detected during one weak (2004–2005) and one strong (2015–2016) EN event, but none were detected during several moderate-strength EN events that induced comparable rainfall deficits (Figure 7). Assuming that the MCD64A1 dataset is a sufficient indicator of large-scale fire occurrences within KNP, we posit that other factors, such as temperature and the incidental presence of ignition sources, have regulated the intensity and spread of fire throughout KNP during severe droughts. However, ongoing deforestation in KNP and predictions of more frequent and extreme super El Niño events [48–51] will worsen the fire hazard of this reserve during future droughts.

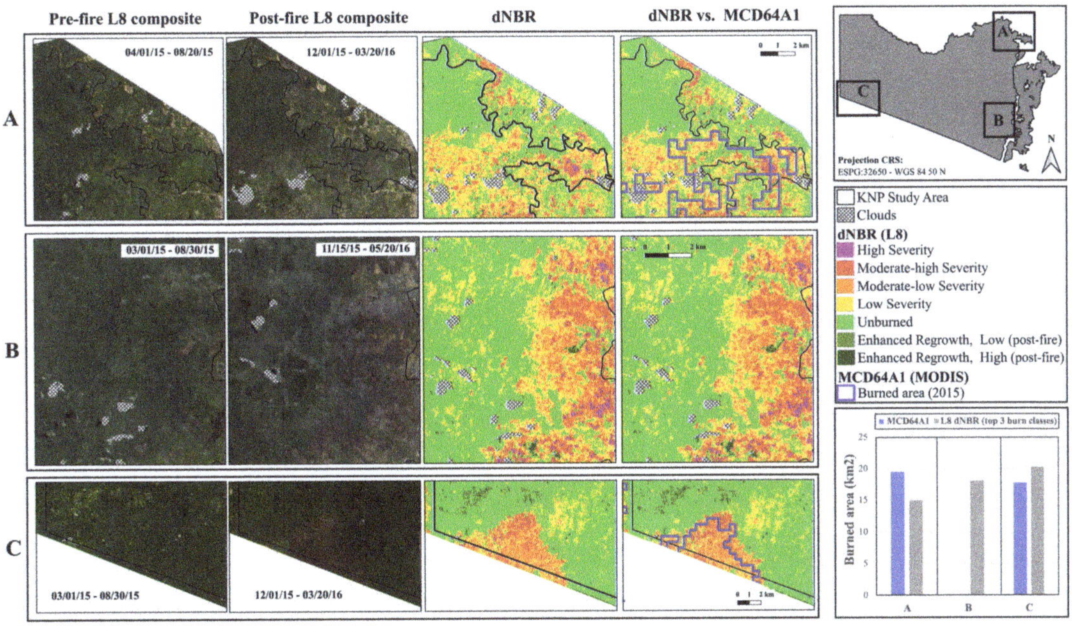

Figure 5. Validation of MCD64A1 burned area in 2015 over northwestern (**A**), southeastern (**B**), and southwestern (**C**) sample regions of Kutai National Park using differenced normalized burn ratio (dNBR) of pre- and post-fire Landsat 8 composite imagery. In the bottom right chart, the spatial extent of the MCD64A1-detected burned areas is compared against the top three classes of dNBR burn severity over each sample region.

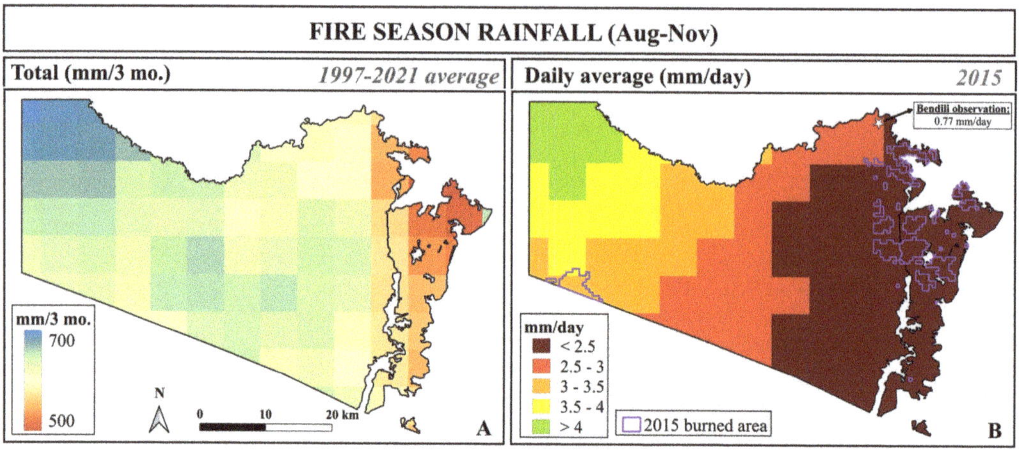

Figure 6. Per-pixel averages (1997–2021) of total rainfall (**A**) and daily average rainfall vs. burned area in 2015 (**B**) during the standard fire season (August–November) within Kutai National Park. Rainfall estimates from CHIRPS Pentad; burned area estimates from MCD64A1.

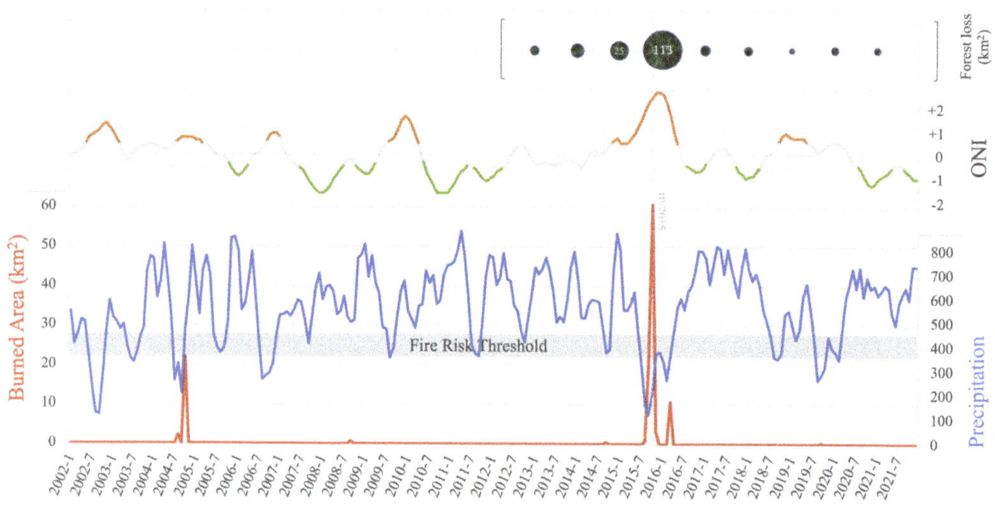

Figure 7. Estimates of monthly burned area (MCD64A1) and three-monthly total precipitation (CHIRPS) within Kutai National Park, as well as the Oceanic Niño Index (ONI) between 2002–2021, with phases of El Niño in orange and La Niña in green. Green bubbles (top) represent annual estimates of forest loss within KNP (HGFC v1.9) between 2013–2021 in square kilometers (aligned to Jan 1 of the detection year). The reported fire risk threshold for this region of 4–5 mm/day (360–450 mm/3 month period) is denoted by the grey band.

4. Conclusions

In this study, we leveraged the extensive collection of historical satellite data and the processing power of Google Earth Engine to investigate trends in and influences on forest loss within Indonesia's Kutai National Park over various time frames since 1997. We found a gradual yet considerable westward progression of forest loss across the eastern (settled/farmed) region of KNP since 2000 and patches of fire-related forest loss along the park's southern border associated with the El Niño-related drought and fires of 2015–2016. In total, the HGFC v1.9 dataset detected 272 km^2 of forest loss (or 15% of baseline forest cover) since 2000, half of which (137 km^2) occurred between 2015–2016 with significant overlap of detected burn scars. Our validation analyses support the detection performance of the HGFC v1.9 over NE KNP but they caution the use of the MCD64A1 burned area dataset over eastern KNP where considerable croplands compromise its accuracy. Our analysis of rainfall estimates also suggests that most El Niño events since 2000 have induced prolonged droughts over KNP below the fire risk threshold (4–5 mm/day) for this region. However, we found that drought severity alone cannot explain fire intensity within KNP, indicating the influences of other modulating factors (e.g., anomalous temperatures, incidental presence of fire sources, recent logging) in this phenomenon. Unlike all other areas of KNP, its eastern settled/farmed region receives average rainfall amounts that approach the fire risk threshold during the standard fire season (August–November), which is potentially (in part) attributable to widespread deforestation as demonstrated elsewhere in Borneo (Chapman et al. 2020; McAlpine et al. 2018). Lastly, we found that the most intense wildfire event of the past two decades—in 2015—coincided with rainfall estimates of <2.5 mm/day over the most fire-affected areas of the park.

Limitations of our analyses include the use of composite imaging in our validation assessments to address the obstacle of high-cloud cover over KNP. With this approach, we were still able to detect clear signals of canopy openings and burn scars in our validation analyses, thus adding confidence to our discussion of forest loss and its relative contributions over our study area. However, precise contributions of fire- and non-fire disturbances to annual forest loss within KNP could not be estimated with our approach, which along with standard validation assessments of the datasets used, should be the subject of future studies within KNP. Future studies of fire dynamics over KNP should also examine mechanisms of fire spread during wildfire years to identify targets of fire management efforts in this region.

The rates and influences of forest loss reported here contribute to undermining tourism-related revenue, degrading essential ecosystem services, heightening fire risk and severity in the region, and threatening the considerable biodiversity that resides within the park, including a stronghold population of the critically endangered Northeast Bornean orangutan. With the relocation of the Indonesian government capital to East Kalimantan scheduled for 2024, unprecedented new human migration into East Kalimantan and subsequent land and building requirements are soon likely to exert added pressures on KNP's forests. Wildfires and illegal logging within KNP must therefore be the priority subject of future conservation, management, and research programs to avoid further loss of forest cover in this protected area.

Author Contributions: Conceptualization, R.G.; formal analysis, R.G.; funding acquisition, X.W.; methodology, R.G.; resources, A.E.R.; supervision, X.W.; writing—original draft, R.G.; writing—review and editing, X.W. and A.E.R. All authors have read and agreed to the published version of the manuscript.

Funding: This research was supported by the Natural Science and Engineering Research Council of Canada and the New Frontiers in Research Fund.

Data Availability Statement: The source codes for the Google Earth Engine scripts used in this paper are available at: https://github.com/ryan-guild/Kutai-National-Park-Deforestation-Fire-Rainfall-Mapping-Study.git. The Hansen Global Forest Change v1.9 dataset is available at: https://developers.google.com/earth-engine/datasets/catalog/UMD_hansen_global_forest_change_2021_v1_9 (accessed on 1 September 2022). The MCD64A1.061 MODIS dataset is available at: https://developers.google.com/earth-engine/datasets/catalog/MODIS_061_MCD64A1 (accessed on 1 September 2022). The CHIRPS Pentad dataset is available at: https://developers.google.com/earth-engine/datasets/catalog/UCSB-CHG_CHIRPS_PENTAD?hl=en (accessed on 1 September 2022).

Conflicts of Interest: The authors declare no conflict of interest.

Appendix A

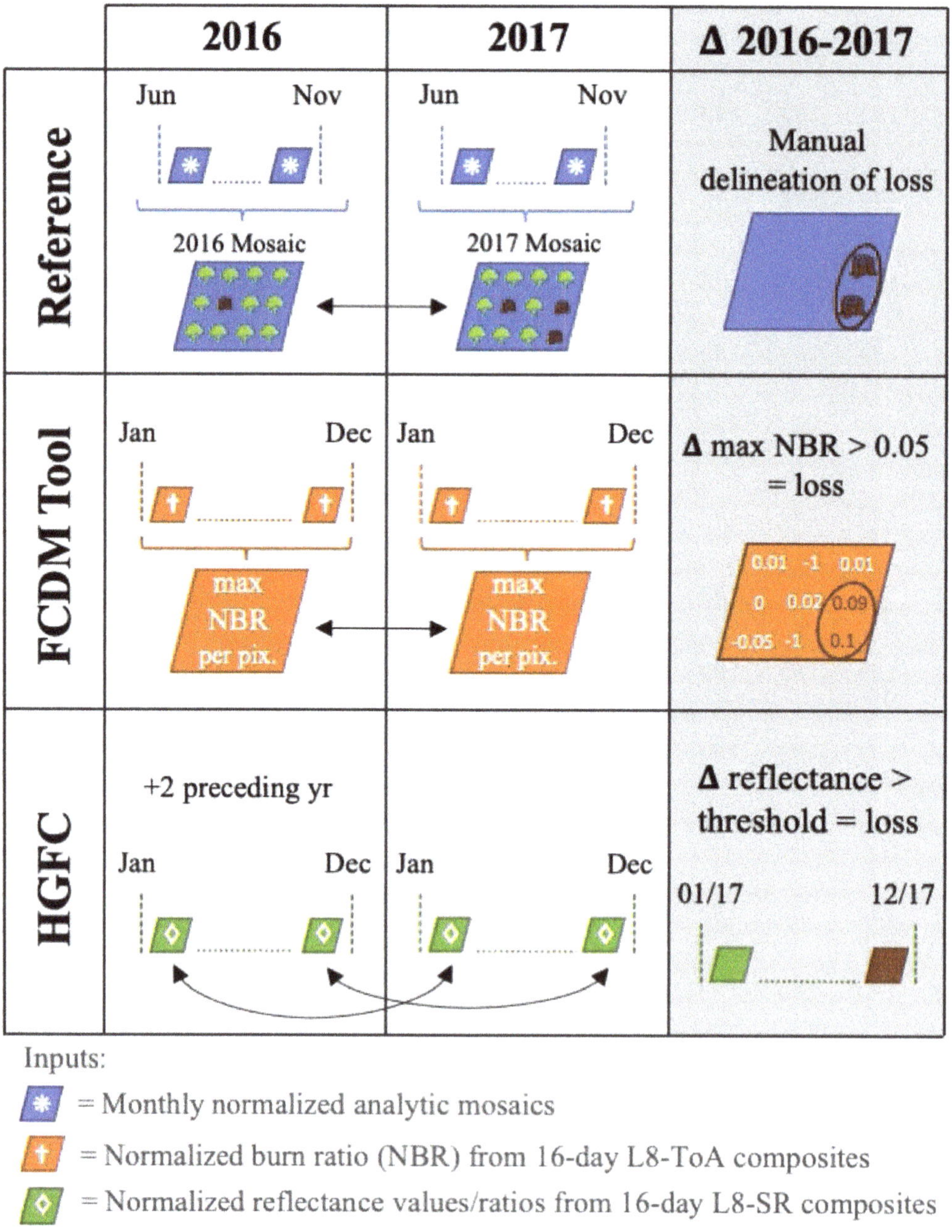

Figure A1. Differences in detection methodology between the three forest loss datasets used in this study, including the reference dataset (Planet/NICFI imagery), FCDM Tool [26], and the Hansen Global Forest Change product [23].

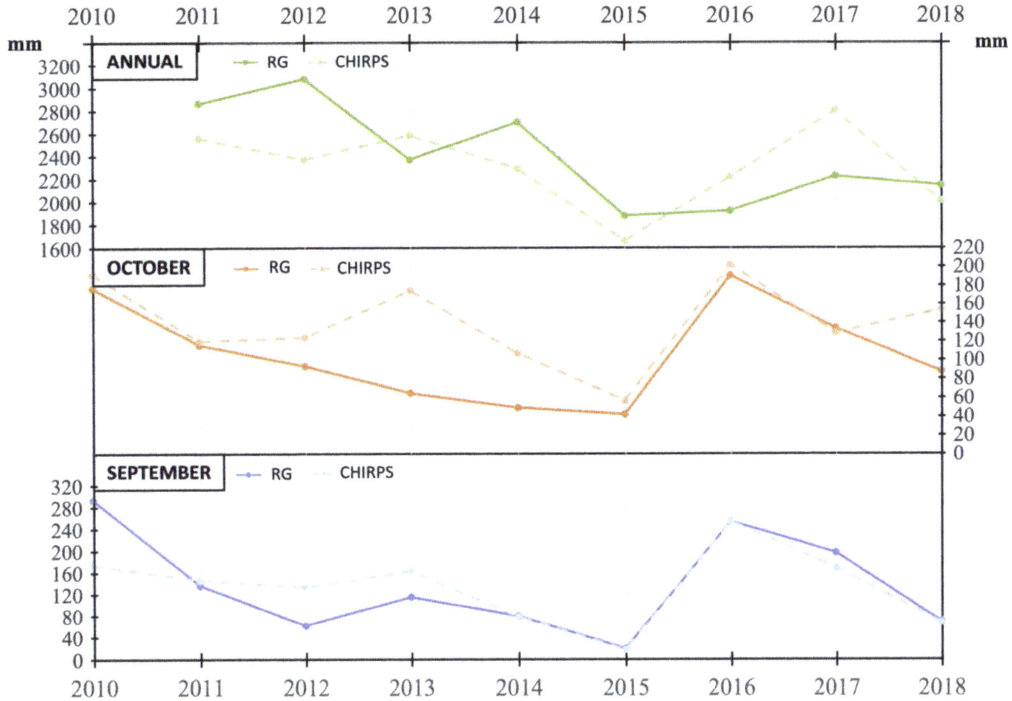

Figure A2. Comparison of rainfall estimates (CHIRPS) and rain gauge observations (RG) at the Bendili research station in Kutai National Park for annual, September, and October time periods between 2010–2018. Bendili RG observations are courtesy of the Orangutan Kutai Project.

References

1. Soehartono, T.; Mardiastuti, A. *The voice of National Parks in Kalimantan, Indonesia: Searching the Truth of Thirty Year National Park Development*; Nata Samastha Foundation: Bogor, Indonesia, 2013; p. 244.
2. Utami-Atmoko, S.; Traylor-Holzer, K.; Rifqi, M.A.; Siregar, P.G.; Achmad, B.; Priadjati, A.; Husson, S.; Wich, S.; Hadisiswoyo, P.; Saputra, F.; et al. *Orangutan Population and Habitat Viability Assessment: Final Report*; IUCN/SSC Conservation Breeding Specialist Group: Bogor, Indonesia, 2017; p. 121.
3. Field, R.D.; van der Werf, G.R.; Fanin, T.; Fetzer, E.J.; Fuller, R.; Jethva, H.; Levy, R.; Livesey, N.J.; Luo, M.; Torres, O.; et al. Indonesian fire activity and smoke pollution in 2015 show persistent nonlinear sensitivity to El Niño-induced drought. *Proc. Natl. Acad. Sci. USA* **2016**, *113*, 9204–9209. [CrossRef] [PubMed]
4. Siegert, F.; Hoffmann, A.A. The 1998 forest fires in East Kalimantan (Indonesia): A quantitative evaluation using high resolution, mulitemporal ERS-2 SAR Images and NOAA-AVHRR hotspot data. *Remote Sens. Environ.* **2000**, *72*, 64–77. [CrossRef]
5. Hadiprakarsa, Y.Y.; Rianta, E.; Christy, L.; Siregar, P.; Kitchener, D.; Hartman, P. *PSSF (Private Sector Sustainability Facility) Site Threats: Evaluation of Threats to Orangutan and Priority Interventions to Abate These Threats at PSSF Focused Sites in North Sumatra and East Kalimantan*; Report for the United States Agency for International Development (USAID); USAID: Washington, DC, USA, 2009; 32p.
6. Guild, R. Threats to wild orangutans: A case study in Kutai National Park of East Kalimantan, Indonesia. Master's Thesis, York University, Toronto, ON, Canada, 2019; p. 79.
7. Ustin, S.L.; Middleton, E.M. Current and near-term advances in Earth observation for ecological applications. *Ecol. Process.* **2021**, *10*, 1–57. [CrossRef] [PubMed]
8. Hasanah, N.I. Perubahan penutupan lahan di Taman Nasional Kutai provinsi Kalimantan Timur (Land cover change in Kutai National Park of East Kalimantan province). Ph.D. Thesis, Bogor Agricultural University, Bogor City, Indonesia, 2011; p. 68.
9. BTNK (Balai Taman National Kutai). *Statistik Taman Nasional Kutai: Tahun 2017 (Kutai National Park Statistics: 2017)*; BTNK (Balai Taman National Kutai): Bontang, Indonesia, 2017; p. 135.
10. BTNK (Balai Taman National Kutai). *Statistik Taman Nasional Kutai: Tahun 2016*; BTNK (Balai Taman National Kutai): Bontang, Indonesia, 2016; p. 173.

11. Rist, L.; Feintrenie, L.; Levang, P. The livelihood impacts of oil palm: Smallholders in Indonesia. *Biodivers. Conserv.* **2010**, *19*, 1009–1024. [CrossRef]
12. Cahyadi, E.R.; Waibel, H. Contract farming and vulnerability to poverty among oil palm smallholders in Indonesia. *J. Dev. Stud.* **2016**, *52*, 681–695. [CrossRef]
13. Naylor, R.L.; Higgins, M.M.; Edwards, R.B.; Falcon, W.P. Decentralization and the environment: Assessing smallholder oil palm development in Indonesia. *Ambio* **2019**, *48*, 1195–1208. [CrossRef]
14. Schweithelm, J.; Glover, D. Causes and impacts of the fires. In *Indonesia's Fires and Haze: The Cost of Catastrophe*; Glover, D., Jessup, T., Eds.; Institute of Southeast Asian Studies: Singapore, Singapore, 2006; pp. 1–13.
15. Goldammer, J.G. Natural rain forest fires in eastern Borneo during the Pleistocene and Holocene. *Naturwissenschaften* **1989**, *76*, 518–520. [CrossRef]
16. Wirawan, N. *Kutai National Park Management Plan 1985–1990*; World Wildlife Fund (WWF): Bogor, Indonesia, 1985.
17. Leighton, M.; Wirawan, N. Catastrophic drought and fire in Bornean tropical rain forest associated with the 1982–1983 ENSO event. In *Tropical Rain Forests and the World Atmosphere*; Prance, G., Ed.; Westview Press: Boulder, CO, USA, 1986; pp. 75–102.
18. Mori, T. Effects of droughts and forest fires on dipterocarp forests in East Kalimantan. In *Rainforest Ecosystems of East Kalimantan*; Guhardja, E., Fatawi, M., Sutisna, M., Mori, T., Ohta, S., Eds.; Springer: Tokyo, Japan, 2000; pp. 29–42.
19. Slik, J.F.W.; Verburg, R.W.; Keßler, P.J.A. Effects of fire and selective logging on tree species composition of lowland dipterocarp forest in East Kalimantan, Indonesia. *Biodivers. Conserv.* **2002**, *11*, 85–98. [CrossRef]
20. Russon, A.E.; Kuncoro, P.; Ferisa, A. Orangutan behavior in Kutai National Park after drought and fire damage: Adjustments to short- and long-term natural forest regeneration. *Am. J. Primatol.* **2015**, *77*, 1276–1289. [CrossRef]
21. Pan, X.; Chin, M.; Ichoku, C.M.; Field, R.D. Connecting Indonesian fires and drought with the type of El Niño and phase of the Indian Ocean Dipole during 1979–2016. *J. Geophisical Res. Atmos.* **2018**, *123*, 7974–7988.
22. Schiavina, M.; Freire, S.; MacManus, K. GHS-POP R2019A—GHS Population Grid Multitemporal (1975–1990–2000–2015). European Commission, Joint Research Centre (JRC) [Dataset]. 2019. Available online: https://doi.org/10.2905/0C6B9751-A71F-4062-830B-43C9F432370F (accessed on 18 September 2022).
23. Hansen, M.C.; Potapov, P.V.; Moore, R.; Hancher, M.; Turubanova, S.A.; Tyukavina, A.; Thau, D.; Stehman, S.V.; Goetz, S.J.; Loveland, T.R.; et al. High-resolution global maps of 21st-century forest cover change. *Science* **2013**, *342*, 850–853. [CrossRef] [PubMed]
24. Potapov, P.; Hansen, M.C.; Kommareddy, I.; Kommareddy, A.; Turubanova, S.; Pickens, A.; Adusei, B.; Tyukavina, A.; Ying, Q. Landsat analysis ready data for global land cover and land cover change mapping. *Remote Sens.* **2020**, *12*, 426. [CrossRef]
25. Potapov, P.; Tyukavina, A.; Turubanova, S.; Talero, Y.; Hernandez-Serna, A.; Hansen, M.; Saah, D.; Tenneson, K.; Poortinga, A.; Aekakkararungroj, A.; et al. Annual continuous fields of woody vegetation structure in the Lower Mekong region from 2000-2017 Landsat time-series. *Remote Sens. Environ.* **2019**, *232*, 111278. [CrossRef]
26. Langner, A.; Miettinen, J.; Kukkonen, M.; Vancutsem, C.; Simonetti, D.; Vieilledent, G.; Verhegghen, A.; Gallego, J.; Stibig, H.-J. Towards Operational Monitoring of Forest Canopy Disturbance in Evergreen Rain Forests: A Test Case in Continental Southeast Asia. *Remote Sens.* **2018**, *10*, 544. [CrossRef]
27. Giglio, L.; Boschetti, L.; Roy, D.P.; Humber, M.L.; Justice, C.O. The collection 6 MODIS burned area mapping algorithm and product. *Remote Sens. Environ.* **2018**, *217*, 72–85. [CrossRef]
28. Hall, J.V.; Argueta, F.; Giglio, L. Validation of MCD64A1 and FireCCI51 cropland burned area mapping in Ukraine. *Int. J. Appl. Earth Obs. Geoinf.* **2021**, *102*, 102443. [CrossRef]
29. Katagis, T.; Gitas, I.Z. Assessing the Accuracy of MODIS MCD64A1 C6 and FireCCI51 Burned Area Products in Mediterranean Ecosystems. *Remote Sens.* **2022**, *14*, 602. [CrossRef]
30. Liu, T.; Marlier, M.E.; Karambelas, A.; Jain, M.; Singh, S.; Singh, M.K.; Gautam, R.; DeFries, R.S. Missing emissions from post-monsoon agricultural fires in northwestern India: Regional limitations of MODIS burned area and active fire products. *Environ. Res. Commun.* **2019**, *1*, 011007. [CrossRef]
31. Liu, T.; Mickley, L.J.; Marlier, M.E.; DeFries, R.S.; Khan, M.F.; Latif, M.T.; Karambelas, A. Diagnosing spatial biases and uncertainties in global fire emissions inventories: Indonesia as regional case study. *Remote. Sens. Environ.* **2020**, *237*, 111557. [CrossRef]
32. Funk, C.; Peterson, P.; Landsfeld, M.; Pedreros, D.; Verdin, J.; Shukla, S.; Husak, G.; Rowland, J.; Harrison, L.; Hoell, A.; et al. The climate hazards infrared precipitation with stations—A new environmental record for monitoring extremes. *Sci. Data* **2015**, *2*, 150066. [CrossRef]
33. Sloan, S.; Tacconi, L.; Cattau, M.E. Fire prevention in managed landscapes: Recent success and challenges in Indonesia. *Mitig. Adapt. Strat. Glob. Chang.* **2021**, *26*, 1–30. [CrossRef]
34. Petersen, R.; Askenov, D.; Esipova, E.; Goldman, E.; Harris, N.; Kuksina, N.; Kurakina, I.; Loboda, T.; Manisha, A.; Sargent, S.; et al. *Mapping Tree Plantations with Multispectral Imagery: Preliminary Results for Seven Tropical Countries*; World Resources Institute: Washington, DC, USA, 2016; p. 18.
35. Runting, R.K.; Ruslandi Griscom, B.; Struebig, M.J.; Satar, M.; Meijaard, E.; Burivalova, Z.; Cheyne, S.M.; Deere, N.J.; Game, E.T.; Putz, F.E.; et al. Larger gains from improved management over sparing-sharing for tropical forests. *Nat. Sustain.* **2019**, *2*, 53–61. [CrossRef]

36. Ministry of Energy and Natural Resources Indonesia. "ESDM One Map Indonesia". 2022. Available online: https://geoportal.esdm.go.id/minerba/ (accessed on 20 September 2022).
37. Rodman, P.S. Synecology of Bornean primates: With special reference to the behavior and ecology of orangutans. Ph.D. Thesis, Harvard University, Cambridge, MA, USA, 1973; p. 218.
38. Leighton, M. Fruit Resources and Patterns of Feeding, Spacing and Grouping Among Sympatric Bornean hornbills (*Bucerotidae*). Ph.D. Thesis, University of California, Davis, CA, USA, 1982.
39. Wilson, C.C.; Wilson, W.L. The influence of selective logging on primates and some other animals in East Kalimantan. *Folia Primatol.* **1975**, *23*, 245–274. [CrossRef] [PubMed]
40. Meijaard, E.; Buchori, D.; Hadiprakarsa, Y.; Utami-Atmoko, S.S.; Nurcahyo, A.; Tiju, A.; Prasetyo, D.; Nardiyono; Christie, L.; Ancrenaz, M.; et al. Quantifying killing of orangutans and human-orangutan conflict in Kalimantan, Indonesia. *PLoS ONE* **2011**, *6*, e27491. [CrossRef] [PubMed]
41. Abram, N.K.; Meijaard, E.; Wells, J.A.; Ancrenaz, M.; Pellier, A.S.; Runting, R.K.; Gaveau, D.; Wich, S.; Nardiyono; Tiju, A.; et al. Mapping perceptions of species' threats and population trends to inform conservation efforts: The Bornean orangutan case study. *Divers. Distrib.* **2015**, *21*, 487–499. [CrossRef]
42. Siegert, F.; Ruecker, G.; Hinrichs, A.; Hoffmann, A.A. Increased damage from fires in logged forests during droughts caused by El Niño. *Lett. Nat.* **2001**, *414*, 437–440. [CrossRef]
43. Van Nieuwstadt, M.G.L.; Sheil, D. Drought, fire and tree survival in a Borneo rain forest, East Kalimantan, Indoensia. *J. Ecol.* **2005**, *93*, 191–201. [CrossRef]
44. Zhang, T.; Wooster, M.J.; de Jong, M.; Xu, W. How well does the "small fire boost" methodology used within the GFED4.1s fire emissions database represent the timing, location and magnitude of agricultural burning? *Remote Sens.* **2018**, *10*, 823. [CrossRef]
45. Chapman, S.; Syktus, J.; Trancoso, R.; Salazar, A.; Thatcher, M.; Watson, J.E.M.; Meijaard, E.; Sheil, D.; Dargusch, P.; McAlpine, C.A. Compounding impact of deforestation on Borneo's climate during El Niño events. *Environ. Res. Lett.* **2020**, *15*, 084006. [CrossRef]
46. Perugini, L.; Caporaso, L.; Marconi, S.; Cescatti, A.; Quesada, B.; De Noblet, N.; House, J.I.; Arneth, A. Biophysical effects on temperature and precipitation due to land cover change. *Environ. Res. Lett.* **2017**, *12*, 1–13. [CrossRef]
47. McAlpine, C.A.; Johnson, A.; Salazar, A.; Syktus, J.; Wilson, K.; Meijaard, E.; Seabrook, L.; Dargusch, P.; Nordin, H.; Sheil, D. Forest loss and Borneo's climate. *Environ. Res. Lett.* **2018**, *13*, 044009. [CrossRef]
48. Cai, W.; Borlace, S.; Lengaigne, M.; van Rensch, P.; Collins, M.; Vecchi, G.; Timmermann, A.; Santoso, A.; McPhaden, M.J.; Wu, L.; et al. Increasing frequency of extreme El Niño events due to greenhouse warming. *Nat. Clim. Change* **2014**, *4*, 111–116. [CrossRef]
49. Cai, W.; Wang, G.; Dewitte, B.; Wu, L.; Santoso, A.; Takahashi, K.; Yang, Y.; Carréric, A.; McPhaden, M.J. Increased variability of eastern Pacific El Niño under greenhouse warming. *Nature* **2018**, *564*, 201–206. [CrossRef] [PubMed]
50. Chen, C.C.; Lin, H.W.; Yu, J.Y.; Lo, M.H. The 2015 Borneo Fires: What have we learned from the 1997 and 2006 El Niños? *Environ. Res. Lett.* **2016**, *11*, 104003. [CrossRef]
51. Wang, B.; Luo, X.; Yang, Y.M.; Sun, W.; Cane, M.A.; Cai, W.; Yeh, S.W.; Liu, J. Historical changes of El Niño properties sheds light on future changes of extreme El Niño. *Proc. Natl. Acad. Sci. USA* **2019**, *116*, 22512–22517. [CrossRef]

remote sensing

MDPI

Article

Weakened Impacts of the East Asia-Pacific Teleconnection on the Interannual Variability of Summertime Precipitation over South China since the Mid-2000s

Wei Lu [1], Yimin Zhu [1,2], Zhong Zhong [1,2], Yijia Hu [1] and Yao Ha [1,2,*]

1 College of Meteorology and Oceanography, National University of Defense Technology, Changsha 410003, China
2 School of Atmospheric Sciences, Jiangsu Collaborative Innovation Center for Climate Change, Nanjing University, Nanjing 210023, China
* Correspondence: hayao86@nudt.edu.cn or hayao1986@yeah.net

Abstract: The current study concentrates on the interdecadal shift in the interannual variability of summertime precipitation (IVSP) over South China (SC). Possible causes for the interdecadal shift are explored. The IVSP on a decadal time scale presents a significant weakening after the mid-2000s. The results show that the variances of the interannual precipitation variability over the SC region between 1993 and 2004 (hereafter S1) and 2005 and 2020 (hereafter S2) are 1.40 mm d^{-1} and 0.58 mm d^{-1}, respectively. The variance of the IVSP has decreased by 58.6% since the mid-2000s. The current study reveals that the reduction in the IVSP over SC after the mid-2000s is prominently attributed to the weakened impact of the East Asia-Pacific (EAP) teleconnection. Before the mid-2000s, the interannual variation of the east-west movement of the western Pacific subtropical high was more significant. The warming over the tropical central-eastern Pacific (CEP) and cooling over the western Pacific (WP) suppress the Walker cell in the tropical Pacific and induce anomalous Hadley cell with its descending branch over the WP in the wet years. The anomalies of SST and atmospheric circulation show opposite phases in the dry years. This SSTA pattern enhances the northward propagation of the EAP teleconnection through a Rossby-wave-type response, which triggers an ascending/descending branch with active/suppressed convection over the northwestern Pacific in the wet/dry years. Therefore, the cooling WP and El Niño in its developing phase provide an ideal condition for more precipitation over SC. However, the above ocean–atmosphere interactions changed after the mid-2000s. The significant SST changes in the tropical CEP and the WP weaken the EAP teleconnection and atmospheric circulation anomalies over SC, leading to a significant interdecadal reduction in the IVSP over SC after the mid-2000s.

Keywords: summertime precipitation; interannual variability; interdecadal shift; South China

Citation: Lu, W.; Zhu, Y.; Zhong, Z.; Hu, Y.; Ha, Y. Weakened Impacts of the East Asia-Pacific Teleconnection on the Interannual Variability of Summertime Precipitation over South China since the Mid-2000s. *Remote Sens.* **2022**, *14*, 5098. https://doi.org/10.3390/rs14205098

Academic Editor: Xander Wang

Received: 19 August 2022
Accepted: 7 September 2022
Published: 12 October 2022

1. Introduction

East Asian summer monsoon (EASM) exhibits significant interannual and interdecadal variabilities, which have a great influence on summertime precipitation in East Asia [1]. Huang et al. gave a comprehensive and systematic summary of the spatiotemporal features, processes, and causes of the EASM variability and pointed out that the EASM has significant variabilities on interannual and interdecadal time scales [2]. Moreover, the EASM experienced an interdecadal weakening around the late 1970s, while summertime precipitation increased in the Yangtze-Huai River Basin (YHRB) and decreased in North China correspondingly. Summertime precipitation in South China (SC) increased after 1992 and decreased in YHRB after 1999 [3]. Correspondingly, the correlation between interannual variations of the EASM and ENSO became unstable [4]. South China (SC) is located at the east of the Qinghai-Tibet Plateau and south of YHRB. Due to the interactions between tropical and middle- to high-latitude weather systems, the weather and climate in SC are complicated and diverse. SC is the area with the amplest precipitation and the longest rainy season in China. Severe drought and flood disasters associated with abnormal precipitation cause huge losses of lives and properties and greatly influence local economies [5–8].

Influenced by the EASM, summertime precipitation over eastern China shows multi-scale variability, including seasonal and intraseasonal variability, as well as interannual and interdecadal variability [2,9–15]. Previous studies have shown that there were three interdecadal variations of summertime precipitation that occurred in the mid-1970s and the early and late 1990s [16]. The interdecadal decrease of summertime precipitation in SC in the 1970s [17,18] is attributed to the weakening of the EASM [19] and the warming of the tropical Indian Ocean (IO) and the tropical central-eastern Pacific (CEP) [20,21].

Summertime precipitation in SC exhibited an interdecadal intensification after the early 1990s [22] and a weakening after the late 1990s [23,24]. Most studies pointed out that uninterrupted warming in the tropical IO and CEP in summertime can cause anomalous Walker Cell and Hadley Cell circulations [25,26]. The interdecadal intensification of precipitation over SC in the early 1990s is attributed to joint effects of the Walker Cell and Hadley Cell anomalies and the reduction in spring snowfall in northern Eurasia [25,27]. Due to the strengthening of the western Pacific subtropical high (WPSH) caused by the tropical Pacific SSTA [23] and the significant reduction in tropical cyclone activity in the western Pacific [28], summertime precipitation in SC decreased in the late 1990s.

Previous studies have shown that there is a link between the East Asian climate and the Northwest Pacific climate, which can be established through atmospheric teleconnection [29]. On the basis of previous studies, Nitta found that there is a seesaw pattern in the atmospheric circulations between regions near Japan and the Philippine Sea and proposed the Pacific-Japan (PJ) teleconnection mode for the first time [30]. Based on the analysis of observations, Huang and Li also found that there is a tele-connected wave train similar to the "tripole" structure of PJ in the atmosphere extending from the Philippines to the Sea of Okhotsk, which is named East Asia-Pacific (EAP) teleconnection type [31]. Many studies focused on the formation mechanism of this teleconnection type. Based on numerical experiments using a barotropic model, Kurihara and Tsuyuki found that convective activities near the Philippines can trigger two-dimensional northward propagating Rossby waves with a horizontal structure similar to the observed PJ/EAP teleconnection [32]. Wakabayashi and Kawamura defined the PJ index based on the decomposition of the orthogonal function through the distribution position of the central node of the graph [33]. Later, the PJ index was widely applied to find the connection between the Northwest Pacific climate anomalies and the East Asian climate anomalies. It is also used in climate forecast, especially the prediction of summertime precipitation anomalies in eastern China. Similar to summertime precipitation in South China, the position and intensity of the EAP teleconnection also underwent significant interdecadal variations. After the late 1970s, the position of the EAP teleconnection pattern changed significantly with an obvious shift to the west and south [34]. Yin et al. pointed out that rainfall anomalies over YHRB associated with the EAP teleconnection were weaker and less evident after the 1980s [35]. Xu et al. revealed that the PJ/EAP (hereafter EAP) teleconnection shifted to the east after the late 1990s, and its intensity weakened significantly [36].

The above studies are helpful for understanding the interdecadal change in the mean state of summertime precipitation over SC. However, less attention has been paid to the study of the interdecadal change of the IVSP in SC [37]. Many studies revealed that the factors which influence the interannual variation of the East Asian summertime climate have changed a lot on the interdecadal time scale in the past decades [38–42]. Therefore, the IVSP could have enormous contributions to the occurrence of extreme precipitation events and should be further investigated. Moreover, interannual changes in various meteorological elements could improve the precision of seasonal prediction by statistical climate prediction models [43]. Therefore, it is necessary to study the interdecadal shift of the IVSP over SC and explore the possible mechanisms behind.

2. Materials and Methods

The precipitation data used in this study are the Precipitation Reconstruction dataset on 2.5 × 2.5° grids provided by NOAA [44–46]. The monthly atmospheric circulation data are derived from the NCEP/NCAR Reanalysis on a 2.5 × 2.5° resolution [47]. The daily outgoing longwave radiation (OLR) dataset produced based on NOAA polar-orbiting satellite remote sensing data with a horizontal resolution of 2.5 × 2.5° grids is used in this study [48]. In addition, the NOAA OI SST V2 High Resolution Dataset on a 0.25 × 0.25° resolution is put to use [49].

The time span of this study is the summertime season (JAS, the second rainy season in SC, which usually has a strong relationship with tropical systems) from 1979 to 2020. A 10-year running *F*-test is performed on the time series of summertime precipitation to assess the interdecadal change. The Student's *t* test is used to examine the confidence level of the composites. The wave activity flux (WAF) is applied to describe the energy propagation of wave trains [50]. The conversion of local kinetic energy (CK) is calculated to depict the conversion of kinetic energy from mean flow to turbulence [51].

3. Results

3.1. Weakened Interannual Variability of Summer Precipitation over SC since the Mid-2000s

To describe the interdecadal shift of the IVSP in the mid-2000s, Figure 1 shows the regional mean time series, interannual difference, and 10-year running *F*-test values of summertime precipitation over SC (20–35°N, 100–125°E) from 1979 to 2020. The time series and interannual difference both reveal a significant weakening of the interannual variability after the mid-2000s. To describe the interdecadal shift of the IVSP over SC in the mid-2000s and determine the interdecadal turning point, the 10-year running *F*-test and 11-year running standard deviation of summertime precipitation over SC are analyzed (Figure 1b). The results suggest that the IVSP over SC experienced a significant weakening around2004/05. Hence, the year 2004 is taken as the shift point and the stage (1993–2004/2005–2020) with strong/weak IVSP is defined as S1/S2. It is found that the variance in S1 is 1.40 mm d^{-1} and that in S2 is 0.58 mm d^{-1}, and the mean values of the interannual difference in summertime precipitation are 1.82 mm d^{-1} in S1 and 0.81 mm d^{-1} in S2. Furthermore, the mean value of the 11-year running standard deviation is 1.33 mm d^{-1} in S1 and the value reduces to 0.64 mm d^{-1} in S2. All the above results support the conclusion that the interdecadal shift occurred around2004/05.

The time series of summertime precipitation with the linear trends in S1 and S2 being removed is shown in Figure 2. It is found that the variance in S1 is 1.32 mm d^{-1} and the value reduces to 0.37 mm d^{-1} in S2. The variance of the detrended time series in S2 decreases by 72.0% compared to that in S1, which is larger than the value of 58.6% for the original time series that includes the linear trend. Based on the detrended time series, the years with detrended summertime main precipitation larger (smaller) than 0 are determined to be wet (dry) years. The wet and dry years during S1 and S2 are listed in Table 1. Then the differences between the wet and dry years are composited to explore the reasons for the interdecadal shift of the IVSP over SC. Linear trends are eliminated for all environmental variables before the composite analysis is conducted.

The composite differences in summertime precipitation between wet and dry years during S1 and S2 are shown in Figure 3b,d, respectively. It is found that summertime precipitation over SC shows positive anomalies for the two periods but presents different spatial distributions and intensities. In S1, the positive anomalies center is located at (26°N, 113°E), and the maximum intensity of anomaly reaches 3.2–3.4 mm d^{-1} (Figure 3b). In S2, the two centers of positive anomalies are located at (29°N, 113°E) and (23°N, 115°E), which are to the north and south of the positive center in S1, respectively. The maximum value at the two positive centers in S2 is around 1.2–1.8 mm d^{-1}, weaker than that in S1 (Figure 3d). Consistent with the composite differences of summertime precipitation, the center of the maximum standard deviation is located at (26°N, 113°E) in S1 but it splits into two centers, which shift to the north and south respectively in S2. The values of standard deviation at the two positive anomaly centers also decrease in S2 compared to that in S1, which agrees with the interdecadal weakening of the IVSP over SC (Figure 3a,c).

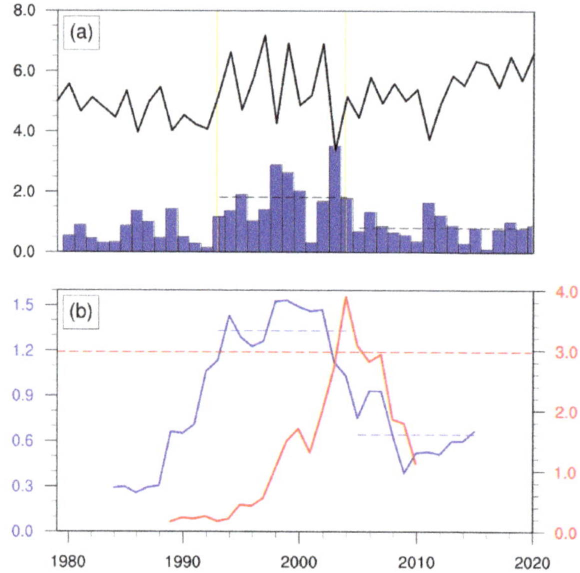

Figure 1. (**a**) Time series (black solid line) and interannual difference (blue bars) of summertime precipitation over SC. (**b**) *F*-test values (red solid line) and 11-year running standard deviations (blue solid lines) of summertime precipitation over SC (black dashed lines in (**a**) are averaged values of interannual variability in S1 and S2, respectively; blue dashed lines in (**b**) are averaged values of running standard deviation in S1 and S2, respectively; the red dashed line in (**b**) denotes the significant value exceeding the 95% confidence level; unit: mm d^{-1}).

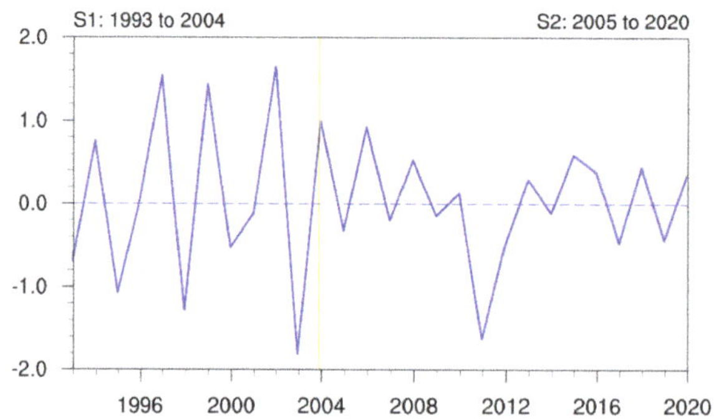

Figure 2. Time series of detrended summertime precipitation over SC during S1 and S2.

Table 1. Wet and dry years picked out in S1 and S2 based on the detrended summertime main precipitation larger and smaller than 0, respectively.

	S1	S2
Wet years	1994, 1996, 1997, 1999, 2002, 2004	2006, 2008, 2010, 2013, 2015, 2016, 2018, 2020
Dry years	1993, 1995, 1998, 2000, 2001, 2003	2005, 2007, 2009, 2011, 2012, 2014, 2017, 2019

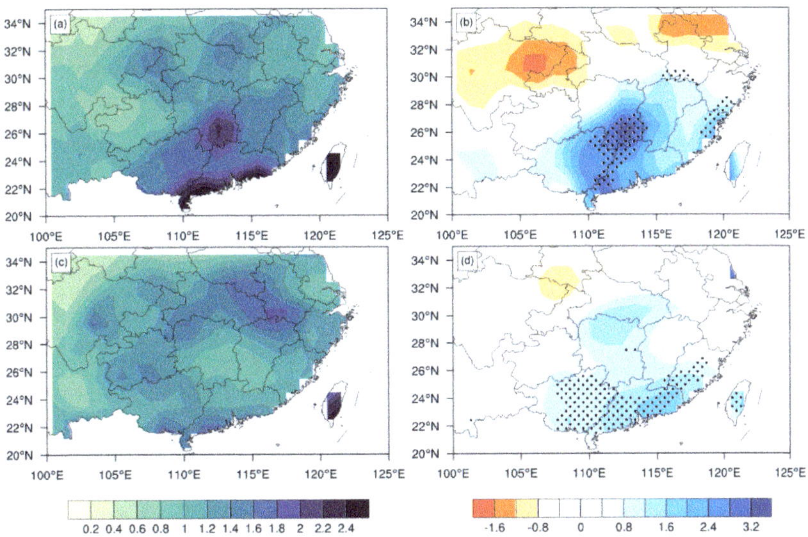

Figure 3. Composite differences in summertime precipitation between wet years and dry years (unit: mm d^{-1}; (**b**,**d**) for S1 and S2, respectively; dots indicate the regions exceeding 95% confidence level.) and the standard deviation of summertime precipitation (unit: mm d^{-1}; (**a**,**c**) for S1 and S2).

3.2. Atmospheric Circulation Anomalies Responsible for the Interdecadal Shift of Summertime Precipitation over SC

To find out the reasons why there are large differences in the IVSP between S1 and S2, atmospheric circulation characteristics are analyzed in this section. Figure 4 shows composite differences of 200 and 500 hPa geopotential height as well as 200 hPa zonal wind between wet and dry years. The center of positive geopotential height anomalies occurs at 45°N in both S1 and S2, but the center in S2 is located at the east of that in S1. The negative geopotential height anomalies are smaller in S2 compared to that in S1 (Figure 4a,b,d,e). Composites of the western Pacific subtropical high (WPSH) feature line (5880 gpm contour) in wet and dry years are displayed in Figure 4b,e, respectively. In S1, west extension of the WPSH is significant in dry years; in contrast, it retreats to the east in wet years (Figure 4b). In S2, the location and intensity of the WPSH show little difference between wet and dry years (Figure 4e). This suggests that the interannual variability of the WPSH significantly weakened in S2 compared to that in S1, which is similar to the variation in summertime precipitation. Consistent with geopotential height anomalies, the positive and negative anomalies centers of zonal wind at 200 hPa are located more eastward in S1 and more westward in S2. The intensity of positive and negative anomalies centers of zonal wind is weaker in S2 than in S1 (Figure 4c,f).

The influence of atmospheric circulation anomalies on the IVSP is analyzed from the perspective of large-scale circulation above. To explore the impact of water vapor transport and local convection on the IVSP, composite differences in moisture flux and its divergence in the middle and lower levels, OLR, and omega in the middle and lower levels, 850 hPa relative vorticity and horizontal wind between wet years and dry years are analyzed. Results are displayed in Figure 5. In S1, negative OLR and omega anomalies are centered over SC (Figure 5a); in S2, however, the centers of the negative OLR and omega anomalies are located more eastward and become weak (Figure 5d). This result indicates that the intensity of convection is weaker in S2 than in S1. From the perspective of moisture transport, strong easterly anomalies pulled lots of moisture from the Northwestern Pacific (NWP) to SC in S1, resulting in significant moisture convergence over SC (Figure 5b). In S2, however, no significant convergence and divergence of moisture flux anomalies can be found over SC (Figure 5e). Positive relative vorticity anomalies are centered over SC in S1 with strong easterly anomalies (Figure 5c). The meridionally distributed centers of anomalies of positive and negative relative vorticity resemble the EAP teleconnection. Compared to that in S1, the EAP teleconnection moves eastward and becomes weak in S2 (Figure 5f). The changes in the location and intensity of the EAP teleconnection could be a key factor that affects the interdecadal shift of the IVSP over SC.

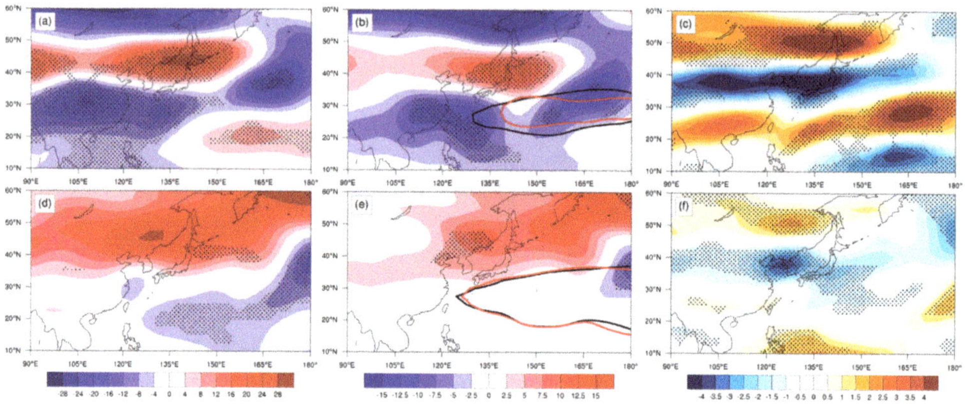

Figure 4. Composite differences of (**a**,**d**) 200 hPa geopotential height (unit: gpm), (**b**,**e**) 500 hPa geopotential height (unit: gpm), and (**c**,**f**) 200 hPa zonal wind (unit: m s^{-1}) between wet years and dry years. (**a**–**c**) for S1 and (**d**–**f**) for S2; dots indicate the regions exceeding 95% confidence level; red and black lines in (**b**,**e**) indicate composites of the WPSH feature line for wet and dry years, respectively).

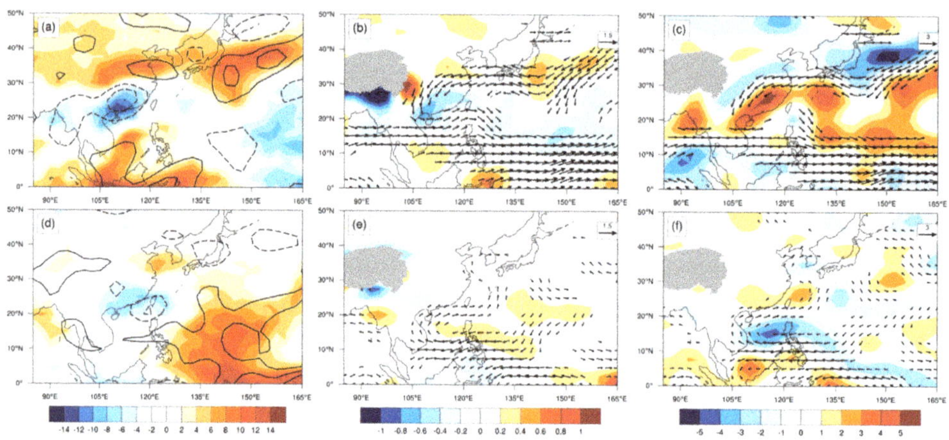

Figure 5. Composite differences of (**a**,**d**) vertically integrated omega from 850 to 500 hPa (contours with the interval of 2, unit : kg (m s)$^{-3}$) and OLR (shading, unit : W m^{-2}); (**b**,**e**) vertically integrated moisture flux from 1000 to 500 hPa (vectors, unit : kg m^{-1} s^{-1}) and its divergence (shading, unit : 10^{-6} kg m^{-2} s^{-1}); and (**c**,**f**) 850 hPa wind (vectors, unit: m s^{-1}) and relative vorticity (shading, unit : 10^{-6} s^{-1}) between wet years and dry years. ((**a**–**c**) for S1; (**d**–**f**) for S2; significant values exceeding the 95% confidence level are displayed; the dashed/solid lines suggest values larger/smaller than zero in (**a**,**d**)).

It is found that the composite differences of summertime precipitation between wet and dry years and standard deviations in the corresponding years over the SC show highly consistent spatial distribution characteristics (Figure 3). To inspect the consistency of the composite differences in moisture transport and local convection between dry and wet years and their standard deviations in the corresponding years, Figure 6 shows the standard deviations of moisture flux divergence, vertically integrated omega from 850 to 500 hPa and OLR in S1 and S2, and the respective differences. Consistent with the results shown in Figure 5, moisture flux divergence, omega, and OLR have larger standard deviations over SC in S1 (Figure 6a–c) than in S2 (Figure 6d–f). Moreover, the differences in moisture flux divergence, omega, and OLR between S1 and S2 (Figure 6g–i) display a similar spatial distribution pattern to that of the standard deviation of summertime precipitation over SC in S1.

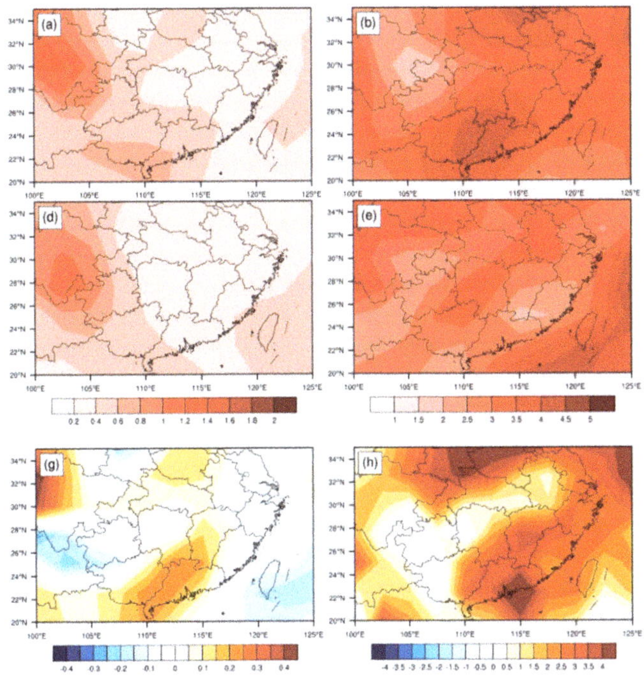

Figure 6. Standard deviations of vertically integrated moisture flux divergence from 1000 to 500 hPa ((**a**,**d**,**g**); unit : 10^{-6} kg m^{-2} s^{-1}), vertically integrated omega from 850 to 500 hPa ((**b**,**e**,**h**); unit : kg (m s)$^{-3}$), OLR ((**c**,**f**,**i**); unit : W m^{-2}) in S1 (**a**–**c**); and S2 (**d**–**f**) and the respective difference between S2 and S1 (**g**–**i**).

3.3. *Weakened Impacts of the EAP Teleconnection on Summertime Precipitation over SC since the Mid-2000s*

As referred to in Section 3.2, the EAP teleconnection is of vital importance for the interdecadal shift of the IVSP over SC. This section explores the possible mechanisms of the interdecadal shift of the EAP teleconnection impact on summertime precipitation over SC around the mid-2000s. The composite differences of OLR and 850 hPa relative vorticity between wet and dry years are shown in Figure 5, which can reflect some features of the EAP teleconnection. In S1, negative differences of OLR (Figure 5a) and positive differences of vorticity (Figure 5c) are centered over SC, while the positive OLR difference center and negative vorticity difference center are distributed to the north and south, respectively. In S2, the centers of negative OLR (Figure 5b) and positive vorticity (Figure 5d) anomalies both move eastward and their intensity weakens. These results suggest the EAP teleconnection may have experienced an interdecadal weakening process after the mid-2000s.

Since the EAP teleconnection is characterized by meridionally distributed centers of relative vorticity anomalies with alternating signs [36], the 200 and 850 hPa relative vorticity anomalies connected with the EAP teleconnection in wet years during S1 and S2 are shown in Figure 7. In S1, the centers of positive and negative relative vorticity anomalies are meridionally distributed along 105–120°E with the maximum positive center in lower levels located over SC (Figure 7b). In S2, the centers of positive and negative relative vorticity anomalies move eastward and are meridionally distributed along 115–130°E with the intensity weakened in lower levels (Figure 7d). Following the approach proposed by Huang to define the EAP index using 500 hPa geopotential height anomalies [52], the EAP index is defined in the present study using 850 hPa relative vorticity anomalies. The EAP index can be expressed as:

$$I_{PJ} = -0.25Vor'(R_1) + 0.5Vor'(R_2) - 0.25Vor'(R_3) \quad (1)$$

where $Vor'(R_1)$, $Vor'(R_2)$, and $Vor'(R_3)$ represent averaged 850 hPa relative vorticity anomalies over (30–36°N, 105–120°E), (16–29°N, 105–120°E), and (8–15°N, 105–120°E), respectively. The above three areas are denoted by the red boxes in Figure 7b,d. The EAP index is 1.187 in S1 and −0.002 in S2, respectively. The large difference in the EAP index between S1 and S2 suggests that the intensity of the EAP teleconnection experienced a significant weakening since the mid-2000s. Furthermore, the weakening of the EAP teleconnection is partly attributed to the weakening of its intensity and partly attributed to its eastward shift.

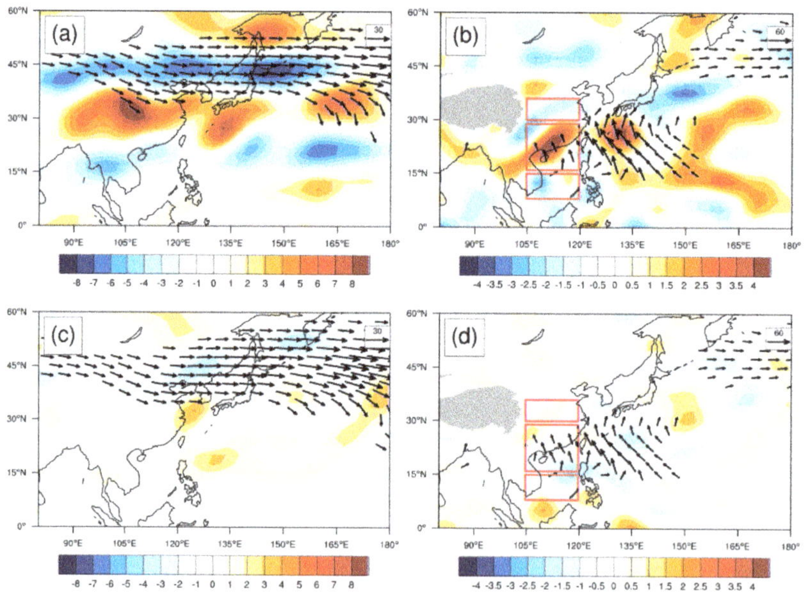

Figure 7. Composite anomalies of relative vorticity (shading, unit : 10^{-6} s^{-1}) and associated WAF (vectors, unit : m^2 s^{-2}) at (**a**,**c**) 200 and (**b**,**d**) 850 hPa in wet years of S1 (**a**,**b**) and S2 (**c**,**d**) (significant values exceeding the 95% confidence level are displayed).

In order to illustrate the propagation of Rossby waves, the wave activity flux (WAF) is also composited. In S1, the WAF connected with the EAP teleconnection emanates from the tropical WP, propagates to the north, and reaches and converges SC and the adjacent sea to its east in the lower levels (Figure 7b). In the higher levels, EAP of the WAF predominantly propagates along the westerly jet stream at the middle- to high- latitudes (Figure 7a). In S2, the propagation paths of WAF in lower (Figure 7d) and higher levels (Figure 7c) both are similar to that in S1, but the intensity of WAF is weaker than that in S1. A previous study indicated that the energy of the EAP teleconnection in the wet years of S1 predominantly hails from a Rossby-wave-type response [53].

Traditional studies regard the EAP teleconnection as a free Rossby wave train propagating from the subtropics to the mid latitudes in the lower troposphere [29,31]. To explain in a clearer way the interdecadal variation of the EAP teleconnection from the perspective of energy, the composite differences in CK vertically integrated from 1000 to 100 hPa between wet years and dry years are displayed in Figure 8. Note that positive CK indicates that fluctuations are easier to develop by extracting kinetic energy from basic flow. In S1, positive CK is centered at 150–160° E, 35–40° N and 105–115° E, 15–20° N (Figure 8a). In S2, the positive CK shift to the east and are centered at 170–180° E, 35–45° N and 125–135° E, 25–35° N (Figure 8b). Traditional studies regard the EAP teleconnection as a free Rossby wave train propagating from the subtropics to the mid latitudes in the lower troposphere [29,31]. Therefore, it is obvious that the CK shift to the east in S2 contributes significantly to the eastward shift of the EAP teleconnection.

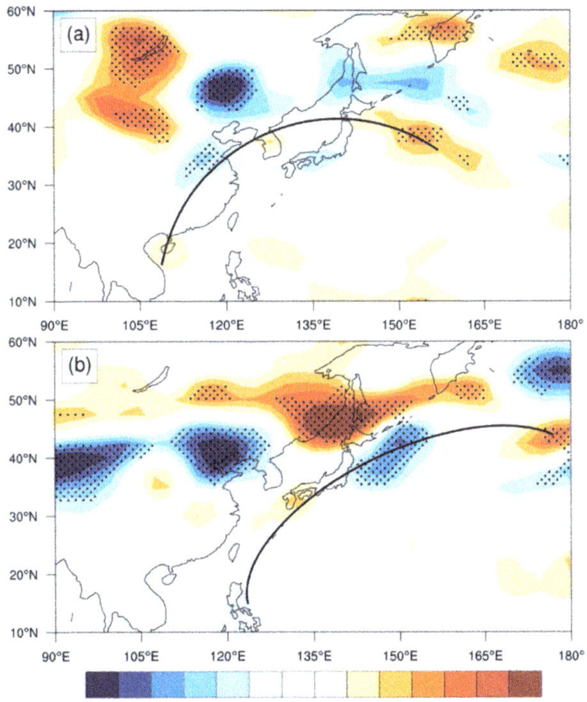

Figure 8. Composite differences of the conversion of local kinetic energy (CK) vertically integrated from 1000 to 100 hPa between wet and dry years in (**a**) S1 and (**b**) S2 (dots indicate the regions exceeding 95% confidence level; black curves in (**a**,**b**) represent the EAP teleconnection).

It has been found that the eastward shift of EAP teleconnection is mainly forced by the shift of related SSTA [54,55]. Figure 9 shows the composite differences of SSTA (unit: K) between wet and dry years from the prior winter (DJF) to the subsequent summertime (JAS). In S1, the EAP teleconnection is connected with the transitional phase from La Niña to El Niño and from "North-warm-South-cold" to "North-cold-South-warm" in the Indian Ocean (IO), as well as a significant cooling phase over the western Pacific (WP) (Figure 9a–c). In S2, the EAP teleconnection is connected with the El Niño Modoki [56] decaying and a cooling phase in the IO, as well as a warming phase over WP (Figure 9d–f).

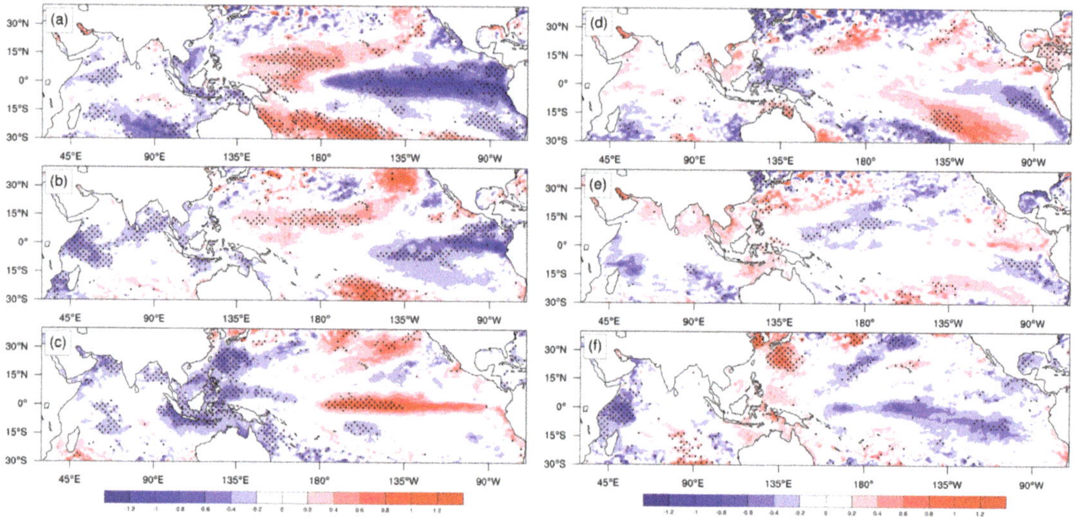

Figure 9. Composite differences of SSTA (unit: K) between wet and dry years (**a**,**d**)/(**b**,**e**)/(**c**,**f**) for pre-winter (DJF)/spring (MAM)/summer (JAS); (**a**–**c**)/(**d**–**f**) for S1/S2; dots indicate the regions exceeding 95% confidence level).

The patterns of composite SSTA in S1 and S2 are significantly different, which is at least partly responsible for the interdecadal shift of the EAP teleconnection. Looking at the situation in the summertime during S2, it is clear that the cooling in the tropical central-eastern Pacific (CEP) and the warming in the WP (Figure 9f) are favorable for the development of convective activity over WP [57,58]. Strong convective activities in the WP strengthen the WPSH by affecting the Hadley cell due to the weakening of EAP teleconnection. Since the enhancement of the IVSP in S1 attracts more attention, the present study focuses on the mechanism for the strengthening of summertime precipitation in S1.

It is found that the CEP warming can induce anomalous convection activities in the NWP by a Rossby-wave-type response and simultaneously destabilize the atmosphere in lower levels [59–61]. To further explore the influence of SSTA pattern on atmospheric circulation anomalies in S1, we analyze the composite differences (shown in Figure 10) in the Walker cell (Figure 10a) and Hadley cell (Figure 10b) between wet and dry years. In the summertime of S1, the updraft in the tropical CEP and downdraft in the tropical WP are significantly enhanced (Figure 10a) due to the warming in the CEP and the cooling in the WP (Figure 9c). This kind of Walker Cell induces an anomalous Hadley cell over the northern WP (Figure 10b). Descending branches appear in the WP around 10S–10N and North China around 35N. An ascending branch occurs near 25°N, which is located at the center of the largest IVSP area (Figure 3a,b). The updraft there suppresses the development of the WPSH (Figure 4b) and promotes more summertime precipitation in the wet years of S1. The Walker Cell and Hadley Cell accompanied by SSTA have opposite impacts on summertime precipitation in SC in the dry years of S1. These two distinctly different processes are responsible for the more significant IVSP in S1.

In a nutshell, the enhanced EAP teleconnection is a Rossby-wave-type response to the warming in the CEP and the cooling in the WP, which result in anomalous Walker Cell and Hadley Cell in the wet years of S1. The northward propagation of Rossby wave energy promotes convective activities and suppresses the WPSH in the NWP, providing an advantageous condition for the increase of summertime precipitation over SC.

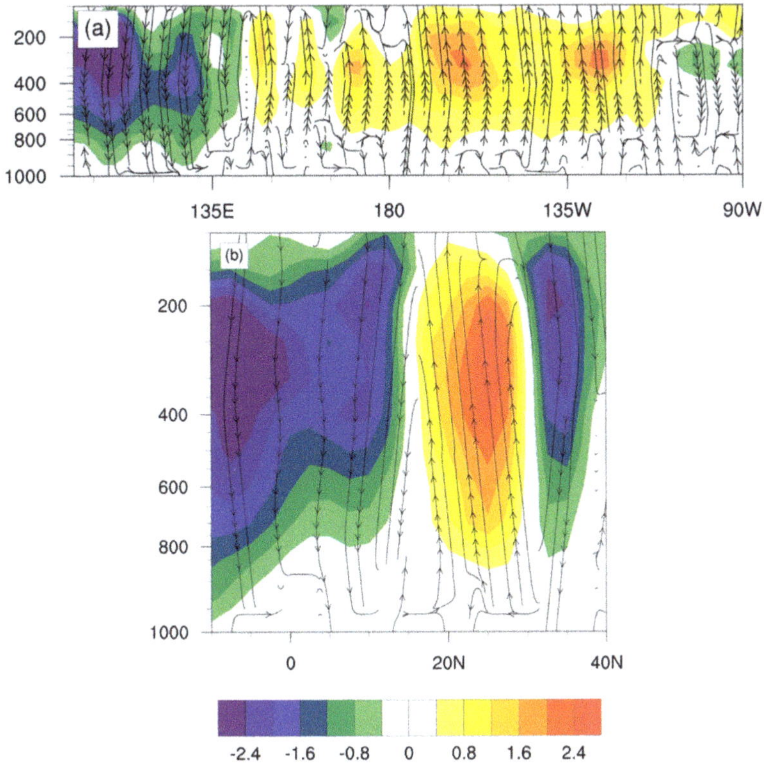

Figure 10. Composite differences of (**a**) Walker cell (averaged over 10°S~10°N) and (**b**) Hadley cell (averaged over 100~120°E) between wet and dry years in S1 (shading regions suggest composite differences of vertical velocity (unit: 10^{-3} m s^{-1}) between wet and dry years. Significant values exceeding the 95% confidence level are displayed).

4. Conclusions

Many studies revealed that the factors that influence the interannual variation of the East Asian summertime climate have changed a lot on interdecadal time scale in the past decades [38–42]. This study reveals that the IVSP over SC experienced a significant weakening after the mid-2000s. Taking 2004 as the shift point, the periods 1993–2004 and 2005–2020 are defined as S1 and S2, respectively. By composite differences of various atmospheric elements between wet and dry years in S1 and S2, the reasons why the interdecadal shift of the IVSP over SC occured are explored.

The composite differences of summertime precipitation between wet and dry years in S1 and S2 are compared. It is found that summertime precipitation over SC shows positive anomalies for both periods but presents distinctly different spatial distributions and intensities. In S1, the center of positive anomalies is located at (26°N, 113°E) and the intensity of the anomalies reaches 3.2–3.4 mm d^{-1}. In S2, two centers of positive anomalies are distributed to the north and south of the positive center in S1. The two centers are respectively located at (29°N, 113°E) and (23°N, 115°E). The intensity of summertime precipitation anomalies in S2 is 1.2–1.8 mm d^{-1} at the two positive centers, weaker than that in S1. Consistent with the composite differences of summertime precipitation, the center of the standard deviation is located at (26°N, 113°E) in S1 but it splits into two centers that shift to the north and south respectively in S2. The magnitude of standard deviation at the positive anomaly centers also decreases in S2 compared to that in S1, which agrees well with the interdecadal weakening of the IVSP over SC. Consistent with summertime precipitation, local moisture flux divergence, omega, and OLR all have larger standard deviations over SC in S1 than in S2.

To find out the reasons for the enormous differences of the IVSP between S1 and S2, atmospheric circulation characteristics are analyzed. From the perspective of large-scale circulation, the center of positive geopotential height anomalies is located at 45N for both S1 and S2, yet the center in S2 is located further westward compared to that in S1. The local negative geopotential height anomalies are weakening from S1 to S2. Moreover, the WPSH elongates to the west in dry years and moves back to the east in wet years during S1. The location and intensity of the WPSH have little difference between wet and dry years during S2. This suggests that the WPSH interannual variability significantly weakens in S2 compared to that in S1, which is similar to the variation in summertime precipitation. Consistent with geopotential height anomalies, the centers of positive and negative anomalies of zonal wind at 200 hPa are located relatively eastward in S1 and westward in S2. The intensity at centers of positive and negative anomalies of zonal wind is weaker in S2 than in S1. From the perspective of moisture transport and local convection, negative OLR and omega anomalies are centered over SC in S1, while the centers of negative anomalies are located eastward and become weak in S2. This result indicates that the intensity of convection weakens from S1 to S2. Moreover, there are strong easterly anomalies that pull huge amounts of moisture from the Northwestern Pacific (NWP) in S1, resulting in significant moisture convergence over SC. However, no significant convergence and/or divergence of moisture flux anomalies can be found over SC in S2. Positive relative vorticity anomalies are centered over SC in S1 accompanied by strong easterly anomalies. The meridional centers of positive and negative relative vorticity anomalies resemble the EAP teleconnection, which moves eastward and becomes weak from S1 to S2.

Since the EAP teleconnection is of vital importance for the interdecadal shift of the IVSP over SC, the reasons for the interdecadal shift of the EAP teleconnection impact on summertime precipitation over SC around the mid-2000s are investigated. The weakening of the EAP teleconnection is partly attributed to the weakening of its intensity and partly attributed to its eastward shift. It was found that the eastward shift of EAP teleconnection is mainly forced by the shift of related SSTA [54,56]. In S1, the EAP teleconnection is connected with the transitional phase from La Niña to El Niño, from the "North-warm-South-cold" pattern to "North-cold-South-warm" pattern in the IO and a significant cooling phase in the WP. In S2, the EAP teleconnection is connected with the Central Pacific El Niño decaying and a cooling phase in the IO, as well as a warming phase over the western Pacific. The patterns of composite SSTA are significantly different between S1 and S2, which is at least partly responsible for the interdecadal shift of the EAP teleconnection. Looking at the situation in the summertime of S2, the cooling in the tropical CEP and the warming in the western Pacific are favorable for the development of convective activity over WP [57,58]. Strong convective activities in the WP strengthen the WPSH by affecting the Hadley cell due to the weakening of EAP teleconnection. In the summertime of S1, the enhanced EAP teleconnection is a Rossby-wave-type response to the warming in the CEP and the cooling in the WP, which results in anomalous Walker Cell and Hadley Cell in the wet years of S1. The northward propagation of Rossby wave energy promotes convective activities and suppresses the WPSH in the NWP, providing an advantageous condition for the increase of summertime precipitation over SC.

The present study concentrates on the interdecadal shift of the IVSP over SC and the possible mechanisms behind it. However, the present study mainly performs some dynamic and statistical analysis on observational datasets. In the following work, it would be better to do numerical experiments to verify the reliability of the results. This study reveals that the interdecadal shift of the EAP teleconnection is significant around the mid-2000s, which is caused by the interdecadal shift of the SSTA patterns in the CEP and the western Pacific. However, there may exist multiple factors responsible for the interdecadal shift of the EAP teleconnection. It is found that the interdecadal change of the PDO phase in the late-2000s could result in the interdecadal shift of the EAP teleconnection and variations of summertime precipitation in eastern China [61]. Further studies are necessary to explore the complex feedback loops that affect summertime precipitation in China.

Author Contributions: Conceptualization, Y.H. (Yao Ha) and Y.Z.; methodology, Y.H. (Yao Ha); validation, Z.Z. and Y.Z.; formal analysis, Y.H. (Yijia Hu); investigation, W.L.; data curation, Y.H. (Yijia Hu); writing—original draft preparation, W.L.; writing—review and editing, Y.H. (Yao Ha); supervision, Z.Z. All authors have read and agreed to the published version of the manuscript.

Funding: This research received no external funding.

Acknowledgments: This work is sponsored jointly by the National Natural Science Foundation of China (41975090); the Natural Science Foundation of Hunan Province, China (2022JJ20043); the Scientific Research Program of National University of Defense Technology (18/19-QNCXJ); and the Jiangsu Collaborative Innovation Center for Climate Change in Nanjing University.

Conflicts of Interest: The authors declare no conflict of interest.

References

1. Chen, G. Comments on "Interdecadal Change of the South China Sea Summer Monsoon Onset". *J. Clim.* **2022**, *28*, 9029–9035. [CrossRef]
2. Huang, R.H.; Chen, J.; Wang, L.; Lin, Z.D. Characteristics, processes, and causes of the spatio-temporal variabilities of the East Asian monsoon system. *Adv. Atmos. Sci.* **2012**, *29*, 910–942. [CrossRef]
3. Yao, R.; Ren, X. Decadal and Interannual Variability of Persistent Heavy Rainfall Events over the Middle and Lower Reaches of the Yangtze River Valley. *J. Meteorol. Res.* **2019**, *33*, 1031–1043. [CrossRef]
4. Wang, H.J.; Fan, K. Recent changes in the East Asian monsoon. *Chin. J. Atmos. Sci.* **2013**, *37*, 313–318. (In Chinese)
5. Ma, W.Q.; Huang, W.Y.; Yang, Z.F.; Wang, B.; Lin, D.Y.; He, X.S. Dynamic and Thermodynamic Factors Associated with Different Precipitation Regimes over South China during Pre-Monsoon Season. *Atmosphere* **2018**, *9*, 219. [CrossRef]
6. Miao, R.; Wen, M.; Zhang, R.; Li, L. The influence of wave trains in mid-high latitudes on persistent heavy rain during the first rainy season over South China. *Clim. Dyn.* **2019**, *53*, 2949–2968. [CrossRef]
7. Li, X.; Wang, C.; Lan, J. Role of the South China Sea in Southern China rainfall: Meridional moisture flux transport. *Clim. Dyn.* **2021**, *56*, 2551–2568. [CrossRef]
8. Ji, Y.; Sun, X.; Xu, Y.; Yao, J.; Yang, X.Q. Summer Regional Pentad Extreme Precipitation in Eastern China and Their Possible Causes. *Front. Earth Sci.* **2021**, *8*, 598025. [CrossRef]
9. Zhang, R.H. Natural and human-induced changes in summer climate over the East Asian monsoon region in the last half century: A review. *Adv. Clim. Change Res.* **2015**, *6*, 131–140. [CrossRef]
10. Zhang, R.H. Changes in East Asian summer monsoon and summer rainfall over eastern China during recent decades. *Sci. Bull.* **2015**, *60*, 1222–1224. [CrossRef]
11. Sun, S.; Ren, Y.; Li, Q.; Zhou, S.; Zhao, C.; Chai, R.; Wang, J. Intra-annual differences of 3-month Standardized Precipitation-Evapotranspiration Index dryness/wetness sensitivity over southwest China. *Atmos. Sci. Lett.* **2018**, *19*, e830. [CrossRef]
12. Cao, X.; Liu, Y.; Wu, R.; Bi, M.; Dai, Y.; Cai, Z. Northwestwards shift of tropical cyclone genesis position during autumn over the western North Pacific after the late 1990s. *Int. J. Climatol.* **2019**, *40*, 1885–1899. [CrossRef]
13. Yang, R.; Wang, J. Interannual variability of the seesaw mode of the interface between the Indian and East Asian summer monsoons. *Clim. Dyn.* **2019**, *53*, 2683–2695. [CrossRef]
14. Zhao, H.; Zhao, K.; Klotzbach, P.J.; Wu, L.; Wang, C. Interannual and interdecadal drivers of meridional migration of Western North Pacific tropical cyclone lifetime maximum intensity location. *J. Clim.* **2022**, *35*, 2709–2722. [CrossRef]
15. Liu, F.; Wang, B.; Ouyang, Y.; Wang, H.; Qiao, S.; Chen, G.; Dong, W. Intraseasonal variability of global land monsoon precipitation and its recent trend. *Npj Clim. Atmos. Sci.* **2022**, *5*, 30. [CrossRef]
16. Zhu, Y.L.; Wang, H.J.; Zhou, W.; Ma, J. Recent changes in the summer precipitation pattern in East China and the background circulation. *Clim. Dyn.* **2011**, *36*, 1463–1473. [CrossRef]
17. Xin, X.; Yu, R.; Zhou, T.; Wang, B. Drought in Late Spring of South China in Recent Decades. *J. Clim.* **2006**, *19*, 3197–3206. [CrossRef]
18. Ye, H.; Lu, R.Y. Dominant patterns of summer rainfall anomalies in East China during 1951–2006. *Adv. Atmos. Sci.* **2012**, *29*, 695–704. [CrossRef]
19. Wang, H.J. The weakening of the Asian monsoon circulation after the end of 1970's. *Adv. Atmos. Sci.* **2001**, *18*, 376–386. [CrossRef]
20. Li, H.; Dai, A.; Zhou, T.; Lu, J. Responses of East Asian summer monsoon to historical SST and atmospheric forcing during 1950–2000. *Clim. Dyn.* **2010**, *34*, 501–514. [CrossRef]
21. Yu, L.; Furevik, T.; Otterå, O.H.; Gao, Y. Modulation of the Pacific Decadal Oscillation on the summer precipitation over East China: A comparison of observations to 600-years control run of Bergen Climate Model. *Clim. Dyn.* **2015**, *44*, 475–494. [CrossRef]
22. Wang, B.; Huang, F.; Wu, Z.; Yang, J.; Fu, X.; Kikuchi, K. Multi-scale climate variability of the South China Sea monsoon: A review. *Dyn. Atmos. Ocean.* **2009**, *47*, 15–37. [CrossRef]

23. Ha, Y.; Zhong, Z.; Chen, H.; Hu, Y. Out-of-phase decadal changes in boreal summer rainfall between Yellow-Huaihe River valley and southern China around 2002/2003. *Clim. Dyn.* **2016**, *47*, 137–158. [CrossRef]
24. Ha, Y.; Zhong, Z.; Hu, Y.; Zhu, Y.; Zang, Z.; Zhang, Y.; Yao, Y.; Sun, Y. Differences between decadal decreases of boreal summer rainfall in southeastern and southwestern China in the early 2000s. *Clim. Dyn.* **2019**, *52*, 3533–3552. [CrossRef]
25. Wu, B.; Li, T.; Zhou, T.J. Relative contributions of the Indian Ocean and local SST anomalies to the maintenance of the western North Pacific anomalous anticyclone during the El Niño decaying summer. *J. Clim.* **2010**, *23*, 2974–2986. [CrossRef]
26. Zhu, Z.W.; Li, T.; He, J.H. Out-of-phase relationship between boreal spring and summer decadal rainfall changes in southern China. *J. Clim.* **2014**, *27*, 1083–1099. [CrossRef]
27. Zhang, Y.S.; Li, T.; Wang, B. Decadal change of the spring snow depth over the Tibetan Plateau: The associated circulation and influence on the East Asian summer monsoon. *J. Clim.* **2004**, *17*, 2780–2793. [CrossRef]
28. Ha, Y.; Zhong, Z.; Sun, Y.; Lu, W. Decadal change of South China Sea tropical cyclone activity in mid-1990s and its possible linkage with intraseasonal variability. *J. Geophys. Res. Atmos.* **2014**, *119*, 5331–5344. [CrossRef]
29. Nitta, T. Convective activities in the tropical western Pacific and their impact on the Northern Hemisphere summer circulation. *J. Meteorol. Soc. Jpn.* **1987**, *65*, 373–390. [CrossRef]
30. Wu, B.; Zhou, T.; Li, T. Impacts of the Pacific–Japan and Circumglobal Teleconnection Patterns on the Interdecadal Variability of the East Asian Summer Monsoon. *J. Clim.* **2016**, *29*, 3253–3271. [CrossRef]
31. Huang, R.H.; Li, W.J. Influence of heat source anomaly over the western tropical Pacific on the subtropical high over East Asia and its physical mechanism. *Chin. J. Atmos. Sci.* **1988**, *12* (Suppl. S1), 107–116. (In Chinese)
32. Kurihara, K.; Tsuyuki, T. Development of the barotropic high around Japan and its association with Rossby wave-like propagations over the North Pacific: Analysis of August 1984. *J. Meteorol. Soc. Jpn.* **1987**, *65*, 237–246. [CrossRef]
33. Wakabayashi, S.; Kawamura, R. Extraction of major teleconnection patterns possibly associated with the anomalous summer climate in Japan. *J. Meteorol. Soc. Jpn.* **2004**, *82*, 1577–1588. [CrossRef]
34. Sun, Y.; Xu, H.M.; Deng, J.C. Interdecadal variation in Pacific−Japan teleconnection patterns and possible causes. *Chin. J. Atmos. Sci.* **2014**, *38*, 1055–1065. (In Chinese)
35. Yin, X.; Zhou, L.; Huangfu, J. Weakened Connection between East China Summer Rainfall and the East Asia-Pacific Teleconnection Pattern. *Atmosphere* **2021**, *12*, 704. [CrossRef]
36. Xu, P.; Wang, L.; Chen, W.; Feng, J.; Liu, Y. Structural Changes in the Pacific–Japan Pattern in the Late 1990s. *J. Clim.* **2019**, *32*, 607–621. [CrossRef]
37. Lu, W.; Zhu, Y.; Ha, Y.; Zhong, Z.; Hu, Y. Recent Decadal Weakening of the Summertime Rainfall Interannual Variability Over Yellow-Huaihe River Valley Attributable to the Western Pacific Cooling. *Front. Earth Sci.* **2022**, *10*, 946252. [CrossRef]
38. Chen, J.; Wang, X.; Zhou, W.; Wen, Z. Interdecadal change in the summer SST-precipitation relationship around the late 1990s over the South China Sea. *Clim. Dyn.* **2018**, *51*, 2229–2246. [CrossRef]
39. Li, X.; Wen, Z.; Chen, D.; Chen, Z. Decadal Transition of the Leading Mode of Interannual Moisture Circulation over East Asia–Western North Pacific: Bonding to Different Evolution of ENSO. *J. Clim.* **2019**, *32*, 289–308. [CrossRef]
40. He, Z.; Wang, W.; Wu, R.; Kang, I.; He, C.; Li, X.; Xu, K.; Chen, S. Change in Coherence of Summer Rainfall Variability over the Western Pacific around the Early 2000s: ENSO Influence. *J. Clim.* **2020**, *33*, 1105–1119. [CrossRef]
41. Wang, Z.; Wen, Z.; Chen, R.; Li, X.; Huang, S. Interdecadal enhancement in the interannual variability of the summer monsoon meridional circulation over the South China Sea around the early 1990s. *Clim. Dyn.* **2020**, *55*, 2149–2164. [CrossRef]
42. Chen, R.; Wen, Z.; Lu, R.; Liu, W. Interdecadal changes in the interannual variability of the summer temperature over Northeast Asia. *J. Clim.* **2021**, *34*, 8361–8376. [CrossRef]
43. Hu, Y.J.; Zhu, Y.M.; Zhong, Z.; Ha, Y. New Predictors and a Statistical Forecast Model for Mei-Yu Onset Date in the Middle and Lower Reaches of the Yangtze River Valley. *Weather. Forecast.* **2014**, *29*, 654–665. [CrossRef]
44. Chen, M.; Xie, P.; Janowiak, J.E.; Arkin, P.A. Global Land Precipitation: A 50-yr Monthly Analysis Based on Gauge Observations. *J. Hydrometeorol.* **2002**, *3*, 249–266. [CrossRef]
45. Şen, Z.; Almazroui, M. Actual Precipitation Index (API) for Drought Classification. *Earth Syst. Environ.* **2021**, *5*, 59–70. [CrossRef]
46. Valipour, M. Calibration of mass transfer-based models to predict reference crop evapotranspiration. *Appl. Water Sci.* **2017**, *7*, 625–635. [CrossRef]
47. Leetmaa, A.; Reynolds, R.; Jenne, R.; Josepht, D. The NCEP/NCAR 40-year reanalysis project. *Bull. Am. Meteor. Soc* **1996**, *77*, 437–471.
48. Liebmann, B.; Smith, C.A. Description of a complete (interpolated) OLR dataset. *Bull. Amer. Meteor. Soc.* **1996**, *77*, 1275–1277.
49. Reynolds, R.W.; Smith, T.M.; Liu, C.; Chelton, D.B.; Casey, K.S.; Schlax, M.G. Daily high-resolution-blended analyses for sea surface temperature. *J. Clim.* **2007**, *20*, 5473–5496. [CrossRef]
50. Takaya, K.; Nakamura, H. A formulation of a phase-independent wave-activity flux for stationary and migratory quasigeostrophic eddies on a zonally varying basic flow. *J. Atmos. Sci.* **2001**, *58*, 608–627. [CrossRef]
51. Kosaka, Y.; Nakamura, H. Structure and dynamics of the summertime Pacific-Japan teleconnection pattern. *Q. J. R. Meteorol. Soc.* **2006**, *132*, 2009–2030. [CrossRef]
52. Huang, G. An index measuring the interannual variation of the East Asian summer monsoon—The EAP index. *Adv. Atmos. Sci.* **2004**, *21*, 41–52. [CrossRef]

53. Wang, L.; Li, T.; Nasuno, T. Impact of Rossby and Kelvin wave components on MJO eastward propagation. *J. Clim.* **2018**, *31*, 6913–6931. [CrossRef]
54. Yim, S.Y.; Wang, B.; Kwon, M. Interdecadal change of the controlling mechanisms for East Asian early summer rainfall variation around the mid-1990s. *Clim. Dyn.* **2014**, *42*, 1325–1333. [CrossRef]
55. Chen, X.; Zhou, T. Relative role of tropical SST forcing in the 1990s periodicity change of the Pacific-Japan pattern interannual variability. *J. Geophys. Res. Atmos.* **2014**, *119*, 13043–13066. [CrossRef]
56. Ashok, K.; Behera, S.K.; Rao, S.A.; Weng, H.; Yamagata, T. El Niño Modoki and its possible teleconnection. *J. Geophys. Res. Ocean.* **2007**, *112*, C11007. [CrossRef]
57. Zhang, J.-Y.; Wang, L.; Yang, S.; Chen, W.; Huangfu, J. Decadal changes of the wintertime tropical tropospheric temperature and their influences on the extratropical climate. *Sci. Bull.* **2016**, *61*, 737–744. [CrossRef]
58. Wu, L.; Zhang, H.; Chen, J.; Feng, T. Impact of Two Types of El Niño on Tropical Cyclones over the Western North Pacific: Sensitivity to Location and Intensity of Pacific Warming. *J. Clim.* **2018**, *31*, 1725–1742. [CrossRef]
59. Gill, A. Some simple solutions for heat-induced tropical circulation. *Q. J. R. Meteorol. Soc.* **1980**, *106*, 447–462. [CrossRef]
60. He, Z.; Wu, R. Indo-Pacific remote forcing in summer rainfall variability over the South China Sea. *Clim. Dyn.* **2014**, *42*, 2323–2337. [CrossRef]
61. Ren, Y.; Song, L.; Wang, Z.; Xiao, Y.; Zhou, B. A possible abrupt change in summer precipitation over eastern China around 2009. *J. Meteorol. Res.* **2017**, *31*, 397–408. [CrossRef]

remote sensing

MDPI

Article

Comparison of Land Use Land Cover Classifiers Using Different Satellite Imagery and Machine Learning Techniques

Sana Basheer [1,2], Xiuquan Wang [1,2,*], Aitazaz A. Farooque [1,2], Rana Ali Nawaz [1,2], Kai Liu [3], Toyin Adekanmbi [1,2] and Suqi Liu [1,4]

1 Canadian Centre for Climate Change and Adaptation, University of Prince Edward Island, St. Peter's Bay, PE C0A 2A0, Canada
2 School of Climate Change and Adaptation, University of Prince Edward Island, Charlottetown, PE C1A 4P3, Canada
3 School of Mathematical and Computational Sciences, University of Prince Edward Island, Charlottetown, PE C1A 4P3, Canada
4 Department of Agriculture and Land, Government of Prince Edward Island, Charlottetown, PE C1A 4N6, Canada
* Correspondence: xxwang@upei.ca

Abstract: Accurate land use land cover (LULC) classification is vital for the sustainable management of natural resources and to learn how the landscape is changing due to climate. For accurate and efficient LULC classification, high-quality datasets and robust classification methods are required. With the increasing availability of satellite data, geospatial analysis tools, and classification methods, it is essential to systematically assess the performance of different combinations of satellite data and classification methods to help select the best approach for LULC classification. Therefore, this study aims to evaluate the LULC classification performance of two commonly used platforms (i.e., ArcGIS Pro and Google Earth Engine) with different satellite datasets (i.e., Landsat, Sentinel, and Planet) through a case study for the city of Charlottetown in Canada. Specifically, three classifiers in ArcGIS Pro, including support vector machine (SVM), maximum likelihood (ML), and random forest/random tree (RF/RT), are utilized to develop LULC maps over the period of 2017–2021. Whereas four classifiers in Google Earth Engine, including SVM, RF/RT, minimum distance (MD), and classification and regression tree (CART), are used to develop LULC maps for the same period. To identify the most efficient and accurate classifier, the overall accuracy and kappa coefficient for each classifier is calculated throughout the study period for all combinations of satellite data, classification platforms, and methods. Change detection is then conducted using the best classifier to quantify the LULC changes over the study period. Results show that the SVM classifier in both ArcGIS Pro and Google Earth Engine presents the best performance compared to other classifiers. In particular, the SVM in ArcGIS Pro shows an overall accuracy of 89% with Landsat, 91% with Sentinel, and 94% with Planet. Similarly, in Google Earth Engine, the SVM shows an accuracy of 87% with Landsat 8 and 92% with Sentinel 2. Furthermore, change detection results show that 13.80% and 14.10% of forest areas have been turned into bare land and urban class, respectively, and 3.90% of the land has been converted into the urban area from 2017 to 2021, suggesting the intensive urbanization. The results of this study will provide the scientific basis for selecting the remote sensing classifier and satellite imagery to develop accurate LULC maps.

Keywords: remote sensing; LULC classification; ArcGIS Pro; Google Earth Engine; machine learning; change detection

Citation: Basheer, S.; Wang, X.; Farooque, A.A.; Nawaz, R.A.; Liu, K.; Adekanmbi, T.; Liu, S. Comparison of Land Use Land Cover Classifiers Using Different Satellite Imagery and Machine Learning Techniques. *Remote Sens.* **2022**, *14*, 4978. https://doi.org/10.3390/rs14194978

Academic Editor: Magaly Koch

Received: 13 September 2022
Accepted: 3 October 2022
Published: 6 October 2022

1. Introduction

Accurate information on land use land cover (LULC) can facilitate various research activities related to floods, droughts, migration, and climate change at several scales. The constant and precise investigation of LULC is a vital aspect of any region's sustainability and development. LULC changes (i.e., deforestation and urbanization, etc.) are one the prominent drivers of climate change around the globe [1]. Furthermore, climate change has significant impacts on water balance [2,3], geomorphology [4], water quality and groundwater management [5–8], resources management and their impacts on humans and their surroundings [9], and land monitoring [10–12], all of which require detailed LULC maps as an essential input [13,14]. LULC maps can assist in identifying which types of land are suitable for agriculture practices and watershed management or useful for urban planning. The most frequent method for this purpose is mapping the LULC and analyzing change over time [15–18].

The severity of extreme events is mainly modulated by the LULC changes. The extreme events of heat waves in 2003 and 2010 in Western Europe and Russia inspired researchers to investigate extreme events by looking at the LULC of the area [19–21]. Therefore, it is crucial to identify the best methodology to analyze the change in LULC to address the growing demands of the population to overcome the issues related to agriculture management, natural resource management, and energy generation. Generating more accurate LULC maps for dense regions requires vast quantities of data. As a result, enormous storage capacities, significant computing power, and the ability to employ various strategies are required [22]. Classification of LULC is the foundation of research on LULC change detection. Traditional visual interpretation and mathematical statistics no longer match the criteria for the accuracy of LULC classification [23]. Training samples, classifiers, and supplementary datasets are the most influential components in supervised LULC classification [24]. However, modern state-of-the-art tools (i.e., ArcGIS Pro and Google Earth Engine) enable us to develop more accurate LULC maps using different LULC classification algorithms.

Analysis of different LULC categories, using machine learning approaches, can be performed in a geographic information system platform (i.e., ArcGIS Pro) or Google Earth Engine using remote sensing datasets. The geographic information system can be used for a better understanding of spatial and temporal analysis related to LULC classification [25]. These analyses can help us to understand the trend of change in LULC. Google Earth Engine is a platform that combines remote sensing data (i.e., satellite imagery from different sources) with high-performance computer service, making satellite imagery processing quicker and simple [26–29]. Google Earth Engine contains satellite imagery from various sources, including Landsat 8, Sentinel 2, MODIS, and many other sources of datasets that are freely accessible. JavaScript is used in Google Earth Engine to develop client libraries, while Python handles code modification [30–34]. Google Earth Engine uses the method of MapReduce architecture for parallel processing, which is used for dividing huge volumes of data into smaller sets and proceeding them across several tools. Then, the data were compiled for output datasets after processing them as several individual components.

Recently, state-of-the-art machine learning algorithms (i.e., random forest/random tree (RF/RT), classification and regression tree (CART), support vector machine (SVM), maximum likelihood (ML), and minimum distance (MD)) have captured the researchers' attention [35,36]. Nowadays, most available research focuses on these LULC classifiers comparisons using different machine learning models. Machine learning has the potential to deal with big historical and present datasets using different algorithms for LULC analysis. In the last few decades, many researchers focused on LULC classification using different remote sensing data with machine learning techniques for different types of study areas to analyze the LULC classification at different platforms [24,30,31]. However, only a few studies focus on using the different platforms for LULC classification and mapping, utilizing different LULC classifiers or approaches to compare LULC maps derived from various multispectral satellite images. On the other hand, the stress for accurate LULC data from satellite imagery around the globe increases. Therefore, it is more critical to compare different classification algorithms and their effectiveness in ArcGIS Pro and Google Earth Engine platforms. The objective of this study is to check the performance evaluation of different algorithms or classifiers for LULC categorization utilizing Landsat

8 surface reflectance Tier 1, Sentinel 2 Level-1C, and Planet imagery data with 30 m, 10 m, and 3–5 m resolutions, respectively, with two different platforms (ArcGIS and Google Earth Engine). Moreover, the accuracy of classifiers, i.e., RF/RT, CART, SVM, ML, and MD [37–39] for both platforms, was determined. Furthermore, the changes in LULC for the city of Charlottetown from 2017 to 2021 were also analyzed utilizing the accurate classifier on the selected platform. The results of this study will provide insights for future studies by identifying the most suitable classifier and platform for the remotely sensed images.

2. Data and Methods

2.1. Study Area

This study focuses on Charlottetown, the capital of Prince Edward Island (PEI), commonly known as the Island, a province of Canada located in the Maritimes (Figure 1). PEI has been facing the adverse effects of climate change in the past few decades [40]. Charlottetown is Prince Edward Island's capital. In the past few years, the population of Charlottetown has increased significantly. As a result, the city is facing deforestation and urbanization. The study area has different land cover types (e.g., forest vegetation, water bodies, and urban areas), and its changing due to human interference. Therefore, it is important to analyze the LULC changes with maximum accuracy using the best machine learning approach. Charlottetown has a total size of 44.29 km^2 [41]. It is located on the province's southern coast and has a population of 80,347. Charlottetown is situated on its namesake harbor. By the convergence of three rivers, this harbor was created on the island's south coast. The entrance to the harbor is the Northumberland Strait. The downtown area of Charlottetown consists of the city's five hundred ancient lots and the waterfront area facing the Hillsborough River. Initially, the downtown area was bordered by four villages (Parkdale, Spring Park, Brighton, and Sherwood). The districts to all sides of downtown except the south area have been developed in recent decades with many commercial developments and an increase in residential areas, while the city's outside areas are still primarily farmland.

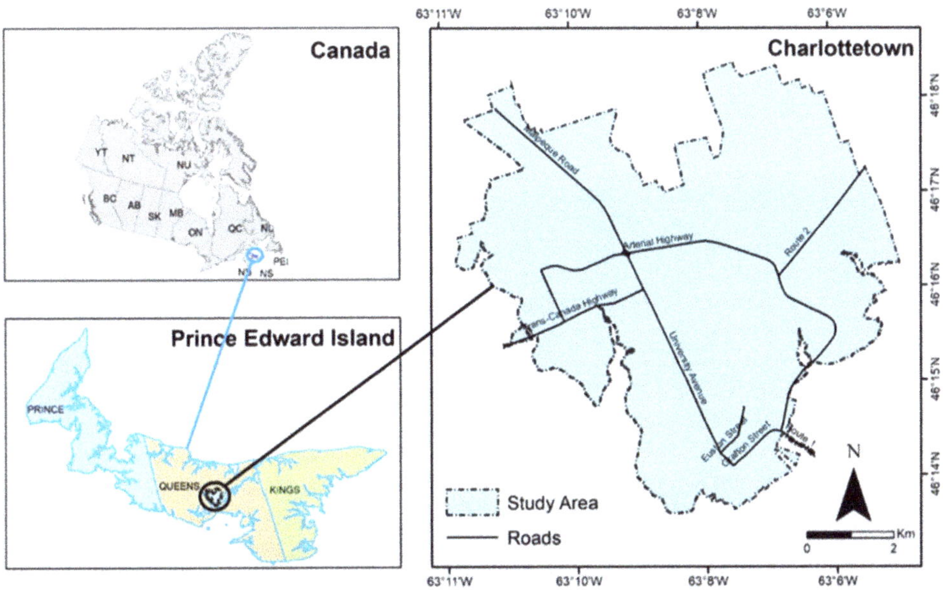

Figure 1. The location of the study area (city of Charlottetown).

2.2. Satellite Data

This project will employ three types of datasets to perform LULC categorization. These datasets include low (Landsat 8), median (Sentinel 2), and high (Planet) resolution satellite images. In this research, all visible and infrared bands (the thermal infrared band was excluded) were used for the LULC analysis [42].

- Landsat 8: the Landsat program is a component of the USGS National Land Imaging (NLI) Program, which has ensured continuity, reliability, and comparability of data since 1974. Landsat has a spatial resolution of 30 m [43].
- Sentinel 2: Sentinel 2 is a high-resolution (10 m), wide-swath, and multispectral imaging system. This aids Copernicus land monitoring studies, such as the examination of water cover and water bodies, vegetation, soil, and the observation related to coastal areas and inland water cover. Sentinel 2 contains thirteen spectral bands [44].
- Planet: since 2016, Planet has captured multispectral photography for consumer use. Planet imagery has high-resolution (3–5 m) imagery that is becoming an essential resource for earth imaging [45]. Planet Scope has more than 150 nanosatellites with a unique combination (coverage of the area, frequency, and high resolution) that can take daily images at a 3 m resolution [25]. These datasets are much more helpful for a better understanding of LULC changes.

2.3. Methodological Framework for LULC Classification

Currently, geospatial Big Data are garnering an abundance of attention and are emphasized globally. ArcGIS Pro and Google Earth Engine are the most popular platforms for large data processing and LULC classification. Several approaches have been established and employed to measure changes in different LULC changes. However, there is currently no research that uses both platforms and includes different input datasets to examine the accurate LULC changes through time in Charlottetown, PEI. In this study, a framework was developed to compare and analyze the performance and viability of different classification methods in ArcGIS Pro and Google Earth Engine, using different datasets. Table 1 shows the major LULC classes scheme for the study area. We use five major classes, e.g., water bodies, urban, forest, bare land, and vegetation, that represent the overall land cover of the city of Charlottetown. Figure 2 shows the methodological frameworks for LULC classification using different classifiers (see Section 2.4) in ArcGIS Pro and Google Earth Engine using Landsat 8, Sentinel 2, and Planet imagery from 2017 to 2021. Image treatment is entirely independent in ArcGIS Pro than in Google Earth Engine. Some data processing factors can affect the outputs, including the algorithms for image selection, cloud masking, and filling the missing pixels. Overall accuracy and kappa coefficient were also calculated to evaluate the performance of each classifier with different input datasets. Maps and graphs were used to display the entire analysis.

Table 1. Name and description of LULC classes scheme.

Class Name	Description of Class
Water Bodies	Lakes, bays, oceans, and estuaries.
Urban	Residential, commercial, industrial, roads and transportation, communications, and utilities.
Forest	Deciduous forest area, evergreen forest, mixed forest, residential forest.
Bare Land	Beaches, sandy areas, open fields without vegetation, exposed rock, gravel pits, transitional areas, and mixed barren land.
Vegetation	Cropland, shrubland, grass.

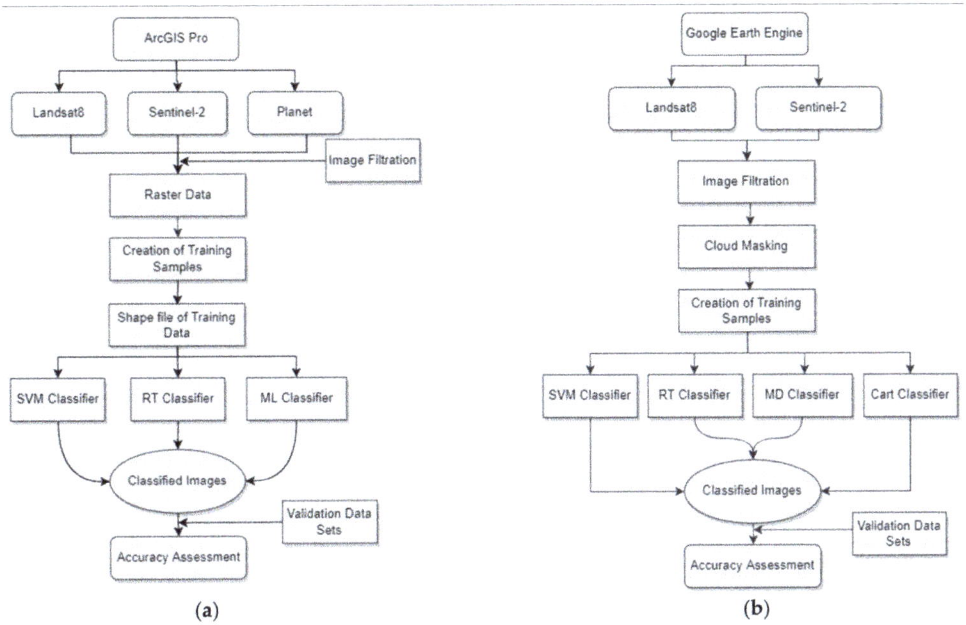

Figure 2. Methodology of LULC classification using (**a**) ArcGIS Pro and (**b**) Google Earth Engine.

2.4. Classification Methods

To perform a pixel-based supervised classification, a set of training samples specific to each year was obtained. Every training sample pixel was assigned to a LULC class on the basis of supplementary data, such as Google Street View and orthophoto images from each year of the study [46]. We use SVM, ML, and RT classier in ArcGIS Pro and CART, SVM, RF, and MD classifiers in the Google Earth Engine platform. In this study, the following classifiers were used for LULC classification:

- Classification and regression tree (CART) is a decision classification tree that facilitates straightforward decision-making and regression analyses [47]. Based on a predetermined threshold, CART runs by separating nodes until it reaches the terminal nodes. This method divides input data into further group sets, and trees are generated using all but one of those group sets. The "classifier.smileCart" approach, which is part of the Google Earth Engine library, was used in this project.
- Random forest/random tree (RF/RT) RF/RT constructs an ensemble classifier and is one of the commonly used classifiers [48] by mixing many CART. RF/RT generates many decision trees using a random selection of training datasets and factors. The two most essential input criteria for this classifier are the size of the training dataset, and the number of trees generated [49,50].
- Support vector machine (SVM). One of the supervised learning techniques is SVM which is used to resolve different problems related to regression and classifications. In the training phase, SVM classifiers generate an ideal hyperplane that splits several classes with the less misclassified pixels from input datasets. The kernel functions, cost parameters, and gamma [51] are key parameters for determining support vectors.
- Minimum distance classifier (MD). The MD classifier classifies imagery datasets into classes that reduce the distance between the class in multi-feature space and image data. This distance between the image data is expressed as a similarity index, such that the minimal distance and highest similarity are equivalent. The main idea behind this process is to calculate the spectral distance between the measurement vector for the candidate pixel and the mean vector for each signature [52].

- Maximum likelihood classifier (ML). Maximum likelihood classifier is a supervised classification method that describes every band by a normal distribution. This supervised classification method is based on the Byes theorem [53].

2.5. Data Processing

In Google Earth Engine, LULC classification maps were generated using Landsat 8 and Sentinel 2 imagery. We use SVM, RF, MD, and CART classifiers in Google Earth Engine. Contaminated pixels caused by cloud cover were eliminated from all images using the cloud mask method provided on the Google Earth Engine [11]. Temporal aggregation approaches such as mean, first, and median was employed to fill the gaps in foggy images.

Using SVM, RT, and ML classifiers, LULC maps were created in ArcGIS Pro. As the major input for classification, images with the least or zero cloud cover were employed. Atmospherically and geometrically corrected images from the three sources (Landsat 8, Sentinel 2, and Planet) with zero cloud cover were utilized as the primary input. The source of data was USGS and Planet Scope sites. Images from the Planet Scope site were mosaiced to cover the whole study area in ArcGIS Pro. Then, all images from all three data sources were clipped to the city of Charlottetown boundary using ArcGIS Pro for LULC classification. Image treatment is entirely independent in ArcGIS Pro than in Google Earth Engine.

A total of 354 training samples were used for LULC classification, from which 284 were used as training samples, and 145 were used as testing samples. Google Earth Pro was used to visually evaluate high-resolution orthophoto images from 2017 to 2021. Training and validation samples were generated with Google Earth Pro. They were then loaded into ArcGIS Pro and the Google Earth Engine platform as shape files to train classifiers. As a rule [54], every classification class should have a minimum of 50 training samples. Training and validation data used for LULC classification was the same in both platforms (ArcGIS Pro and Google Earth Engine). LULC was divided into five broad classes: forest, urban, bare land, water bodies, and vegetation.

For the CART classifier: the optimal cross-validation factor was estimated to be ten based on the research of Kohavi [44] and utilized as an input. For the RT classifier: more trees within the range of 50–100 demonstrated greater classification accuracy and performance [55]. In the present investigation, 100 trees produced positive outcomes. For the SVM classifier: cost, kernel type, and gamma are crucial parameters. The linear kernel is suitable for large datasets [56]. The gamma value is unnecessary for linear kernels. The cost parameter establishes the degree of the punishment for misclassified data, if the C value is greater, it means fewer data misclassifications. For SVM classification, the C-SVC method with a value of 10 as a cost parameter and a linear kernel is also utilized. After LULC classification using post-change detection methodologies, the result of each categorization for the study period was estimated, and a change detection analysis from 2017 to 2021 was also conducted.

2.6. Accuracy Assessment

Methods, procedures, time, and space affected the classification's precision [57–59]. Several research [42,60] found slight to significant variations in the classification accuracy of LULC using different LULC classifiers. The efficacy of various classifiers was evaluated based on their accuracy. The most frequent indicator for measuring the accuracy and also efficiency of all classifiers is overall accuracy, which indicates the percentage of correctly classified testing data. The accuracy evaluation in this investigation reveals a modest discrepancy between the outputs of the classifiers utilized in this instance. The precision of a LULC classification varies not just by classifier but also by location and time. This might be caused by atmospheric, surface, and light fluctuations [61]. Using stratified random sampling, a dataset was generated [62]. Stratified sampling is to divide the dataset or strata according to the characteristics of its attribute. After randomly selecting samples from the types or strata through stratification, these samples in each stratum have a certain commonality. The main purpose of the stratified sampling method is to obtain the sample data with high efficiency. It is generally believed that stratified sampling makes it easy to extract representative samples [63]. These sampling datasets were separated into validation and training sets. Seventy percent of the entire datasets were used for training, while thirty percent were used as validation datasets [64]. Some auxiliary datasets such as Google Street view were used as a reference to collect the sampling datasets. ArcGIS Pro and Google Earth Engine offer different methods to check the accuracy of many LULC classifiers, and some of these accuracy evaluations were utilized to determine the accuracy of each categorized map from 2017 to 2021.

$$Overall\ Accuracy = \frac{T_{C.P}}{T_{S.P}} \times 100 \tag{1}$$

Whereas $T_{C.P}$ is the total number of correctly classified pixels and $T_{S.P}$ is the total number of sample pixels.

$$kappa\ coefficient = \frac{O.A - C.A}{1 - C.A} \tag{2}$$

Whereas $O.A$ is the overall accuracy and $C.A$ is chance agreement. Equations (1) and (2) show the calculation method of overall accuracy and kappa coefficient. The validation data gathered for each year was compared to each map to generate an error matrix. In addition, the kappa coefficient was determined to evaluate the accuracy of each used classifier for every year throughout the study period in ArcGIS pro and Google Earth Engine platforms for their respective classifiers [65]. We tested all classified results from ArcGIS pro on the ArcGIS Pro platform and tested all Google Earth Engine classified results on the Google Earth Engine platform by the same methods (overall accuracy, kappa coefficient) to check the accuracy. A higher value of the kappa coefficient shows the perfect arrangement of the raster data [66], e.g., 0.8 to 1.0 value shows higher efficiency of LULC classification.

2.7. Change Detection

Numerous techniques exist to identify the LULC change from different imagery datasets [67], but this process is not always convenient or straightforward. Comparing other datasets from different satellites obtained on many dates is relatively straightforward yet effective for change detection in the LULC of any area. Some techniques based on this methodology [68,69] calculate descriptive statistics for the vegetation index difference between two different time points. A threshold of the spectral difference is utilized to distinguish between pixels with and without change. In this study, after the classification and accuracy assessment of all three datasets, we choose the dataset and the classifier with the higher value of overall accuracy and also the kappa coefficient value to analyze the change in LULC using ArcGIS Pro. After that, area change in every class throughout the study period was calculated.

3. Results and Discussion

3.1. LULC Classification Maps

3.1.1. LULC Classification of Landsat 8 Imagery in ArcGIS Pro

From Figure 3, we can observe that the LULC classification of Landsat 8 resulted in most of the forest area being misclassified as an urban class or bare land in the years 2020 and 2021 for RT. Vegetation class was misclassified as bare land as water bodies in 2018 for RT. For the SVM classifier, forest and vegetation were misclassified as urban in 2020 to some extent. Vegetation was also misclassified as bare land or forest in 2018 and 2019. For 2020, SVM classified the image well except for bare land and some urban areas. For all Landsat 8 images from all five years, the SVM classifier performed well in comparison to the other two classifiers in terms of overall accuracy. For 2017 and 2019, the ML classifier misclassified vegetation as a forest. This may be because of identical reflectance, the same as forests [65].

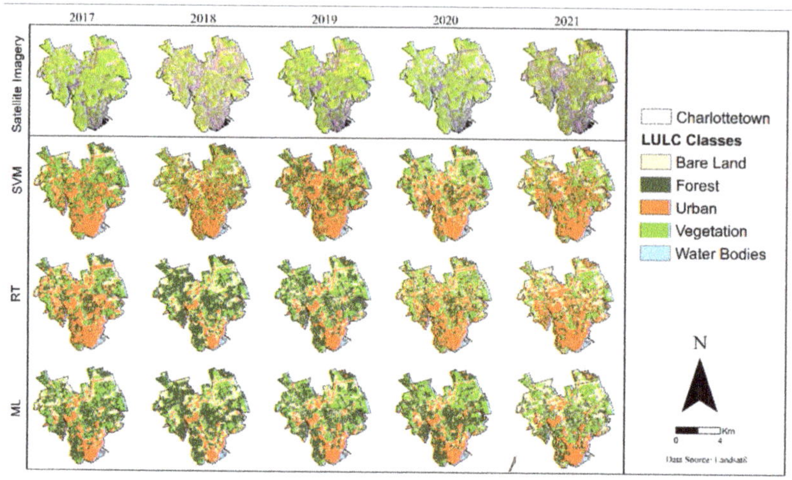

Figure 3. LULC classification maps of Landsat 8 images using SVM, RT, and ML classifiers for the years 2017 to 2021 in ArcGIS Pro.

3.1.2. LULC Classification of Sentinel 2 Imagery in ArcGIS Pro

Figure 4 shows the LULC classification maps using SVM, RT, and ML classifiers for the years 2017 to 2021. From Figure 4, we can see that classification of Sentinel 2 using an SVM classifier resulted in more accuracy as compared to the other two classifiers [11]. For SVM, forest and vegetation were misclassified as urban in 2017 and 2018. For 2021, SVM classified the input imagery data well except for forest and urban classes. For 2020, the RT classifier misclassified the urban and bare land as forest and vegetation in the years 2020 and 2021. Forest was misclassified as urban and, to a lesser extent, as water bodies in 2021 for ML. Vegetation was misclassified as bare land to some extent or forest in 2018 and 2019 using the ML classifier. For 2017 and 2020, the ML classifier misclassified vegetation as bare land and forest, respectively.

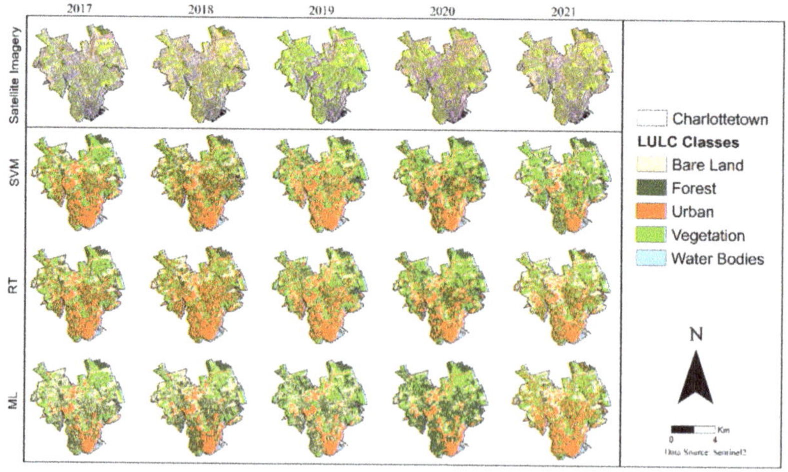

Figure 4. LULC classification maps of Sentinel 2 images using SVM, RT, and ML classifiers for the years 2017 to 2021 in ArcGIS Pro.

3.1.3. LULC Classification of Planet Imagery in ArcGIS Pro

Results show that the classification output of Planet imagery resulted in most of the LULC classes being well classified for the study area with the SVM classifier (Figure 5). For SVM, vegetation was slightly misclassified as bare land and urban in 2020 and 2021. For 2017, the RT classifier misclassified the bare land as vegetation. Forest was misclassified as vegetation and, to some extent, as urban in 2020 and 2021, respectively. For all years, Sentinel 2 images throughout the study period, the ML classifier misclassified the forest and urban areas as bare land in some areas.

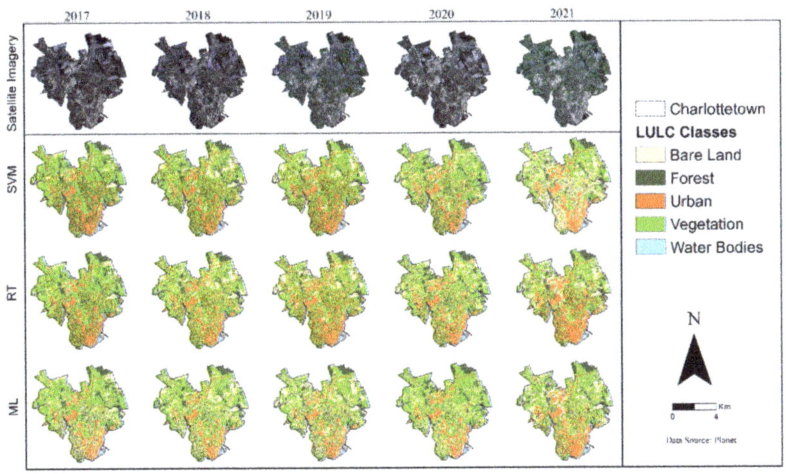

Figure 5. LULC classification maps of Planet images using SVM, RT, and ML classifiers for the years 2017 to 2021 in ArcGIS Pro.

3.1.4. LULC Classification of Landsat 8 Imagery in Google Earth Engine

Figure 6 shows the LULC classification maps using Landsat 8 imagery. In most of the forest area, this class was misclassified as vegetation for the study area with the SVM and CART classifiers in 2020 and 2021. For the SVM classifier, bare land was misclassified as vegetation and urban for 2017 and 2021. For 2017, the RF classifier misclassified the bare land as vegetation. In 2021, vegetation was classified as bare land in some areas. Urban was misclassified as forest in 2017, 2018, and 2019 in the MD classifier. Throughout the study period, the CART classifier misclassified the forest as urban areas. Overall, the SVM classifier performs better than the other three classifiers using Landsat 8 imagery in the Google Earth Engine.

3.1.5. LULC Classification of Sentinel 2 Imagery in Google Earth Engine

Figure 7 shows the classification output of Sentinel 2 imagery, which shows that SVM classifiers perform well as compared to CART, MD, and RF classifiers. For SVM, the forest was misclassified as vegetation and bare land for 2019 and 2020, respectively. For 2017, the RF classifier misclassified the bare land and forest area as urban. In 2020, vegetation was classified as water bodies in some areas. Urban was misclassified as a forest for all five years in the MD classifier [51]. Throughout the study period, the CART classifier misclassified most classes as urban areas [70]. Overall SVM classifier performs better than the other three classifiers as urban and vegetation areas are well classified in the Google Earth Engine using Sentinel 2 imagery.

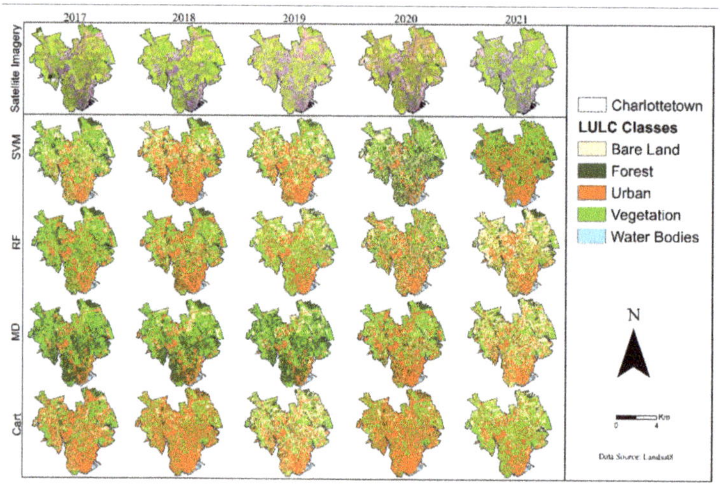

Figure 6. LULC classification maps of Landsat 8 images using SVM, RF, MD, and CART classifiers for the years 2017 to 2021 in Google Earth Engine.

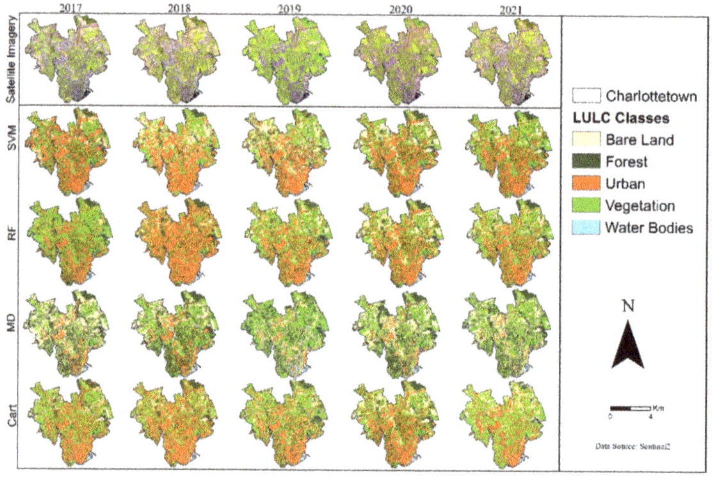

Figure 7. LULC classification maps of Sentinel 2 images using SVM, RF, MD, and CART classifiers for the years 2017 to 2021 in Google Earth Engine.

3.2. Accuracy Assessment

In this research, to check the accuracy of each classifier, we used a validation dataset that was different from the training datasets. By stratified random sampling, we distributed these points over the study area [64] to ensure that all LULC classes accurately and uniformly were represented. After LULC classification using all classifiers, overall accuracy and kappa coefficient were calculated to check the accuracy of these LULC classified maps in the ArcGIS Pro and Google Earth Engine. Tables 2 and 3 show the overall accuracy of different classifiers in ArcGIS Pro and Google Earth Engine platform using Landsat 8, Sentinel 2, and Planet imagery. From Table 2, we can observe the overall accuracy of the SVM classifier with Planet imagery was higher than the other classifiers. The SVM classifier performs higher accuracy with an overall accuracy of 96% in 2017 and 2019 using Planet imagery in ArcGIS Pro. For Sentinel 2 dataset, the higher value of overall accuracy was for the year 2018, with a value of 87%. The lowest value of overall accuracy (74%) for Sentinel

2 imagery was for the year 2019 using the RT classifier. Landsat 8 shows the lowest overall accuracy for all tree classifiers using the ArcGIS Pro platform throughout the study period compared to the other two datasets. Using Landsat 8 imagery, the SVM classifier shows a higher overall accuracy throughout the study period, with the highest value of 85% in 2021.

Table 2. Kappa coefficient and overall accuracy of Landsat 8, Sentinel 2, and Planet imagery for SVM, RT, and ML classifiers using ArcGIS Pro.

Year	Classifier	Landsat 8		Sentinel 2		Planet	
		Overall Accuracy (%)	Kappa Coefficient	Overall Accuracy (%)	Kappa Coefficient	Overall Accuracy (%)	Kappa Coefficient
2017	SVM	84	0.79	86	0.82	96	0.95
	ML	82	0.76	85	0.81	89	0.87
	RT	76	0.72	81	0.75	88	0.85
2018	SVM	83	0.78	87	0.83	90	0.87
	ML	81	0.75	84	0.79	88	0.85
	RT	78	0.71	81	0.77	84	0.84
2019	SVM	80	0.82	85	0.87	96	0.95
	ML	82	0.77	76	0.71	95	0.94
	RT	71	0.65	74	0.67	94	0.92
2020	SVM	83	0.78	84	0.81	94	0.92
	ML	82	0.75	83	0.81	92	0.90
	RT	81	0.72	80	0.77	90	0.87
2021	SVM	85	0.81	86	0.83	93	0.91
	ML	81	0.76	83	0.79	92	0.90
	RT	80	0.73	82	0.77	92	0.90

Table 3. Kappa coefficient and overall accuracy of Landsat 8, Sentinel 2 imagery for SVM, RF, MD, and CART classifiers.

Year	Classifier	Landsat 8		Sentinel 2	
		Overall Accuracy (%)	Kappa Coefficient	Overall Accuracy (%)	Kappa Coefficient
2017	CART	85	0.59	91	0.72
	SVM	88	0.67	92	0.78
	MD	83	0.54	90	0.68
	RF	84	0.55	91	0.73
2018	CART	85	0.64	92	0.80
	SVM	87	0.67	93	0.83
	MD	83	0.53	90	0.68
	RF	84	0.56	91	0.79
2019	CART	88	0.68	91	0.75
	SVM	90	0.80	92	0.82
	MD	86	0.62	89	0.64
	RF	87	0.64	90	0.80
2020	CART	81	0.55	90	0.81
	SVM	83	0.57	92	0.89
	MD	77	0.47	85	0.65
	RF	78	0.49	91	0.83
2021	CART	87	0.70	89	0.72
	SVM	90	0.76	91	0.86
	MD	81	0.64	87	0.74
	RF	85	0.69	88	0.80

Using the Google Earth Engine, the SVM classifier performed the highest overall accuracy of 92% and 90% with Sentinel 2 and Landsat 8 imagery, respectively, in the year 2019 compared with the other three classifiers [71]. In 2021, the MD classifier showed the lowest overall accuracy value, 87%, with Sentinel 2 imagery. Throughout the study, the SVM classifier performs well compared to the other classifiers with Sentinel 2 data. Using Landsat 8 data, the SVM classifier shows the highest overall accuracy of 90% in 2019 and 2021. The lowest overall accuracy of 77% was observed for the MD classifier in 2018. Throughout the study period, the MD classifier showed the lowest overall accuracy with both datasets in the Google Earth Engine platform.

Figure 8a shows the kappa coefficient values for all tested classifiers using three different imagery (Landsat 8, Sentinel 2, and Planet) at the ArcGIS Pro platform. Values of the kappa coefficient from Figure 8a show that Planet imagery shows the highest accuracy with the SVM classifier compared to the other two datasets throughout the study period. Figure 8b shows the values of the kappa coefficient for all tested classifiers using Landsat 8 and Sentinel 2 imagery at the Google Earth Engine platform. The Sentinel 2 dataset shows a higher value of the kappa coefficient as compared to the Landsat 8 dataset from 2017 to 2021 with the SVM classifier. Results show that SVM performs well as compared to CRAT and other classifiers on both platforms. Similar results are reported in the literature [47,72]. At the same time, the lowest value for the kappa coefficient was observed for Landsat 8 data using the RF classifier.

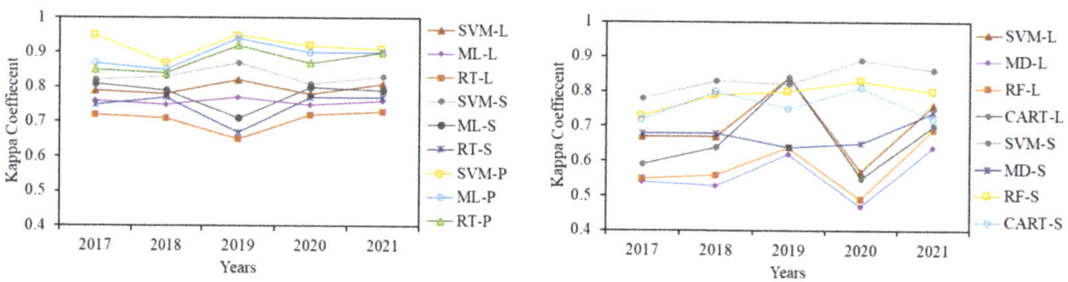

(a) (b)

Figure 8. Kappa coefficient (**a**) using SVM, ML, and RT classifiers in ArcGIS Pro, (**b**) using SVM, MD, RF, and CART classifiers in Google Earth Engine.

The average overall accuracy for the SVM, ML, and RT classifiers was 94%, 91%, and 89%, respectively, using Planet imagery. While using Sentinel 2 and Landsat imagery, the overall accuracy for SVM, ML, and RT classifiers was 85%, 82%, 82.2% and 83%, 81%, and 77%, respectively, at the ArcGIS Pro platform. In ArcGIS Pro, the average value of the kappa coefficient from 2017 to 2021 was 0.92 for the SVM classifier using Planet imagery (Table 4), which was higher than the other classifiers. The lowest value for the average kappa coefficient value was observed for the RT classifier using Landsat 8 data, which was 0.70, respectively. In Google Earth Engine, the average value of overall accuracy for the SVM classifier was 0.83 with Sentinel 2 (Table 5), which was higher than the other classifiers for the Sentinel 2 dataset. Using Landsat 8 data, the highest value of average overall accuracy and kappa coefficient with a value of 83% and 0.79 was for the SVM classifier. MD and RF classifiers show almost close values of average overall accuracy and kappa coefficient in the Google Earth Engine platform.

Table 4. Average overall accuracy and average kappa coefficient of Landsat 8, Sentinel 2, and Planet imagery for SVM, RT, and ML classifiers using ArcGIS Pro.

Classifier	Landsat 8		Sentinel 2		Planet	
	Overall Accuracy (%)	Kappa Coefficient	Overall Accuracy (%)	Kappa Coefficient	Overall Accuracy (%)	Kappa Coefficient
SVM	83	0.79	85	0.83	94	0.92
ML	81	0.75	82	0.78	91	0.89
RT	77	0.70	79	0.74	89	0.87

Table 5. Average overall accuracy and average kappa coefficient of Landsat 8, Sentinel 2 imagery for SVM, RF, MD, and CART classifiers using Google Earth Engine.

Classifier	Landsat 8		Sentinel 2	
	Overall Accuracy (%)	Kappa Coefficient	Overall Accuracy (%)	Kappa Coefficient
CART	83	0.63	90	0.76
SVM	87	0.69	92	0.83
MD	84	0.56	88	0.68
RF	82	0.56	89	0.79

3.3. Change Detection

There are a variety of approaches for detecting LULC changes using different satellite image data [68], but this process is not always simple. Comparing remotely sensed data obtained on multiple dates is a simple yet efficient method for change detection [73]. Figure 9 shows the significant changes in the LULC of Charlottetown from 2017 to 2021 in all LULC classes in the city of Charlottetown. A significant change was observed in forest and vegetation classes, which were converted into other LULC classes during the study period. Bare land areas were also converted to urban areas due to rapid urbanization in the study area.

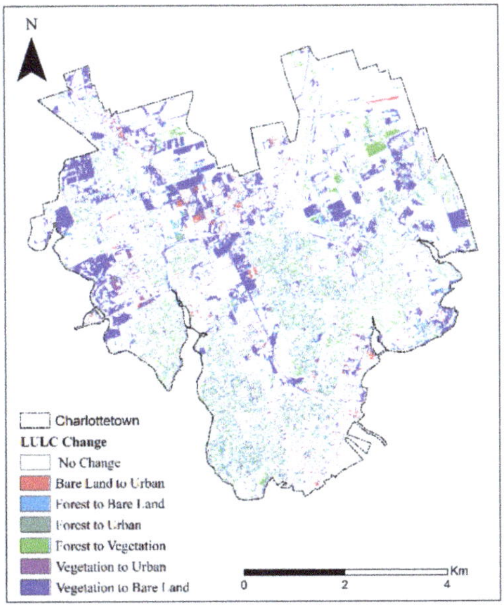

Figure 9. Change in LULC over study area from 2017 to 2019.

The LULC changes can also be determined by using a transition matrix. The transition matrix can explain the change in every LULC class over a specific period [74]. This matrix describes the changed and unchanged amount of area of each LULC class [75]. Change in every class can be observed by a transition matrix in Table 6 from 2017 to 2021 in percentage and km^2. The bare land class changed to the urban class with an increase of 38.81%, showing an urban area expansion from bare land, forest, and vegetation areas as per the literature review. At the same time, another significant change was observed in the vegetation class, which had a decrease of 32.24% in the bare land class and 1.63% in the urban class. Overall, 3.9% of the total study area was converted to an urban area from 2017 to 2021, which shows rapid urbanization in the study area. Forest areas changed into vegetation by 21.3%. Forest area was also converted into bare land by 13.80% due to growth in the lumber industry and into urban by 14.10%, whereas the urban class and water bodies class shows minor changes in their respective area of interest.

Table 6. Transition matrix for area change in different LULC classes from 2017 to 2021 in km^2.

	Bare Land	Forest	Urban	Vegetation	Water Bodies	Area Changes	Area Unchanged	Total Area
Bare Land	0.15	0.33	1.25	0.49	0.00	+2.07	1.15	3.22
	4.65%	10.24%	38.81%	15.21%	0%			
Forest	1.38	5.18	1.41	2.13	0.00	−4.92	5.18	10.10
	13.80%	51.80%	14.10%	21.30%	0%			
Urban	1.29	0.24	9.00	0.014	0.00	+1.54	9.00	10.54
	12.23%	2.29%	86.12%	0.13%	0%			
Vegetation	7.50	0.59	0.33	11.74	0.00	−7.42	12.74	20.16
	37.20%	7.80%	1.63%	58.23%	0%			
Water Bodies	0.01	0.00	0.01	0.00	0.44	−0.011	0.44	0.46
	1.08%	0%	1.30%	0%	95.65%			

4. Conclusions

The climate around the globe has been modified by human activities through the LULC changes. The LULC changes (i.e., deforestation, urbanization, etc.) played an important role in climate change since the pre-industrial era [76]. There are many studies conducted to measure the LULC changes. However, there is no study, to the best of the authors' knowledge, which compares the performance of different approaches available on other platforms. In this study, the performance of different classifiers was observed on two different platforms (ArcGIS Pro and Google Earth Engine) using different datasets, e.g., Landsat 8, Sentinel 2, and Planet imagery. This study aimed to identify the best-performing classifier with different input data. To check the accuracy of different classifiers, the accuracy assessment was performed for every classifier using the error matrix and kappa coefficient for all three satellite imageries from 2017 to 2021. The SVM showed higher accuracy as compared to other classifiers within both platforms. The SVM classifier performs the highest overall accuracy using the Planet imagery on ArcGIS Pro and using Sentinel 2 imagery in the Google Earth Engine. Due to the high resolution of Planet data, it performs well with an average overall accuracy of 94% compared to Sentinel 2 with 85% and Landsat 8 with 83% in the ArcGIS Pro platform. On the Google Earth Engine platform, the SVM classifier performs well with Sentinel 2 imagery with an average overall accuracy of 92% compared to Landsat 8 imagery, which shows 87% accuracy. For this type of study area, which includes different LULC types, e.g., the forest, vegetation, urban, and water bodies, the SVM classifier shows promising results [11].

Change detection analysis shows deforestation and rapid urbanization as 18.80% of forest area was converted to bare land, and 38.81% of bare land was converted to urban area throughout the study period. In addition, some classes may be categorized using expert knowledge and supplementary data [46]. The most accurate and suitable classifier can also affect by the region of interest, number, and quality of training samples. Similarly, identical reflectance of satellite imagery was another limitation for LULC classification. Some pixels in the LULC maps represent different classes after the LULC classification from the training due to the fact that some of the pixels correspond to different spectral responses. These LULC classifications may be modified based on the study area, data, etc. Furthermore, many studies suggested that the different classifiers show different results in different climatic and geographic situations. Therefore, there is a dire need to further explore the LULC classification accuracy by adding more detailed LULC classes and more training samples with high-resolution input datasets.

Author Contributions: Conceptualization, X.W.; Data curation, S.B.; Formal analysis, R.A.N.; Investigation, A.A.F. and K.L.; Project administration, X.W.; Supervision, X.W.; Validation, S.B. and T.A.; Visualization, A.A.F., R.A.N. and S.L.; Writing—original draft, S.B.; Writing—review and editing, R.A.N., K.L. and T.A. All authors have read and agreed to the published version of the manuscript.

Funding: This research was supported by the Natural Science and Engineering Research Council of Canada, the New Frontiers in Research Fund, Atlantic Canada Opportunities Agency, and Agriculture and Agri-Food Canada.

Data Availability Statement: Publicly available data sets were used in this study. These can be found at: https://earthexplorer.usgs.gov/ (Landsat8 and Sentinel-2 datasets) and https://www.planet.com/explorer/ (Planet datasets).

Conflicts of Interest: The authors declare no conflict of interest.

References

1. Li, Y.; Brando, P.M.; Morton, D.C.; Lawrence, D.M.; Yang, H.; Randerson, J.T. Deforestation-induced climate change reduces carbon storage in remaining tropical forests. *Nat. Commun.* **2022**, *13*, 1964. [CrossRef] [PubMed]
2. Sridhar, V.; Kang, H.; Ali, S.A. Human-Induced Alterations to Land Use and Climate and Their Responses for Hydrology and Water Management in the Mekong River Basin. *Water* **2019**, *11*, 1307. [CrossRef]
3. Sridhar, V.; Jin, X.; Jaksa, W.T.A. Explaining the hydroclimatic variability and change in the Salmon River basin. *Clim. Dyn.* **2012**, *40*, 1921–1937. [CrossRef]
4. Sujatha, E.; Sridhar, V. Spatial Prediction of Erosion Risk of a Small Mountainous Watershed Using RUSLE: A Case-Study of the Palar Sub-Watershed in Kodaikanal, South India. *Water* **2018**, *10*, 1608. [CrossRef]
5. Sridhar, V.; Billah, M.M.; Hildreth, J.W. Coupled Surface and Groundwater Hydrological Modeling in a Changing Climate. *Groundwater* **2017**, *56*, 618–635. [CrossRef] [PubMed]
6. Xiao, Y.; Liu, K.; Yan, H.; Zhou, B.; Huang, X.; Hao, Q.; Zhang, Y.; Zhang, Y.; Liao, X.; Yin, S. Hydrogeochemical constraints on groundwater resource sustainable development in the arid Golmud alluvial fan plain on Tibetan plateau. *Environ. Earth Sci.* **2021**, *80*, 750. [CrossRef]
7. Xiao, Y.; Xiao, D.; Hao, Q.; Liu, K.; Wang, R.; Huang, X.; Liao, X.; Zhang, Y. Accessible Phreatic Groundwater Resources in the Central Shijiazhuang of North China Plain: Perspective From the Hydrogeochemical Constraints. *Front. Environ. Sci.* **2021**, *9*, 475. [CrossRef]
8. Xiao, Y.; Hao, Q.; Zhang, Y.; Zhu, Y.; Yin, S.; Qin, L.; Li, X. Investigating Sources, Driving Forces and Potential Health Risks of Nitrate and Fluoride in Groundwater of a Typical Alluvial Fan Plain. *Sci. Total Environ.* **2022**, *802*, 149909. [CrossRef]
9. Sridhar, V.; Ali, S.A.; Sample, D.J. Systems Analysis of Coupled Natural and Human Processes in the Mekong River Basin. *Hydrology* **2021**, *8*, 140. [CrossRef]
10. Jamali, B.; Bach, P.M.; Cunningham, L.; Deletic, A. A Cellular Automata Fast Flood Evaluation (CA-ffé) Model. *Water Resour. Res.* **2019**, *55*, 4936–4953. [CrossRef]
11. Rahman, A.; Abdullah, H.M.; Tanzir, M.T.; Hossain, M.J.; Khan, B.M.; Miah, M.G.; Islam, I. Performance of Different Machine Learning Algorithms on Satellite Image Classification in Rural and Urban Setup. *Remote Sens. Appl. Soc. Environ.* **2020**, *20*, 100410. [CrossRef]
12. Sridhar, V.; Anderson, K.A. Human-Induced Modifications to Land Surface Fluxes and Their Implications on Water Management under Past and Future Climate Change Conditions. *Agric. For. Meteorol.* **2017**, *234–235*, 66–79. [CrossRef]
13. Cihlar, J. Land Cover Mapping of Large Areas from Satellites: Status and Research Priorities. *Int. J. Remote Sens.* **2000**, *21*, 1093–1114. [CrossRef]
14. Renschler, C.S.; Harbor, J. Soil Erosion Assessment Tools from Point to Regional Scales—The Role of Geomorphologists in Land Management Research and Implementation. *Geomorphology* **2002**, *47*, 189–209. [CrossRef]
15. Roy, D.P.; Wulder, M.A.; Loveland, T.R.; Woodcock, C.E.; Allen, R.G.; Anderson, M.C.; Helder, D.; Irons, J.R.; Johnson, D.M.; Kennedy, R.; et al. Landsat-8: Science and product vision for terrestrial global change research. *Remote Sens. Environ.* **2014**, *145*, 154–172. [CrossRef]

16. Wulder, M.A.; White, J.C.; Loveland, T.R.; Woodcock, C.E.; Belward, A.S.; Cohen, W.B.; Fosnight, E.A.; Shaw, J.; Masek, J.G.; Roy, D.P. The Global Landsat Archive: Status, Consolidation, and Direction. *Remote Sens. Environ.* **2016**, *185*, 271–283. [CrossRef]
17. Gómez, C.; White, J.C.; Wulder, M.A. Optical Remotely Sensed Time Series Data for Land Cover Classification: A Review. *ISPRS J. Photogramm. Remote Sens.* **2016**, *116*, 55–72. [CrossRef]
18. Phan, T.N.; Kuch, V.; Lehnert, L.W. Land Cover Classification Using Google Earth Engine and Random Forest Classifier—The Role of Image Composition. *Remote Sens.* **2020**, *12*, 2411. [CrossRef]
19. Fischer, E.M.; Schär, C. Consistent Geographical Patterns of Changes in High-Impact European Heatwaves. *Nat. Geosci.* **2010**, *3*, 398–403. [CrossRef]
20. Seneviratne, S.I.; Lüthi, D.; Litschi, M.; Schär, C. Land–Atmosphere Coupling and Climate Change in Europe. *Nature* **2006**, *443*, 205–209. [CrossRef]
21. Stott, P.A.; Stone, D.A.; Allen, M.R. Human Contribution to the European Heatwave of 2003. *Nature* **2004**, *432*, 610–614. [CrossRef] [PubMed]
22. Xie, S.; Liu, L.; Zhang, X.; Yang, J.; Chen, X.; Gao, Y. Automatic Land-Cover Mapping Using Landsat Time-Series Data Based on Google Earth Engine. *Remote Sens.* **2019**, *11*, 3023. [CrossRef]
23. Gibril, M.B.A.; Idrees, M.; Yao, K.; Shafri, H.Z.M. Integrative Image Segmentation Optimization and Machine Learning Approach for High Quality Land-Use and Land-Cover Mapping Using Multisource Remote Sensing Data (Erratum). *J. Appl. Remote Sens.* **2018**, *12*, 019902. [CrossRef]
24. Johnson, B. Scale Issues Related to the Accuracy Assessment of Land Use/Land Cover Maps Produced Using Multi-Resolution Data: Comments on "the Improvement of Land Cover Classification by Thermal Remote Sensing". Remote Sens. **2015**, *7*, 8368–8390. *Remote Sens.* **2015**, *7*, 13436–13439. [CrossRef]
25. Szostak, M.; Likus-Cieślik, J.; Pietrzykowski, M. PlanetScope Imageries and LiDAR Point Clouds Processing for Automation Land Cover Mapping and Vegetation Assessment of a Reclaimed Sulfur Mine. *Remote Sens.* **2021**, *13*, 2717. [CrossRef]
26. Gorelick, N.; Hancher, M.; Dixon, M.; Ilyushchenko, S.; Thau, D.; Moore, R. Google Earth Engine: Planetary-Scale Geospatial Analysis for Everyone. *Remote Sens. Environ.* **2017**, *202*, 18–27. [CrossRef]
27. Sidhu, N.; Pebesma, E.; Câmara, G. Using Google Earth Engine to Detect Land Cover Change: Singapore as a Use Case. *Eur. J. Remote Sens.* **2018**, *51*, 486–500. [CrossRef]
28. Tamiminia, H.; Salehi, B.; Mahdianpari, M.; Quackenbush, L.; Adeli, S.; Brisco, B. Google Earth Engine for Geo-Big Data Applications: A Meta-Analysis and Systematic Review. *ISPRS J. Photogramm. Remote Sens.* **2020**, *164*, 152–170. [CrossRef]
29. Kolli, M.K.; Opp, C.; Karthe, D.; Groll, M. Mapping of Major Land-Use Changes in the Kolleru Lake Freshwater Ecosystem by Using Landsat Satellite Images in Google Earth Engine. *Water* **2020**, *12*, 2493. [CrossRef]
30. Shelestov, A.; Lavreniuk, M.; Kussul, N.; Novikov, A.; Skakun, S. Exploring Google Earth Engine Platform for Big Data Processing: Classification of Multi-Temporal Satellite Imagery for Crop Mapping. *Front. Earth Sci.* **2017**, *5*, 17. [CrossRef]
31. Patel, N.N.; Angiuli, E.; Gamba, P.; Gaughan, A.; Lisini, G.; Stevens, F.R.; Tatem, A.J.; Trianni, G. Multitemporal Settlement and Population Mapping from Landsat Using Google Earth Engine. *Int. J. Appl. Earth Obs. Geoinf.* **2015**, *35*, 199–208. [CrossRef]
32. Mateo-García, G.; Gómez-Chova, L.; Amorós-López, J.; Muñoz-Marí, J.; Camps-Valls, G. Multitemporal Cloud Masking in the Google Earth Engine. *Remote Sens.* **2018**, *10*, 1079. [CrossRef]
33. Pimple, U.; Simonetti, D.; Sitthi, A.; Pungkul, S.; Leadprathom, K.; Skupek, H.; Som-ard, J.; Gond, V.; Towprayoon, S. Google Earth Engine Based Three Decadal Landsat Imagery Analysis for Mapping of Mangrove Forests and Its Surroundings in the Trat Province of Thailand. *J. Comput. Commun.* **2018**, *6*, 247–264. [CrossRef]
34. Nery, T.; Sadler, R.; Solis-Aulestia, M.; White, B.; Polyakov, M.; Chalak, M. Comparing Supervised Algorithms in Land Use and Land Cover Classification of a Landsat Time-Series. In Proceedings of the 2016 IEEE International Geoscience and Remote Sensing Symposium (IGARSS), Beijing, China, 10–15 July 2016. [CrossRef]
35. Maxwell, A.E.; Warner, T.A.; Fang, F. Implementation of Machine-Learning Classification in Remote Sensing: An Applied Review. *Int. J. Remote Sens.* **2018**, *39*, 2784–2817. [CrossRef]
36. Adam, E.; Mutanga, O.; Odindi, J.; Abdel-Rahman, E.M. Land-Use/Cover Classification in a Heterogeneous Coastal Landscape Using RapidEye Imagery: Evaluating the Performance of Random Forest and Support Vector Machines Classifiers. *Int. J. Remote Sens.* **2014**, *35*, 3440–3458. [CrossRef]
37. Bar, S.; Parida, B.R.; Pandey, A.C. Landsat-8 and Sentinel-2 Based Forest Fire Burn Area Mapping Using Machine Learning Algorithms on GEE Cloud Platform over Uttarakhand, Western Himalaya. *Remote Sens.Appl. Soc. Environ.* **2020**, *18*, 100324. [CrossRef]
38. Liu, D.; Chen, N.; Zhang, X.; Wang, C.; Du, W. Annual Large-Scale Urban Land Mapping Based on Landsat Time Series in Google Earth Engine and OpenStreetMap Data: A Case Study in the Middle Yangtze River Basin. *ISPRS J. Photogramm. Remote Sens.* **2020**, *159*, 337–351. [CrossRef]
39. Tassi, A.; Vizzari, M. Object-Oriented LULC Classification in Google Earth Engine Combining SNIC, GLCM, and Machine Learning Algorithms. *Remote Sens.* **2020**, *12*, 3776. [CrossRef]
40. *Prince Edward Island (PEI) Climate Change Risk Assessment*; Department of Environment, Energy and Climate Action: Charlottetown, PE, Canada, 2021.
41. Wikipedia Contributors. Charlottetown. Available online: https://en.wikipedia.org/wiki/Charlottetown (accessed on 1 June 2022).

42. Reis, S. Analyzing Land Use/Land Cover Changes Using Remote Sensing and GIS in Rize, North-East Turkey. *Sensors* **2008**, *8*, 6188–6202. [CrossRef]
43. Wahab, N.A.; Shafri, H.Z. *Utilization of Google Earth Engine (GEE) for Land Cover Monitoring over Klang Valley, Malaysia*; IOP Conference Series: Earth and Environmental Science; IOP Publishing: Bristol, UK, 2020; Volume 540, p. 012003.
44. Ustin, S.L.; Middleton, E.M. Current and near-term advances in Earth observation for ecological applications. *Ecol. Process.* **2021**, *10*, 1. [CrossRef]
45. Frazier, A.E.; Hemingway, B.L. A Technical Review of Planet Smallsat Data: Practical Considerations for Processing and Using PlanetScope Imagery. *Remote Sens.* **2021**, *13*, 3930. [CrossRef]
46. Gordon, A.D.; Breiman, L.; Friedman, J.H.; Olshen, R.A.; Stone, C.J. Classification and Regression Trees. *Biometrics* **1984**, *40*, 874. [CrossRef]
47. Shetty, S. Analysis of Machine Learning Classifiers for LULC Classification on Google Earth Engine. Master's Thesis, University of Twente, Enschede, The Netherlands, 2019.
48. Breiman, L. Random Forests. *Mach. Learn.* **2001**, *45*, 5–32. [CrossRef]
49. Xie, Z.; Chen, Y.; Lu, D.; Li, G.; Chen, E. Classification of Land Cover, Forest, and Tree Species Classes with ZiYuan-3 Multispectral and Stereo Data. *Remote Sens.* **2019**, *11*, 164. [CrossRef]
50. Pal, M. Random forest classifier for remote sensing classification. *Int. J. Remote Sens.* **2005**, *26*, 217–222. [CrossRef]
51. Talukdar, S.; Singha, P.; Mahato, S.; Shahfahad; Pal, S.; Liou, Y.-A.; Rahman, A. Land-Use Land-Cover Classification by Machine Learning Classifiers for Satellite Observations—A Review. *Remote Sens.* **2020**, *12*, 1135. [CrossRef]
52. Murtaza, K.O.; Romshoo, S.A. Determining the suitability and accuracy of various statistical algorithms for satellite data classification. *Int. J. Geomat. Geosci.* **2014**, *4*, 585–599.
53. Norovsuren, B.; Tseveen, B.; Batomunkuev, V.; Renchin, T.; Natsagdorj, E.; Yangiv, A.; Mart, Z. Land Cover Classification Using Maximum Likelihood Method (2000 and 2019) at Khandgait Valley in Mongolia. *IOP Conf. Ser. Earth Environ. Sci.* **2019**, *381*, 012054. [CrossRef]
54. Thomas, M.L.; Ralph, W.; Kiefer, J.C. Remote Sensing and Image Interpretation (Fifth Edition). *Geogr. J.* **2004**, *146*, 448–449. [CrossRef]
55. Kohavi, R. *A Study of Cross-Validation and Bootstrap for Accuracy Estimation and Model Selection*; Ijcai: Montreal, QC, Canada, 1995; Volume 14, pp. 1137–1145.
56. Belgiu, M.; Drăguţ, L. Random Forest in Remote Sensing: A Review of Applications and Future Directions. *ISPRS J. Photogramm. Remote Sens.* **2016**, *114*, 24–31. [CrossRef]
57. Hsu, C.W.; Chang, C.C.; Lin, C.J. *A Practical Guide to Support Vector Classification*; Technical Report; Department of Computer Science and Information Engineering, University of National Taiwan: Taipei, Taiwan, 2003; pp. 1396–1400.
58. Camargo, F.F.; Sano, E.E.; Almeida, C.M.; Mura, J.C.; Almeida, T. A Comparative Assessment of Machine-Learning Techniques for Land Use and Land Cover Classification of the Brazilian Tropical Savanna Using ALOS-2/PALSAR-2 Polarimetric Images. *Remote Sens.* **2019**, *11*, 1600. [CrossRef]
59. Rodriguez-Galiano, V.F.; Chica-Rivas, M. Evaluation of Different Machine Learning Methods for Land Cover Mapping of a Mediterranean Area Using Multi-Seasonal Landsat Images and Digital Terrain Models. *Int. J. Digit. Earth* **2012**, *7*, 492–509. [CrossRef]
60. Foody, G.M. Harshness in Image Classification Accuracy Assessment. *Int. J. Remote Sens.* **2008**, *29*, 3137–3158. [CrossRef]
61. Islam, K.; Jashimuddin, M.; Nath, B.; Nath, T.K. Land Use Classification and Change Detection by Using Multi-Temporal Remotely Sensed Imagery: The Case of Chunati Wildlife Sanctuary, Bangladesh. *Egypt. J. Remote Sens. Space Sci.* **2018**, *21*, 37–47. [CrossRef]
62. Manandhar, R.; Odeh, I.; Ancev, T. Improving the Accuracy of Land Use and Land Cover Classification of Landsat Data Using Post-Classification Enhancement. *Remote Sens.* **2009**, *1*, 330–344. [CrossRef]
63. Li, C.; Ma, Z.; Wang, L.; Yu, W.; Tan, D.; Gao, B.; Feng, Q.; Guo, H.; Zhao, Y. Improving the Accuracy of Land Cover Mapping by Distributing Training Samples. *Remote Sens.* **2021**, *13*, 4594. [CrossRef]
64. Zhao, Y.; Gong, P.; Yu, L.; Hu, L.; Li, X.; Li, C.; Zhang, H.; Zheng, Y.; Wang, J.; Zhao, Y.; et al. Towards a common validation sample set for global land-cover mapping. *Int. J. Remote Sens.* **2014**, *35*, 4795–4814. [CrossRef]
65. Loukika, K.N.; Keesara, V.R.; Sridhar, V. Analysis of Land Use and Land Cover Using Machine Learning Algorithms on Google Earth Engine for Munneru River Basin, India. *Sustainability* **2021**, *13*, 13758. [CrossRef]
66. McHugh, M.L. Interrater Reliability: The Kappa Statistic. *Biochem. Med.* **2012**, *22*, 276–282. [CrossRef]
67. Costa, H.; Almeida, D.; Vala, F.; Marcelino, F.; Caetano, M. Land Cover Mapping from Remotely Sensed and Auxiliary Data for Harmonized Official Statistics. *ISPRS Int. J. Geo-Inf.* **2018**, *7*, 157. [CrossRef]
68. Zhu, Z. Change Detection Using Landsat Time Series: A Review of Frequencies, Preprocessing, Algorithms, and Applications. *ISPRS J. Photogramm. Remote Sens.* **2017**, *130*, 370–384. [CrossRef]
69. Lu, D.; Mausel, P.; Brondízio, E.; Moran, E. Change Detection Techniques. *Int. J. Remote Sens.* **2004**, *25*, 2365–2401. [CrossRef]
70. Ramachandra, T.; Kumar, U. Geographic Resources Decision Support System for Land Use, Land Cover Dynamics Analysis. In Proceedings of the FOSS/GRASS Users Conference, Bangkok, Thailand, 12–14 September 2004; pp. 12–14.
71. Pu, R.; Gong, P.; Tian, Y.; Miao, X.; Carruthers, R.I.; Anderson, G.L. Using Classification and NDVI Differencing Methods for Monitoring Sparse Vegetation Coverage: A Case Study of Saltcedar in Nevada, USA. *Int. J. Remote Sens.* **2008**, *29*, 3987–4011. [CrossRef]

72. Shao, Y.; Lunetta, R.S. Comparison of Support Vector Machine, Neural Network, and CART Algorithms for the Land-Cover Classification Using Limited Training Data Points. *ISPRS J. Photogramm. Remote Sens.* **2012**, *70*, 78–87. [CrossRef]
73. Jin, S.; Yang, L.; Danielson, P.; Homer, C.; Fry, J.; Xian, G. A Comprehensive Change Detection Method for Updating the National Land Cover Database to circa 2011. *Remote Sens. Environ.* **2013**, *132*, 159–175. [CrossRef]
74. Shi, G.; Jiang, N.; Yao, L. Land Use and Cover Change during the Rapid Economic Growth Period from 1990 to 2010: A Case Study of Shanghai. *Sustainability* **2018**, *10*, 426. [CrossRef]
75. Hu, Y.; Batunacun; Zhen, L.; Zhuang, D. Assessment of Land-Use and Land-Cover Change in Guangxi, China. *Sci. Rep.* **2019**, *9*, 2189. [CrossRef]
76. Solomon, S.; On, P.; Al, E. *Climate Change 2007: The Physical Science Basis*; Cambridge University Press: Cambridge, UK, 2007.

remote sensing MDPI

Article

Climate Sensitivity of the Arid Scrublands on the Tibetan Plateau Mediated by Plant Nutrient Traits and Soil Nutrient Availability

Ben Chen [1,2,†], Hui Chen [1,†], Meng Li [3], Sebastian Fiedler [4], Mihai Ciprian Mărgărint [5], Arkadiusz Nowak [6,7], Karsten Wesche [8,9,10], Britta Tietjen [11,12] and Jianshuang Wu [2,5,*]

1 Hebei Key Laboratory of Environmental Change and Ecological Construction, Hebei Technology Innovation Center for Remote Sensing Identification of Environmental Change, School of Geographical Sciences, Hebei Normal University, Shijiazhuang 050024, China
2 Institute of Environment and Sustainable Development in Agriculture, Chinese Academy of Agricultural Sciences, Beijing 100081, China
3 School of Geographic Sciences, Nantong University, Nantong 226007, China
4 Department of Ecosystem Modelling, Büsgen-Institute, Universität Göttingen, 37077 Göttingen, Germany
5 Department of Geography, Geography and Geology Faculty, Alexandru Ioan Cuza University of Iaşi, 700505-RO Iaşi, Romania
6 Botanical Garden Center for Biological Diversity Conservation in Powsin, Polish Academy of Sciences, 02-973 Warsaw, Poland
7 Department of Botany and Nature Conservation, University of Warmia and Mazury in Olsztyn, Plac Łódzki 1, 10-727 Olsztyn, Poland
8 Department of Botany, Senckenberg Museum of Natural History Görlitz, 02826 Görlitz, Germany
9 International Institute (IHI) Zittau, Technische Universität Dresden, 02763 Zittau, Germany
10 German Centre for Integrative Biodiversity Research (iDiv) Halle-Jena-Leipzig, 04103 Leipzig, Germany
11 Institute of Biology, Theoretical Ecology, Freie Universität Berlin, 14195 Berlin, Germany
12 Berlin-Brandenburg Institute of Advanced Biodiversity Research, 14195 Berlin, Germany
* Correspondence: wujianshuang@caas.cn
† These authors contributed equally to this work.

Citation: Chen, B.; Chen, H.; Li, M.; Fiedler, S.; Mărgărint, M.C.; Nowak, A.; Wesche, K.; Tietjen, B.; Wu, J. Climate Sensitivity of the Arid Scrublands on the Tibetan Plateau Mediated by Plant Nutrient Traits and Soil Nutrient Availability. *Remote Sens.* **2022**, *14*, 4601. https://doi.org/10.3390/rs14184601

Academic Editor: Xander Wang

Received: 25 July 2022
Accepted: 6 September 2022
Published: 15 September 2022

Highlights:

What are the main findings?

- Principal component regressions revealed the climate sensitivity of Tibetan arid desert scrubs.
- Plant nutrient traits and soil nutrient availability regulate desert scrubs' climate sensitivity.

What is the implication of the main finding?

- Scrubs' sensitivity to temperature is mainly regulated by the nitrogen contents of soils and leaves.
- Scrubs' sensitivity to precipitation is affected by the leaf carbon content of dominant species.
- Neither soil nor plant nutritional properties alone can well explain scrubs' sensitivity to droughts.

Abstract: Climate models predict the further intensification of global warming in the future. Drylands, as one of the most fragile ecosystems, are vulnerable to changes in temperature, precipitation, and drought extremes. However, it is still unclear how plant traits interact with soil properties to regulate drylands' responses to seasonal and interannual climate change. The vegetation sensitivity index (VSI) of desert scrubs in the Qaidam Basin (NE Tibetan Plateau) was assessed by summarizing the relative contributions of temperature (S_{GST}), precipitation (S_{GSP}), and drought (temperature vegetation dryness index, S_{TVDI}) to the dynamics of the normalized difference vegetation index (NDVI) during plant growing months yearly from 2000 to 2015. Nutrient contents, including carbon, nitrogen, phosphorus, and potassium in topsoils and leaves of plants, were measured for seven types of desert scrub communities at 22 sites in the summer of 2016. Multiple linear and structural equation models were used to reveal how leaf and soil nutrient regimes affect desert scrubs' sensitivity to climate variability. The results showed that total soil nitrogen (STN) and leaf carbon content (LC), respectively, explained 25.9% and 17.0% of the VSI variance across different scrub communities. Structural equation modeling (SEM) revealed that STN and total soil potassium (STK) mediated desert scrub's

VSI indirectly via S_{GST} (with standardized path strength of −0.35 and +0.32, respectively) while LC indirectly via S_{GST} and S_{GSP} (with standardized path strength of −0.31 and −0.19, respectively). Neither soil nor leave nutrient contents alone could explain the VSI variance across different sites, except for the indirect influences of STN and STK via S_{TVDI} (−0.18 and 0.16, respectively). Overall, this study disentangled the relative importance of plant nutrient traits and soil nutrient availability in mediating the climatic sensitivity of desert scrubs in the Tibetan Plateau. Integrating soil nutrient availability with plant functional traits together is recommended to better understand the mechanisms behind dryland dynamics under global climate change.

Keywords: climate change; dryland ecosystem; leaf nutrient traits; Qaidam Basin; soil nutrient availability; vegetation sensitivity index

1. Introduction

Drylands cover approximately 41% of the Earth's land surface [1] and support more than 38% of the global population [2], although they are sparsely vegetated and with limited productivity. Recently, drylands have been degraded due to ongoing climate change and intensifying human disturbance [3,4]. For example, increased droughts mainly caused by warming are predicted to accelerate the degradation of the Mediterranean drylands under human overexploitation [5]. More than 75% of drylands in developing countries are also likely to expand due to further global warming [6]. Consequently, poverty alleviation will become more challenging in arid undeveloped countries [7]. Therefore, a better understanding of how drylands respond to changes in climate and other potential regulators is urgently needed.

Climate change drives dryland dynamics across different spatial scales, including warming, shifting precipitation regimes, and intensifying droughts. Warming can limit plant growth and survival via increasing evapotranspiration and intensifying water deficit in soils [8,9]. Experimental studies have reported that increased precipitation can improve nutrient availability and facilitate plant growth, development, and reproduction in drylands via enhancing moisture content in topsoils and vice versa [10,11]. Meanwhile, remote-sensing-related technologies help monitor large-scale vegetation dynamics [12]. However, current studies on ecosystem responses to climate change are mainly based on average climate states but usually ignore the effects of climate variability and extremes [13].

Recently, Seddon et al. [14] have proposed a framework to assess the vegetation sensitivity index (VSI) to short-term variability in temperature, precipitation, and radiation (cloud mask) from a global perspective. Such an algorithm has been increasingly applied to reveal the relative importance of different climatic variables in driving the dynamics in vegetation coverage, productivity, and phenology within and across various ecosystems [15–18]. For example, Li et al. [15] assessed the variability of VSI with altitude gradient in alpine grasslands on the Tibetan Plateau and found that vegetation in higher altitude regions is more sensitive to climate variability than that in lower altitudes. In another study, Yuan et al. [18] found that vegetation in Central Asia was more sensitive to climate variability in spring than other three seasons. These studies have mapped the differential spatiotemporal patterns of alpine and arid vegetation in response to climate variability and greatly enhanced our understanding of the vegetation–climate relationship. However, they failed to provide mechanistic explanations for these patterns from the perspective of plant and soil nutrient properties. Many in situ manipulative experiments have reported that the vegetation response to climate change can be largely influenced by plant functional traits and soil nutrient availability. For example, in a climate simulation study of alpine grasslands, Henn et al. [19] found that alpine plants can adapt to climate change by shaping leaf traits. In a recent meta-analysis, soil nitrogen availability was found to be a determinant of terrestrial ecosystem productivity under changing climates [20]. However, the findings from local in situ experiments are always hard to be directly applied to large-scale ecosystem management. Therefore, it is necessary to combine remote sensing technologies and field observations to gain insights into the mechanisms of how vegetation adapts to changing climate via plant and soil nutritional properties.

Plants can respond specifically and actively to environmental changes rather than only passively withstand external stresses [21]. According to the community functional ecology theories [22–24], plant seed germination, seedling growth, and individual development and survival, along with environmental gradients, are mainly regulated by functional traits at species and community levels [25–30]. Plant individuals can modify their water requirements by reducing leaf size, stomatal conductance, and photosynthetic rates in response to droughts [3,31,32]. At the community level, plants can alter the C:N:P stoichiometry among organs above- and belowground to offset physical inhibitions caused by warming or cooling [33,34], declined moisture [35–37], and changed nutrient availability [38]. For example, Jiang et al. [39] found that warming could substantially enhance N and P mineralization and consequently improve nutrient provision to plants in the Arctic tundra. Recently, Delgado-Baquerizo et al. [40] and Jiao et al. [41] also found that increased aridity, mainly due to warming, could reduce the concentrations of C and N but increase that of P in drylands. Fiedler et al. [42] also found that climate change could indirectly affect the trade-offs and synergies among different ecosystem functions by affecting soil nutrients and plant traits. These studies indicated that ecosystem responses to climate change are complicated and depend on abiotic and biotic factors [43]. Network analyses that include plant functional traits and physical environments together may improve our prediction of ecosystem functional changes across space and over time [44–46]. However, our understanding of the interactions of plant functional traits with soil nutrients of high-elevated drylands in response to climate variability is still limited.

In this study, we aimed to improve our understanding of how plant nutrient traits interact with low soil nutrient availability to regulate the sensitivity of desert scrubs in the Qaidam Basin, northeastern Qinghai-Tibetan Plateau, in response to climate variability. Specifically, we have (i) explored the patterns and trends of climate change and desert scrub coverage in the last decade, (ii) examined the differences in climatic sensitivity and its components for different desert scrub communities, and finally, (iii) investigated the networking paths of interactions between plant nutrient traits and soil nutrient availability to the components of desert scrubs' sensitivity to climate variability across the entire Qaidam Basin.

2. Materials and Methods

2.1. Study Area

The Qaidam Basin, with elevations ranging from 2640 m to 6000 m, is surrounded by the Kunlun, Qilian, and Altun Mountains and located in the northeastern Qinghai-Tibetan Plateau (90°16′E~99°16′E, 35°00′N~39°20′N, Figure 1). The area is characterized by an alpine arid continental climate, where the mean annual temperature (MAT) is generally below 5 °C and mean annual precipitation (MAP) is less than 200 mm [47]. The air relative humidity is between 30% and 40% throughout the year, while the average yearly sunshine duration can reach 3000 h in the Qaidam Basin. The annual average evaporation in the basin is between 1900 mm and 2600 mm [48]. Soil nutrients and moisture availability are low in desert scrublands in this area [49]. The vegetation has co-evolved with harsh physical conditions and is dominated by scrubs and semi-scrubs, such as *Kalidium foliatum*, *Salsola abrotanoides*, and *Ephedra sinica* [50]. Desert plants sprout in May and defoliate in early October; therefore, this period was usually defined as the plant growing season in recent studies [51].

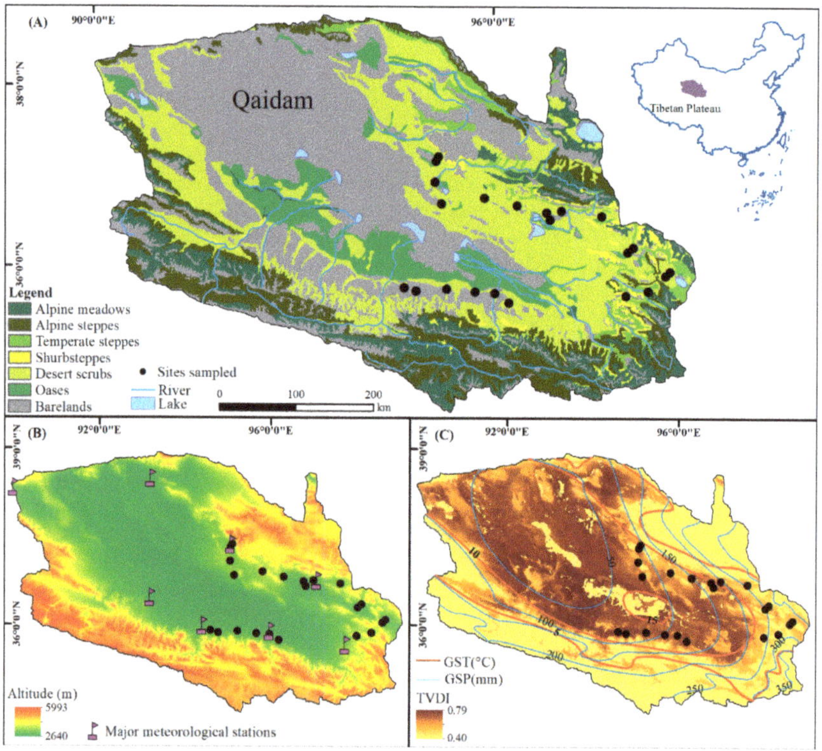

Figure 1. Vegetation (**A**), elevation and major meteorological stations (**B**), climate (**C**), and site locations (black dots) in the Qaidam Basin, Qinghai-Tibetan Plateau. In Panel C, GST and GSP refer to the average temperature and sum precipitation during the plant growing season, respectively, from May to October during 2000–2015. TVDI means the temperature vegetation dryness index.

2.2. Remote Sensing Data Collection and Processing

It is the key to choosing a suitable vegetation index for accurately examining and assessing vegetation dynamics under climate change. Recently, Zhang et al. [50] have evaluated the correlation between different vegetation indices and vegetation cover observed with UAV in the Qaidam Basin. They found that the Normalized Difference Vegetation Index (NDVI) performed much better than other indices such as ratio vegetation index (RVI), difference vegetation index (DVI), modified vegetation index (MVI), modified soil-adjusted vegetation index (MSAVI), and normalized difference greenness index (NDGI)) and normalized difference greenness index (NDGI) in assessing vegetation cover. Therefore, in this study, the NDVI of the Moderate Resolution Imaging Spectroradiometer (MODIS) MOD13Q1 Products was used as a proxy for vegetation productivity of desert scrubs. We downloaded the NDVI data with a temporal resolution of 16 days and spatial one of 250 m for the Qaidam Basin for the period from 2000 to 2015 (https://ladsweb.modaps.eosdis.nasa.gov/, accessed on 1 October 2021). As NDVI is sensitive to soil background information, therefore, we used the Savitzky–Golay filter [52] to smooth the time-series NVDI and eliminate the contaminations caused by clouds, snow, and ice cover, following Li et al. [15]. The pixels whose peak NDVI value was less than 0.05 were classified as extremely sparse vegetation or barren lands. Those pixels were manually excluded from further analyses following the recent study of Yuan et al. [18] in Central Asia. Finally, the peak monthly NDVI generated by maximum value composites [53] was used to describe the sensitivity and its components of desert scrub vegetation in the Qaidam Basin, Qinghai-Tibetan Plateau, to the variability of different climate variables.

2.3. Weather Data Collection and Processing

The monthly data of temperature and precipitation of the main meteorological stations within the Qaidam Basin (Figure 1B) yearly from May to September were downloaded and mapped for the period of 2000 to 2015. Seasonal and inter-annual fluctuations in temperature and precipitation, as well as their extreme values, were clearly shown in the supplementary Figure S1. It makes sense that the weather data from 2000–2015 are sufficient to reflect the climate fluctuations in the region. Therefore, we downloaded monthly temperature and precipitation of 200 meteorological stations located within and around the Qinghai-Tibetan Plateau from the China Meteorological Data Service Centre (http://data.cma.cn/en, accessed on 5 October 2021) for the plant growing seasons (yearly from May to October) between 2000 and 2015. First, the average temperature and sum precipitation during the plant growing seasons (termed as GST and GSP, respectively) were calculated for each station. In addition to monthly weather records, we interpolated GSP and GST into rasters with ANUSPLIN (The software is developed by Australian National University in Australia and the version is 4.3) [54] at a spatial resolution of 250 m to match the remote sensing data (see details below). A digital elevation model (DEM) with a 30 m resolution was used as the covariate, following Li et al. [55], who also confirmed that the interpolated climate rasters matched well the field observations across the northern Tibetan Plateau.

In this study, we used the temperature vegetation dryness index (TVDI) to explore the vegetation-drought relationship as recommended by Sandholt et al. [56] because it combines vegetation greenness (MODIS MOD13A2) and land surface temperature (MOD11A2). TVDI ranges from 0 to 1, with higher values indicating drier soils. We downloaded monthly TVDI data from the National Earth System Science Data Center (http://www.geodata.cn, accessed on 12 October 2021) and resampled it to match the spatial-temporal resolutions of climate and NDVI data. Finally, we used the Mann–Kendall tests [57,58] to examine the trends of GST, GSP, NDVI, and TVDI at the pixel scale between 2000 and 2015 (see more details in Figures S2–S5). The MK test outputs Z-values and slope, with positive and negative slope values representing increasing and decreasing trends. $|Z| > 1.96$ reaches a significant level ($\alpha = 0.05$).

2.4. Climatic Sensitivity Index of Desert-Scrubs

In this study, we used NDVI as an agent of desert-scrub coverage to explore its sensitivity to changes in GST, GSP, and TVDI. First, we transformed the time-series NDVI and climate variables into z-score anomalies using each variable's mean and standard deviation (Equation (1)).

$$Z = \frac{X - \overline{X}}{\sigma} \tag{1}$$

where Z is the standard score, X is the raw data, $-\overline{X}$ is the mean value of data, and σ is the standard deviation of data. Then, we used a principal component regression (PCR) to determine the relative importance of each climate variable in driving the monthly NDVI dynamic at the pixel scale. Principal components with significant relationships to climatic variables ($p < 0.1$) were retained for the subsequent VSI calculations.

The climate weights (W_i) were obtained by multiplying the loading scores of each variable with the corresponding PCR coefficients. Here, the weight of each climate variable was also rescaled to be between 0 and 1. Then we detrended each time-series variable and extracted its weight variance. The residuals of the mean-variance relationship between NDVI and climate variables were fitted using a linear-quadratic model, following Seddon et al. [14]. The residuals were normalized between 0 and 100. We take the $\log_{10}$-transformed ratios between NDVI variability and climate variables as the sensitivity metrics

(S_i). The NDVI logged for one month was used as the fourth variable to explore the time-lag effect of vegetation response to climate variability, following Seddon et al. [14].

$$VSI = \sum_{i=1}^{3} W_i \times S_i \quad (2)$$

where VSI is the sum of the product of the weights of each climate variable (W_i) and its sensitivity metrics (S_i). We rescaled the VSI between 0 and 100 (unitless), and the larger the value, the more sensitive to climate variability. We constructed a 500 m buffer centered on each sample point. Thus, it makes sense that the specific VSI values of the pixels in the buffer area can serve as repeats for multiple comparisons among different scrub communities. Moreover, we calculated the VSI average of the pixels in the buffer area for each site and used it in multiple linear regressions and structural equation modeling across sites within the study area.

2.5. Plant Nutrient Traits and Soil Nutrient Availability

According to the 1:1,000,000 China Vegetation Atlas, which is available at the Resource and Environmental Science Data Center, Chinese Academy of Sciences (http://www.resdc.cn/data.aspx?DATAID=122, accessed on 1 May 2016), our field observations in 2016 at 22 sites covered seven main types of desert scrubs within the Qaidam Basin (Table 1). Plant nutrient traits and soil nutrient properties were measured to examine their potential regulating effects on the desert scrubs' sensitivity to climate variability.

We laid five 5 m × 5 m sample squares evenly along the diagonal in a relatively homogeneous area of 250 m × 250 m at each site, where the plant community was intact and undisturbed by human activities while the proportion of desert scrub plants was higher than 90% in the total vegetation cover. Healthy leaves were collected from at least five mature individuals for each scrub/semi-scrub species in each quadrat. Plant leaves were stored in a separate bag for each species during the field campaign and then dried at 70 °C for 48 h in the lab to a constant weight. Before chemical analyses, all dried leaves were grounded into fine powder through a 0.15 mm sieve and kept in brown glass bottles. Leaf total P (LTP, g kg^{-1}) and K (LTK, g kg^{-1}), respectively, were determined by Mo-Sb colorimetric method and flame spectrophotometer [59].

We randomly sampled three profiles of the 0–50 cm depth at each site, about a half kilogram for each sample. Soil samples were air-dried and sieved through a 100-mesh sieve to remove root fractions and gravel. Then, soils were grounded into fine powders and stored in brown glasses before chemical analyses [60]. Total soil and leaf nitrogen (STN and LTN, g kg^{-1}) were determined by Kjeldahl method. Total soil and leaf phosphorus (STP and LTP, g kg^{-1}) were measured with the Mo-Sb anti-spectrophotometry method, while total soil and leaf Kalium (STK and LTK, g kg^{-1}) with the flame photometry method. Soil organic carbon (SOC, g kg^{-1}) and leaf carbon (LC, g kg^{-1}) were determined by the potassium dichromate oxidation–ferrous sulfate titrimetric method.

Table 1. Locations, the most dominant species and their relative coverage (DSC), climate conditions, soil properties, and plant nutrient traits at each site sampled in 2016 across the Qaidam Basin. GST and GSP refer to the average temperature and precipitation during the plant growing season. Soil properties included the contents of organic carbon (SOC), total nitrogen (STN), total phosphorus (STP), and total potassium (STK) of topsoils at the 0–50 cm depth. Plant nutrient traits were the community weighted means (CWMs) of leaf carbon (LC), total nitrogen (LTN), total phosphorus (LTP), and total potassium (LTK) in the leaves of all plants at each sampling site.

Site	Long (°E)	Lat. (°N)	Alt. (m)	Dominant Species	DSC (%)	GST (°C)	GSP (mm)	SOC (g kg^{-1})	STN (g kg^{-1})	STK (g kg^{-1})	STP (g kg^{-1})	LC (g kg^{-1})	LTN (g kg^{-1})	LTP (g kg^{-1})	LTK (g kg^{-1})
Site1	95.091	36.346	2890	*Sympegma regelii*	53.0	16.4	66	1.48	0.09	14.00	0.45	326.60	20.95	1.00	14.72
Site2	95.267	36.325	2970	*Ephedra sinica*	65.0	15.8	72	1.21	0.11	14.67	0.43	311.36	16.67	0.45	8.16
Site3	95.703	36.381	2850	*Ephedra sinica*	48.0	16.8	73	1.76	0.09	15.27	0.46	397.40	15.09	0.65	11.13
Site4	96.126	36.378	2765	*Tamarix chinensis*	100.0	17.3	79	3.91	0.14	20.46	0.35	348.00	19.25	0.92	7.88
Site5	96.413	36.382	2855	*Krascheninnikovia ceratoides*	50.0	16.7	92	2.20	0.12	17.29	0.49	307.20	21.28	2.10	18.18
Site6	96.630	36.293	2850	*Haloxylon ammodendron*	100.0	16.4	101	3.52	0.07	16.27	0.43	132.00	30.45	1.36	28.97
Site7	98.329	36.469	3320	*Kalidium foliatum*	80.0	12.2	224	3.23	0.44	19.09	0.49	576.00	23.10	1.28	15.68
Site8	98.648	36.536	3530	*Kalidium foliatum*	59.0	10.4	264	6.16	0.51	18.74	0.46	353.51	22.87	1.63	17.19
Site9	98.883	36.723	3175	*Kalidium foliatum*	100.0	12.3	261	7.26	1.13	19.84	0.54	375.00	22.20	1.72	16.88
Site10	98.947	36.773	3070	*Kalidium foliatum*	65.0	13.1	257	4.25	0.43	18.14	0.53	451.50	33.84	1.88	9.71
Site11	98.312	36.962	2945	*Sympegma regelii*	27.0	14.5	214	4.58	0.40	17.65	0.52	338.68	20.31	1.94	23.31
Site12	98.390	37.017	3130	*Kalidium foliatum*	61.0	13.0	236	4.58	0.40	17.92	0.49	492.66	21.99	0.80	14.55
Site13	97.886	37.344	3205	*Salsola abrotanoides*	56.0	12.8	222	6.38	0.58	15.77	0.48	233.33	30.24	1.37	16.91
Site14	97.284	37.366	2970	*Sympegma regelii*	70.0	14.6	172	5.35	0.38	17.12	0.37	307.80	28.65	2.33	13.93
Site15	97.127	37.258	2860	*Krascheninnikovia ceratoides*	56.0	16.3	146	2.20	0.29	18.29	0.42	440.00	23.90	1.86	16.50
Site16	97.068	37.333	2905	*Kalidium foliatum*	83.33	15.8	149	4.90	0.46	19.04	0.44	384.00	26.95	1.83	13.69
Site17	96.623	37.384	2980	*Sympegma regelii*	60.00	15.7	126	3.30	0.34	19.08	0.47	403.80	23.42	1.36	23.54
Site18	96.135	37.440	3620	*Salsola abrotanoides*	66.67	11.2	135	6.71	0.58	20.40	0.66	346.67	23.57	1.02	20.15
Site19	95.395	37.855	3310	*Sympegma regelii*	40.00	13.1	91	7.33	0.39	22.71	0.58	321.60	14.70	2.90	21.44
Site20	95.364	37.800	3190	*Krascheninnikovia ceratoides*	100.00	13.8	85	12.38	0.29	23.09	0.49	528.00	25.67	1.37	34.63
Site21	95.381	37.565	3185	*Krascheninnikovia ceratoides*	50.00	14.1	82	3.25	0.13	18.56	0.50	330.00	20.65	3.40	17.60
Site22	95.506	37.328	3035	*Ephedra sinica*	40.00	15.3	76	2.31	0.12	15.52	0.56	444.60	16.19	2.17	10.66

2.6. Plant Functional Trait Diversity

Our plant nutrient traits referred to the contents of N, P, K, and C in leaves (LTN, LTP, LTK, and LC) for all species sampled (Table 1). We calculated community-weighted means (CWMs) to describe the regulating effects of plant nutrient trait diversity on desert scrubs' sensitivity to climate variability (see Equations (2) and (3)). CWMs are essential for understanding community reorganization in response to environmental filtering [61] and are widely recommended and used in plant functional ecology research [62].

$$CWM_j = \sum_{i=1}^{n} P_{ij} T_{ij} \tag{3}$$

where P_{ij} is the relative cover of species i in sample site j; T_{ij} is the mean of trait values of species i in sample site j; CWM_j is the community-weighted mean of traits of each species in sample site j.

2.7. Statistical Analyses

First, with the Mann–Kendall test [57,58], we examined the trends of GST, GSP, TVDI, and NDVI between 2000 and 2015 at the pixel scale. Moreover, we extracted the yearly value and trends of GST, GSP, TVDI, and NDVI based on each site's geographical coordinate records at the pixels where we performed field surveys. Thus, we compared the medians of climate and vegetation variables and their trends among the seven types of desert scrubs (see more details in Figures S2–S5).

Second, we calculated the VSI of 2000–2015 based on principal component regressions and used a red–green–blue (RGB) composition map to show the relative contribution of GST, GSP, and TVDI to the climatic sensitivity of desert scrubs at the pixel scale. We also extracted the VSI and its components to each site and examined the difference in them at the community level with the Wilcoxon test [63].

Next, we divided the data into three groups: VSI as the response variable and leaf traits (LTN, LTP, LTK, LC) and soil attributes (STN, STP, STK, SOC) as potential explanatory variables. We examined the normality of the data containing response and predictor variables and normalized these data in R using a scale function (see Equation (1)). Moreover, we examined the correlation between the responders and predictors using the corrplot package (version 0.92) [64] in R. The multivariate linear model was applied to investigate the main effects of leaf and soil nutrient attributes on the variation of VSI. In this step, we also determined the multicollinearity of each factor based on the variance inflation factor (VIF) [65]. Thus, the relative importance of each explanatory can be disentangled as the fraction of the variance VSI explained in the best-fitted model.

Finally, structural equation modeling (SEM) [66] was used to explore how vegetation traits and soil nutrients drive changes in VSI using the lavaan package (version 0.6.1) [67]. Stepwise backward selections were used to remove the least significant term until the best-fitted model was picked out. All statistical analyses were performed in R (The software is developed by R Core Team in Austria and the version is 4.1.1) [68], and maps were generated with ArcGIS (The software is developed by the Environment System Research Institute in America and the version is 10.2) [69].

3. Results

3.1. Vegetation Sensitivity Index and Its Contributors

The overall VSI of desert scrubs is low (15.0) and distributed unevenly across the Qaidam Basin. The VSI was greater than 30 at only 1.18% pixels scattered near the foothills and riverbanks (Figures 1A and 2A, Table 2). The relative contribution of each climatic variable also varies heterogeneously across space. Specifically, in the western and northeastern areas, vegetation is more sensitive to changes in GSP. However, the desert scrubs in the southern and northern areas were mainly affected by temperature dynamics. In the eastern part, desert scrubs were jointly controlled by GSP and TVDI (Figure 2B). Moreover, the desert scrubs in the central region were driven by GST and TVDI together.

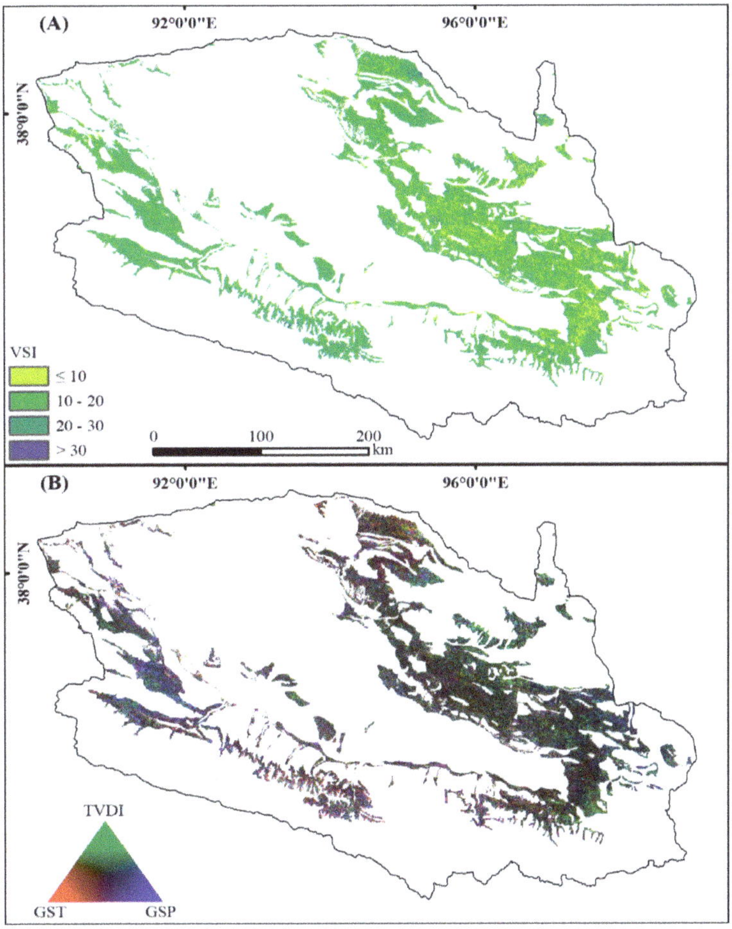

Figure 2. Vegetation Sensitivity Index (VSI, (**A**)) and its contributions (**B**) from the temperature and precipitation during the plant growing season (GST and GSP, respectively, in red and blue) and temperature vegetation dryness index (TVDI, in green).

Table 2. Vegetation Sensitivity Index (VSI) among different desert scrub communities across the Qaidam Basin.

	Haloxylon ammodendron (%)	*Tamarix chinensis* (%)	*Krascheninnikovia ceratoides* (%)	*Sympegma regelii* (%)	*Ephedra sinica* (%)	*Salsola abrotanoides* (%)	*Kalidium foliatum* (%)	Others (%)	Total (%)
VSI ≤ 10	3.24	1.10	4.57	0.32	0.69	4.07	0.83	2.45	17.32
10 < VSI ≤ 20	8.00	6.96	16.64	1.39	2.88	20.70	5.46	7.88	70.27
20 < VSI ≤ 30	0.96	1.28	2.78	0.16	0.41	3.64	0.80	1.09	11.23
VSI > 30	0.09	0.10	0.36	0.01	0.04	0.40	0.06	0.11	1.18

Communities dominated by *Haloxylon ammodendron* and *Tamarix chinensis* were more sensitive to climate variability than others (Figure 3A). The climatic weights vary among desert-scrub types. Communities dominated by *Tamarix chinensis* were more sensitive to GST than others (Figure 3B), while communities dominated by *Haloxylon ammodendron* were more vulnerable to GSP (Figure 3C). In response to changes in TVDI, communities dominated by *Tamarix chinensis* were slightly more vulnerable than those dominated by

Haloxylon ammodendron. However, both communities were more sensitive than other desert scrub communities across the Qaidam Basin (Figure 3D).

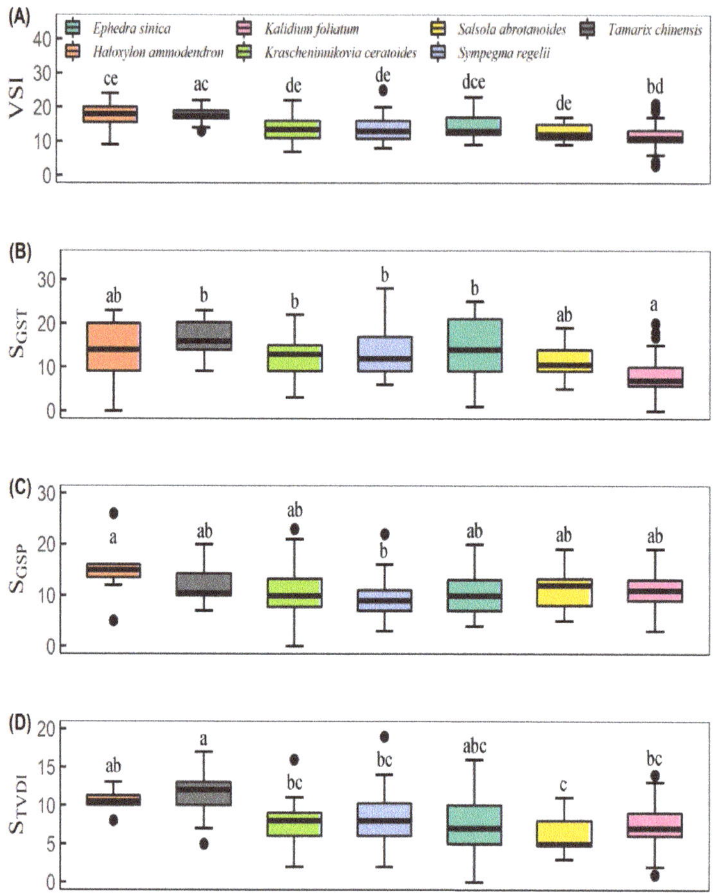

Figure 3. Comparisons of VSI (**A**) and its climatic weights among desert scrub communities. Panels (**B–D**) displayed the weights of mean temperature and sum precipitation during the plant growing season (noted as S_{GST} and S_{GSP}, respectively) as well as the temperature vegetation dryness index (S_{TVDI}). Different lowercase letters indicate significant differences among species.

3.2. Effects of Soil and Plant Properties on VSI

We found that VSI strongly correlated with the weights of GST (S_{GST}, r = 0.81) and droughts (S_{TVDI}, r = 0.65, Figure S6). S_{GST} was negatively correlated with STN (r = −0.53) and LTN (r = −0.51). The weight of GSP (S_{GSP}) was negatively correlated with LC (r = −0.49). Moreover, SOC had strong correlations with S_{GST} (r = −0.53), STK (r = 0.74), LTK (r = 0.59), and STN (r = −0.48), respectively. Across different dominant species of the 22 sites surveyed in this study, we only found that S_{GST} declined linearly with increasing STN and LTN (Figures 4 and 5) while S_{GSP} declined linearly with increasing STP and LC (Figures 4 and 5).

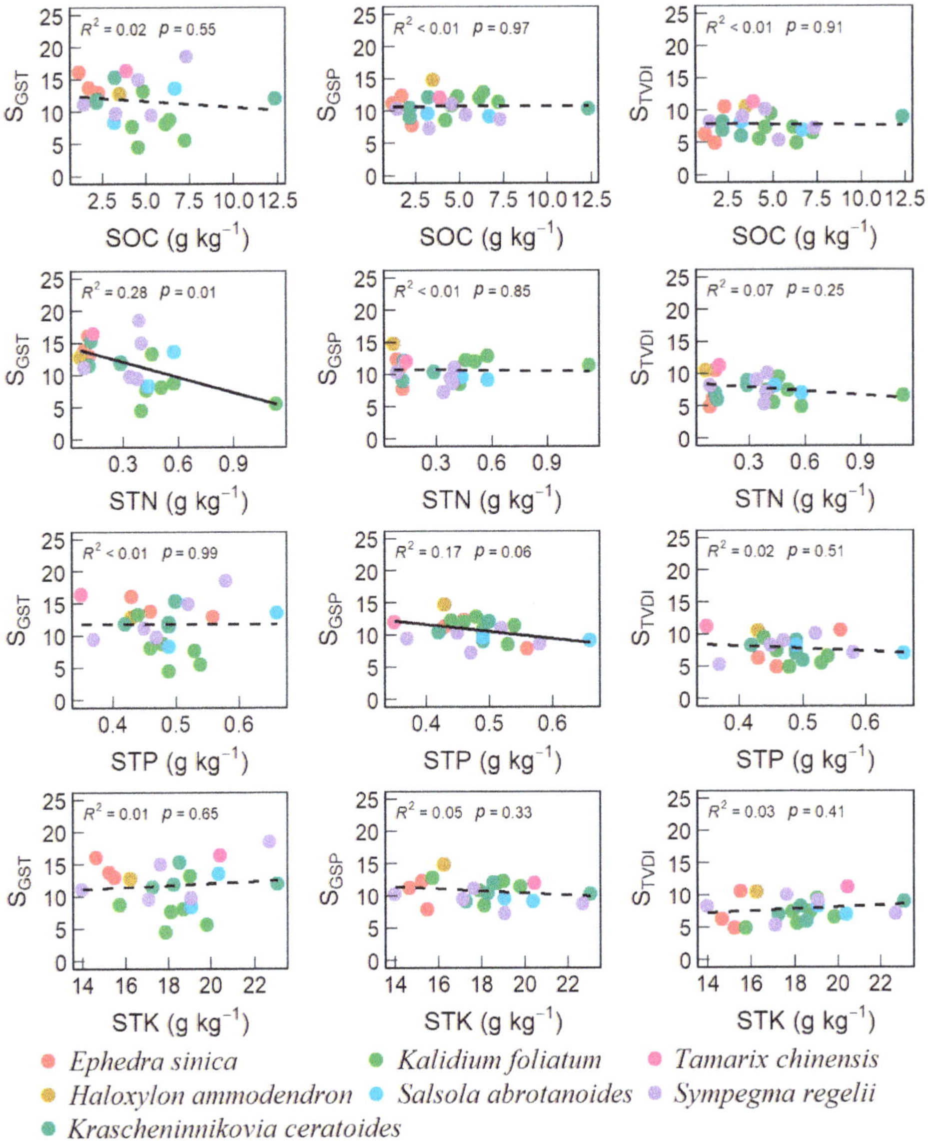

Figure 4. Trends of vegetation sensitivity index and its contributions with soil nutrients.

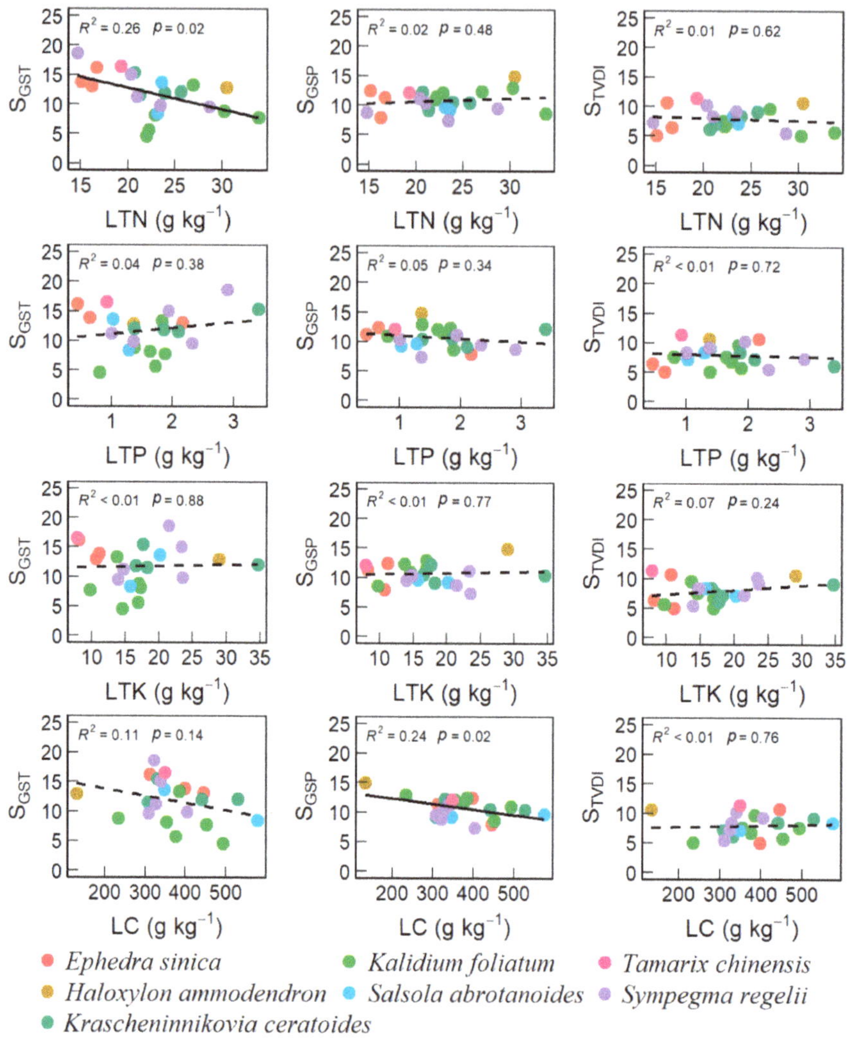

Figure 5. Trends of vegetation sensitivity index and its contributions with leaf nutrients.

However, we also found differential responses, linear, U-shaped, and unimodal, for different top-dominant species along the soil and foliar nutrient gradients (Figures S7 and S8). When examining the relationships of sensitivity components with soil nutrients, we only found that the S_{GST} of *Kalidium foliatum* and S_{TVDI} of *Sympegma regelii* had unimodal relationships with SOC and STP, respectively, at a marginal significance level ($p < 0.1$, Figure S7). When examining the relationships of sensitivity components with foliar nutrimental traits, we found a significant unimodal relationship between S_{TVDI} and LC for *Sympegma Regeli* ($p < 0.05$, Figure S8), a marginally positive linear relationship between S_{GSP} and LTK for *Kalidium foliatum* ($p < 0.1$, Figure S8), a significant unimodal relationship between S_{GSP} and LTN for *Kalidium foliatum*, and a marginally positive linear relationship between S_{GST} and LTP for *Krascheninnikovia ceratoides* ($p < 0.1$, Figure S8).

STN and LC explained 42.9% of the variance in VSI across different desert scrubs ($p < 0.05$, Table 3). S_{GST} was significantly affected by STN and LC, which accounted for 49.0% of the variance in S_{GST} ($p < 0.01$, Table 3). In addition to STN and LC, LTN and STK affected S_{GST} marginally, respectively, to explain 9.11% and 7.62% of the variance in S_{GST} (p = 0.053 and 0.073, respectively, Table 3). S_{GSP} was affected by LC and STP marginally, with 14.0% and 18.9% of its variance explained by LC and STP ($p < 0.1$, Table 3). Neither soil nutrients nor plant leaf traits influenced S_{TVDI} in this study ($p > 0.05$, Table 3).

Table 3. Main effects of soil nutrient availability and plant nutrient traits on vegetation sensitivity index (VSI) of desert scrubs across the Qaidam Basin. Abbreviations were the same as in Table 1. d.f., the degree of freedom; F, variance ratio; p, significance level; η^2, Eta squared, the percentage of sum squares explained. The difference significant at the 0.01 and 0.05 levels were labeled with ** and *, respectively.

Predictor	d.f.	VSI, $R^2 = 0.57$			S_{GST}, $R^2 = 0.74$			S_{GSP}, $R^2 = 0.45$			S_{TVDI}, $R^2 = 0.21$		
		F	p	η^2 (%)	F	p	η^2 (%)	F	p	η^2 (%)	F	p	η^2 (%)
SOC	1	0.29	0.597	0.98	0.89	0.362	1.80	0.00	0.970	0.01	0.01	0.910	0.07
STN	1	7.74	0.016 *	25.88	14.87	0.002 **	29.90	0.07	0.795	0.30	1.54	0.236	8.34
STP	1	0.14	0.710	0.48	2.01	0.180	4.04	4.43	0.055	18.88	0.09	0.764	0.51
STK	1	1.31	0.273	4.39	3.79	0.073	7.62	1.94	0.188	8.25	1.82	0.200	9.83
LTN	1	1.35	0.266	4.53	4.53	0.053	9.11	0.04	0.850	0.16	0.00	0.963	0.01
LTP	1	0.00	0.946	0.02	1.11	0.311	2.23	0.08	0.788	0.32	0.45	0.516	2.41
LTK	1	0.96	0.345	3.22	0.05	0.831	0.10	0.63	0.443	2.67	1.57	0.232	8.50
LC	1	5.09	0.042 *	17.02	9.48	0.009 **	19.06	3.27	0.094	13.96	0.02	0.903	0.08

3.3. Causal Network of Plant Traits and Soil Nutrients to VSI under Climate Variability

Structural equation modeling revealed the causal networking paths among plant and soil nutrient traits and climate weights to the sensitivity of desert scrubs across the Qaidam Basin (Figure 6). STN, LC, and LTN had direct adverse effects on S_{GST}. In contrast, STK had a significant positive effect on S_{GST}. LC had direct adverse effects on S_{GSP}, accounting for about 24% of its total variance. STN had direct adverse effects on S_{TVDI}, while STK had a significant positive effect on S_{TVDI}. However, STN and STK accounted for only 16% of the variance in S_{TVDI}.

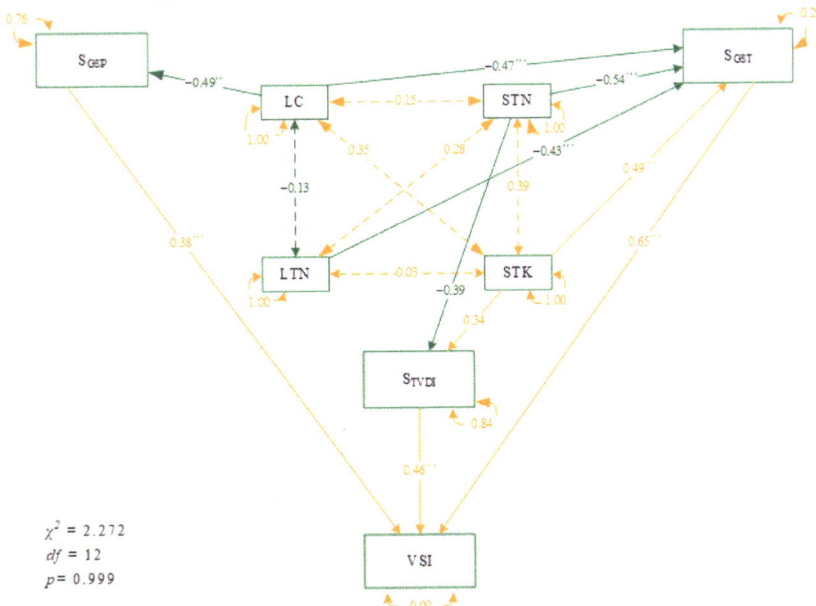

Figure 6. Structural equation modeling reveals the direct and indirect effects of leaf traits and soil properties on VSI and its components of desert scrubs across the Qaidam Basin. Notes are the same as

in Table 1. The orange and dark-green lines represent the positive and negative influences, respectively. The difference significant at the 0.001 and 0.01 levels were indicated with *** and **, respectively.

4. Discussion

4.1. Climate Sensitivity among Different Desert Scrubs

In this study, we mapped the climate sensitivity of desert scrubs in the Qaidam Basin using principal component regressions (Figure 2). We found that the climatic sensitivity of desert scrubs in this study was not high as expected. However, this finding is consistent with Seddon et al. [14] and Yuan et al. [18], and they found that arid deserts across the globe and Central Asia, respectively, were not as sensitive to climate dynamics. This phenomenon may have resulted from the strong memory effect of desert plants (Figure S9). The vegetation memory effect generally refers to the fact that previous environmental conditions can strongly influence the current ecosystem [70]. Ogle et al. [71] quantified ecological memory in vegetation and ecosystem processes in arid and semi-arid regions and found when the vegetation's memory effects were considered, 18–28% more of the variance in a given responsible variable could be explained. Seddon et al. [14] also found such a strong memory effect in global drylands' sensitivity to climate change. Moreover, the low climatic sensitivity of desert scrubs in the Qaidam Basin was also likely due to the relatively slow warming during the last decade (Figure S2), a rate of only 0.06 °C per decade. This finding was also confirmed by Easterling and Wehner [72] that at the beginning of the 21st century, there would be a 10- to 20-year warming hiatus. Overall, the stable climatic conditions may be another explanation for the low climate sensitivity of desert scrubs in the Qaidam Basin.

Although the desert scrubs were not sensitive as expected to climate change, we also found high species-dependency for the seven community types involved in this study. Specifically, we found that scrub communities with tall plant individuals, such as *Tamarix chinensis* communities and *Haloxylon ammodendron* communities, were more sensitive to climate change (Figure 3). Similarly, Zhu et al. [73] have found that northern China's trees and tall shrubs are more vulnerable to climate change. This is because taller plants have predictably wider water-conducting conduits, so they are more vulnerable to conduction-blocking embolisms [74]. Therefore, the response of different plant communities to climate change is influenced by the functional traits evolved to adapt to long-term environmental changes. However, vegetation also reacts and adjusts positively to environmental stresses [21]. Environmental-vegetation interactions have shaped scrub communities with different tolerance, and the response of these scrub communities to climate variability varies widely in the Qaidam Basin. The environmental filtering hypothesis predicts that, excluding biotic interactions, only the species that have already evolved with specific functional or phenotypic traits for survival, such as tolerance or avoidance of dryland plants to extreme water deficits [21,75]. This might be the reason for the differences in the relative weight of temperature (S_{GST}), precipitation (S_{GSP}), and drought (S_{TVDI}) found among the top-dominant species (Figure 3).

4.2. Desert-Scrubs' VSI Correlated with Soil and Leaf Nutrients

Warming is one of the important drivers for community structural and functional changes in drylands. Here, we found that STN, STK, LTN, and LC were closely related to the S_{GST} of vegetation (Table 3) and together explained more than 65% of the variance in S_{GST} (Table 3). This might be because warming can enhance soil microorganism activities, promote SOM decomposition and N mineralization, and thereby increase the supply and availability of soil nutrients [76–78]. In addition to STN and STK, warming can also enhance the activity and supply of K ions in soil solution [79].

In addition to soil nutrients, we also found that S_{GST} generally declined with increasing LTN and LC (Figure 5), indicating a high photosynthetic rate with high LTN and LC can mitigate the negative effects of warming. LTN and LC are usually used to descript the photosynthetic activity under manipulated stresses in experimental studies [80,81]. Our findings were partly consistent with Li et al. [15] that high-elevated grasslands have greater S_{GST} than those in low-elevations on the Tibetan Plateau. Dryland plants can increase leaf photosynthetic enzymes by enhancing LTN to promote photosynthetic rate and water use efficiency [82–84]. Moreover, our findings confirmed that the relationships of sensitivity climatic components with soil and foliar nutrients varied among different top-dominant species (Figures S7 and S8), likely due to species niche overlapping and differentiation.

Water availability is the main limiting factor for desert vegetation, and we found that STP and LC were the main factors affecting vegetation S_{GSP}, explaining 18.8% and 13.96% of the variation in S_{GSP}, respectively. P is an important element that affects vegetation growth, development, and metabolism; however, it can be readily combined with Ca^{2+} in arid soils and then is hard to be uptaken by plants. Pulsed precipitation in drylands can influence P transport, transformation, and availability through biochemical processes that control organophosphorus mineralization [85,86]. This might be the reason for the declining S_{GSP} with increasing STP (Figure 4) in our study.

LC is closely related to the photosynthesis of vegetation. Under drought conditions, vegetation reduces its water loss by decreasing leaf area and lowering light saturation point, which results in impaired ATP synthesis and reduced carboxylase activity, resulting in impaired leaf photosynthesis and thus reduced LC fixation [87]. Pulsed precipitation can briefly mitigate vegetation drought conditions and provide good moisture conditions for vegetation photosynthesis, thus increasing LC accumulation. This is why we also found that S_{GSP} declined with increasing LC content across different species (Figure S8). In addition, we also found that the LC of the seven arid scrub species was concentrated in the 300–400 g kg^{-1} interval, which may imply the coexistence of species has evolved with overlapped niches to successfully survive in the Qaidam Basin, Qinghai-Tibetan Plateau.

It is worthy to note that neither soil nor plant nutritional properties alone can well explain the variance in S_{TVDI}, which describes the availability of soil moisture, in other words, the combining effects of temperature and precipitation. In this study, we found that S_{TVDI} could be associated with STN and STK at the same time (Figure 6). Both STN and STK could explain 16% of the variance in S_{TVDI} across different species and sites. Soil water deficit caused by decoupling temperature and precipitation can limit soil microbial activity and slow down soil N mineralization [88]. Moreover, a drying environment with high temperatures and limited precipitation can also exacerbate the volatilization of gaseous N and lead to high evaporative demand [89], which in turn reduces the availability of STN [90]. Dry conditions can enhance soil inorganic N content [91]; however, they are hardly taken up by plants due to diffusion limitations [92]. The uptake of potassium by vegetation is dependent on soil moisture [93]; soil water deficit reduces soil potassium release capacity and availability [94] and inhibits soil potassium mobility, limiting potassium uptake by vegetation.

4.3. The Networks of Direct and Indirect Environmental Influences on VSI

The response of vegetation to climate change is characterized by the integrated effect of multiple factors and processes, forming complex and specific climate-species relationships [95,96]. In this study, we found that soil and foliar nutrient properties affect the scrub communities' climate sensitivity by regulating different climate sensitivity components. Even more, such regulating effects are dependent on the spatial scale, across sites, or for a given species. Finally, we found that the effect of soil nutrient effectiveness on VSI was greater compared to leaf nutrient effectiveness. The path strength of STK and STN on S_{GST} were greater than that of LC and LTN (Figure 6), and the explanatory power of STK and STN on the variance in S_{GST} (37.52%) was greater than that of LC and LTN (28.17%, Table 3). Soil nutrients provide indispensable nutrients for the growth and development of vegetation, and the decoupling of soil nutrient cycles due to climate warming may negatively affect dryland ecosystem services [40]. We should pay attention to the role of soil nutrients in drylands to maintain ecosystem stability. Therefore, in the future, the dependent and combining effects of biotic and abiotic factors on vegetation's sensitivity and vulnerability should be well examined and disentangled.

4.4. Limitations and Future Work

In this study, we integrated remote sensing data and field measurements together to investigate the mechanisms behind desert scrub's sensitivity to climate change in the Qaidam Basin, Qinghai-Plateau. However, there are still several uncertainties in this study. First, there might be a scaling-mismatch problem between remote sensing data and field measurements. Indeed, NDVI can be used to reflect the information on all vegetation in the whole region. Limited by budgets, only dominant desert scrub plants were sampled and measured for functional traits. Herbaceous plants were not fully considered in field observations. This may affect the accuracy of CWMs of plant functional traits and lead to some bias in the interpretation of community climate sensitivity. In further studies, both scrubs and herbaceous plants should be well considered because they have differential strategies to adapt to changes in physical circumstances.

In addition, this study only considered several soil nutrients and failed to take into account the effects of microorganisms and trace elements in soils. A recent review suggests that soil microbes dominate soil life activities by mediating nutrient cycling, decomposing organic matter, inhibiting plant diseases, influencing soil structure, and maintaining vegetation productivity, are major drivers of soil ecosystem change, and have game-changing potential in restoring soil function [97]. Moreover, soil trace elements are directly involved in the metabolic activities of organisms and influence the growth and development of vegetation, and have received increasing attention globally [98]. Therefore, future studies should also integrate the effects of multiple factors to obtain a more comprehensive understanding of the vegetation–climate relationship.

5. Conclusions

This study assessed and disentangled the sensitivity and its components of desert scrubs to climate variability with the time series of NDVI and weather data at the large scale crossing the Qaidam Basin. The regulatory mechanisms of plant and soil functional/nutrimental traits behind the desert scrub's climate sensitivity were investigated and compared among multiple communities dominated by different plant species. The results confirmed that the sensitivity of vegetation to temperature change was mainly regulated by the contents of leaf carbon and nitrogen as well as soil nitrogen and potassium. In contrast, the sensitivity of vegetation to precipitation change was regulated by the contents of leaf carbon and soil total nitrogen. Neither soil nutrients nor plant foliar traits alone can well explain scrubs' sensitivity to droughts. Plant foliar traits and soil nutrient properties indirectly regulate the different components of vegetation's climate sensitivity. Due to harsh physical circumstances in the Qaidam Basin and limited available funds, a scaling-mismatch problem might still exist between remote-sensing data and field measurements in this study. Anyway, this study combined soil and plant functional traits together to provide a new perspective to investigate the mechanisms behind desert vegetation dynamics under climate change when assessed with large-scale remote-sensing data. These findings also highlighted the necessity to make management and conservation strategies specifically according to the different factors that regulate the vegetation–climate relationship among different plant communities.

Supplementary Materials: The following supporting information can be downloaded at: https://www.mdpi.com/article/10.3390/rs14184601/s1.

Author Contributions: J.W. and H.C. conceived ideas and designed field protocols. B.C., J.W. and M.L. collected data. B.C. and H.C. performed chemical analyses. J.W. and B.C. analyzed data, prepared figures, and led the writing. M.L., S.F., M.C.M., A.N., K.W., B.T. and H.C. interpreted the results. All authors have read and agreed to the published version of the manuscript.

Funding: This research was jointly supported by the Second Tibetan Plateau Scientific Expedition and Research (STEP, 2019QZKK1002), the Innovation Talent Exchange of Foreign Expert Program under the Belt and Road Initiative (DL2021056001L), the Key Project of the Hebei Normal University (L2021Z05) and the National Natural Sciences Foundation of China (41877448 and 40971118).

Acknowledgments: J.W. had been funded with a two-year scholarship from January 2017 to April 2019 by the Alexander von Humboldt Foundation in Germany. J.W. has been supported by the Young Talent Scientist Program of the Chinese Academy of Agricultural Sciences since December 2019. S.F. was funded by the Deutsche Forschungsgemeinschaft (DFG, project number 192626868–SFB 990/2-3). We appreciated the valuable comments and suggestions given by Ben Niu and Zhipeng Wang, and Yun Jäschke on the first draft of this manuscript.

Conflicts of Interest: The authors declare that they have no known competing financial interests or personal relationships that could have influenced the work reported in this paper.

References

1. Reynolds, J.F.; Smith, D.M.; Lambin, E.F.; Turner, B.L., 2nd; Mortimore, M.; Batterbury, S.P.; Downing, T.E.; Dowlatabadi, H.; Fernandez, R.J.; Herrick, J.E.; et al. Global desertification: Building a science for dryland development. *Science* **2007**, *316*, 847–851. [CrossRef] [PubMed]
2. Millennium Ecosystem Assessment. *Ecosystems and Human Well-Being: Wetlands and Water*; Island Press: Washington, DC, USA, 2005; Volume 5.
3. Berdugo, M.; Delgado-Baquerizo, M.; Soliveres, S.; Hernandez-Clemente, R.; Zhao, Y.; Gaitan, J.J.; Gross, N.; Saiz, H.; Maire, V.; Lehmann, A.; et al. Global ecosystem thresholds driven by aridity. *Science* **2020**, *367*, 787–790. [CrossRef] [PubMed]
4. Huang, J.; Li, Y.; Fu, C.; Chen, F.; Fu, Q.; Dai, A.; Shinoda, M.; Ma, Z.; Guo, W.; Li, Z.; et al. Dryland climate change: Recent progress and challenges. *Rev. Geophys.* **2017**, *55*, 719–778. [CrossRef]
5. Vicente-Serrano, S.M.; Zouber, A.; Lasanta, T.; Pueyo, Y. Dryness is accelerating degradation of vulnerable shrublands in semiarid Mediterranean environments. *Ecol. Monogr.* **2012**, *82*, 407–428. [CrossRef]
6. Huang, J.P.; Yu, H.P.; Guan, X.D.; Wang, G.Y.; Guo, R.X. Accelerated dryland expansion under climate change. *Nat. Clim. Chang.* **2016**, *6*, 166–171. [CrossRef]
7. Barbier, E.B.; Hochard, J.P. Land degradation and poverty. *Nat. Sustain.* **2018**, *1*, 623–631. [CrossRef]
8. Tietjen, B.; Schlaepfer, D.R.; Bradford, J.B.; Lauenroth, W.K.; Hall, S.A.; Duniway, M.C.; Hochstrasser, T.; Jia, G.; Munson, S.M.; Pyke, D.A.; et al. Climate change-induced vegetation shifts lead to more ecological droughts despite projected rainfall increases in many global temperate drylands. *Glob. Chang. Biol.* **2017**, *23*, 2743–2754. [CrossRef]
9. Hantson, S.; Huxman, T.E.; Kimball, S.; Randerson, J.T.; Goulden, M.L. Warming as a driver of vegetation loss in the Sonoran desert of California. *J. Geophys. Res.-Biogeo.* **2021**, *126*, e2020JG005942. [CrossRef]
10. Gherardi, L.A.; Sala, O.E. Effect of interannual precipitation variability on dryland productivity: A global synthesis. *Glob. Chang. Biol.* **2019**, *25*, 269–276. [CrossRef]
11. Hoover, D.L.; Pfennigwerth, A.A.; Duniway, M.C. Drought resistance and resilience: The role of soil moisture-plant interactions and legacies in a dryland ecosystem. *J. Ecol.* **2021**, *109*, 3280–3294. [CrossRef]
12. Pettorelli, N.; Laurance, W.F.; O'Brien, T.G.; Wegmann, M.; Nagendra, H.; Turner, W. Satellite remote sensing for applied ecologists: Opportunities and challenges. *J. Appl. Ecol.* **2014**, *51*, 839–848. [CrossRef]
13. Thomas, C.D.; Cameron, A.; Green, R.E.; Bakkenes, M.; Beaumont, L.J.; Collingham, Y.C.; Erasmus, B.F.; De Siqueira, M.F.; Grainger, A.; Hannah, L.; et al. Extinction risk from climate change. *Nature* **2004**, *427*, 145–148. [CrossRef] [PubMed]
14. Seddon, A.W.; Macias-Fauria, M.; Long, P.R.; Benz, D.; Willis, K.J. Sensitivity of global terrestrial ecosystems to climate variability. *Nature* **2016**, *531*, 229–232. [CrossRef]
15. Li, L.; Zhang, Y.; Wu, J.; Li, S.; Zhang, B.; Zu, J.; Zhang, H.; Ding, M.; Paudel, B. Increasing sensitivity of alpine grasslands to climate variability along an elevational gradient on the Qinghai-Tibet plateau. *Sci. Total Environ.* **2019**, *678*, 21–29. [CrossRef] [PubMed]
16. Li, M.; Wu, J.S.; Song, C.Q.; He, Y.T.; Niu, B.; Fu, G.; Tarolli, P.; Tietjen, B.; Zhang, X.Z. Temporal variability of precipitation and biomass of alpine grasslands on the northern tibetan plateau. *Remote Sens.* **2019**, *11*, 360. [CrossRef]
17. Yao, Y.; Wang, X.; Li, Y.; Wang, T.; Shen, M.; Du, M.; He, H.; Li, Y.; Luo, W.; Ma, M.; et al. Spatiotemporal pattern of gross primary productivity and its covariation with climate in China over the last thirty years. *Glob. Chang. Biol.* **2018**, *24*, 184–196. [CrossRef]
18. Yuan, Y.; Bao, A.; Liu, T.; Zheng, G.; Jiang, L.; Guo, H.; Jiang, P.; Yu, T.; de Maeyer, P. Assessing vegetation stability to climate variability in central Asia. *J. Environ. Manag.* **2021**, *298*, 113330. [CrossRef]
19. Henn, J.J.; Buzzard, V.; Enquist, B.J.; Halbritter, A.H.; Klanderud, K.; Maitner, B.S.; Michaletz, S.T.; Potsch, C.; Seltzer, L.; Telford, R.J.; et al. Intraspecific trait variation and phenotypic plasticity mediate alpine plant species response to climate change. *Front. Plant Sci.* **2018**, *9*, 1548. [CrossRef]
20. Liu, Y.; Men, M.; Peng, Z.; Houx, J.H., 3rd; Peng, Y. Nitrogen availability determines ecosystem productivity in response to climate warming. *Ecology* **2022**, e3823. [CrossRef]

21. Palmquist, K.A.; Schlaepfer, D.R.; Renne, R.R.; Torbit, S.C.; Doherty, K.E.; Remington, T.E.; Watson, G.; Bradford, J.B.; Lauenroth, W.K. Divergent climate change effects on widespread dryland plant communities driven by climatic and ecohydrological gradients. *Glob. Chang. Biol.* **2021**, *27*, 5169–5185. [CrossRef]
22. Violle, C.; Navas, M.L.; Vile, D.; Kazakou, E.; Fortunel, C.; Hummel, I.; Garnier, E. Let the concept of trait be functional! *Oikos* **2007**, *116*, 882–892. [CrossRef]
23. McGill, B.J.; Enquist, B.J.; Weiher, E.; Westoby, M. Rebuilding community ecology from functional traits. *Trends Ecol. Evol.* **2006**, *21*, 178–185. [CrossRef] [PubMed]
24. Kearney, M.; Porter, W.P. Ecologists have already started rebuilding community ecology from functional traits. *Trends Ecol. Evol.* **2006**, *21*, 481–482; author reply 482–483. [CrossRef]
25. Bruelheide, H.; Dengler, J.; Purschke, O.; Lenoir, J.; Jimenez-Alfaro, B.; Hennekens, S.M.; Botta-Dukat, Z.; Chytry, M.; Field, R.; Jansen, F.; et al. Global trait-environment relationships of plant communities. *Nat. Ecol. Evol.* **2018**, *2*, 1906–1917. [CrossRef]
26. De Bello, F.; Lepš, J.A.N.; Sebastià, M.-T. Predictive value of plant traits to grazing along a climatic gradient in the Mediterranean. *J. Appl. Ecol.* **2005**, *42*, 824–833. [CrossRef]
27. Guo, X.; Liu, H.; Ngosong, C.; Li, B.; Wang, Q.; Zhou, W.; Nie, M. Response of plant functional traits to nitrogen enrichment under climate change: A meta-analysis. *Sci. Total Environ.* **2022**, *834*, 155379. [CrossRef]
28. Midolo, G.; Kuss, P.; Wellstein, C. Land use and water availability drive community-level plant functional diversity of grasslands along a temperature gradient in the Swiss Alps. *Sci. Total Environ.* **2021**, *764*, 142888. [CrossRef]
29. Prieto, I.; Roumet, C.; Cardinael, R.; Dupraz, C.; Jourdan, C.; Kim, J.H.; Maeght, J.L.; Mao, Z.; Pierret, A.; Portillo, N.; et al. Root functional parameters along a land-use gradient: Evidence of a community-level economics spectrum. *J. Ecol.* **2015**, *103*, 361–373. [CrossRef]
30. Wang, H.; Wang, R.; Harrison, S.P.; Prentice, I.C. Leaf morphological traits as adaptations to multiple climate gradients. *J. Ecol.* **2022**, *110*, 1344–1355. [CrossRef]
31. Balachowski, J.A.; Volaire, F.A. Implications of plant functional traits and drought survival strategies for ecological restoration. *J. Appl. Ecol.* **2018**, *55*, 631–640. [CrossRef]
32. Lian, X.; Piao, S.L.; Chen, A.P.; Huntingford, C.; Fu, B.J.; Li, L.Z.X.; Huang, J.P.; Sheffield, J.; Berg, A.M.; Keenan, T.F.; et al. Multifaceted characteristics of dryland aridity changes in a warming world. *Nat. Rev. Earth Environ.* **2021**, *2*, 232–250. [CrossRef]
33. Woods, H.A.; Makino, W.; Cotner, J.B.; Hobbie, S.E.; Harrison, J.F.; Acharya, K.; Elser, J.J. Temperature and the chemical composition of poikilothermic organisms. *Funct. Ecol.* **2003**, *17*, 237–245. [CrossRef]
34. Zhang, Q.; Zhou, J.; Li, X.; Yang, Z.; Zheng, Y.; Wang, J.; Lin, W.; Xie, J.; Chen, Y.; Yang, Y. Are the combined effects of warming and drought on foliar c:N:P:K stoichiometry in a subtropical forest greater than their individual effects? *For. Ecol. Manag.* **2019**, *448*, 256–266. [CrossRef]
35. Sun, Y.; Liao, J.H.; Zou, X.M.; Xu, X.; Yang, J.Y.; Chen, H.Y.H.; Ruan, H.H. Coherent responses of terrestrial c:N stoichiometry to drought across plants, soil, and microorganisms in forests and grasslands. *Agric. For. Meteorol.* **2020**, *292–293*, 108104. [CrossRef]
36. Niu, D.; Zhang, C.; Ma, P.; Fu, H.; Elser, J.J. Responses of leaf c:N:P stoichiometry to water supply in the desert shrub *Zygophyllum xanthoxylum*. *Plant Biol.* **2019**, *21*, 82–88. [CrossRef]
37. Castellanos, A.E.; Llano-Sotelo, J.M.; Machado-Encinas, L.I.; Lopez-Pina, J.E.; Romo-Leon, J.R.; Sardans, J.; Penuelas, J. Foliar c, n, and p stoichiometry characterize successful plant ecological strategies in the Sonoran desert. *Plant Ecol.* **2018**, *219*, 775–788. [CrossRef]
38. Huang, J.; Wang, P.; Niu, Y.; Yu, H.; Ma, F.; Xiao, G.; Xu, X. Changes in c:N:P stoichiometry modify n and p conservation strategies of a desert steppe species *Glycyrrhiza uralensis*. *Sci. Rep.* **2018**, *8*, 12668. [CrossRef]
39. Jiang, Y.Y.; Rocha, A.V.; Rastetter, E.B.; Shaver, G.R.; Mishra, U.; Zhuang, Q.L.; Kwiatkowski, B.L. C-n-p interactions control climate driven changes in regional patterns of c storage on the north slope of Alaska. *Landsc. Ecol.* **2016**, *31*, 195–213. [CrossRef]
40. Delgado-Baquerizo, M.; Maestre, F.T.; Gallardo, A.; Bowker, M.A.; Wallenstein, M.D.; Quero, J.L.; Ochoa, V.; Gozalo, B.; Garcia-Gomez, M.; Soliveres, S.; et al. Decoupling of soil nutrient cycles as a function of aridity in global drylands. *Nature* **2013**, *502*, 672–676. [CrossRef]
41. Jiao, F.; Shi, X.R.; Han, F.P.; Yuan, Z.Y. Increasing aridity, temperature and soil pH induce soil c-n-p imbalance in grasslands. *Sci. Rep.* **2016**, *6*, 19601. [CrossRef] [PubMed]
42. Fiedler, S.; Monteiro, J.A.F.; Hulvey, K.B.; Standish, R.J.; Perring, M.P.; Tietjen, B. Global change shifts trade-offs among ecosystem functions in woodlands restored for multifunctionality. *J. Appl. Ecol.* **2021**, *58*, 1705–1717. [CrossRef]
43. Jäschke, Y.; Heberling, G.; Wesche, K. Environmental controls override grazing effects on plant functional traits in Tibetan rangelands. *Funct. Ecol.* **2020**, *34*, 747–760. [CrossRef]
44. Gross, N.; Bagousse-Pinguet, Y.L.; Liancourt, P.; Berdugo, M.; Gotelli, N.J.; Maestre, F.T. Functional trait diversity maximizes ecosystem multifunctionality. *Nat. Ecol. Evol.* **2017**, *1*, 132. [CrossRef] [PubMed]
45. He, N.; Li, Y.; Liu, C.; Xu, L.; Li, M.; Zhang, J.; He, J.; Tang, Z.; Han, X.; Ye, Q.; et al. Plant trait networks: Improved resolution of the dimensionality of adaptation. *Trends Ecol. Evol.* **2020**, *35*, 908–918. [CrossRef] [PubMed]
46. He, N.; Liu, C.; Piao, S.; Sack, L.; Xu, L.; Luo, Y.; He, J.; Han, X.; Zhou, G.; Zhou, X.; et al. Ecosystem traits linking functional traits to macroecology. *Trends Ecol. Evol.* **2019**, *34*, 200–210. [CrossRef]
47. Jin, X.M.; Liu, J.T.; Wang, S.T.; Xia, W. Vegetation dynamics and their response to groundwater and climate variables in Qaidam basin, China. *Int. J. Remote Sens.* **2016**, *37*, 710–728. [CrossRef]

48. Han, J.J.; Wang, J.P.; Chen, L.; Xiang, J.Y.; Ling, Z.Y.; Li, Q.K.; Wang, E.L. Driving factors of desertification in Qaidam basin, China: An 18-year analysis using the geographic detector model. *Ecol. Indic.* **2021**, *124*, 107404. [CrossRef]
49. Shi, X.; Yu, D.; Warner, E.; Pan, X.; Petersen, G.; Gong, Z.; Weindorf, D. Soil database of 1:1,000,000 digital soil survey and reference system of the Chinese genetic soil classification system. *Soil Surv. Horiz.* **2004**, *45*, 129–136. [CrossRef]
50. Zhang, S.Q.; Chen, H.; Fu, Y.; Niu, H.H.; Yang, Y.; Zhang, B.X. Fractional vegetation cover estimation of different vegetation types in the Qaidam basin. *Sustainability* **2019**, *11*, 864. [CrossRef]
51. Fu, Y.; Chen, H.; Niu, H.H.; Zhang, S.Q.; Yang, Y. Spatial and temporal variation of vegetation phenology and its response to climate changes in Qaidam basin from 2000 to 2015. *J. Geogr. Sci.* **2018**, *28*, 400–414. [CrossRef]
52. Savitzky, A.; Golay, M.J. Smoothing and differentiation of data by simplified least squares procedures. *Anal. Chem.* **1964**, *36*, 1627–1639. [CrossRef]
53. Holben, B.N. Characteristics of maximum-value composite images from temporal AVHRR data. *Int. J. Remote Sens.* **1986**, *7*, 1417–1434. [CrossRef]
54. Hutchinson, M. *ANUSPLIN*, version 4.3; Centre for Resource and Environmental Studies, The Australian National University: Canberra, Australia, 2004.
55. Li, M.; Wu, J.S.; He, Y.T.; Wu, L.; Niu, B.; Song, M.H.; Zhang, X.Z. Dimensionality of grassland stability shifts along with altitudes on the Tibetan plateau. *Agric. For. Meteorol.* **2020**, *291*, 108080. [CrossRef]
56. Sandholt, I.; Rasmussen, K.; Andersen, J. A simple interpretation of the surface temperature/vegetation index space for assessment of surface moisture status. *Remote Sens. Environ.* **2002**, *79*, 213–224. [CrossRef]
57. Kendall, M.G. *Rank Correlation Methods*, 4th ed.; Charles Grifin: London, UK, 1975.
58. Mann, H.B. Nonparametric tests against trend. *Econometrica* **1945**, *13*, 245–259. [CrossRef]
59. Akhtar, K.; Wang, W.Y.; Ren, G.X.; Khan, A.; Feng, Y.Z.; Yang, G.H. Changes in soil enzymes, soil properties, and maize crop productivity under wheat straw mulching in Guanzhong, China. *Soil Till. Res.* **2018**, *182*, 94–102. [CrossRef]
60. Wu, J.S.; Song, M.H.; Ma, W.L.; Zhang, X.Z.; Shen, Z.X.; Tarolli, P.; Wurst, S.; Shi, P.L.; Ratzmann, G.; Feng, Y.F.; et al. Plant and soil's δ n-15 are regulated by climate, soil nutrients, and species diversity in alpine grasslands on the northern Tibetan plateau. *Agric. Ecosyst. Environ.* **2019**, *281*, 111–123. [CrossRef]
61. Gauzere, P.; Doulcier, G.; Devictor, V.; Kefi, S. A framework for estimating species-specific contributions to community indicators. *Ecol. Indic.* **2019**, *99*, 74–82. [CrossRef]
62. Violle, C.; Reich, P.B.; Pacala, S.W.; Enquist, B.J.; Kattge, J. The emergence and promise of functional biogeography. *Proc. Natl. Acad. Sci. USA* **2014**, *111*, 13690–13696. [CrossRef]
63. Wilcoxon, F. Individual comparisons by ranking methods. In *Breakthroughs in Statistics*; Springer: Berlin/Heidelberg, Germany, 1992; pp. 196–202.
64. Wei, T.; Simko, V. R Package "Corrplot": Visualization of a Correlation Matrix (Version 0.92). 2021. Available online: https://github.com/taiyun/corrplot (accessed on 26 February 2022).
65. Niu, W.; Chen, H.; Wu, J. Soil moisture and soluble salt content dominate changes in foliar delta(13)c and delta(15)n of desert communities in the Qaidam basin, Qinghai-Tibetan plateau. *Front. Plant Sci.* **2021**, *12*, 675817. [CrossRef]
66. Wu, J.; Li, M.; Fiedler, S.; Ma, W.; Wang, X.; Zhang, X.; Tietjen, B. Impacts of grazing exclusion on productivity partitioning along regional plant diversity and climatic gradients in Tibetan alpine grasslands. *J. Environ. Manag.* **2019**, *231*, 635–645. [CrossRef]
67. Rosseel, Y. Lavaan: An R package for structural equation modeling. *J. Stat. Softw.* **2012**, *48*, 1–36. [CrossRef]
68. R Core Team. *R: A Language and Environment for Statistical Computing*; R Foundation for Statistical Computing: Vienna, Austria, 2021.
69. ESRI. *ArcGIS Desktop: Release 10.2*; Environmental Systems Research Institute: Redlands, CA, USA, 2014.
70. Kusch, E.; Davy, R.; Seddon, A.W.R. Vegetation-memory effects and their association with vegetation resilience in global drylands. *J. Ecol.* **2022**, *110*, 1561–1574. [CrossRef]
71. Ogle, K.; Barber, J.J.; Barron-Gafford, G.A.; Bentley, L.P.; Young, J.M.; Huxman, T.E.; Loik, M.E.; Tissue, D.T. Quantifying ecological memory in plant and ecosystem processes. *Ecol. Lett.* **2015**, *18*, 221–235. [CrossRef] [PubMed]
72. Easterling, D.R.; Wehner, M.F. Is the climate warming or cooling? *Geophys. Res. Lett.* **2009**, *36*. [CrossRef]
73. Zhu, Y.K.; Zhang, J.T.; Zhang, Y.Q.; Qin, S.G.; Shao, Y.Y.; Gao, Y. Responses of vegetation to climatic variations in the desert region of northern China. *Catena* **2019**, *175*, 27–36. [CrossRef]
74. Olson, M.E.; Soriano, D.; Rosell, J.A.; Anfodillo, T.; Donoghue, M.J.; Edwards, E.J.; Leon-Gomez, C.; Dawson, T.; Camarero Martinez, J.J.; Castorena, M.; et al. Plant height and hydraulic vulnerability to drought and cold. *Proc. Natl. Acad. Sci. USA* **2018**, *115*, 7551–7556. [CrossRef]
75. Kraft, N.J.B.; Adler, P.B.; Godoy, O.; James, E.C.; Fuller, S.; Levine, J.M. Community assembly, coexistence and the environmental filtering metaphor. *Funct. Ecol.* **2015**, *29*, 592–599. [CrossRef]
76. Aerts, R.; Chapin, F.S. The mineral nutrition of wild plants revisited: A re-evaluation of processes and patterns. In *Advances in Ecological Research*; Fitter, A.H., Raffaelli, D.G., Eds.; Academic Press: Cambridge, MA, USA, 2000; Volume 30, pp. 1–67.
77. Hobbie, S.E.; Nadelhoffer, K.J.; Hogberg, P. A synthesis: The role of nutrients as constraints on carbon balances in boreal and arctic regions. *Plant Soil* **2002**, *242*, 163–170. [CrossRef]
78. Kirschbaum, M.U.F. The temperature-dependence of soil organic-matter decomposition, and the effect of global warming on soil organic-c storage. *Soil Biol. Biochem.* **1995**, *27*, 753–760. [CrossRef]

79. Sparks, D.L.; Liebhardt, W.C. Temperature effects on potassium exchange and selectivity in Delaware soils. *Soil Sci.* **1982**, *133*, 10–17. [CrossRef]
80. Lewis, J.D.; Lucash, M.; Olszyk, D.M.; Tingey, D.T. Relationships between needle nitrogen concentration and photosynthetic responses of douglas-fir seedlings to elevated CO_2 and temperature. *New Phytol.* **2004**, *162*, 355–364. [CrossRef]
81. Berry, J.; Bjorkman, O. Photosynthetic response and adaptation to temperature in higher plants. *Annu. Rev. Plant Physiol.* **1980**, *31*, 491–543. [CrossRef]
82. Field, C. The photosynthesis-nitrogen relationship in wild plants. In *On the Economy of Form and Function*; Cambridge University Press: Cambridge, UK, 1986.
83. Wright, I.J.; Reich, P.B.; Westoby, M. Strategy shifts in leaf physiology, structure and nutrient content between species of high- and low-rainfall and high- and low-nutrient habitats. *Funct. Ecol.* **2001**, *15*, 423–434. [CrossRef]
84. Wright, I.J.; Reich, P.B.; Cornelissen, J.H.; Falster, D.S.; Garnier, E.; Hikosaka, K.; Lamont, B.B.; Lee, W.; Oleksyn, J.; Osada, N.; et al. Assessing the generality of global leaf trait relationships. *New Phytol.* **2005**, *166*, 485–496. [CrossRef]
85. Blackwell, M.S.A.; Brookes, R.C.; de la Fuente-Martinez, N.; Gordon, H.; Murray, P.J.; Snars, K.E.; Williams, J.K.; Bol, R.; Haygarth, P.M. Chapter 1—Phosphorus solubilization and potential transfer to surface waters from the soil microbial biomass following drying–rewetting and freezing–thawing. In *Advances in Agronomy*; Sparks, D.L., Ed.; Elsevier: Amsterdam, The Netherlands, 2010; Volume 106, pp. 1–35.
86. Wang, R.Z.; Creamer, C.A.; Wang, X.; He, P.; Xu, Z.W.; Jiang, Y. The effects of a 9-year nitrogen and water addition on soil aggregate phosphorus and sulfur availability in a semi-arid grassland. *Ecol. Indic.* **2016**, *61*, 806–814. [CrossRef]
87. Farooqui, A.; Pillai, S.K.; Agnihotri, D.; Khan, S.; Tewari, R.; Shukla, S.K.; Ali, S.; Trivedi, A.; Pandita, S.K.; Kumar, K.; et al. Impact of climate on the evolution of vegetation in tectonically active Karewa basin, Kashmir Himalayas. *J. Earth Syst. Sci.* **2021**, *130*, 93. [CrossRef]
88. Li, Z.; Tian, D.; Wang, B.; Wang, J.; Wang, S.; Chen, H.Y.H.; Xu, X.; Wang, C.; He, N.; Niu, S. Microbes drive global soil nitrogen mineralization and availability. *Glob. Chang. Biol.* **2019**, *25*, 1078–1088. [CrossRef] [PubMed]
89. Trenberth, K.E.; Dai, A.G.; van der Schrier, G.; Jones, P.D.; Barichivich, J.; Briffa, K.R.; Sheffield, J. Global warming and changes in drought. *Nat. Clim. Chang.* **2014**, *4*, 17–22. [CrossRef]
90. Brenner, D.L.; Amundson, R.; Baisden, W.T.; Kendall, C.; Harden, J. Soil n and n-15 variation with time in a California annual grassland ecosystem. *Geochim. Cosmochim. Acta* **2001**, *65*, 4171–4186. [CrossRef]
91. Evans, S.E.; Burke, I.C. Carbon and Nitrogen Decoupling Under an 11-Year Drought in the Shortgrass Steppe. *Ecosystems* **2013**, *16*, 704–705. [CrossRef]
92. Körner, C.; Kèorner, C. *Alpine Plant Life: Functional Plant Ecology of High Mountain Ecosystems*; Springer: Berlin/Heidelberg, Germany, 1999.
93. Kuchenbuch, R.; Claassen, N.; Jungk, A. Potassium availability in relation to soil moisture. *Plant Soil* **1986**, *95*, 233–243. [CrossRef]
94. Bargagli, R.; Brown, D.H.; Nelli, L. Metal biomonitoring with mosses: Procedures for correcting for soil contamination. *Environ. Pollut.* **1995**, *89*, 169–175. [CrossRef]
95. Mahlstein, I.; Daniel, J.S.; Solomon, S. Pace of shifts in climate regions increases with global temperature. *Nat. Clim. Chang.* **2013**, *3*, 739–743. [CrossRef]
96. Moles, A.T.; Perkins, S.E.; Laffan, S.W.; Flores-Moreno, H.; Awasthy, M.; Tindall, M.L.; Sack, L.; Pitman, A.; Kattge, J.; Aarssen, L.W.; et al. Which is a better predictor of plant traits: Temperature or precipitation? *J. Veg. Sci.* **2014**, *25*, 1167–1180. [CrossRef]
97. Coban, O.; de Deyn, G.B.; van der Ploeg, M. Soil microbiota as game-changers in restoration of degraded lands. *Science* **2022**, *375*, abe0725. [CrossRef]
98. Watanabe, T.; Broadley, M.R.; Jansen, S.; White, P.J.; Takada, J.; Satake, K.; Takamatsu, T.; Tuah, S.J.; Osaki, M. Evolutionary control of leaf element composition in plants. *New Phytol.* **2007**, *174*, 516–523. [CrossRef]

remote sensing

MDPI

Article

Satellite Fog Detection at Dawn and Dusk Based on the Deep Learning Algorithm under Terrain-Restriction

Yinze Ran [1], Huiyun Ma [1], Zengwei Liu [1], Xiaojing Wu [2], Yanan Li [1] and Huihui Feng [1,*]

1 School of Geosciences and Info-Physics, Central South University, Changsha 410083, China
2 National Satellite Meteorological Center, China Meteorological Administration, Beijing 100081, China
* Correspondence: hhfeng@csu.edu.cn

Abstract: Fog generally forms at dawn and dusk, which exerts serious impacts on public traffic and human health. Terrain strongly affects fog formation, which provides a useful clue for fog detection from satellite observation. With the aid of the advanced Himawari-8 imager data (H8/AHI), this study develops a deep learning algorithm for fog detection at dawn and dusk under terrain-restriction and enhanced channel domain attention mechanism (DDF-Net). The DDF-Net is based on the traditional U-Net model, with the digital elevation model (DEM) data acting as the auxiliary information to separate fog from the low stratus. Furthermore, the squeeze-and-excitation networks (SE-Net) is integrated to optimize the information extraction for eliminating the influence of solar zenith angles (SZA) on the spectral characteristics over a large region. Results show acceptable accuracy of the DDF-Net. The overall probability of detection (POD) is 84.0% at dawn and 83.7% at dusk. In addition, the terrain-restriction strategy improves the results at the edges of foggy regions and reduces the false alarm rate (FAR) for low stratus. The accuracy is expected to be improved when training at a season or month scale, rather than at a longer temporal scale. Results of our study help to improve the accuracy of fog detection, which could further support the relevant traffic planning or healthy travel.

Keywords: H8/AHI; fog detection at dawn and dusk; DEM; U-Net; SE-Net

Citation: Ran, Y.; Ma, H.; Liu, Z.; Wu, X.; Li, Y.; Feng, H. Satellite Fog Detection at Dawn and Dusk Based on the Deep Learning Algorithm under Terrain-Restriction. *Remote Sens.* **2022**, *14*, 4328. https://doi.org/10.3390/rs14174328

Academic Editor: Xander Wang

Received: 30 June 2022
Accepted: 29 August 2022
Published: 1 September 2022

Publisher's Note: MDPI stays neutral with regard to jurisdictional claims in published maps and institutional affiliations.

1. Introduction

Fog refers to the suspension of microscopic water droplets in the atmosphere [1]. Besides during the night, solar radiation is also weak during the dawn and dusk, resulting in a relatively high probability of fog formation because of the low surface air temperature and high vapor saturation [2]. The dense fog seriously reduces horizontal visibility, which adversely affects public traffic and human health (particularly in rush hours at dawn and dusk) [3]. In addition, anthropogenically generated chemicals dissolve in the foggy water, which would strongly worsen air pollution [4]. Therefore, fog detection at dawn and dusk is crucial to effectively support traffic planning and to provide information and bulletins for reducing risks to human health [5].

Traditional terrestrial fog detection mainly relies on meteorological observation stations. However, it is difficult to be applied at a large-scale due to the limited spatial and temporal resolutions. With the high temporal resolution and spatial continuity observation, satellite remote sensing shows great potential for fog detection on a large scale [6]. Physically, the satellite fog detection algorithm follows the fact that the emissivity of opaque water clouds is lower in the mid-infrared (MIR) band than that in the thermal infrared (TIR) band [7]. Based on this theory, Eyre et al. [8] used the 3.7 μm and 11 μm brightness temperature (BT) of the Advanced Very High-Resolution Radiometer (AVHRR) to identify the nighttime fog and low stratus. Considering the great difference of the dual-channel brightness temperature difference (BTD) between fog, surface and clouds, the BTD is widely used in nighttime fog detection [9–14]. However, the method is only suitable for fog detection at nighttime, for which it is difficult to isolate the fog and low stratus (FLS) simultaneously [15]. The MIR band is affected by the solar radiation reflection of the target in the daytime, which causes the dual-channel differential separation detection threshold to vary with the solar zenith angle (SZA) [2]. At dawn and dusk, the MIR reflection is relatively weak due to the high SZA. In regions near to the terminator, the BTD between the fog and surface is similar or consistent, resulting in large uncertainty in fog detection. Gurka et al. [16] analyzed the fog dissipation process and its correlation with the thickness of the fog layer based on the visible light (VIS) band of the Synchronous Meteorological Satellite (SMS-1). Rao et al. [17] discussed the feasibility of identifying FLS in the VIS band of 0.64 μm. At present, the daytime fog detection algorithms are mainly based on the threshold of the spectral and textural differences between cloud, fog, and surface in the 0.67 μm, 3.7 μm, and 11 μm bands [18,19]. However, the MIR and VIS bands are affected by the high SZA at dawn and dusk and the spectral characteristics difference is weak between fog and surface, leaving low confidence in fog detection at dawn and dusk. Using the difference of BTD and reflectivity at 0.67 μm, Yoo et al. [5] proposed a FLS detection algorithm at dawn based on the Communication, Ocean and Meteorological Satellite (COMS) and Fengyun-2D (FY-2D) satellite. However, the method is difficult to monitor fog areas in real-time due to the low temporal resolution of the polar-orbiting satellites. Yang et al. [20] used the geostationary Fengyun-4A (FY-4A) and Himawari-8 (H8) with a higher temporal resolution to retrieve the probability index of Japan/surrounding FLS at dawn, which provides a methodological reference for fog detection at dawn and dusk.

The algorithms mentioned above mainly rely on the spectral difference between fog and other ground objects. Surface environments (particularly the terrain) also strongly affect fog formation and detection [21], which are lacking in most previous algorithms. In general, fog forms near the ground, while clouds form at high altitudes [22]. Furthermore, the edge of fog is restricted by terrain [13], which presents unique spatial variation, vertical structure, and optical properties. By adding terrain information to the fog detection algorithm, it is expected to further improve the accuracy of fog detection. For example, Shang et al. [22] combined DEM with traditional cloud/haze detection indicators to improve the accuracy of cloud, haze, and clear recognition. However, the terrain features of fog classification are the same as the edge elevation of objects, while the spectral features of remote sensing images are mainly at pixel scale. Therefore, it is difficult to fuse the two features in traditional fog detection methods.

The deep learning approach realizes the mapping of input data to detection target results (that is, the detection of fog) through a series of data transformations (layers), replacing the complex multi-stage process with a simple, end-to-end learning model. By using a series of nonlinear functions, the deep learning model describes and classifies the characteristics of the targets [23], which provides a potentially efficient way for fog detection at dawn or dusk. Based on the U-Net model, Jacob et al. [24] proposed an RS-Net cloud detection algorithm that effectively improved the efficiency of cloud detection. Wu et al. [25] proposed a geographic information-driven network (GeoInfo-Net) for cloud and snow detection. In addition to using remote sensing data, the network would encode the auxiliary data sets of the elevation, latitude, and longitude information for network training, which helps to improve the accuracy significantly. However, fog detection is sensitive to the remote sensing bands under different terrain conditions and SZA, resulting in various weight values of each band during deep learning. Meanwhile, fog is a small probability event. The model training strategy needs to be adjusted to speed up the convergence process of model training and avoid overfitting. The problems above leave great uncertainty for the application of deep learning in satellite fog detection. These problems can be optimized by improving the application of the deep learning model. For example, the sensitivity of different SZA and terrain conditions to fog detection can be effectively alleviated by introducing the Squeeze-and-Excitation Networks (SE-Net), which could automatically obtain the weights of the contributions of the parameters [26]. Uneven distribution of training data samples can be solved through the batch normalization of input data. Additionally, the training process could be optimized by designing the independent adaptive learning rates for different parameters and by using efficient optimization algorithms [27]. However, the relevant research is still rare, thus further investigation is warranted.

This study aims to develop a deep learning-based algorithm for satellite fog detection under the terrain-restriction. The DEM data is added into the information input layer to distinguish fog and low cloud. Furthermore, the Squeeze-and-Excitation Networks (SE-Net) is integrated to optimize the model's information extraction to eliminate the influence of different SZA over a large region. Finally, a deep learning fog detection algorithm under terrain constraints and enhanced channel domain attention mechanism (DDF-Net) is developed for fog detection at dawn and dusk. This study is organized as follows: Section 2 describes the study area, data, details regarding pre-processing methodologies, and the realization of the algorithm. Section 3 includes the results of quantitative and qualitative validation. Finally, the performance and limitation of the DDF-Net are concluded and discussed in Section 4.

2. Materials and Methods

2.1. Study Area and Data Sources

Figure 1 shows an overview of the study area, which is located in northern China (104°30′E~136°00′E and 33°00′N~47°30′N). The western part of the study area includes the Inner Mongolia Plateau and Loess Plateau, with an average altitude of more than 1500 m. The eastern of the study area is located in the third step of the flat terrain of the Chinese topographic demarcation line, where the average altitude drops to 500–1000 m. The study area is located in the eastern of the Henhe-Tengchong Line and has a developed economy and large population. The climate is characterized by a rich source of water vapor from the Western Pacific, which helps to form fog for more than 30 days per year [28]. The frequent fog at dawn and dusk exerts a serious impact on local traffic, especially aviation and highways, and also the economic development and people's physical and mental health [29].

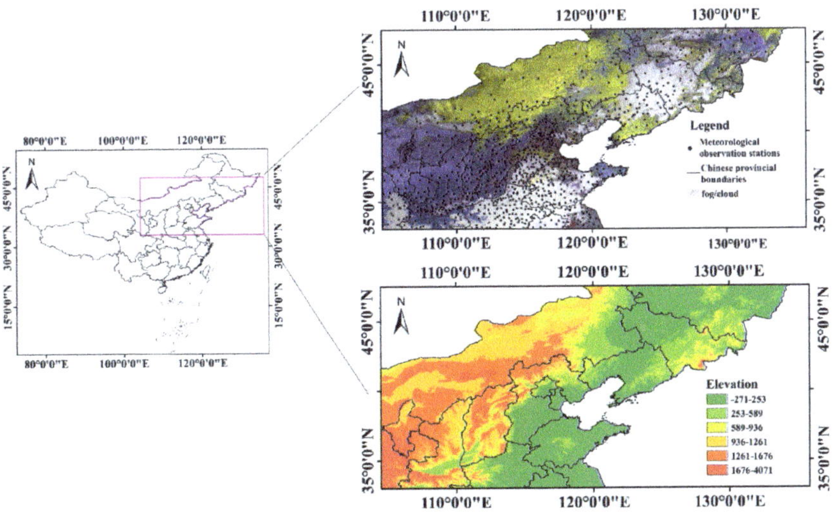

Figure 1. The study area.

(1) Satellite data

We adopt the Advanced Himawari-8 Imager (AHI) data with a temporal resolution of 10 min for the investigation. The data set has 16 channels ranging from 0.47 μm to 13.3 μm, with the details could be seen in ref [30]. By referencing the previous studies [22,31–33], we select the data sets of bands 3, 5, 6, 7, 11, 13, and 14 for model training and validation. AHI data in foggy days from November to December during 2015~2017 are selected to build 3122 fog mask images (size: 256 × 256) and then divided into 70% for training and 30% for validating.

(2) Ground observation data

The ground observation data was used to execute and validate the algorithm, these data were available from the China Meteorological Administration ground observation stations. The data include observations 8 times per day: 02:00 local time (LT), 05:00 LT, 08:00 LT, 11:00 LT, 14:00 LT, 17:00 LT, 20:00 LT, and 23:00 LT. Among them, the data at 8:00 and 17:00 are used to evaluate the accuracy of the algorithm. The climate would be marked as foggy when the "visibility" is less than 1 km in most criteria. In this study, we followed the criteria and classified the fog as strong, dense and hazy as visibility ranges between 0~200, 200~500 and 500~1000 m, respectively. Meanwhile, the China Meteorological Administration (CMA) promulgated the national standard of Grade of fog forecast (GB/T 27964-2011GB/T) [34], which also defined haze when the horizontal visibility ranges from 1.0 to 10.0 km. Furthermore, it is less harmful to traffic when the visibility is greater than 3 km. Therefore, we defined the haze as visibility ranges from 500~3000 m.

(3) Terrain data

The terrain data used in this study came from the SRTM1-DEM data with a spatial resolution of 30 m released by the United States Geological Survey (USGS). To match the spatial resolution of the H8/AHI data, the DEM data was resampled to 2 km.

2.2. Deep Learning Algorithm

Deep learning realizes the mapping of input data to target results through a series of data transformations (layers) (Figure 2). The basic principle is to output the image as a target result according to the mapping function f of the initial network model [23]. The model compares the prediction with the truth, and then adjusts the parameters and weights through an iterative method to capture the optimal prediction result. The components of the deep learning model generally include the mapping function f, convolutional layers (CL) and pooling layers (PL) [23]. Specifically, f defines the functional relationship between the input x and the target mask classification map y; CL uses a specified number of convolution kernels in each layer to extract the feature information of each pixel of x. The shallow layer includes the textural and spectral features that are used as the basis classification, and the deeper layer represents the higher-level semantics, which is the feature information obtained after several convolutions (feature extraction) [35]. The PL is implemented to sample the channel feature information downward, which divides the input image into several rectangular areas to get the maximum value of each sub-area. Through this way, it will continuously reduce the spatial size of the data, which could reduce the parameters and improve the computation speed [23]. Furthermore, the PL can ensure that the network continuously extracts and utilizes the information through the multi-layer structure of biological neurons. The activation function (Relu) is located between the CL and the PL, which is used to achieve nonlinear transformation. The sigmoid function maps the output of the last CL to values from 0 and 1, and then realizes the classification of the input x for each pixel by setting an appropriate threshold.

In the execution of the deep learning model, the target probability is generated according to the last CL and the sigmoid function. It is defined as the target pixel when the probability is greater than a certain threshold. The deviation between the actual fog coverage and the predicted target mask is calculated using a binary cross-entropy loss function, and the parameter gradient of the mapping function f is calculated and updated by the Backpropagation algorithm (BP) [36] until the optimal parameters of the model are obtained. Specifically, the actual fog coverage is generated by the combination of visual interpretation and ground observation, which is used as the label data for training or validating the algorithm.

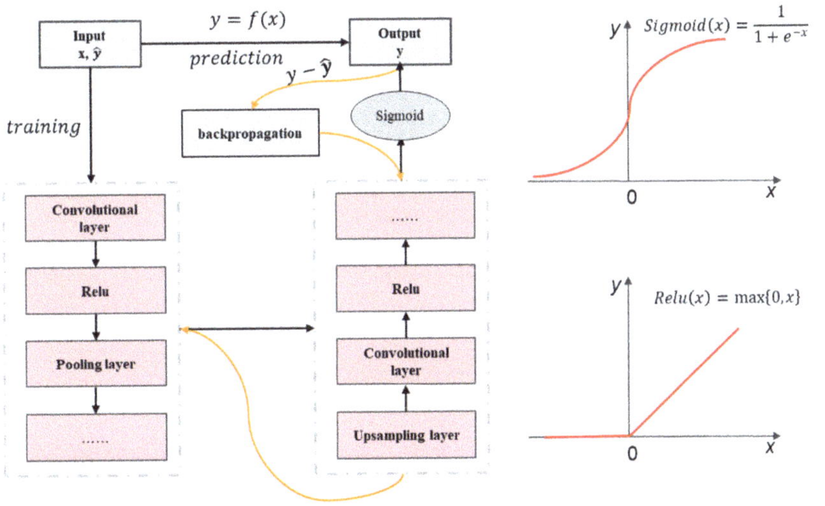

Figure 2. The Schematic diagram of the deep learning model. Relu is used to achieve non-linear transformation between the CL and PL layers.

2.3. The SE-Net Module

The output of the CL does not consider the dependencies of each channel. SE-Net allows the network to selectively enhance features and suppress useless features. We choose to adopt this approach in our study. The SE-Net module contains 1 global average pooling operation (*GP*), 2 small fully connected layers (*FC*), 1 Sigmoid function and 1 channel weighted operation, with the details could be seen in ref [26]. Specifically, the Global pooling (*GP*) refers to the average of all the pixels of the feature map on each channel. The FC includes two sub-modes of FC_1 and FC_2. FC_1 consists of fully-connected layers with C/16 filters, which is calculated by the weighted summation to reduce the number of parameters and improve the calculation efficiency. The activation function (Relu) is located between the FC_1 and the FC_2, which is used to achieve nonlinear transformation. FC_2 consists of fully-connected layers with C filters, which assure the same numbers of outputs and channels. Finally, the sigmoid normalizes these learned weights to be between 0–1 for dimensionless processing of different features.

Execution steps of the SE-Net module are as follows:

(1) Obtaining the global information from the feature channels. The information of the DEM-AHI fog image at dawn and dusk is extracted through the CL formed by *C* convolution kernels, and image *X* with a characteristic channel number of *C* is formed. C_p is calculated by GP in each feature channel of the image X (Equation (1)):

$$C_p = \frac{1}{W \times H} \sum_{i=1}^{W} \sum_{j=1}^{H} x_p(i,j) \tag{1}$$

where C_p is the global information of the feature channel *p* after GP, *W* and *H* are the length and width of the image *X*, respectively, x_p (*i,j*) represents the pixel value of X at point (*i,j*) in channel *p*.

(2) Calculating the correlation between each feature channel. According to the calculated correlation results, the weight of the feature channel that improves the accuracy is increased, while the suppressed or ineffective feature channel's weight is decreased (Equation (2)):

$$F_{P\_F_1} = ReLU(FC_1(C_P)) \tag{2}$$

where $F_{\_P\_F1}$ is the feature of output channel p after FC_1 and Relu activation function, the FC_1 is the first fully connected layer [23,26], the size of $F_{\_P\_F1}$ is $C/16 \times 1 \times 1$.

(3) Calculate the weight of each feature channel. The Sigmoid function is used to calculate the weight factor W_p of the channel p (Equation (3)):

$$W_P = Sigmoid(FC_2(F_{P\_F_1})) \tag{3}$$

where the FC_2 is the second fully connected layer. W_p represents the weight factor of channel p after function calculation of FC_2 and Sigmoid function.

(4) Attention learning of channels. The weighted attention feature x'_p is acquired through the weight factor Wp and the pixel value x_p of X at channel p (Equation (4)):

$$x'_P = W_p \cdot x_p \tag{4}$$

The x'_p represents the attention feature of channel p learned by SE-Net.

2.4. Deep Learning-Based Algorithm of Fog Detection under Terrain Restriction

The spatio-spectral characteristics of fog, medium/high clouds and surfaces provide the physical basis for the algorithm of fog detection, while the weak solar radiation severely hinders the separation of fog from the surface and clouds at dawn and dusk. The U-net model had been proven to be a robust and effective method due to its excellent performance and transferability in image segmentation [24,37], which is used as the basic model in our algorithm. Furthermore, terrain strongly affects the formation and movement of fog [21], which should be taken into consideration in the algorithm of fog detection. To overcome these challenges above, this study develops a deep learning-based algorithm for satellite fog detection at dawn and dusk under terrain constraints. The improvement of our method refers to the adaptive algorithm under different SZA and terrain conditions by introducing channel attention mechanisms and adding DEM data to remote sensing images. Specifically, our method extracts fog from the AHI-DEM datasets by integrating the SE-Net module in the CL of the U-net model. Figure 3 depicts the flowchart of the algorithm. The inputs include the ground observation data, satellite data of H8/AHI (bands 3, 5, 6, 7, 11, 13, and 14) and the terrain data (DEM), while the outputs refer to spatial coverage of fog detection under different times.

The main structure of the DDF-Net is shown in Figure 4. The input side includes 5 CL, 4 PL and 3 SE-Net modules. The CL can extract the feature information of fog, cloud and surface at dawn or dusk through designing different numbers of 3×3 convolution kernel. The BN operation is executed to normalize the input data and the Relu function acts by increasing the model's ability to fit nonlinear relationships. The PL selects the maximum value of the 4 pixels to reduce the dimension of the feature information and complete the down-sampling of the feature map, which can reduce the parameters and accelerate the training process of the model. The output side includes 4 CL and 4 up-sampling layers. The up-sampling layer restores the dimension of the feature map by duplicating each pixel in the feature map as 4 pixels, which generates less local information in the feature map. Alternatively, the skip connection fuses the feature map to establish a contextual feature information connection and compensate local information, which helps to obtain the semantic space context information of the model [37]. The DDF-Net model composed of these multi-layer structures can approximately fit all complex functions. This is very beneficial to our research, because the state of fog is quite different under different terrains and SZA.

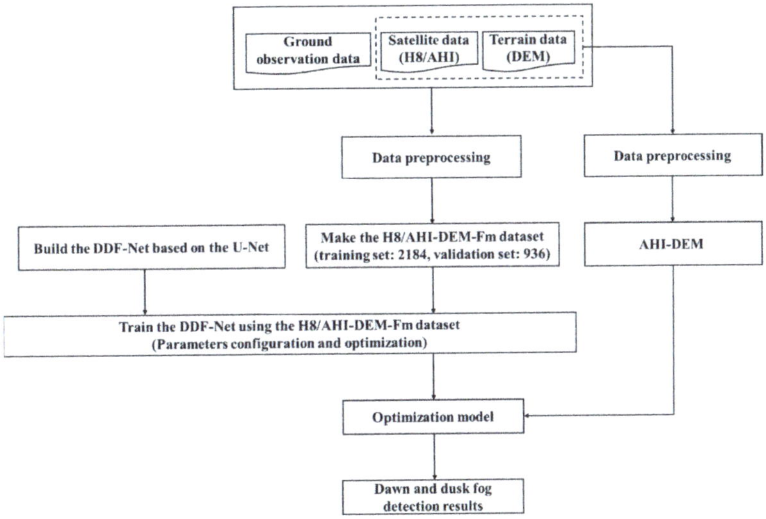

Figure 3. Flow chart of the methodology.

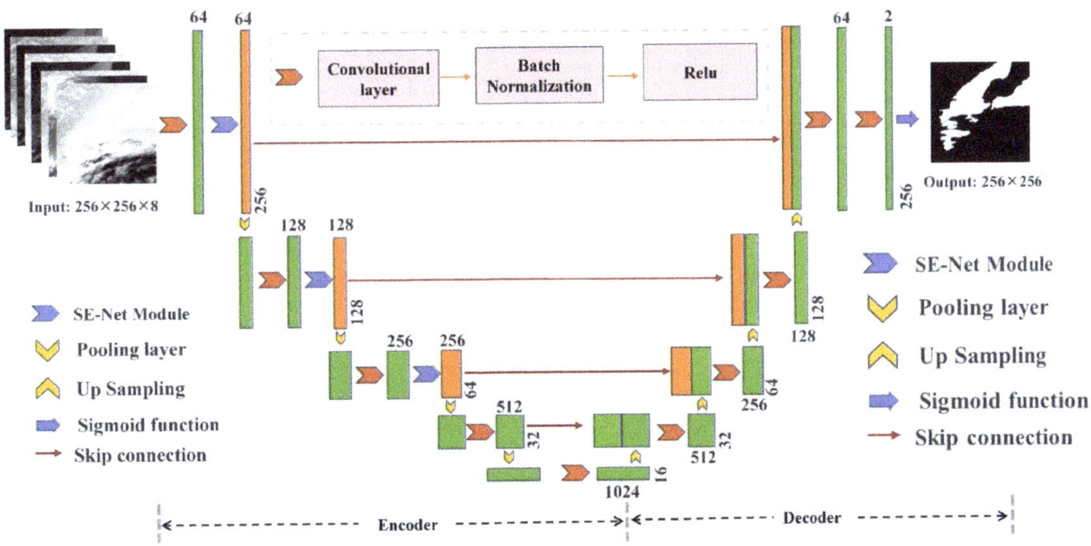

Figure 4. The DDF-Net's architecture. The top/bottom numbers of the rectangle represent the channels of the feature map, and the right numbers are the size of the feature map.

2.4.1. Building the H8/AHI-DEM-Fm Fog Mask Dataset

Based on the H8/AHI data, this study firstly builds the training data set (Figure 5). The DEM data was added to the AHI data to refine the model's detection accuracy at the edge of the foggy area. Fog pixels are labeled by the spectral, textural and movement characteristics of fog, surface and cloud.

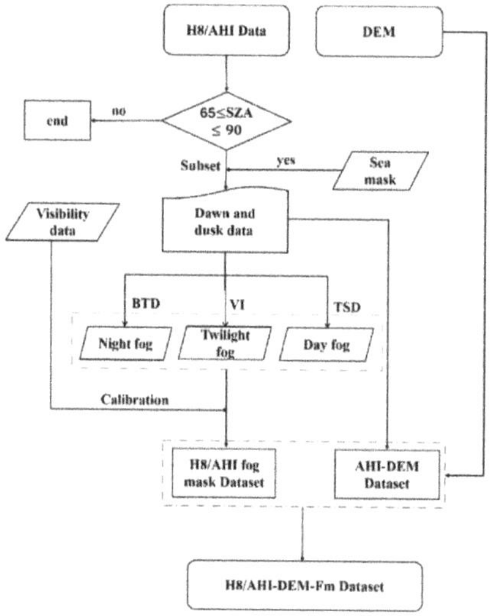

Figure 5. The process of making the H8/AHI-DEM-Fm dataset.

(1) Building the AHI-DEM dataset at dawn and dusk.

The remote sensing images including dawn and dusk are screened according to the SZA. Meanwhile, band clipping and sea-land mask clipping are performed. The DEM is combined with AHI data as a channel to construct the AHI-DEM dataset containing the time of dawn and dusk.

(2) Building the fog mask dataset Fm.

The fog pixels marked on the AHI-DEM data are used to build the fog mask data Fm. The pixels are marked as daytime, nighttime and twilight fog through the textural and spectral difference (TSD), bright temperature difference (BTD) and visual interpretation (VI). TSD refers to the differences in the texture spectrum of daytime fog, clouds, and surfaces. The BTD of fog is lower than surfaces and medium/clouds at nighttime. The criterion of VI is that the fog has the characteristics of spatial and temporal consistency within the adjacent 30 min. According to the labeling of the image at the previous moment, it is determined whether there are fog pixels at the current moment. VI is mainly used to identify fog that is similar or consistent with surface spectral information at dawn and dusk. Meanwhile, the ground observation data is used to strictly calibrate the Fm.

(3) Building the H8/AHI-DEM-Fm fog mask dataset.

AHI-DEM and Fm are simultaneously cropped into sub-images of 256 $\times$ 256 sizes with 30% overlap to improve the training efficiency of the model and the ability to capture edge information. Finally, H8/AHI-DEM-Fm is produced through the above process.

2.4.2. Integration of the SE-Net Module

SZA strongly affects the fog spectral characteristics, even with the same remote sensing band. For example, the VIS band has a strong effect on the low SZA regions (due to fog being more reflective than the surface and less reflective than clouds), while it is ineffective in the nighttime regions. Therefore, the influence of each band on fog detection under different SZA should be taken into account, which is very important for large-scale dawn and dusk fog detection. Meanwhile, the terrain strongly affects the fog formation and development, which would act as one of the key parameters of fog detection [21]. The SE-Net module can automatically learn the weight of each band in the fog detection task at different conditions of SZA terrain in a data-driven manner, which shows the great potential for fog detection under different SZA.

Assuming that X is the feature map of the c channel output by the Relu function of a certain layer in the encoder, then X can be expressed as:

$$X = \{x1, x2, \ldots, xc\} \tag{5}$$

SE-Net automatically learns the weights W of feature maps on each channel by training:

$$W = \{w1, w2, \ldots, wc\} \tag{6}$$

Finally, the feature map X' output by SE-net can be expressed as:

$$X' = X \cdot W = \{x1 \times w1, x2 \times w2, \ \ldots, \ xc \times wc\} \tag{7}$$

The SE-Net module is embedded after the first three Relu (rather than after each Relu in the encoder) to further improve the extraction of fog information under different terrain and SZA conditions, and also avoids the introduction of too many parameters to slow down the model training.

2.4.3. Adjust the Training Strategy of the Model

To optimize the network model, the Adam gradient descent algorithm (Adam) [27,38,39] and BN [40] are introduced to adjust the training strategy. The Adam is introduced to speed up the training process of the model by automatically obtaining the direction of optimized parameters [27,38,39]. The BN is embedded after each CL to normalize the input data to overcome the sudden change in the distribution of a certain batch of training data (Figure 4). In addition, it can also improve the training speed and shortens the training time [38]. The configuration parameters (such as software, graphics card, programming framework, etc.) are shown in Table 1.

Table 1. Training platform and parameter configuration information.

Parameter Configuration	Platform and Version
GPU	NVIDIA GeForce RTX 2080Ti (NVIDIA Corporation, Santa Clara, CA, United States)
Memory	11GB
Operating system	Windows 10
Programming language	Python3.6.6
Framework	PyTorch1.5.1
Cuda	10.0
Batch size	8
Epoch	1500
Decaying Epoch	1440
$\alpha_{initial}$	0.0035

(1) By setting the initial parameters of the model (Table 1), the encoder extracts the feature map of the input image. Batch size represents the number of images inputted to the network during model training. Epoch represents the maximum number of rounds of model iteration. $\alpha_{initial}$ indicates the initial learning rate. Decaying Epoch indicates that the model adjusts the learning rate dynamically according to formula (8) after the studies number of Epoch minus Decaying Epoch.

$$\alpha = \alpha_{Last-round} \times \left(\frac{Decaying\ \ Epoch}{Epoch} \right)^{0.9} \tag{8}$$

(2) The decoder combines high-level semantic and low-level spatial information to complete the initial classification of pixels. The model finally outputs the binary result of

0 (non-fog) and 1 (fog). Then, the error between the predicted result and the actual fog coverage is calculated (Equations (9)–(11)). BP assigns the error to each parameter and the Adam is used to update the parameters by using the learning rate provided by the model.

(3) Processes 1–2 are iterated until the indicators are no longer significantly changing and fluctuate around a certain value.

(4) The new remote sensing data are input to the trained DDF-Net model to realize fog detection over the whole study area. The performance of the algorithm could be finally evaluated with the aid of ground observation.

2.5. Evaluation Metrics

This study adopts several metrics to evaluate the accuracy of the training model and the performance accuracy of our algorithm, which are described as follows:

(1) Model Evaluation Metrics

We use the metric of semantic segmentation [33,37], namely Accuracy (*Acc*), F1-Score (*F1*) and Intersection-over-Union (*IoU*) to evaluate the accuracy of the training model. Each indicator is defined as follows:

$$Acc = \frac{N_{correct}}{N_{total}} \tag{9}$$

$$F1 = \frac{2p \cdot r}{p + r} \tag{10}$$

$$IoU = \frac{N_{correct}}{N_{ground-truth} \cup N_{false}} \tag{11}$$

where *Acc* is the ratio of the number of correctly classified pixels ($N_{correct}$) to the total number of pixels in the scene (N_{total}). p represents the proportion of the $N_{correct}$ to the sum of $N_{correct}$ and the number of falsely classified pixels (N_{false}), *r* represents the proportion of the Ncorrect to the total number of pixels that are foggy in the corresponding label of the images ($N_{ground\text{-}truth}$). *IoU* represents the ratio of fog pixels in the image predicted by the model to the sum of total number of pixels that are foggy in the corresponding label of the images ($N_{ground\text{-}truth}$) and N_{false}. All indicators are scaled from 0 to 1, the larger the *IoU*, the higher the segmentation accuracy.

(2) Quantitative evaluation metrics of fog detection accuracy

The probability of detection (*POD*), false alarm ratio (*FAR*) and critical success index (*CSI*) [11] are used for the validation of fog detection, which are described as follows:

$$POD = \frac{N_H}{N_H + N_M}, \tag{12}$$

$$FAR = \frac{N_F}{N_H + N_F}, \tag{13}$$

$$CSI = \frac{N_H}{N_H + N_M + N_F}, \tag{14}$$

where N_H, N_M and N_F are the pixel numbers of hitting, missing and false fog detections. All indicators are scaled from 0 to 1, with high *POD* and *CSI* and low *FAR* indicating great performance of the algorithm.

3. Results

The fog is detected at dawn (7:00–8:30) and dusk (15:30–17:00) on 18 November 2016 over northern China (Figures 6 and 7) through our algorithm, and ground observation sites at 8:00 or 17:00 are shown in Figures 6c and 7d. The detection results show that the fog area increases from 7:00 to 7:30 and gradually decreases from 7:30–8:30. Comparing the position of the terminator line, it can be seen that the fog gradually dissipated after sunrise. With the aid of our algorithm, large areas of fog are detected in the central and western parts of the study area. Furthermore, the detection results of satellite data can capture the generation and disappearance of fog, which is highly consistent with the ground observations. Figure 6a shows that the fog was in the formation stage. The fog area was fragmented because it was interspersed with low cloud, resulting in incomplete fog detection results. Figure 6b shows that a relatively stable fog zone has formed, with fog covering the largest region before sunrise. As the terminator line continues to move westwards, the surface temperature in the eastern portion of the figure increases rapidly due to solar radiation, which causes fog to dissipate gradually (Figure 6c,d). The ground observation data show a high spatial consistence with the results of satellite fog detection (Figure 6c). The rectangular box in Figure 6c is located in the Chinese topographic transition zone according to the terrain of the study area (Figure 1), whose foggy area's boundary is defined well because of the application of terrain data. The missing detection is primarily located in the eastern portion of the figure (circular box in Figure 6c). The main reason for this missing detection might originate from the influence of low clouds. Influenced by the characteristics of the passive sensor images, it is relatively difficult to detect the under-cloud fog for the algorithm.

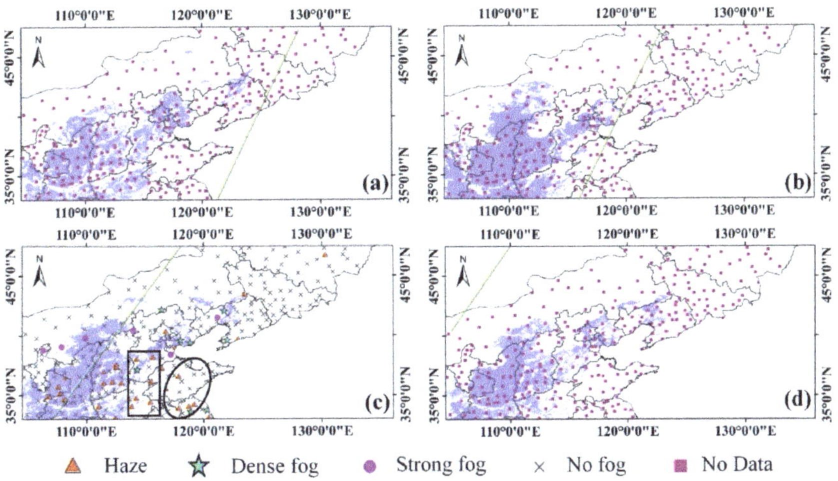

Figure 6. The fog detection results at (**a**) 7:00, (**b**) 7:30, (**c**) 8:00 and (**d**) 8:30 on 18 November 2016. The triangle, star, circle, and cross symbols represent haze, dense fog, strong fog, and non-fog stations, respectively. The square symbol indicates that there is no observation data because of the limit monitoring frequency. The blue area is the fog detected by our algorithm. All the times are at the Beijing Time (8 h later than the Universal Time Coordinated (UTC)) and the green line is the terminator line.

Figure 7 shows the fog detection results at dusk from 15:30 to 17:00. The ground observation data show large areas of haze at 17:00. It also shows a high spatial consistence with the results of satellite fog detection (Figure 7d), which supports the performance of our fog detection approach (Figure 7d). Satellite detection results show that the area of fog takes on a relatively stable trend from 15:30 to 17:00 (Figure 7a–d). The false detection occurred to the left of the terminator line at 16:30 (circular box in Figure 7c), which is characterized by high topographic undulations and low cloud coverage. Overall, our algorithm successfully detected more than 80% of fog in the study area.

We validate the algorithm with the aid of 10 fog cases randomly selected from 18th November in 2016 to 3 January in 2017. The statistical evaluation shows that the overall *POD*, *FAR* and *CSI* of the algorithm are 84.0%, 16.4% and 72.0% at dawn (Table 2), respectively. The highest *FAR* and lowest *CSI* occurred on January 3, 2017. The main reason is attributed to the low clouds in the Inner Mongolia autonomous region and Northeast Plain, which are difficult to be isolated by infrared remote sensing at nighttime. Table 3 shows the accuracy at dusk, with the mean *POD*, *FAR* and *CSI* values are 83.7%, 15.8% and 72.6%, respectively. Overall, our algorithm demonstrates the broad potential of using deep learning for fog detection at dawn and dusk. We sum all the fog categories to validate the algorithm. It is very useful to evaluate the accuracies under different fog types. However, the observation points are insufficient to be divided into so many sub-datasets. Therefore, a small disturbance of the ground samples would generate great uncertainty on the validation. We will continue the collection of ground observation to validation the performance of our algorithm in our future research.

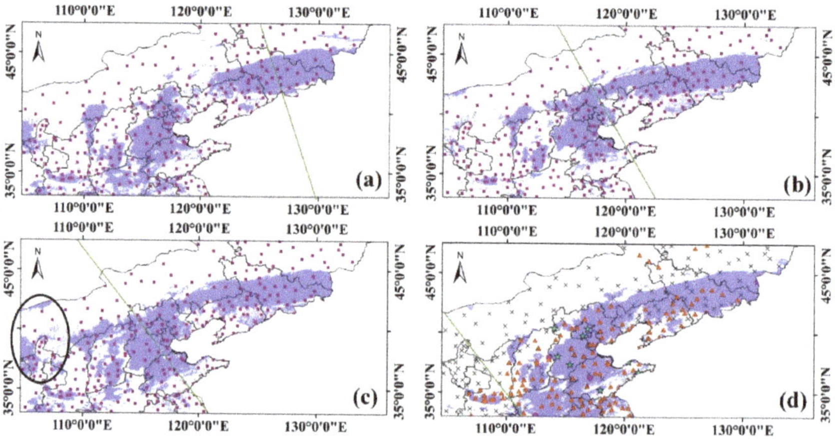

Figure 7. The fog detection results at (**a**) 15:30, (**b**) 16:00, (**c**) 16:30 and (**d**) 17:00 on 18 November 2016. The triangle, star, circle, and cross symbols represent haze, dense fog, strong fog, and non-fog stations, respectively. The square symbol indicates that there is no observation data because of the limit monitoring frequency. The blue area is the fog detected by our algorithm. All the times are at the Beijing Time (8 h later than the Universal Time Coordinated (UTC)) and the green line is the terminator line.

Table 2. Fog detection accuracy at 08:00 (BT).

Date	Satellite Detection	Ground Observation (Fog)	Ground Observation (Nonfog)	*POD*	*FAR*	*CSI*
18 November 2016	Fog Nonfog	34 12	7 210	0.739	0.171	0.642
12 December 2016	Fog Nonfog	44 3	6 255	0.936	0.120	0.830
18 December 2016	Fog Nonfog	16 3	3 276	0.842	0.158	0.727
20 December 2016	Fog Nonfog	19 4	2 322	0.826	0.095	0.760
3 January 2017	Fog Nonfog	18 3	7 381	0.857	0.280	0.643
	mean			0.840	0.164	0.720

Table 3. Fog detection accuracy at 17:00 (BT).

Date	Satellite Detection	Ground Observation (Fog)	Ground Observation (Nonfog)	*POD*	*FAR*	*CSI*
17 November 2016	Fog Nonfog	11 2	2 209	0.846	0.154	0.733
18 November 2016	Fog Nonfog	42 3	3 238	0.933	0.067	0.875
11 December 2016	Fog Nonfog	44 8	10 372	0.846	0.185	0.710
18 December 2016	Fog Nonfog	12 3	4 277	0.800	0.250	0.632
28 December 2016	Fog Nonfog	19 6	3 307	0.760	0.136	0.679
	mean			0.837	0.158	0.726

To evaluate the seasonal adaptability of the algorithm, eight dawn fog cases and five dusk fog cases were randomly selected from February to October 2017 (few fogs occur at dusk from April to July). Ground observations in 08:00 (February to March) and 05:00 (May to September) are adopted for the evaluation at dawn and dusk. Results are shown in Tables 4 and 5. Overall, the *POD*, *FAR* and *CSI* are 66.8%, 45.1% and 43.6% at dawn, and 68.0%, 32.3% and 49.8% at dusk, which show the acceptable accuracy over different seasons. On the other hand, the detection accuracy in summer and autumn was significantly influenced by cloudiness. In summer, the algorithm's performance is robust due to the apparent spectral and textural difference of the surface and fog/cloud. However, low clouds (liquid water clouds) in summer are more likely to be misclassified as fog. The FAR also increased due to the frequent cloud covering in autumn.

Table 4. Fog detection accuracy at dawn over different seasons.

Ground Observation Time	Cases	Satellite Detection	Ground Observation (Fog)	Ground Observation (Nonfog)	*POD*	*FAR*	*CSI*
8:00	5 February 2017, 8:00	Fog Nonfog	21 3	9 402	0.875	0.300	0.636
8:00	17 March 2017, 7:00	Fog Nonfog	5 2	4 413	0.714	0.444	0.455
5:00	5 May 2017, 6:00	Fog Nonfog	8 5	15 409	0.615	0.652	0.286
5:00	27 June 2017, 6:00	Fog Nonfog	24 11	13 392	0.686	0.351	0.500
5:00	29 July 2017, 6:00	Fog Nonfog	6 5	13 426	0.545	0.684	0.250
5:00	31 August 2017, 5:00	Fog Nonfog	12 9	14 399	0.571	0.538	0.343
5:00	17 September 2017, 5:00	Fog Nonfog	5 3	3 377	0.625	0.375	0.455
5:00	12 October 2017, 6:00	Fog Nonfog	22 9	8 415	0.710	0.267	0.564
		mean			0.668	0.451	0.436

Table 5. Fog detection accuracy at dusk over different seasons.

Ground Observation Time	Cases	Satellite Detection	Ground Observation (Fog)	Ground Observation (Nonfog)	*POD*	*FAR*	*CSI*
17:00	21 February 2017, 17:00	Fog	12	3	0.462	0.200	0.414
		Nonfog	14	395			
17:00	13 March 2017, 17:00	Fog	24	11	0.774	0.314	0.571
		Nonfog	7	417			
20:00	22 August 2017, 19:00	Fog	0	0	–	–	–
		Nonfog	7	487			
17:00	15 September 2017, 18:00	Fog	19	12	0.704	0.387	0.487
		Nonfog	8	377			
17:00	5 October 2017, 18:00	Fog	14	9	0.778	0.391	0.519
		Nonfog	4	437			
		mean			0.680	0.323	0.498

4. Discussion

We further compare the performances of our algorithm with those of the previous studies. Specifically, fog detection results from the U-Net and DDF-Net are compared at 08:00 on 18 in November and 12 December in 2016, and at 17:00 on 18 November and 11 December in 2016. The comparisons are shown in Figures 8 and 9. The ground observation indicates that fog area exceeding 85% has been successfully detected in the study area (Figure 8c,d and Figure 9a,b). Specifically, both algorithms detected fog more accurately at dawn (Figure 8c,d), and haze is detected more accurately by both algorithms at dusk (Figure 9a,b). On the other hand, both algorithms have missed detections due to medium and low cloud contamination (circular box in Figure 8a,b and Figure 9c,d). The false detection occurred at 8:00 (rectangular box in Figure 8c,d), which are mainly due to the interference of the low liquid water cloud. Compared with the U-Net, the DDF-Net shows more accurate detection accuracy. For example, the circular box in Figure 8a,b shows that fog edges of the U-Net are slightly coarse with a small proportion of low stratus being misclassified as fog, but the edge detection results of DDF-Net are more refined and the boundary of the foggy area is clearer. Secondly, U-Net detects low cloud as fog mistakenly in the daytime area with high SZA (rectangular box in Figure 8a,c), while the misclassified medium/high clouds are smaller in the DDF-Net (rectangular box in Figure 8b,d).

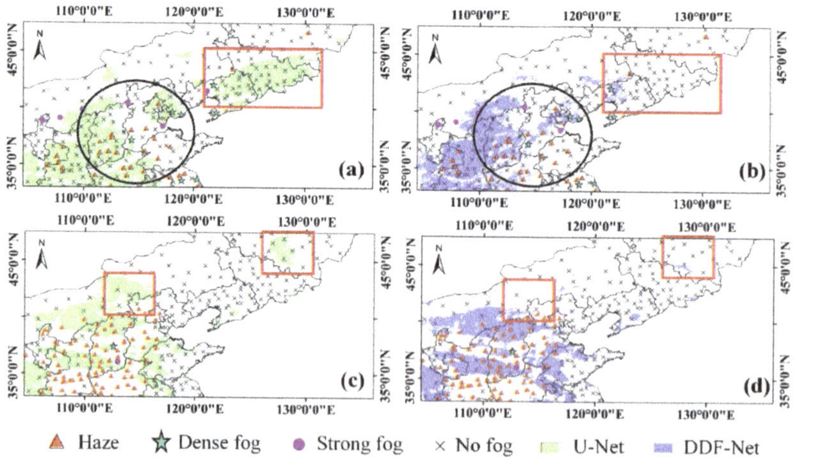

Figure 8. The fog detection results of U-Net and DDF-Net at 08:00 (18 November 2016 (**a**,**b**) and 12 December 2016 (**c**,**d**)). The triangle, star, circle, and cross symbols represent haze, dense fog, strong fog,

and non-fog stations, respectively. The green area is the fog detected by U-Net. The blue area is the fog detected by DDF-Net. All the times are at the Beijing Time (8 h later than the Universal Time Coordinated (UTC)).

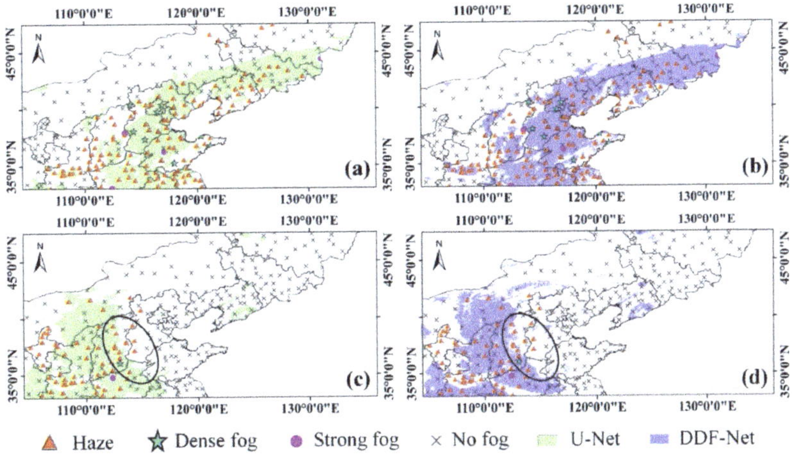

Figure 9. The fog detection results of U-Net and DDF-Net at 17:00 (18 November 2016 (**a**,**b**) and 11 December 2016 (**c**,**d**)). The triangle, star, circle, and cross symbols represent haze, dense fog, strong fog, and non-fog stations, respectively. The green area is the fog detected by U-Net. The blue area is the fog detected by DDF-Net. All the times are at the Beijing Time (8 h later than the Universal Time Coordinated (UTC)).

5. Conclusions

In this study, we developed a novel algorithm for fog detection at dawn and dusk, which is based on the U-Net model and incorporates a channel attention mechanism under-restriction. Several conclusions could be drawn after the execution and validation of the algorithm:

(1) The DDF-Net could detect large-scale fog and has a more refined identification of the edge of the fog area. Both the DDF-Net and U-net are challenged in detecting fog during the periods of time when the fog is dissipating and when there are clouds, while the FAR of DDF-Net is lower than the latter.

(2) The fog near the terminator line can be effectively separated from the surface, which is effective for identifying fog from the surface at dawn and dusk. Specially, the DDF-Net can effectively improve the misdetection of medium/low clouds near and far from the terminator line. The fog detection results are relatively stable with fewer false detections.

(3) The algorithm is simple and efficient, with little human intervention. As a result, the fog detection results can be obtained in near real time and are well suited to the requirements for real-time detection of dawn and dusk fog over large areas of land.

Overall, this study provides a new deep learning-based approach for fog detection at dawn and dusk. Our algorithm shows great advantages and potential for overcoming the limitations of minor spectral differences for fog detection at dawn and dusk, which would significantly improve the satellite application in climate forecast. The limitation refers to the low performance of fog formation and dissipation in mountainous areas with highly complex terrain. Two issues would be addressed in our future research. Firstly, other geographical parameters (i.e., latitude and surface temperature, etc.) would also affect the fog formation, which should be addressed in the algorithm to improve the accuracy of satellite fog detection. Secondly, a public fog dataset for each season would be developed and published for supporting the management of traffic and public health.

Author Contributions: Conceptualization, Y.R.; methodology, Y.R. and H.F.; software, Y.R.; validation, Y.R., H.M. and Z.L.; formal analysis, Y.R.; investigation, X.W. and Y.L.; resources, X.W.; data curation, X.W.; writing—original draft preparation, Y.R.; writing—review and editing, H.F. and H.M.; visualization, Y.R. and Z.L.; supervision, H.M. and H.F.; project administration, H.M.; funding acquisition, H.M. and H.F. All authors have read and agreed to the published version of the manuscript.

Funding: This research was funded by the National Natural Science Foundation of China, grant number 42071334 and 42071378.

Data Availability Statement: The Himawari-8 data supporting the findings of this study are openly available at https://www.eorc.jaxa.jp/ptree/ (accessed on 17 April 2020). The ground observation data can be archived at the following website https://data.cma.cn/ (accessed on 20 April 2020). The SRTM1-DEM data can be archived at the following website https://www.usgs.gov/ (accessed on 5 May 2020).

Acknowledgments: The authors are grateful to JMA for making the Himawari-8 data available. We are also grateful to the National Satellite Meteorological Center for providing access to the Meteorological station visibility data of our research.

Conflicts of Interest: The authors declare no conflict of interest.

References

1. World Meteorological Organization (WMO). *International Meteorological Vocabulary*; Secretariat of the WMO: Geneva, Switzerland, 1992; p. 782.
2. Ma, H.; Li, Y.; Wu, X.; Feng, H.; Ran, Y.; Jiang, B.; Wang, W. A large-region fog detection algorithm at dawn and dusk for high-frequency Himawari-8 satellite data. *Int. J. Remote Sens.* **2022**, *43*, 2620–2637. [CrossRef]
3. Nilo, S.T.; Romano, F.; Cermak, J.; Cimini, D.; Ricciardelli, E.; Cersosimo, A.; Di Paola, F.; Gallucci, D.; Gentile, S.; Geraldi, E. Fog detection based on meteosat second generation-spinning enhanced visible and infrared imager high resolution visible channel. *Remote Sens.* **2018**, *10*, 541. [CrossRef]
4. Kang, L.; Miao, Y.Y.; Miao, M.S.; Hu, Z.Y. The Harm of Fog and Thick Haze to the Body and the Prevention of Chinese Medicine. In Proceedings of the 2nd International Conference on Biomedical and Biological Engineering 2017 (BBE 2017), Guilin, China, 26–28 May 2017; Atlantis Press: Amsterdam, The Netherlands, 2017; pp. 119–124. [CrossRef]
5. Yoo, J.M.; Choo, G.H.; Lee, K.H.; Wu, D.L.; Yang, J.H.; Park, J.D.; Choi, Y.S.; Shin, D.B.; Jeong, J.H.; Yoo, J.M. Improved detection of low stratus and fog at dawn from dual geostationary (COMS and FY-2D) satellites. *Remote Sens. Environ.* **2018**, *211*, 292–306. [CrossRef]
6. Lee, J.R.; Chung, C.Y.; Ou, M.L. Fog detection using geostationary satellite data: Temporally continuous algorithm. *Asia-Pac. J. Atmos. Sci.* **2011**, *47*, 113–122. [CrossRef]
7. Hunt, G.E. Radiative properties of terrestrial clouds at visible and infra-red thermal window wavelengths. *Q. J. R. Meteorol. Soc.* **1973**, *99*, 346–369. [CrossRef]
8. Eyre, J. Detection of fog at night using Advanced Very High Resolution Radiometer (AVHRR) imagery. *Meteorol. Mag.* **1984**, *113*, 266–271.
9. Turner, J.; Allam, R.; Maine, D. A case-study of the detection of fog at night using channels 3 and 4 on the Advanced Very High-Resolution Radiometer (AVHRR). *Meteorol. Mag.* **1986**, *115*, 285–290.
10. Bendix, J. Fog climatology of the Po Valley. *Riv. Meteorol. Aeronaut.* **1994**, *54*, 25–36.
11. Bendix, J. A case study on the determination of fog optical depth and liquid water path using AVHRR data and relations to fog liquid water content and horizontal visibility. *Int. J. Remote Sens.* **1995**, *16*, 515–530. [CrossRef]
12. Cermak, J.; Bendix, J. Dynamical Nighttime Fog/Low Stratus Detection Based on Meteosat SEVIRI Data: A Feasibility Study. *Pure Appl. Geophys.* **2007**, *164*, 1179–1192. [CrossRef]
13. Cermak, J.; Bendix, J. A novel approach to fog/low stratus detection using Meteosat 8 data. *Atmos. Res.* **2008**, *87*, 279–292. [CrossRef]
14. Chaurasia, S.; Sathiyamoorthy, V.; Paul Shukla, B.; Simon, B.; Joshi, P.C.; Pal, P.K. Night time fog detection using MODIS data over Northern India. *Met Apps.* **2011**, *18*, 483–494. [CrossRef]
15. Weston, M.; Temimi, M. Application of a Nighttime Fog Detection Method Using SEVIRI Over an Arid Environment. *Remote Sens.* **2020**, *12*, 2281. [CrossRef]
16. Gurka, J.J. The Role of Inward Mixing in the Dissipation of Fog and Stratus. *Mon. Weather Rev.* **1978**, *106*, 1633–1635. [CrossRef]
17. Rao, P.K.; Holmes, S.J.; Anderson, R.K.; Winston, J.S.; Lehr, P.E. *Weather Satellites: Systems, Data, and Environmental Applications*; American Meteorological Society: Boston, MA, USA, 1990; p. 516.
18. Chaurasia, S.; Gohil, B.S. Detection of Day Time Fog Over India Using INSAT-3D Data. *IEEE J. Sel. Top. Appl. Earth Obs. Remote Sens.* **2015**, *8*, 4524–4530. [CrossRef]

19. Drönner, J.; Egli, S.; Thies, B.; Bendix, J.; Seeger, B. FFLSD—Fast Fog and Low Stratus Detection tool for large satellite time-series. *Comput. Geosci.* **2019**, *128*, 51–59. [CrossRef]
20. Yang, J.H.; Yoo, J.M.; Choi, Y.S. Advanced Dual-Satellite Method for Detection of Low Stratus and Fog near Japan at Dawn from FY-4A and Himawari-8. *Remote Sens.* **2021**, *13*, 1042. [CrossRef]
21. Hůnová, I.; Brabec, M.; Malý, M.; Dumitrescu, A.; Geletič, J. Terrain and its effects on fog occurrence. *Sci. Total Environ.* **2021**, *768*, 144359. [CrossRef]
22. Shang, H.; Chen, L.; Letu, H.; Zhao, M.; Li, S.; Bao, S. Development of a daytime cloud and haze detection algorithm for Himawari-8 satellite measurements over central and eastern China. *J. Geophys. Res. Atmos.* **2017**, *122*, 3528–3543. [CrossRef]
23. LeCun, Y.; Bengio, Y.; Hinton, G. Deep learning. *Nature* **2015**, *521*, 436–444. [CrossRef]
24. Jeppesen, J.H.; Jacobsen, R.H.; Inceoglu, F.; Toftegaard, T.S. A cloud detection algorithm for satellite imagery based on deep learning. *Remote Sens. Environ.* **2019**, *229*, 247–259. [CrossRef]
25. Wu, X.; Shi, Z.; Zou, Z. A geographic information-driven method and a new large scale dataset for remote sensing cloud/snow detection. *Isprs. J. Photogramm.* **2021**, *174*, 87–104. [CrossRef]
26. Hu, J.; Shen, L.; Sun, G. Squeeze-and-excitation networks. In Proceedings of the IEEE Conference on Computer Vision and Pattern recognition, Salt Lake City, UT, USA, 18–23 June 2018; pp. 7132–7141. [CrossRef]
27. Kingma, D.P.; Ba, J. Adam: A method for stochastic optimization. *arXiv* **2014**, arXiv:1412.6980. [CrossRef]
28. Ding, Y.; Liu, Y. Analysis of long-term variations of fog and haze in China in recent 50 years and their relations with atmospheric humidity. *Sci. China Earth Sci.* **2014**, *57*, 36–46. [CrossRef]
29. Chen, H.; Wang, H. Haze Days in North China and the associated atmospheric circulations based on daily visibility data from 1960 to 2012. *J. Geophys. Res. Atmos.* **2015**, *120*, 5895–5909. [CrossRef]
30. Bessho, K.; Date, K.; Hayashi, M.; Ikeda, A.; Imai, T.; Inoue, H.; Kumagai, Y.; Miyakawa, T.; Murata, H.; Ohno, T. An introduction to Himawari-8/9—Japan's new-generation geostationary meteorological satellites. *J. Meteorol. Soc. Jpn.* **2016**, *94*, 151–183. [CrossRef]
31. Ishida, H.; Nakajima, T.Y. Development of an unbiased cloud detection algorithm for a spaceborne multispectral imager. *J. Geophys. Res.* **2009**, *114*. [CrossRef]
32. Knudby, A.; Latifovic, R.; Pouliot, D. A cloud detection algorithm for AATSR data, optimized for daytime observations in Canada. *Remote Sens. Environ.* **2011**, *115*, 3153–3164. [CrossRef]
33. Cermak, J.; Bendix, J. Detecting ground fog from space—A microphysics-based approach. *Int. J. Remote Sens.* **2011**, *32*, 3345–3371. [CrossRef]
34. *GB/T 27964-2011GB/T*; Grade of Fog Forecast. China Meteorological Administration (CMA): Beijing, China, 2011. (In Chinese)
35. Zeiler, M.; Fergus, R. Visualizing and understanding convolutional networks. In Proceedings of the European Conference on Computer Vision, Zurich, Switzerland, 6–12 September 2014; pp. 818–833.
36. Rumelhart, D.E.; Hinton, G.E.; Williams, R.J. Learning representations by back-propagating errors. *Nature* **1986**, *323*, 533–536. [CrossRef]
37. Ronneberger, O.; Fischer, P.; Brox, T. U-Net: Convolutional Networks for Biomedical Image Segmentation. In Proceedings of the Medical Image Computing and Computer-Assisted Intervention—MICCAI 2015, Munich, Germany, 5–9 October 2015; Springer International Publishing: Cham, Switzerland, 2015; pp. 234–241.
38. Shao, Z.; Pan, Y.; Diao, C.; Cai, J. Cloud Detection in Remote Sensing Images Based on Multiscale Features-Convolutional Neural Network. *IEEE Trans. Geosci. Remote* **2019**, *57*, 4062–4076. [CrossRef]
39. Mohajerani, S.; Krammer, T.A.; Saeedi, P. Cloud detection algorithm for remote sensing images using fully convolutional neural networks. *arXiv* **2018**, arXiv:1810.05782. [CrossRef]
40. Ioffe, S.; Szegedy, C. Batch Normalization: Accelerating Deep Network Training by Reducing Internal Covariate Shift. In Proceedings of the 32nd International Conference on Machine Learning, Lille, France, 6–11 July 2015; Francis, B., David, B., Eds.; PMLR: San Diego, CA, USA, 2015; pp. 448–456.

remote sensing

Article

Projection of Future Extreme Precipitation in China Based on the CMIP6 from a Machine Learning Perspective

Yilin Yan [1], Hao Wang [1,2,*], Guoping Li [3,4], Jin Xia [5], Fei Ge [4], Qiangyu Zeng [1], Xinyue Ren [1] and Linyin Tan [1]

1 College of Atmospheric Sounding, Chengdu University of Information Technology, Chengdu 610225, China
2 Key Laboratory of Atmosphere Sounding, China Meteorological Administration, Chengdu 610225, China
3 Collaborative Innovation Center on Forecast and Evaluation of Meteorological Disasters, Nanjing 210044, China
4 School of Atmospheric Sciences/Plateau Atmosphere and Environment Key Laboratory of Sichuan Province/Joint Laboratory of Climate and Environment Change, Chengdu University of Information Technology, Chengdu 610225, China
5 Shiyan City Meteorological Bureau, Shiyan 442000, China
* Correspondence: wh@cuit.edu.cn; Tel.: +86-028-8596-6501

Abstract: In recent years, China has suffered from frequent extreme precipitation events, and predicting their future trends has become an essential part of the current research on this issue. Because of the inevitable uncertainties associated with individual models for climate prediction, this study uses a machine learning approach to integrate and fit multiple models. The results show that the use of several evaluation metrics provides better results than the traditional ensemble median method. The correlation coefficients with the actual observations were found to improve from about 0.8 to 0.9, while the correlation coefficients of the precipitation amount (PRCPTOT), very heavy precipitation days (R20mm), and extreme precipitation intensity (SDII95) reached 0.95. Based on this, the precipitation simulations of moderate forced scenario for sharing socio-economic path (SSP2-4.5) from 27 coupled models in the Coupled Model Intercomparison Project Phase 6 (CMIP6) were used to explore potential changes in future extreme precipitation events in China and to calculate the distribution and trends of the PRCPTOT, extreme precipitation amount (R95pTOT), maximum consecutive 5-day precipitation (Rx5day), precipitation intensity (SDII), SDII95, and R20mm for the early 21st century (2023–2050), mid-21st century (2051–2075), and late 21st century (2076–2100), respectively. The results showed that the most significant increase in extreme precipitation indices is expected to occur by the end of the century, with the R95pTOT, Rx5day, and SDII95 increasing by 13.73%, 9.43%, and 9.34%, respectively, from the base period. The remaining three precipitation indexes, the PRCPTOT, SDII, and R20mm, also showed increases of 8.77%, 6.84%, and 4.02%, respectively. Additionally, there were apparent differences in the spatial variation of extreme precipitation. There were significant increasing trends of extreme precipitation indexes in central China and northeast China in the three periods, among which the total annual precipitation showed an increasing trend in central and northern China and a decreasing trend in western and south China. An increasing trend of annual precipitation intensity was found to be mainly concentrated in central China and south China, and the annual precipitation frequency showed a larger increasing trend at the beginning of this century. The annual precipitation frequency showed an increasing trend in the early part of this century. In general, all the indices showed an overall increasing trend in the future period, with the PRCPTOT, Rx5day, and SDII95 showing the most significant overall increasing trends.

Keywords: machine learning; extreme precipitation indices; climate change; CMIP6; China

Citation: Yan, Y.; Wang, H.; Li, G.; Xia, J.; Ge, F.; Zeng, Q.; Ren, X.; Tan, L. Projection of Future Extreme Precipitation in China Based on the CMIP6 from a Machine Learning Perspective. *Remote Sens.* **2022**, *14*, 4033. https://doi.org/10.3390/rs14164033

Academic Editor: Xander Wang

Received: 14 July 2022
Accepted: 14 August 2022
Published: 18 August 2022

1. Introduction

In recent years, the impacts of extreme weather events such as continuous heat waves, extreme precipitation events and droughts [1,2] have become more prominent and more widespread [3]. Extreme precipitation, in addition to its severe impact and related secondary disasters such as floods, landslides, debris flows, and other natural disasters [4], also brings a great challenge to people's lives as well as to economic development [5,6]. The Sixth Assessment Report of the Intergovernmental Panel on Climate Change (IPCC6), led by the United Nations, notes that continued warming has intensified water cycle processes in the context of widespread global warming, resulting in an increase in atmospheric water content, which, in turn, will lead to enhanced precipitation and the chance of intense precipitation events over most of the globe [7,8]. The scale and frequency of extreme precipitation events will be larger and more frequent in the future as temperatures continue to rise, raising concerns about the changing features of extreme precipitation events [9,10].

China has diverse climatic conditions and complex geography, and is subject to a perennial monsoon climate with frequent extreme precipitation events [11,12]. From the late 1970s to the present, economic losses due to extreme precipitation events in Chinese regions have increased nearly tenfold. Historically, there have also been multiple floods caused by extreme precipitation, such as the 1998 Yangtze River basin-wide mega-flood caused by a multi-day extreme precipitation event [13,14]; in 2012, Beijing was hit by the heaviest rainfall in nearly 61 years, resulting in massive property damage and casualties; the August 2020 Sichuan flood that was a rare event in China's climate history; and the 2012 Sichuan flood that resulted in the loss of life. However, current studies on extreme precipitation have focused more on small regional scales, such as precipitation distribution in urban and watershed areas [15] and trends or changes in extreme precipitation in typical areas [16,17], while there are fewer studies on extreme precipitation at large regional scales.

Since the 1990s, the Coupled Model Intercomparison Project (CMIP) has promoted the simulation of hundreds of climate models from dozens of institutes around the world so as to improve the simulation capability of climate models and enhance a more comprehensive scientific knowledge of weather processes. Most past work on the assessment and prediction of temperature and precipitation in China through CMIP Phase 5 (CMIP5) [18]. Jiang et al. [19] analyzed extreme precipitation using the CMIP5. Simulations show moisture bias in northwestern China as well as in the north, accompanied by a dry tendency in southeastern China. Future projections for the CMIP5 show that the increase in extreme global precipitation may be faster than the increase in total precipitation on rainy days [20]. In addition, China is expected to experience more frequent regional heavy precipitation events while being more destructive [19,21]. With the improvement of the model's simulation performance, the current research focus has gradually shifted to research on the new-generation CMIP6 model [22]. The experimental output of the CMIP6 model contains parameterized settings for more complex dynamic climate processes [23] and more different emission scenarios designed to improve prediction accuracy [24]. Chinese scholars have paid more attention to its performance in China and the new CMIP6 scenarios (i.e., Shared Socio-Economic Pathways, SSPs) regarding how China's climate will change. Recent studies have also begun using CMIP6 data to evaluate and predict the East Asian monsoon climate [25]. They have also shown that the CMIP6 model is compatible with the CMIP5 model. Models simulating precipitation and climate extremes have generally improved [26,27], and there are more models and larger simulation ensembles.

Current predictions of China's climate mainly utilize multi-model ensemble approaches, with some studies exploring the best individual models and using the multi-model ensemble medians approaches [14], and, although these approaches provide some improvements to prediction accuracy, they are still affected by the limitations of individual models. The fields of artificial intelligence (AI) and machine learning (ML) are currently evolving, allowing meteorological researchers to integrate machine learning knowledge into related fields. The biggest advantage of machine learning (ML) methods is that more useful information can be extracted from multi-model data to achieve a closer match with real observations. Numerous studies have shown that machine learning can be better used in weather forecasting, long-term future climate prediction, and supplementing missing climate elements, and significant progress has been made [28]. Since machine learning has greater advantages in solving nonlinear and high-dimensional problems, it allows for the extraction of dynamics and physical processes present in the climate, among other reliable information [28,29].

Based on this, we used ML to integrate the simulation results of multiple models in the next-generation Coupled Model Intercomparison Project Phase 6 (CMIP6) while establishing the nonlinear relationship between them with the objective of real observational data. To better represent the integration effect, we compared ML with the ensemble median model to explore whether ML can effectively improve the ability to simulate observed data and further establish reliable future climate prediction results based on it. The remainder of this study is organized as follows: Section 2 introduces the research area, research data, evaluation indices, and research methods. Section 3 presents the main findings of current and future climate prediction methodology assessments. Finally, Section 4 summarizes and discusses the spatial distribution and temporal variation trend of extreme precipitation in China in the future.

2. Datasets and Methodology

2.1. Study Area

China is located in eastern Asia, with significant differences in winter and summer precipitation, most of which is concentrated in summer [30,31]. China's geographical environment is changeable with regional differences in topography. Due to the role of the East Asian monsoon, the eastern summer rainfall is sufficient. Northwest China is mainly influenced by a westerly, dry climate. Due to the vast size of China, its complex topography and strong monsoon characteristics [32], and the existence of many different climate types, precipitation varies among regions, decreasing from southeast to northwest [33].

In order to further study the future trends of precipitation and to quantify the regional differences, in this study, we divided the Chinese regions into eight climatic zones [34], specifically, the western arid (semiarid) zone (A), the Qinghai–Tibet Plateau (B), the eastern arid zone (C), southwest China (D), northeast China (E), north China (F), central China (G), and south China (H). The specific divisions are shown in Figure 1.

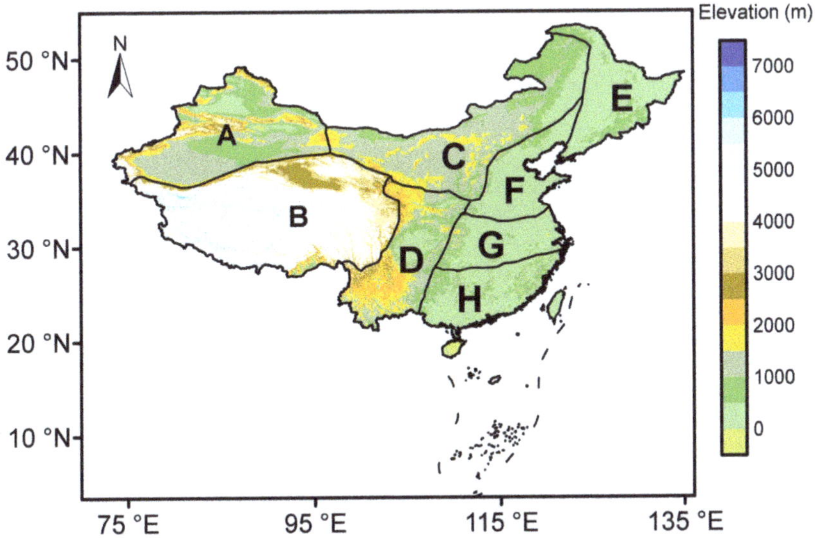

Figure 1. Research area and climate zoning. Among them (A) western arid (semiarid) zone, (B) Qinghai–Tibet Plateau, (C) eastern arid zone, (D) southwest China, (E) northeast China, (F) north China, (G) central China, (H) south China.

2.2. Data

The daily precipitation observations CN05.1 used in the study are from the grid point dataset provided by the China Meteorological Administration. This dataset was obtained by Wu et al. [35] using 2416 national meteorological stations from 1961 to 2014 in the National Meteorological Information Center, and the daily observation data of precipitation with a spatial resolution of $0.25° \times 0.25°$ were obtained by interpolation and superposition of a thin-disk spline function (ANUSPLIN) and the angular distance weighting method (ADW), respectively. Additionally, these data have been used extensively in numerous climate change studies in China [36,37]. The dataset was used as the "true value" for training and verification to verify and evaluate the established model.

We used daily precipitation simulations from 27 CMIP6 global climate models. The numerical simulation experiments available for study in CMIP6 consist of two main parts: (1) historical climate simulation experiments (Historica); and (2) 23 Model Comparison Sub-Plans (MIPs) for future prognosis. Both provide a variety of meteorological variables including precipitation, where the former is based on a large number of observations, including historical climate simulation experiments driven by external forcing of data from ground-based observations and remote sensing observations, and also serves as a reference benchmark experiment, mainly to assess the model's ability to simulate climate change. Secondly, among the many MIPs, we selected the Scenario Model Comparison Program (ScenarioMIP), the most recently included in CMIP6, whose experimental results were used as the initial data for the future prediction part of this study. The study uses historical simulation data from 1961–2014 and future projection data for 2023–2100 from SSP2-4.5 in the Shared Socioeconomic Pathway. SSP2-4.5 was used as the updated RCP4.5 scenario representing the medium socio-economic development path and the medium forcing scenario, which simulates medium land use and aerosol paths [38,39], which can reflect the development scenario under normal development conditions. Before the analysis, all the data were unified to a $1° \times 1°$ resolution using a bilinear interpolation scheme [19].

2.3. Methods

2.3.1. Climate Indices

The six precipitation indices used in the study were defined by the Expert Team on Climate Change Detection and Indices (ETCCDI, Table 1) [40]. They were used to quantify the characteristics of future precipitation characteristics [33,41]. The metrics included the following precipitation indices: total precipitation on rainy days (daily precipitation >1 mm) (PRCPTOT), annual extreme precipitation total above the 95th percentile threshold (R95pTOT), and 5-day maximum precipitation (Rx5day); and the following precipitation intensity indices: precipitation intensity (SDII), extreme precipitation intensity above the 95th percentile threshold (SDII95), and daily precipitation >20 mm of heavy rainfall days (R20 mm). These indices are widely used to identify and monitor climate extremes [42]. In this study, we first calculated the indices of all models and observations and then performed machine learning multi-model integration and calculated the median of the multi-model ensemble on this basis.

Table 1. Precipitation indices and definitions (recommended by the ETCCDI).

Index	Description	Definition	Unit
PRCPTOT	Precipitation amount	Annual total precipitation of daily precipitation >1 mm	mm
R95pTOT	Extreme precipitation amount	Annual total precipitation with daily precipitation >95% threshold	mm
Rx5day	Maximum consecutive 5-day precipitation	Annual maximum consecutive 5-day precipitation	mm
SDII	Precipitation intensity	Precipitation intensity of days with daily precipitation >1 mm	mm/day
SDII95	Extreme precipitation intensity	Precipitation intensity of days with daily precipitation >95% threshold of the entire time series for the grid point	mm/day

Table 1. *Cont.*

Index	Description	Definition	Unit
R20mm	Very heavy precipitation days	Days of annual daily precipitation >20 mm	day

2.3.2. Artificial Neural Networks

The back propagation (BP) learning algorithm is a multilayer feedforward network, an artificial neural network algorithm based on error back propagation with forwarding multilayer feedback. Using the gradient descent method, the error squared minima of the actual output value and the target value were searched for so that the threshold value of the network as well as the connection weights could be adjusted without interruption so as to achieve the desired output for each set of inputs in the adjusted network model [43].

The operation of the BP neural network consists of three main parts, namely, the forward propagation process from the input layer pathway hidden layer to the output layer; the reverse correction process from the output layer to the input layer is determined by the error between the predicted output of the network before the actual observed data; and the training process in which the forward and backward processes are alternated. The BP neural network topology includes an input layer, a hidden layer, and an output layer; $x_1, x_2, \ldots, x_m$ are the input data; $y_1, y_2, \ldots, y_n$ are the output data; and w is the network weights, as shown in Figure 2.

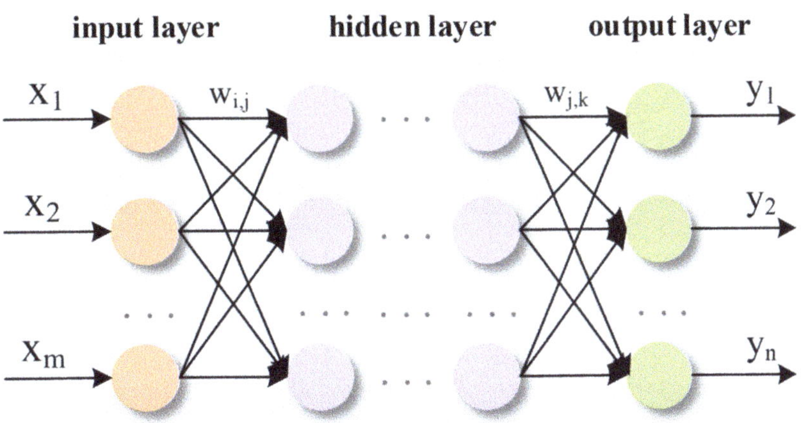

Figure 2. BP neural network topology diagram.

2.3.3. Multi-Model Integrated

The overall flow chart of the multi-model integration processing method is shown in Figure 3. The model data and observed data for the historical period (1961–2014) were divided into two parts: the training set representing 1961–1998 (38 years) and the test set representing 1999–2014 (16 years). Among these, the test set mainly served as an assessment of whether the results of the evaluation after multi-model integration were closer to the observed data. In this study, we integrated six indices separately with the overall goal of integrating the advantages of the individual models' simulation capabilities in different regions to the greatest extent possible so as to obtain the best simulation results for the actual observed data in the Chinese regions. The scheme was eventually applied to the prediction of future periods as a way to achieve the best precipitation forecast. A large number of previous works have used the multi-model ensemble median for future precipitation predictions [26,44]. The CMIP6 multi-model ensemble median method has generally improved the simulation of precipitation and precipitation extremes trends on a global scale compared to its predecessor (CMIP5) [27], so it was used as a reference for the machine learning fitting results.

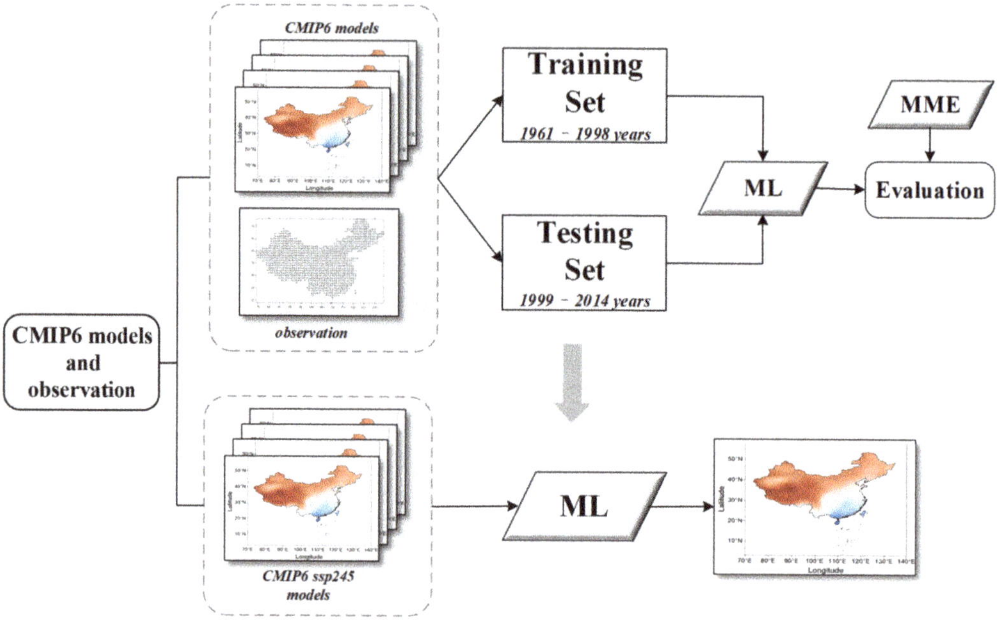

Figure 3. ML model flow chart.

2.3.4. Evaluation Method

In this study, the Taylor diagram was used to evaluate the simulation results of 27 climate models, the multi-model ensemble median, and the integrated multi-model ML in CMIP6; the spatial distribution of each index of model simulation results before and after processing can be compared more intuitively with the consistency of the observed data. Taylor diagrams can integrate multiple evaluation metrics for presentation at the same time, indicating the accuracy of the model matching, the observed data in terms of standard deviation, central root mean squared error (RMSE), and the correlation coefficient [45].

The standard deviation of the observation data and the model is calculated as follows:

$$\sigma_{obs} = \sqrt{\frac{1}{N}\sum_{n=1}^{N}\left(X_{obsn} - \overline{X_{obs}}\right)^2} \tag{1}$$

$$\sigma_{model} = \sqrt{\frac{1}{N}\sum_{n=1}^{N}\left(X_n - \overline{X}\right)^2} \tag{2}$$

Among them, $\overline{X_{obs}}$ and $\overline{X}$ respectively is the average of the observed data and model data.

The central root mean square error, *RMSE*, is defined as:

$$RMSE = \sqrt{\frac{1}{N}\sum_{n=1}^{N}\left[\left(X_n - \overline{X}\right) - \left(X_{obsn} - \overline{X_{obs}}\right)\right]^2} \tag{3}$$

The correlation coefficient r of the observed data and the model data is defined as:

$$r = [\frac{1}{N}\sum_{n=1}^{N}(X_n - \overline{X}) - (X_{obsn} - \overline{X_{obs}})]/(\sigma_{model}\sigma_{obs}) \tag{4}$$

The key to the Taylor diagram is that the standard deviation, correlation coefficient, and central root mean square error of the simulated and observed data satisfy the following relationships:

$$RMSE^2 = \sigma_{obs}{}^2 + \sigma_{model}{}^2 - 2\sigma_{obs}\sigma_{model}r \tag{5}$$

Because of the large uncertainty in future precipitation predictions, We also applied the relative root mean square error ($RMSE'$) for a more accurate quantitative assessment of individual models, integrated models, and median multi-model integration to re-validate the correlation with the observed data results [46,47].

The $RMSE'$ for the assessment of the climate simulation capabilities of the CMIP6 models is defined as follows [48]: first, the RMSE of each model relative to the observed data needs to be calculated as in Equation (6), and, second, using the calculated $RMSE$ for each mode in turn, the $RMSE'$ for each mode is calculated as follows:

$$RMSE' = \frac{\text{RMSE} - \text{RMSE}_{Median}}{\text{RMSE}_{Median}} \tag{6}$$

where RMSE_{Median} is the set median of the RMSE of all the models. In general, a negative (positive) value of $RMSE'$ is better (worse) than half (50%) of all model results for the model.

3. Results and Discussion

3.1. Evaluations

We obtained the PRCPTOT, R95pTOT, Rx5day, SDII, SDII95, and R20mm for each grid point by calculating the actual observations, 27 pattern data, the multi-model ensemble median data, and the ML output data for the validation of the dataset (1999–2014), respectively. On this basis, the central root mean square error (RMSE) and the ratio of the spatial correlation coefficient (R) to the standard deviation (STD) of the above and the observed values were calculated, and the Taylor diagram for a summary of the statistics was obtained. The results are shown in Figure 4. According to the calculation principles of the three indexes, the higher the spatial correlation coefficient, the closer the standard deviation ratio to 1, and the closer the central root mean square error to 0, the better the data simulation effect (the smaller the distance between the simulated data points and observation points, as expressed in the Taylor diagram). It can be seen from the Taylor diagram of each index that for 27 different models, there are great differences between them. For the four precipitation indices, PRCPTOT, R95pTOT, and Rx5day, most of the patterns had correlation coefficients between 0.6 and 0.85, while the RMSE values was mostly between 0.5 and 1.0. The correlation coefficient of the precipitation frequency index R20mm was between 0.3 and 0.85, while the RMSE values were between 0.75 and 1.5. The two intensity indices, SDII and SDII95, performed better with correlation coefficients between 0.7 and 0.9 and the RMSE values between 0.5 and 1.0. In contrast, the effect of the multi-model ensemble medians on all the indices was slightly better than the 27-model data. The ML performed much better than the 27 patterns and the multi-model ensemble medians, which also reflects the ability of machine learning methods to better integrate the advantages of different models. A further analysis showed that the correlation coefficients of the PRCPTOT, R20rmm, and SDII95 all reached about 0.95, while the R95pTOT Rx5day and SDII were slightly different, with correlation coefficients of 0.86, 0.85, and 0.92. The central root mean square errors were all around 0.5, and the final standard deviation ratio was between 0.75 and 1.0 for all the evaluations. The best performance was obtained for the PRCPTOT, R20mm, and SDII95. It was found that ML has the smallest distance between each index and the observed values, indicating that the simulation effect is closest to the observed values.

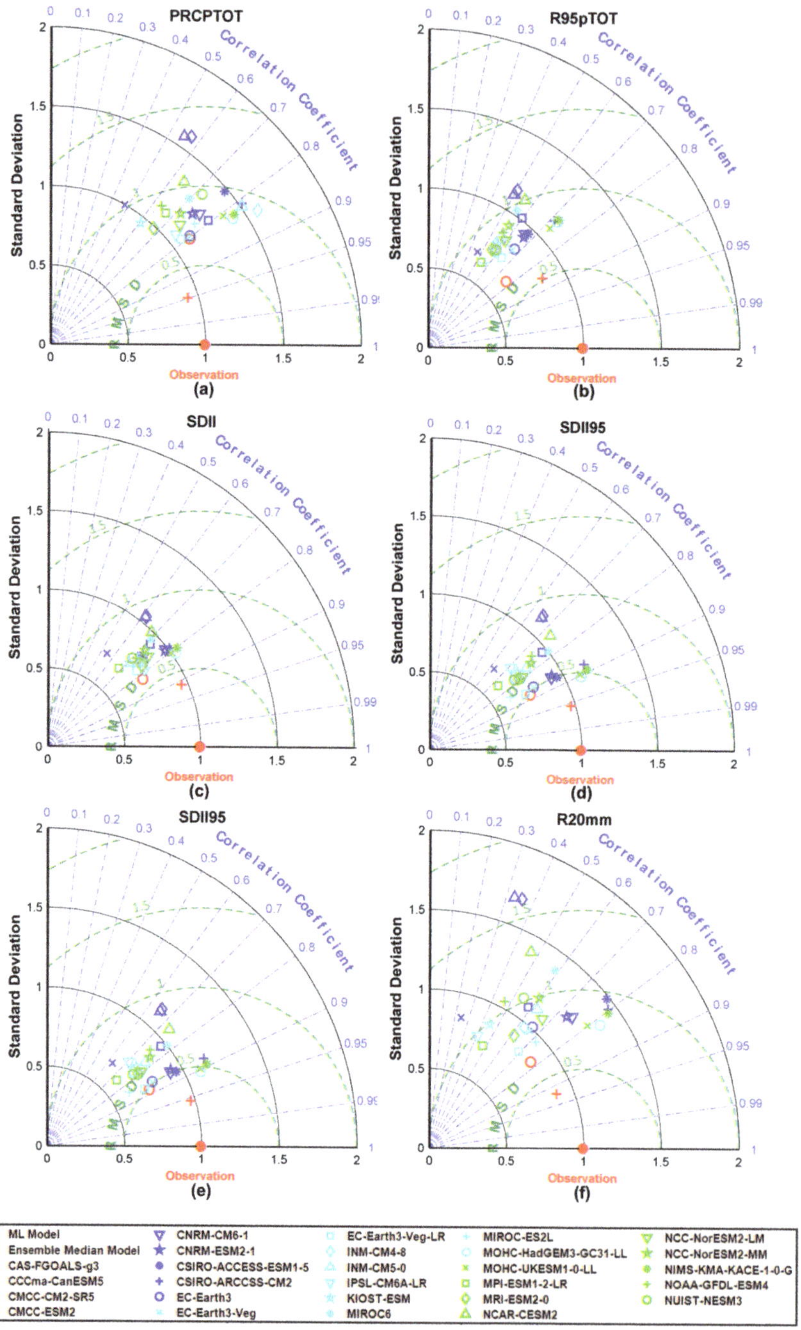

Figure 4. Taylor diagram of PRCPTOT (**a**), R95pTOT (**b**), Rx5day (**c**), SDII (**d**), SDII95 (**e**), and R20mm (**f**). In the figure, the circle's angle corresponds to the spatial correlation coefficient; the black arc represents the standard deviation ratio; the green arc represents the central root mean square error; and the purple dot is the observation representing the observed data.

Figure 5 visualizes the distribution of the *RMSE′* magnitude for each model simulating extreme precipitation indices as compared to the real observations. A larger value indicates a worse simulation performance, while a lower value indicates a better simulation performance.

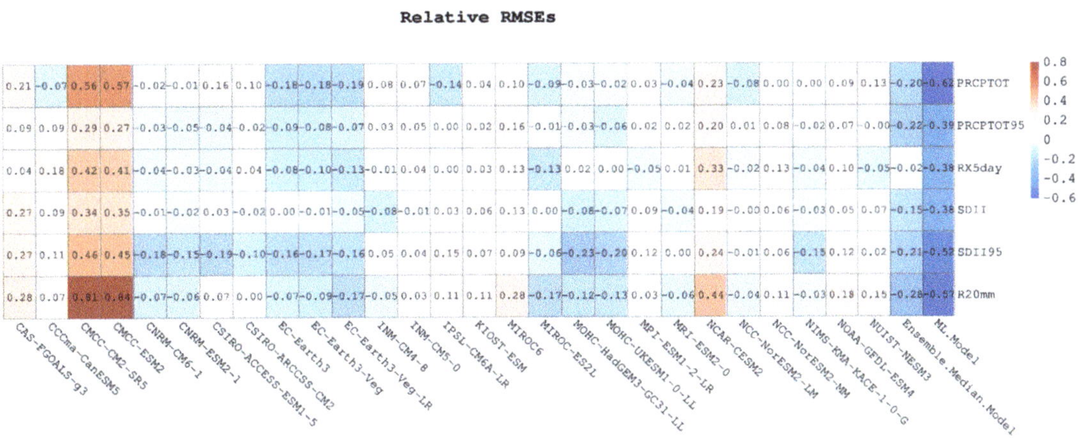

Figure 5. Heat map of the relative RMSE (RMSE′) of the precipitation extremes index simulated by the CMIP6 model versus observed data for the period 1999–2014. The colors range from red to blue. The deeper the blue, the better the simulation ability. The deeper the red, the worse the simulation ability.

The results show that the ability of each model to simulate each index of precipitation varies greatly, among which EC-Earth3, EC-Earth3-Veg-LR, and EC-Earth3-Veg performed better, and the RMSE′ of the different indices was mainly negative, indicating that the three models can simulate the study area well. CAS-FGOALS-g3, CMCC-ESM2, and CMCC-CM2-SR5 performed poorly, while EC-Earth3, EC-Earth3-Veg-LR, and EC-Earth3-Veg were found to be able to better simulate the interannual variability of extreme precipitation events in the Asian region [49]. Since the models exhibited large uncertainties in different metrics, this also reaffirms the importance of model integration through machine learning when studying climate simulations and predictions. The last two columns of Figure 5 show the ensemble median approach and the machine learning model fit, respectively, and it can be analyzed that, for each precipitation index, the integrated median multi-model approach performs better overall than the results of the 27 models simulated individually, but the ML in the last column was even more comprehensive, far outperforming all the models including the multi-model ensemble median approach according to the six extreme precipitation metrics. The results also showed that for all indices, the RMSE′ of ML was the minimum, and the results were better than those of any individual model, which largely eliminates the uncertainty of the structural model, and thus the model can be considered to be reasonably well established for predictive simulations of future climate change.

Based on the above evaluation results, that is, the performance of ML fitting being significantly better than that of a single model, we finally decided to evaluate the spatial simulation capability of ML for each of the six precipitation indices and to better understand the regional climate bias of the model. Figure 6 shows the deviation and relative deviation distribution of the ML fitted values relative to the observed precipitation indices, where a relative deviation greater than 100% indicates a multiplicative overestimation of the ML model mean for the simulated values of the indices. From the figure, we can see that the simulations of the six precipitation indices by the future climate model were underestimated in most areas of central and southern China and the Qinghai–Tibet Plateau, and overestimated in some areas of the western and eastern arid regions, which may be due to the limitations of the climate model itself and its simulation of complex terrain [50]. Some scholars have previously found that the CMIP6 model output for western China had a low agreement for observed precipitation [51], and the use of ML is thus effective in improving the error in this part of the region. Specifically from the spatial point of view, ML was found to have over 50% overestimation for the three extreme precipitation indices in southern Xinjiang and parts of Inner Mongolia; 10%–40% underestimation in parts of Tibet, southwest, central, and southern China; and over 40% underestimation in parts of Qinghai and Xinjiang. ML is relatively good for the simulation of the two extreme precipitation frequency indexes; for SDII especially, the relative deviation was found to not be more than 40%, and for SDII95, a deviation of more than 40% was found to be mainly distributed in parts of Xinjiang. ML overestimated the frequency index R20mm by more than 40% in Xinjiang, Tibet, Qinghai, and parts of Inner Mongolia, and underestimated by more than 40% in Xinjiang and Tibet. In general, ML simulation is relatively accurate for most parts of China, and the biggest deviation was mainly around Xinjiang and the Qinghai–Tibet Plateau, where the resolution of the climate model was relatively low, and it was difficult to completely reproduce the local circulation of complex terrain. In addition, these regions usually lack climate stations, including gridded precipitation data obtained from climate stations, making it hard to accurately simulate precipitation features in complex terrain [52]. Secondly, the simulated precipitation was only slightly underestimated (<10%) in the vast majority of South China compared to the observed data, and comparable results were obtained by Sun et al. [53]; it is also consistent with the underestimation of extreme precipitation in southern China [54]. However, the overall distribution was approximately the same as the previously assessed distribution compared to the multi-model ensemble median used in previous studies, but the simulation effect was greatly improved [55]. The results of this study also differ from some previous studies, and part of the reason for the difference may be that this study is a comparison of climate model ML with gridded precipitation data, rather than the station data typically used in the past. Most regions in China were found to have better performance with the ML model. The uncertainty in northwest China was the smallest. The uncertainty in northeast China was greater than that in northwest China, and the uncertainty in south China was small. The uncertainty in northwest China was low, likely because it is a particularly arid region.

3.2. Projected Changes

3.2.1. Future Changes in Spatial Distribution and Boxplot

Due to the long time projection range of the model by which to facilitate analysis, we divided the future projection time into three periods starting from 2023: 2023–2050 (early 21st century), 2051–2075 (mid-21st century), and 2076–2100 (late 21st century), while the precipitation observation data from 1990 to 2014 were used as the reference period to facilitate the analysis of the distribution of the six precipitation indices in eight climate zones of China in different future periods compared with the base period and the change trends.

As shown in Figures 7 and 8, the spatial distribution of the relative changes of the three precipitation indices (PRCPTOT, R95pTOT, and Rx5day) and the box line plot of the absolute changes under the SSP2-4.5 scenario are shown. It can be seen that all precipitation, in general, showed an increasing trend in this century, and each precipitation index also increased with time. Relative to the base period, the changes in the three indices are more dramatic at the end of the 21st century, with the change rates of 5.69%, 7.47%, and 7.89%, respectively. Therefore, in the context of future climate change, precipitation in China will not increase significantly. Among these findings, 30.85% to 37.68% of the study area is expected to experience a significant increase (>15%) in R95pTOT by the end of the century, mainly in northern China and the southern part of the western arid zone, while the maximum rate of change will reach 40% in some parts of northern China. The largest absolute changes also occur in north China, with weaker sub-increases in northeast China and southwest China. For example, the absolute change in R95pTOT is expected to increase by an average of about 100 mm in north China by the end of the century, while the increase in the southern part of the western arid zone is expected to be about 1 mm. However, the relative rates of change are inversely distributed, with larger percentages of change occurring in the southern part of the western arid zone as well as in the eastern arid zone. This is because the total precipitation in arid and semi-arid regions is quite low, and small changes in precipitation can cause large differences in these regions [56]. In contrast, in central China and southern China, even though there is abundant rainfall and large absolute variability in some areas, its relative variability is not significant. In general, the box line plots of the absolute variation and relative variation distribution of PRCPTOT, R95pTOT, and Rx5day are generally similar, while the relative variation rates of PRCPTOT and Rx5day are more significant in south China than those of R95pTOT.

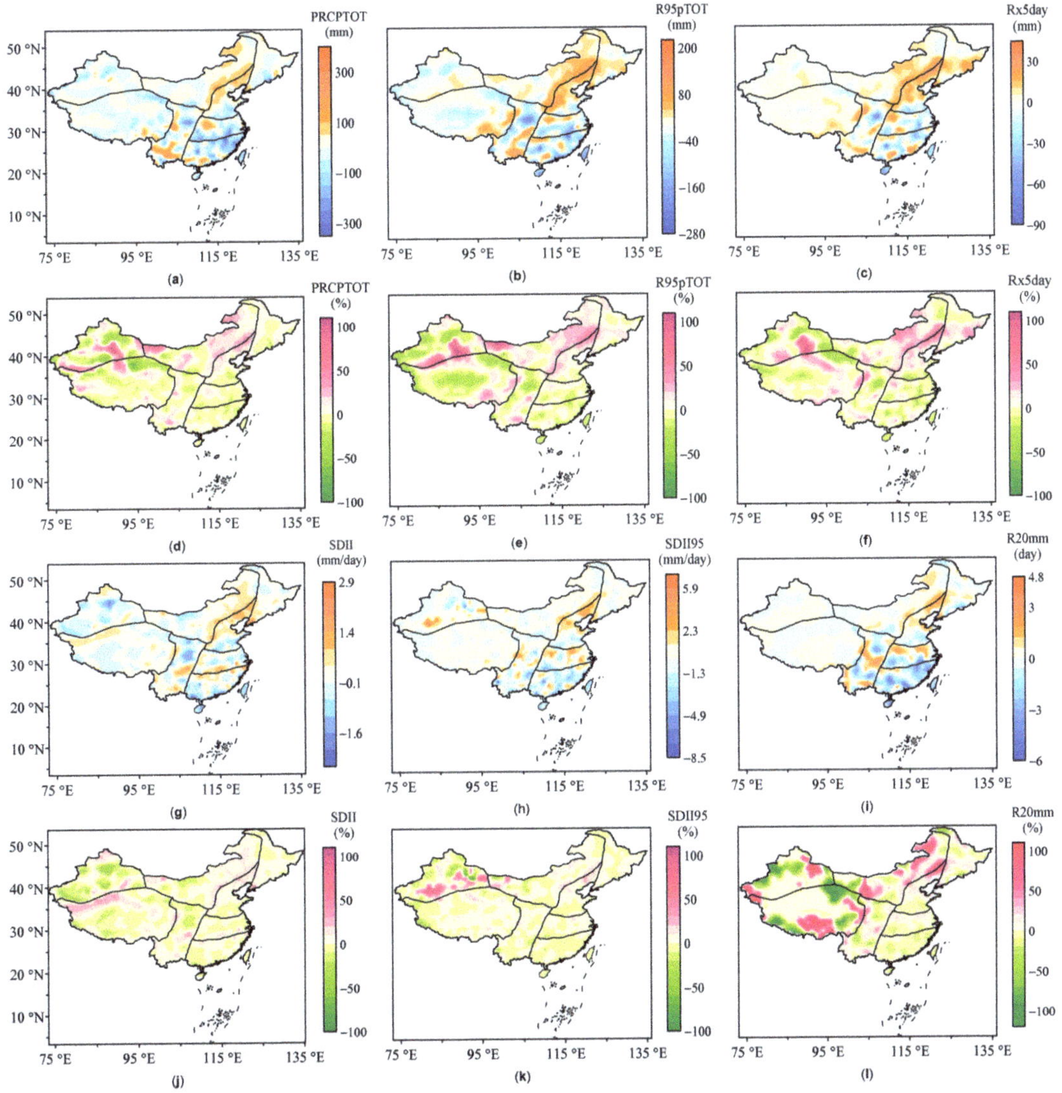

Figure 6. Absolute and relative deviations (%) of ML from observations for PRCPTOT (**a**,**d**), R95pTOT (**b**,**e**), Rx5day (**c**,**f**), SDII (**g**,**j**), SDII95 (**h**,**k**), and R20mm (**i**,**l**) for the period 1999–2014.

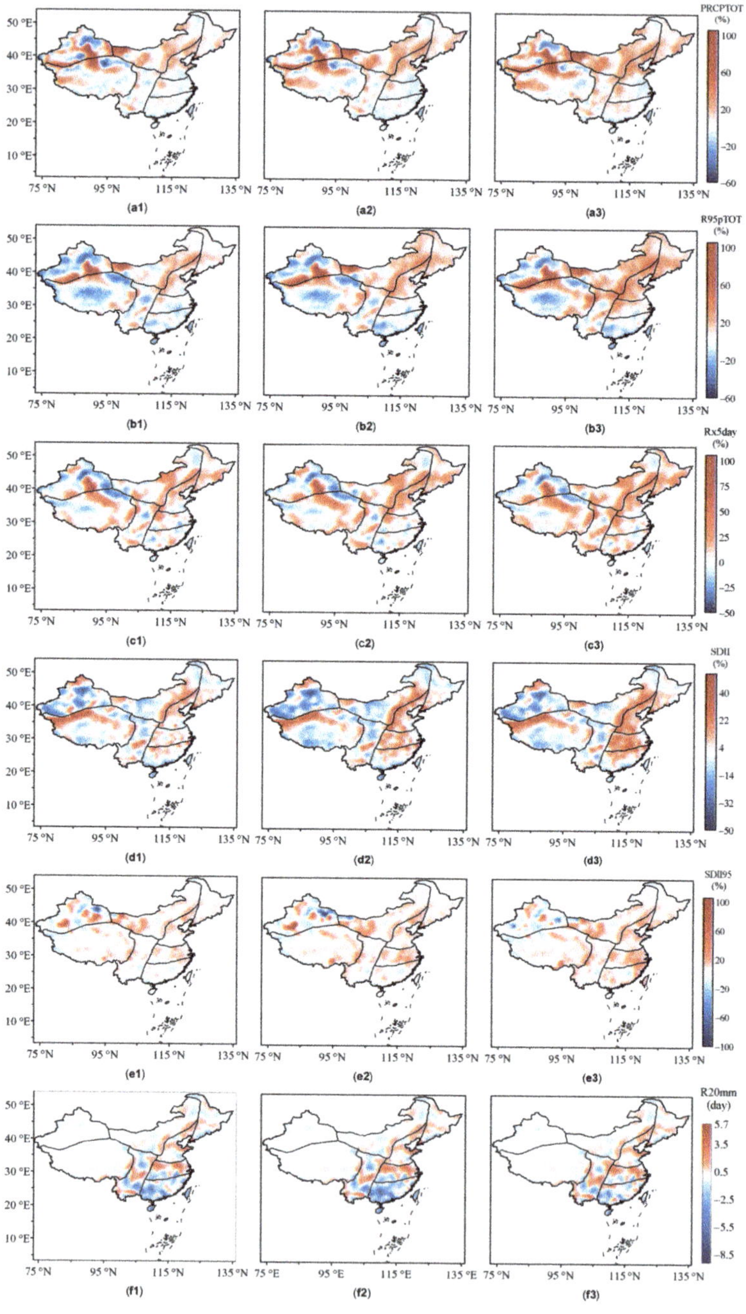

Figure 7. Relative change maps (%) of PRCPTOT (**a1–a3**), R95pTOT (**b1–b3**), Rx5day (**c1–c3**), SDII (**d1–d3**), and SDII95 (**e1–e3**) and absolute change maps of R20mm (**f1–f3**) for the three periods of 2023–2100 compared to the base period of 1990–2014.

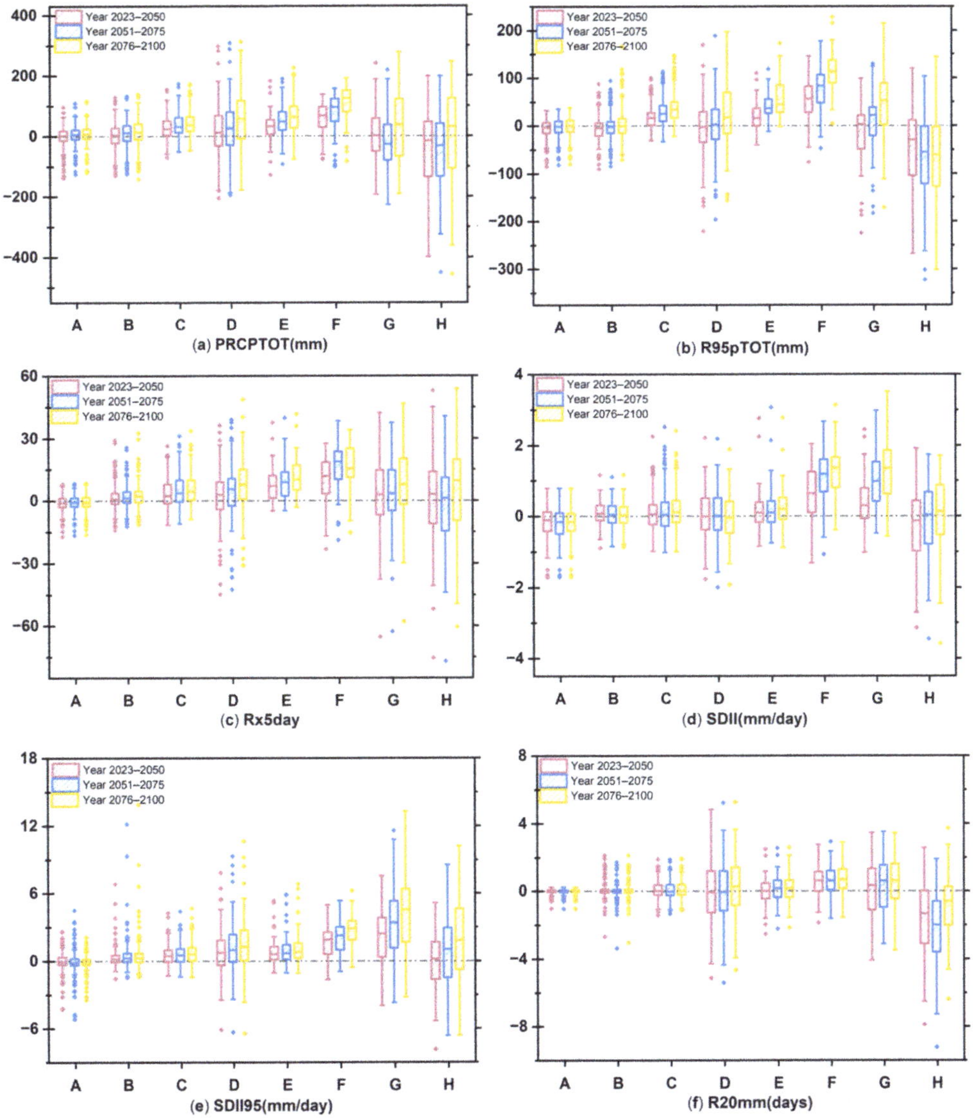

Figure 8. Boxplots of absolute changes from the base period.

Figure 7(d1,e3) and Figure 8d,e show the two precipitation intensity indices SDII and SDII95. It can be seen that there were some differences in their distributions, but the rate of change in the three phases increased gradually with time, as did the amount of precipitation. By the end of the century, SDII and SDII95 are expected to change by about 3.78% and 7.30%, respectively, compared to the base period, it is not difficult to find a more significant increasing trend in the extreme index SDII95. Precipitation intensity over about 14.9–17.5% of China, with significant variation (>15%) over the forecast period, and the areas with small changes (−5–5%) accounted for 37.06% of the total area, indicating that the extreme precipitation intensity increase is concentrated in the local areas of eastern China with a maximum change rate of about 30%, while the remaining areas are not expected to have large changes. Among these, as shown in Figure 8d,e, the main regions that were found to have an impact on the values were central China and north China, and the largest absolute changes were found to occur in central China, for which the SDII and SDII95 were 3.521 mm/day and 13.31 mm/day, respectively, followed by southwest China and south China, while the relative change rates of the western arid zone and the Qinghai–Tibet Plateau regions with less perennial precipitation and absolute changes were not obvious.

Figure 7(f1–f3) and Figure 8f show the distribution and box plots of the annual number of days of intense rainfall (R20mm). Since torrential rain is not common in western China and the probability of occurrence is very low, small changes will all result in high relative rates of change and will also affect the analysis of other regional characteristics. After weighting, we choose to show the absolute change distribution. The R20mm mainly showed a continuous increase in north China and central China, and most of the regions showed an increase as compared with the base period. In contrast, parts of south China experienced a significant decrease in the middle of the 21st century, and even the number of rainstorm days was below the base period until it increased at the end of the century.

In conclusion, all the indices were found to increase with time during the forecast period, and by the end of the 21st century, PRCPTOT is expected to increase steadily in most parts of the country, with changes in heavy precipitation concentrated in north China as well as northeast China. In particular, the three extreme precipitation indices, R95pTOT, Rx5day, and SDII95, all showed more prominent and significant changes. Additionally, the projected percentage increase in northern China was higher than that in southern China, with the largest increases mainly in north China and northeast China. The increase of PRCPTOT in north China is greater than that in south China, but this did not change the distribution in that precipitation in southern China remains higher than in northern China [57]. Due to the perennially dry climate in the northern and northwestern regions, the largest relative increase is expected in the western arid zone with a large relative change, a phenomenon consistent with CMIP5 projections [58].

3.2.2. Future Trend Distribution

Figure 9 shows the spatial distribution of the annual trend changes of the six indices in the three early, middle and end periods in the future. Among these, all the indices passed the 95% MK reliability test of the trend significance level.

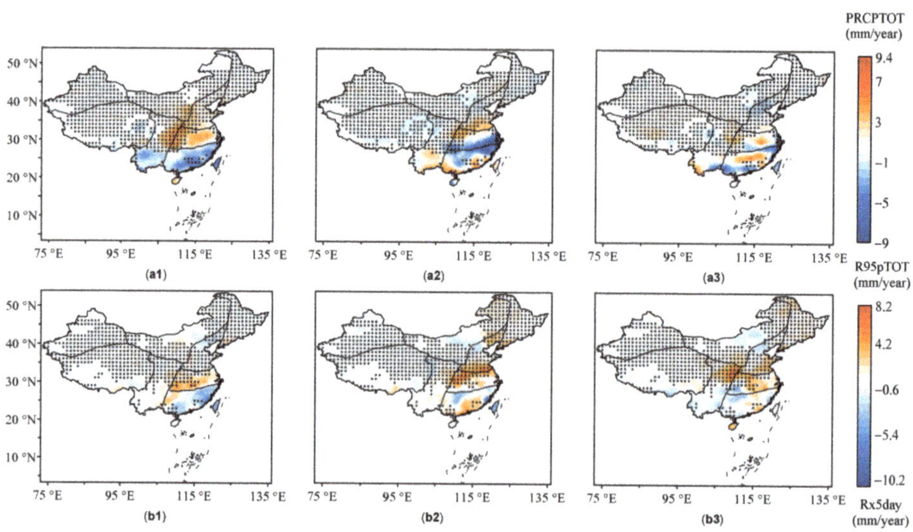

Figure 9. *Cont.*

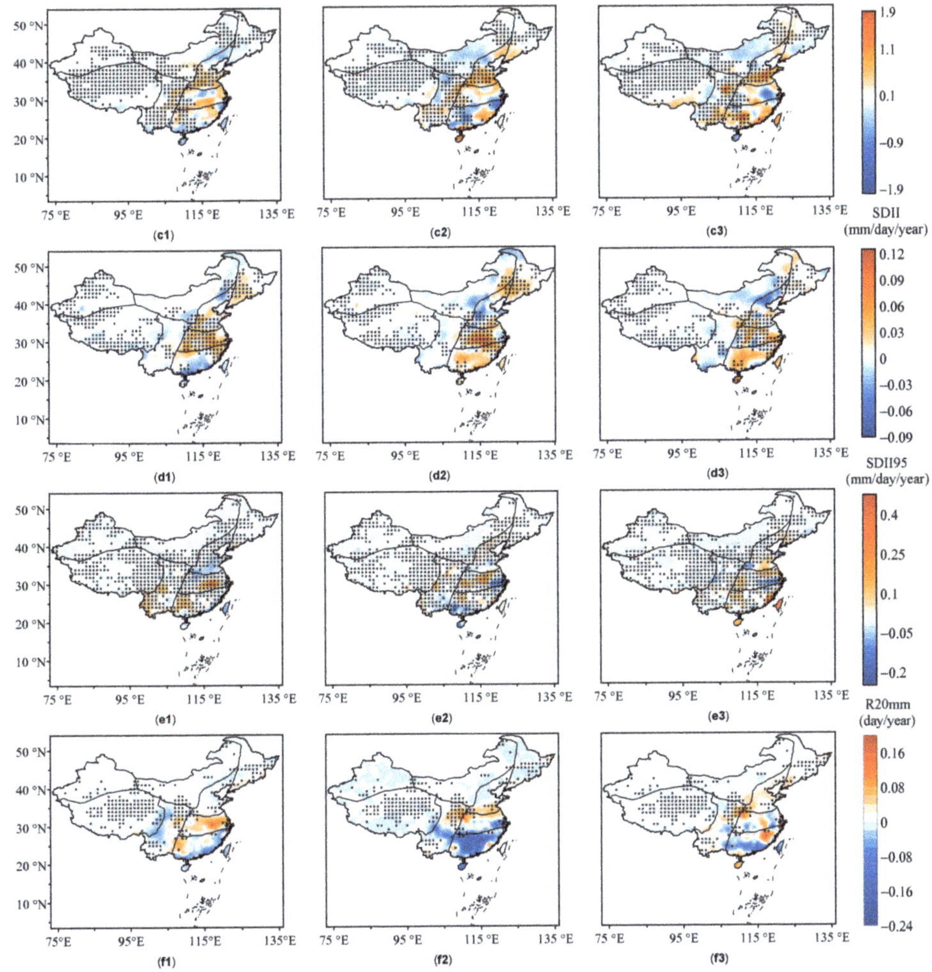

Figure 9. Spatial distribution of trend changes in each phase for PRCPTOT (**a1–a3**), R95pTOT (**b1–b3**), Rx5day (**c1–c3**), SDII (**d1–d3**), SDII95 (**e1–e3**), and R20mm (**f1–f3**). The columns from left to right are the three predicted periods before, during, and at the end of the future period, respectively. The gray dots are marked as the grid points where the trend significance level passed the 95% MK confidence test.

As shown in Figure 9(a1–c3) for the trend distribution of the three precipitation indices PRCPTOT, R95pTOT, and Rx5day, it can be seen that most of the regional grid points were significant nationwide except for south China and some parts of central China. For PRCPTOT, the largest trend of most regions in the three stages was in the early 21st century, except for a few western regions and south China. Among these, southwest China, as well as north China, reached the trend of 0.512 mm/year and 0.491 mm/year in the early stage, respectively, and this trend slowed down in the middle and at the end of the period, but was still in an upward trend. In particular, south China showed a decreasing trend of −0.219 mm/year in the early 21st century, and then showed a significant increasing trend in some areas in the middle and late stages, but, according to the overall trend, there was still a slight decline of 0.056 mm/year. Finally, most of the western regions of China showed an upward trend at all stages, with a significant one in the Qinghai–Tibet Plateau where the overall trend was found to reach 0.354 mm/year. Secondly, the two extreme precipitation indices R95pTOT and Rx5day differed from the former in that the upward trend was most significant in the middle or late stages across China. In the mid-century, R95pTOT in central China reached the maximum trend value of 0.803 mm/year nationwide, while Rx5day in north China also reached the maximum trend of 0.116 mm/year, followed by northeast China and southwest China, which did not have a trend as obvious as that in the abovementioned regions, but still showed a strong upward trend of change. The R95pTOT in south China had a brief upward trend in the middle term, but the overall trend did not fluctuate much.

For the precipitation intensity indexes SDII and SDII95 (Figure 9(d1–e3)), they have different spatial trend distributions. The trend of SDII was more drastic at all stages, but the grid point of SDII95 passing the 95% confidence test was much larger than that of SDII. It is expected that, except for south China, most of the areas on the central map will still show an upward trend in general, especially north China, which has a relatively obvious trend of change where the two indices reached a peak of 0.018 mm/day/year and 0.014 mm/day/year, respectively. In contrast, SDII in north China, the eastern arid zone, and parts of western China showed almost no significant fluctuations. SDII in southwest China showed a slight downward trend in the middle and late 21st century, but SDII95 showed a continuous upward trend. About R20mm generation frequency is shown in Figure 9(f1–f3). Combining the grid points with trend significance using MK95, it can be clearly observed that the maximum trend occurs in southwest China in all three periods, while the maximum decreasing trend is concentrated in parts of central China and south China. It is not difficult to find that the area with an increasing trend in the southwest region increased in the middle and late 21st century, and only a few areas in the region showed a decreasing trend in the later period. The same phenomenon was also observed in north China, where the majority of the areas remained unchanged and decreased at the beginning, and the majority of the areas with increasing trends occurred at the end, which indicates that R20mm in southwest China and north China will increase significantly with time, and the precipitation in this region will also increase.

From this, it can be seen that the area with an increasing trend in the southwest region increased in the middle and late 21st century, and only a few areas in the region showed a decreasing trend in the later period.

In general, the trend distribution of these indices is not consistent, and the value of PRCPTOT is expected to continue to rise in most regions of China in the future, but a decreasing trend will be observed in some parts of south China.

3.2.3. Future Interannual Variation

Figure 10 shows the evolution characteristics of five precipitation indexes over time in the forecast period of 2023–2100 compared with the reference period of 1990–2014. Table 2 shows the statistical data on the change in the time trend of the indicators in Figure 10. Among the grid point statistics, we kept only the grid points whose trend significance levels passed the 95% MK confidence test as statistical objects.

Table 2. Proportion of different intensity echoes of two radars.

Year	PRCPTOT (%/Year)	R95pTOT (%/Year)	Rx5day (%/Year)	SDII (%/Year)	SDII95 (%/Year)	R20mm (%/Year)
2023–2100	0.110	0.219	0.129	0.108	0.087	0.102
2023–2050	0.1666	0.083	0.130	0.112	0.084	0.066
2051–2075	0.0577	0.259	0.145	0.196	0.050	0.161
2076–2100	0.0610	0.387	0.182	0.177	0.049	0.177

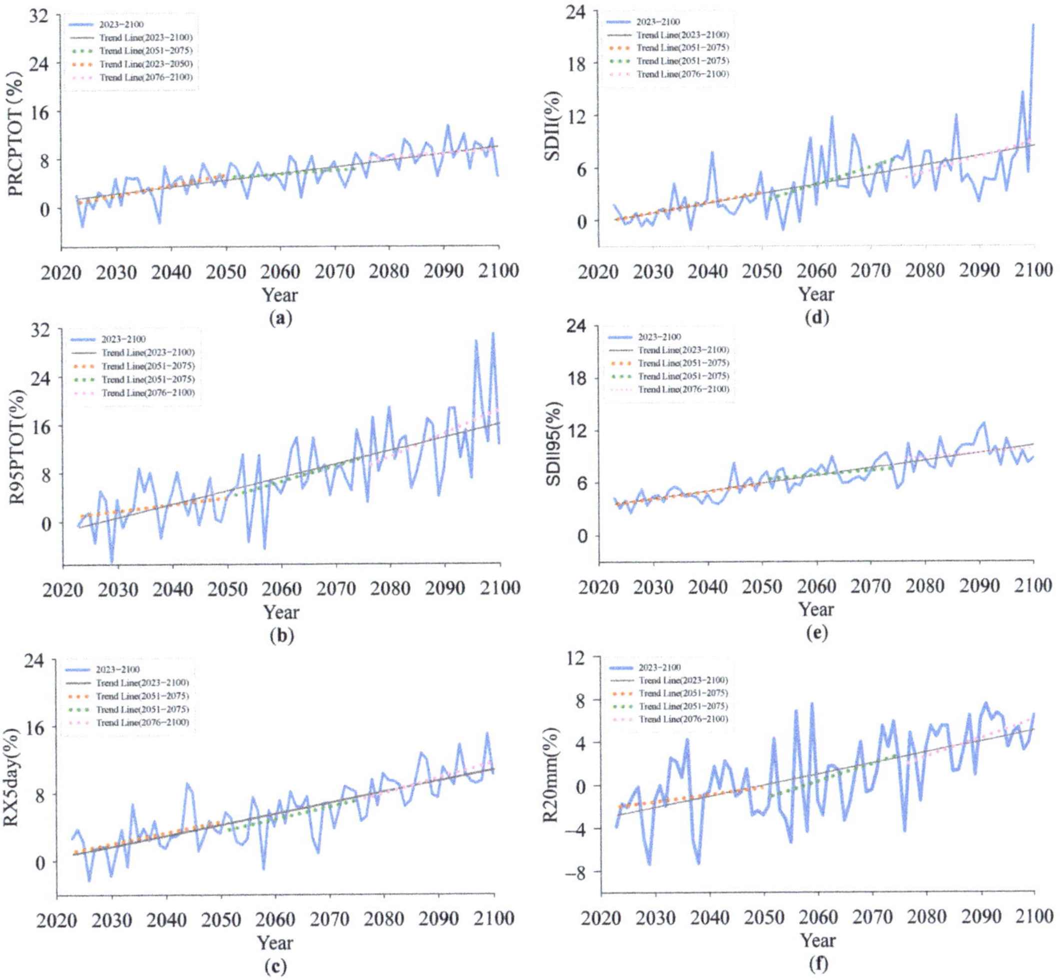

Figure 10. Annual average changes in six national precipitation indices: PRCPTOT (**a**), R95pTOT (**b**), Rx5day (**c**), SDII (**d**), SDII95 (**e**), and R20mm (**f**), all calculated from the base period 1990–2014. The colored dotted lines represent the trend for the three future periods, and the gray lines represent the overall trend.

Compared with the baseline period, the evolution trends of PRCPTOT, R95pTOT, Rx5day, SDII, SDII95, and R20mm in 2023–2100 can all be seen to show an overall increasing trend, and the rate of change of future precipitation for each index was found to mostly increase with time in the middle and end of this century.

Figure 10a–c show the temporal evolution characteristics of three precipitation indices. PRCPTOT (Figure 10a) showed the flattest trend of the three indices, reaching a maximum trend of 0.17%/year at the beginning of this century, and then maintaining a relatively stable trend around 0.06%/year. Throughout the study time period, the average trend of PRCPTOT, compared to the base period, reached 0.11%/year. Similarly, as shown in Figure 10b, the trend of the rate of change of R95pTOT in the future 2023–2100 period showed a clear trend and increased gradually over the period, reaching an average trend of 0.39%/year at the end of this century. In Figure 10c, the curve of Rx5day steadily increases, and its change rate is relatively stable in the future period, with the overall average trend reaching 0.182%/year.

Figure 10d,e show the temporal changes in the rate of change of the two intensity indices. The values of SDII and SDII95 are expected to rise in this century, with SDII showing a more pronounced trend. As shown in Figure 10d, the average trend of SDII reached a maximum of 0.20%/year in the middle of the century and decreased towards the end of the century, but still maintained an increase, with an average trend of 0.11%/year throughout the study period. Finally, the change of SDII95 was the most gentle, showing almost no obvious change trend, with an average trend of 0.09%/year, indicating that China's future extreme precipitation intensity will not change much as a whole, and, from the above, we can see that SDII changes mostly exist in local areas. Figure 10f shows the evolution of the annual change rate of heavy rainfall frequency (R20mm). The evolution trend of the R20mm in the three periods of the 21st century was seen to continue to increase; the frequency of extreme precipitation was found to maintain an increasing trend, and the trend gradually expanded, reaching 0.177%/year in the late 21st century.

4. Discussion

The total amount and intensity of heavy rainfall is expected to increase across the country in the future, which also implies that continuous flooding will be a major problem in China's development. Future changes in precipitation in China are mainly influenced by temperature in addition to the general humidification of the atmosphere, and many studies have shown that precipitation in China is highly susceptible to climate warming, and rising temperatures lead to changes in precipitation levels in China [59]. On the other hand, potential changes in future large-scale monsoon circulation systems also play a key role in the frequency and intensity of precipitation in China, with some studies suggesting a strong influence of the East Asian summer circulation in eastern China and the westerly circulation in northwestern China and the northern Qinghai–Tibet Plateau. Precipitation in eastern China is more frequent in summer and is easily influenced by the East Asian summer winds, leading to changes in future precipitation and spatial distribution [60,61]. Studies have also shown that the expected future intensification of the East Asian summer winds [29,62] may lead to larger increases in precipitation in north China and smaller increases (or even decreases) in southeastern China, especially in the middle and lower reaches of the Yangtze River Basin, and may also affect the timing of precipitation [63]. In addition, the winter northwest wind will also strengthen in the future [64,65], which is also the cause of the projected increase in precipitation in the northwest territories. The above has focused on two factors, namely, thermal and dynamic factors, to explore some possibilities regarding future increases in precipitation intensity.

We used ML to integrate the CMIP6 multi-model, effectively reducing its uncertainty and improving its prediction ability, and revealing the future temporal and spatial changes of precipitation in China, which will allow us to improve our knowledge of the precipitation evolution in China, and the world, in the future. In addition, in the future, we will explore more advanced machine learning techniques to retrieve more information from multiple perspectives and large amounts of data, thereby improving the reliability of future predictions. Additionally, the significant increases in severe extreme precipitation events may have a strong impact on human society and ecosystems. We will also further explore the contribution and impact of extreme precipitation events on future precipitation. The results of this study bring important guidance to the development and design of future remote sensing equipment, as well as guiding directions for the planning and layout of future multi-source observation networks.

5. Conclusions

In this study, we used a machine learning approach to integrate the CMIP6 multi-models to better capture the nonlinear and complex relationships among climate models compared with the traditional ensemble median approach and achieved a more accurate model-based prediction of future precipitation. On this basis, the temporal changes and spatial distribution of indices of eight climate zones in China during three periods of the 21st century, from 2023 to 2100, were analyzed. The study findings are as follows:

(a) In the validation assessment period (1999–2014), the ML integration treatment performed well, and the correlation coefficients of all the indices improved from the multi-model ensemble median to above 0.9 in general, with some reaching 0.95. In general, all the ML-treated indices improved the accuracy of the spatial pattern of precipitation in most regions of China. The improvement was more significant in areas with complex topography, such as around the Qinghai–Tibet Plateau. Due to the uncertainty of the climate model, there were still some errors in some areas, including the western arid zone and eastern arid zone, which were overestimated, while the negative deviations were mainly concentrated in some regions in southern China. The prediction performance of ML for the precipitation intensity index was better than that of the precipitation index, especially for SDII95.

(b) In the SSP2-4.5 scenario, the PRCPTOT, R95pTOT, Rx5day, SDII, SDII95, and R20mm precipitation indices continued to increase in mainland China, and by the end of the 21st century, will increase 8.77%, 13.73%, 9.43%, 6.84%, 9.34%, and 4.02%, respectively. Changes in extreme precipitation indices were the most prominent, with extreme precipitation reaching more than half of the total precipitation at the end of the period. China may experience more frequent and intense extreme precipitation events in the future, and the risk of future flooding throughout China will be greater.

(c) The increase in precipitation in southern China is not apparent. Even in the middle of the 21st century, precipitation in central and south China has decreased. In contrast, there are certain differences in the spatial distribution, especially precipitation likely to increase more in northern China, with the most significant precipitation change occurring at the end of the 21st century when PRCPTOT, R95pTOT, and Rx5day increase significantly in northern China. Additionally, the frequency and intensity of precipitation are expected to increase more sharply mainly in north China and central China, and slightly in western and southern China. The increase in precipitation intensity is expected to be greater in north China than in south China. This change may alleviate droughts and water shortages in some parts of northern China to some extent, but this pattern will not change for more rainfall in the south and less rainfall in the north [19].

(d) The temporal evolution of all the indices showed an increasing trend, with precipitation increasing more rapidly at the end of this century and SDII having a greater rate of increase in the first and middle of the century. The contribution of R95pTOT to PRCPTOT also increased with time, reaching 0.387%/year at the end of the century, while Rx5day also increased at a rate of 0.182%/year.

Author Contributions: Conceptualization, Y.Y., H.W., G.L., and J.X.; methodology, Y.Y., H.W., and F.G.; software, Y.Y., H.W., and X.R.; formal analysis, Y.Y., H.W., and L.T.; writing—original draft preparation, Y.Y.; writing—review and editing, Y.Y., H.W., and Q.Z.; visualization, Y.Y., H.W., and J.X. All authors have read and agreed to the published version of the manuscript.

Funding: This work was supported by the National Natural Science Foundation of China (42175002), the Project of the Sichuan Department of Science and Technology (2022YFS0541), the Key Laboratory of Atmospheric Sounding Program of China Meteorological Administration (2021KLAS02M), the National Key R&D Program of China (2018YFC1506104), Special Funds for the Central Government to Guide Local Technological Development (2020ZYD051), and Application Basic Research of Sichuan Department of Science and Technology (2019YJ0316).

Data Availability Statement: Publicly available datasets were analyzed in this study. These data can be found here: The data used in this study are the coupled model intercomparison project phase 6 where the SSP2-4.5 scenarios at https://esgf-data.dkrz.de/projects/cmip6-dkrz/ (accessed on 1 October 2021).

Acknowledgments: The authors sincerely thank the Coupled Model Intercomparison Project for providing the data used in this paper, and also thank the reviewers for their constructive comments and editorial suggestions, which have greatly contributed to the quality of the paper.

Conflicts of Interest: The authors declare no conflict of interest.

References

1. Meehl, G.A.; Tebaldi, C.; Walton, G.; Easterling, D.; McDaniel, L. Relative increase of record high maximum temperatures compared to record low minimum temperatures in the US. *Geophys. Res. Lett.* **2009**, *36*, L23701. [CrossRef]
2. Trenberth, K.E.; Fasullo, J.T. Climate extremes and climate change: The Russian heat wave and other climate extremes of 2010. *J. Geophys. Res. Atmos.* **2012**, *117*, D17103. [CrossRef]
3. Lu, X.X.; Ran, L. China flood havoc highlights poor urban planning. *Nat. Hazards* **2011**, *56*, 575. [CrossRef]
4. Root, T.L.; Price, J.T.; Hall, K.R.; Schneider, S.H.; Rosenzweig, C.; Pounds, J.A. Fingerprints of global warming on wild animals and plants. *Nature* **2003**, *421*, 57–60. [CrossRef] [PubMed]
5. Changnon, S.A.; Pielke, R.A., Jr.; Changnon, D.; Sylves, R.T.; Pulwarty, R. Human factors explain the increased losses from weather and climate extremes. *Bull. Am. Meteorol. Soc.* **2000**, *81*, 437–442. [CrossRef]
6. Peterson, T.C.; Zhang, X.; Brunet-India, M.; Vázquez-Aguirre, J.L. Changes in North American extremes derived from daily weather data. *J. Geophys. Res. Atmos.* **2008**, *113*, D07113. [CrossRef]
7. Westra, S.; Fowler, H.J.; Evans, J.P.; Alexander, L.V.; Berg, P.; Johnson, F.; Kendon, E.J.; Lenderink, G.; Roberts, N. Future changes to the intensity and frequency of short-duration extreme rainfall. *Rev. Geophys.* **2014**, *52*, 522–555. [CrossRef]
8. Donat, M.G.; Lowry, A.L.; Alexander, L.V.; O'Gorman, P.A.; Maher, N. More extreme precipitation in the world's dry and wet regions. *Nat. Clim. Change* **2016**, *6*, 508–513. [CrossRef]
9. Croitoru, A.-E.; Chiotoroiu, B.-C.; Todorova, V.I.; Torică, V.; Change, P. Changes in precipitation extremes on the Black Sea Western Coast. *Glob. Planet. Change* **2013**, *102*, 10–19. [CrossRef]
10. Kunkel, K.E.; Easterling, D.R.; Redmond, K.; Hubbard, K. Temporal variations of extreme precipitation events in the United States: 1895–2000. *Geophys. Res. Lett.* **2003**, *30*, 1326–1350. [CrossRef]
11. Zhang, Q.; Gemmer, M.; Chen, J. Climate changes and flood/drought risk in the Yangtze Delta, China, during the past millennium. *Quat. Int.* **2008**, *176*, 62–69. [CrossRef]
12. Zhang, Q.; Sun, P.; Singh, V.P.; Chen, X.; Change, P. Spatial-temporal precipitation changes (1956–2000) and their implications for agriculture in China. *Glob. Planet. Change* **2012**, *82*, 86–95. [CrossRef]
13. Min, S.-K.; Zhang, X.; Zwiers, F.W.; Hegerl, G.C. Human contribution to more-intense precipitation extremes. *Nature* **2011**, *470*, 378–381. [CrossRef] [PubMed]
14. Yu, F.; Chen, Z.; Ren, X.; Yang, G. Analysis of historical floods on the Yangtze River, China: Characteristics and explanations. *Geomorphology* **2009**, *113*, 210–216. [CrossRef]
15. Wang, Q.-X.; Fan, X.-H.; Qin, Z.-D.; Wang, M.-B.; Change, P. Change trends of temperature and precipitation in the Loess Plateau Region of China, 1961–2010. *Glob. Planet. Change* **2012**, *92*, 138–147. [CrossRef]
16. Yuan, Z.; Yang, Z.; Yan, D.; Yin, J. Historical changes and future projection of extreme precipitation in China. *Theor. Appl. Climatol.* **2017**, *127*, 393–407. [CrossRef]
17. Wu, S.; Hu, Z.; Wang, Z.; Cao, S.; Yang, Y.; Qu, X.; Zhao, W.J. Spatiotemporal variations in extreme precipitation on the middle and lower reaches of the Yangtze River Basin (1970–2018). *Quat. Int.* **2021**, *592*, 80–96. [CrossRef]
18. Jia, W.; Bo-Tao, Z.; Ying, X. Response of precipitation and its extremes over China to warming: CMIP5 simulation and projection. *Chin. J. Geophys.* **2015**, *58*, 461–473. [CrossRef]
19. Jiang, Z.; Li, W.; Xu, J.; Li, L. Extreme precipitation indices over China in CMIP5 models. Part I: Model evaluation. *J. Clim.* **2015**, *28*, 8603–8619. [CrossRef]
20. Sillmann, J.; Kharin, V.V.; Zwiers, F.; Zhang, X.; Bronaugh, D. Climate extremes indices in the CMIP5 multimodel ensemble: Part 2. Future climate projections. *J. Geophys. Res. Atmos.* **2013**, *118*, 2473–2493. [CrossRef]
21. Peng, Y.; Zhao, X.; Wu, D.; Tang, B.; Xu, P.; Du, X.; Wang, H. Spatiotemporal variability in extreme precipitation in China from observations and projections. *Water* **2018**, *10*, 1089. [CrossRef]
22. Simpkins, G. Progress in climate modelling. *Nat. Clim. Change* **2017**, *7*, 684–685. [CrossRef]
23. Eyring, V.; Bony, S.; Meehl, G.A.; Senior, C.A.; Stevens, B.; Stouffer, R.J.; Taylor, K.E. Overview of the Coupled Model Intercomparison Project Phase 6 (CMIP6) experimental design and organization. *Geosci. Model Dev.* **2016**, *9*, 1937–1958. [CrossRef]
24. Gusain, A.; Ghosh, S.; Karmakar, S. Added value of CMIP6 over CMIP5 models in simulating Indian summer monsoon rainfall. *Atmos. Res.* **2020**, *232*, 104680. [CrossRef]
25. Nie, S.; Fu, S.; Cao, W.; Jia, X.J.T.; Climatology, A. Comparison of monthly air and land surface temperature extremes simulated using CMIP5 and CMIP6 versions of the Beijing Climate Center climate model. *Theor. Appl. Climatol.* **2020**, *140*, 487–502. [CrossRef]
26. Xu, H.; Chen, H.; Wang, H. Future changes in precipitation extremes across China based on CMIP6 models. *Int. J. Climatol.* **2022**, *42*, 635–651. [CrossRef]

27. Chen, H.; Sun, J.; Lin, W.; Xu, H. Comparison of CMIP6 and CMIP5 models in simulating climate extremes. *Sci. Bull.* **2020**, *65*, 1415–1418. [CrossRef]
28. Reichstein, M.; Camps-Valls, G.; Stevens, B.; Jung, M.; Denzler, J.; Carvalhais, N. Deep learning and process understanding for data-driven Earth system science. *Nature* **2019**, *566*, 195–204. [CrossRef]
29. Wang, T.; Miao, J.-P.; Sun, J.-Q.; Fu, Y.-H. Intensified East Asian summer monsoon and associated precipitation mode shift under the 1.5 C global warming target. *Adv. Clim. Change Res.* **2018**, *9*, 102–111. [CrossRef]
30. Wang, H.; Wang, L.; He, J.; Ge, F.; Chen, Q.; Tang, S.; Yao, S.J.E.; Science, S. Can the GPM IMERG hourly products replicate the variation in precipitation during the wet season over the Sichuan Basin, China? *Earth Space Sci.* **2020**, *7*, e2020EA001090. [CrossRef]
31. Wang, H.; Yan, Y.; Keji, L.; Chen, Q.; Fan, X.; Zhang, F.; Tan, L. Relationships between Rapid Urbanization and Extreme Summer Precipitation over the Sichuan–Chongqing Area of China. *Front. Earth Sci.* **2022**, *10*, 909547. [CrossRef]
32. Miao, J.; Lin, Z.-S. Study on the characteristics of the precipitation of nine regions in China and their physical causes II—The relation between the precipitation and physical causes. *J. Trop. Meteorol.* **2004**, *20*, 64–72.
33. Zhang, X.; Alexander, L.; Hegerl, G.C.; Jones, P.; Tank, A.K.; Peterson, T.C.; Trewin, B.; Zwiers, F.W. Indices for monitoring changes in extremes based on daily temperature and precipitation data. *Wiley Interdiscip. Rev. Clim. Change* **2011**, *2*, 851–870. [CrossRef]
34. Zhang, Q.; Kong, D.; Singh, V.P.; Shi, P.J.G.; Change, P. Response of vegetation to different time-scales drought across China: Spatiotemporal patterns, causes and implications. *Glob. Planet. Change* **2017**, *152*, 1–11. [CrossRef]
35. Wu, J.; Gao, X.-J. A gridded daily observation dataset over China region and comparison with the other datasets. *Chin. J. Geophys.* **2013**, *56*, 1102–1111.
36. Ji, Z.; Kang, S. Evaluation of extreme climate events using a regional climate model for China. *Int. J. Climatol.* **2015**, *35*, 888–902. [CrossRef]
37. Wu, J.; Gao, X.; Giorgi, F.; Chen, D. Changes of effective temperature and cold/hot days in late decades over China based on a high resolution gridded observation dataset. *Int. J. Climatol.* **2017**, *37*, 788–800. [CrossRef]
38. O'Neill, B.C.; Tebaldi, C.; Van Vuuren, D.P.; Eyring, V.; Friedlingstein, P.; Hurtt, G.; Knutti, R.; Kriegler, E.; Lamarque, J.-F.; Lowe, J.; et al. The scenario model intercomparison project (ScenarioMIP) for CMIP6. *Geosci. Model Dev.* **2016**, *9*, 3461–3482. [CrossRef]
39. Riahi, K.; Van Vuuren, D.P.; Kriegler, E.; Edmonds, J.; O'neill, B.C.; Fujimori, S.; Bauer, N.; Calvin, K.; Dellink, R.; Fricko, O.; et al. The shared socioeconomic pathways and their energy, land use, and greenhouse gas emissions implications: An overview. *Glob. Environ. Change* **2017**, *42*, 153–168. [CrossRef]
40. Zhao, Y.; Zou, X.; Cao, L.; Xu, X. Changes in precipitation extremes over the Pearl River Basin, southern China, during 1960–2012. *Quat. Int.* **2014**, *333*, 26–39. [CrossRef]
41. Karl, T.R.; Nicholls, N.; Ghazi, A. Clivar/GCOS/WMO workshop on indices and indicators for climate extremes workshop summary. In *Weather and Climate Extremes*; Springer: Berlin/Heidelberg, Germany, 1999; pp. 3–7.
42. Zhu, H.; Jiang, Z.; Li, J.; Li, W.; Sun, C.; Li, L. Does CMIP6 inspire more confidence in simulating climate extremes over China? *Adv. Atmos. Sci.* **2020**, *37*, 1119–1132. [CrossRef]
43. Jiao, L.-C.; Yang, S.-Y.; Liu, F.; Wang, S.-G.; Feng, Z.-X. Seventy years beyond neural networks: Retrospect and prospect. *Chin. J. Comput.* **2016**, *39*, 1697–1716.
44. Lin, W.; Chen, H.J.A.; Letters, O.S. Assessment of model performance of precipitation extremes over the mid-high latitude areas of Northern Hemisphere: From CMIP5 to CMIP6. *Atmos. Ocean. Sci. Lett.* **2020**, *13*, 598–603. [CrossRef]
45. Taylor, K.E. Summarizing multiple aspects of model performance in a single diagram. *J. Geophys. Res. Atmos.* **2001**, *106*, 7183–7192. [CrossRef]
46. Ge, F.; Zhu, S.; Luo, H.; Zhi, X.; Wang, H. Future changes in precipitation extremes over Southeast Asia: Insights from CMIP6 multi-model ensemble. *Environ. Res. Lett.* **2021**, *16*, 024013. [CrossRef]
47. Han, T.; Chen, H.; Hao, X.; Wang, H. Projected changes in temperature and precipitation extremes over the Silk Road Economic Belt regions by the Coupled Model Intercomparison Project Phase 5 multi-model ensembles. *Int. J. Climatol.* **2018**, *38*, 4077–4091. [CrossRef]
48. Dong, S.; Xu, Y.; Zhou, B.; Shi, Y. Assessment of indices of temperature extremes simulated by multiple CMIP5 models over China. *Adv. Atmos. Sci.* **2015**, *32*, 1077–1091. [CrossRef]
49. Dong, T.; Dong, W. Evaluation of extreme precipitation over Asia in CMIP6 models. *Clim. Dyn.* **2021**, *57*, 1751–1769. [CrossRef]
50. Jiang, D.; Wang, H.; Lang, X. Evaluation of East Asian climatology as simulated by seven coupled models. *Adv. Atmos. Sci.* **2005**, *22*, 479–495.
51. Chen, L.; Frauenfeld, O.W. A comprehensive evaluation of precipitation simulations over China based on CMIP5 multimodel ensemble projections. *J. Geophys. Res. Atmos.* **2014**, *119*, 5767–5786. [CrossRef]
52. Torma, C.; Giorgi, F.; Coppola, E. Added value of regional climate modeling over areas characterized by complex terrain—Precipitation over the Alps. *J. Geophys. Res. Atmos.* **2015**, *120*, 3957–3972. [CrossRef]
53. Sun, Q.; Miao, C.; Duan, Q. Projected changes in temperature and precipitation in ten river basins over China in 21st century. *Int. J. Climatol.* **2015**, *35*, 1125–1141. [CrossRef]
54. Ou, T.; Chen, D.; Linderholm, H.W.; Jeong, J.-H. Evaluation of global climate models in simulating extreme precipitation in China. *Tellus A Dyn. Meteorol. Oceanogr.* **2013**, *65*, 19799. [CrossRef]

55. Su, F.; Duan, X.; Chen, D.; Hao, Z.; Cuo, L. Evaluation of the global climate models in the CMIP5 over the Tibetan Plateau. *J. Clim.* **2013**, *26*, 3187–3208. [CrossRef]
56. Guo, J.; Huang, G.; Wang, X.; Li, Y.; Yang, L. Future changes in precipitation extremes over China projected by a regional climate model ensemble. *Atmos. Environ.* **2018**, *188*, 142–156. [CrossRef]
57. Tian, J.; Zhang, Z.; Ahmed, Z.; Zhang, L.; Su, B.; Tao, H.; Jiang, T. Projections of precipitation over China based on CMIP6 models. *Stoch. Environ. Res. Risk Assess.* **2021**, *35*, 831–848. [CrossRef]
58. Zhou, B.; Wen, Q.H.; Xu, Y.; Song, L.; Zhang, X. Projected changes in temperature and precipitation extremes in China by the CMIP5 multimodel ensembles. *J. Clim.* **2014**, *27*, 6591–6611. [CrossRef]
59. Gu, X.; Zhang, Q.; Singh, V.P.; Shi, P. Changes in magnitude and frequency of heavy precipitation across China and its potential links to summer temperature. *J. Hydrol.* **2017**, *547*, 718–731. [CrossRef]
60. Fasullo, J. A mechanism for land–ocean contrasts in global monsoon trends in a warming climate. *Clim. Dyn.* **2012**, *39*, 1137–1147. [CrossRef]
61. Loo, Y.Y.; Billa, L.; Singh, A. Effect of climate change on seasonal monsoon in Asia and its impact on the variability of monsoon rainfall in Southeast Asia. *Geosci. Front.* **2015**, *6*, 817–823. [CrossRef]
62. Xin, X.; Zhang, L.; Zhang, J.; Wu, T.; Fang, Y. Climate change projections over East Asia with BCC_CSM1. 1 climate model under RCP scenarios. *J. Meteorol. Soc. Jpn. Ser. II* **2013**, *91*, 413–429. [CrossRef]
63. An, Z.; Wu, G.; Li, J.; Sun, Y.; Liu, Y.; Zhou, W.; Cai, Y.; Duan, Y.; Li, L.; Mao, J.; et al. Global monsoon dynamics and climate change. *Annu. Rev. Earth Planet. Sci.* **2015**, *43*, 29–77.
64. Sun, Y.; Ding, Y. A projection of future changes in summer precipitation and monsoon in East Asia. *Sci. China Earth Sci.* **2010**, *53*, 284–300. [CrossRef]
65. Ham, S.; Lee, J.-W.; Yoshimura, K. Assessing future climate changes in the East Asian summer and winter monsoon using regional spectral model. *J. Meteorol. Soc. Jpn. Ser. II* **2016**, *94*, 69–87. [CrossRef]

remote sensing

MDPI

Article

Temperature Contributes More than Precipitation to Runoff in the High Mountains of Northwest China

Mengtian Fan [1,*], Jianhua Xu [2], Yaning Chen [3], Meihui Fan [4], Wenzheng Yu [1] and Weihong Li [3]

1 School of Geographical Sciences, Nanjing University of Information Science & Technology, Nanjing 210044, China
2 Key Laboratory of Geographic Information Science (Ministry of Education), School of Geographic Sciences, East China Normal University, Shanghai 200241, China
3 State Key Laboratory of Desert and Oasis Ecology, Xinjiang Institute of Ecology and Geography, Chinese Academy of Sciences, Urumqi 830011, China
4 School of Foreign Languages, Dezhou University, Dezhou 253023, China
* Correspondence: mtfan@nuist.edu.cn

Abstract: In alpine areas in Northwest China, such as the Tianshan Mountains, the lack of climate data (because of scarce meteorological stations) makes it difficult to assess the impact of climate change on runoff. The main contribution of this study was to develop an integrated method to assess the impact of climate change on runoff in data-scarce high mountains. Based on reanalysis products, this study firstly downscaled climate data using machine learning algorithms, then developed a Batch Gradient Descent Linear Regression to calculate the contributions of temperature and precipitation to runoff. Applying this method to six mountainous basins originating from the Tianshan Mountains, we found that climate changes in high mountains are more significant than in lowlands. In high mountains, the runoff changes are mainly affected by temperature, whereas in lowlands, precipitation contributes more than temperature to runoff. The contributions of precipitation and temperature to runoff changes were 20% and 80%, respectively, in the Kumarik River. The insights gained in this study can guide other studies on climate and hydrology in high mountain basins.

Keywords: runoff changes; climate downscaling; data-scarce mountainous basins; quantitative assessment

Citation: Fan, M.; Xu, J.; Chen, Y.; Fan, M.; Yu, W.; Li, W. Temperature Contributes More than Precipitation to Runoff in the High Mountains of Northwest China. *Remote Sens.* **2022**, *14*, 4015. https://doi.org/10.3390/rs14164015

Academic Editor: Xander Wang

Received: 19 July 2022
Accepted: 16 August 2022
Published: 18 August 2022

1. Introduction

Global warming has exacerbated the uncertainty of runoff in mountainous rivers [1,2]. Mountainous rivers provide water for people and support lowland industries and agriculture [3–5]. However, in the alpine areas of Northwest China, such as the Tianshan mountains, limited climate data is available (because of the scarcity of meteorological stations), which makes it difficult to calculate the contribution of climate change to runoff [6,7]. New methods need to be developed in order to address this knowledge gap [8].

In recent years, various methods have been applied to calculate the contribution of climate change to runoff, including correlation analysis [9], sensitivity analysis [10], nonparametric Mann–Kendall tests [11], water–energy balance equations [12,13], two-parameter climate elasticity [14], weight connections [15], multiple linear regression (MLR) [16], Budyko curves [17], intelligent water drop algorithms [18], and hydrological models, such as VIC [19] and SWAT [20,21]. Using weight connections and artificial neural networks, Wang et al. [15] calculated the contributions of precipitation and temperature to runoff, which were 48% and 52% in the Toshkan River and 36% and 64% in the Kumarik River. Climate and streamflow processes have non-linear characteristics [22]. Improved, complete ensemble empirical mode decomposition, with adaptive noise (ICEEMDAN), can effectively decompose signals [23], and can be applied to analyze the impacts of climate change on runoff at different scales [14]. However, existing methods are mostly based on observed data [24], so they cannot be easily applied to data-scarce mountainous basins.

Precipitation and temperature are the main variables affecting runoff changes in mountainous basins [25,26]. Various studies have shown that precipitation changes have led to runoff changes in many rivers around the world [27,28]. Increasing temperatures can accelerate the melting of snow and glaciers [29], providing large amounts of runoff to mountainous rivers [30]. For data-scarce areas, reanalysis products spatially distributed climate data [31]. The ERA5 precipitation (ERAP) and ERA-interim temperature (ERAT) show climate change at the regional scale and are strongly correlated with empirical observations in China [32–36]. However, when applied to basins, the reanalysis products need to be downscaled, in order to improve spatial resolution [37,38].

The purpose of this study is to develop a method to calculate the contributions of temperature and precipitation to runoff in data-scarce mountainous rivers. Selecting six mountainous basins originating from the Tianshan Mountains, we first downscaled temperature and precipitation using machine learning algorithms, and we then developed a Batch Gradient Descent Linear Regression (BGDLR) model to calculate the contributions of temperature and precipitation to runoff. This research can provide references for hydrological forecasts in mountainous areas.

2. Materials and Methods

2.1. Study Area

To calculate the contributions of temperature and precipitation to runoff changes in mountainous rivers, we selected as study areas the Manas River basin (MRB) (84°96′E–86°31′E; 43°07′N–43°98′N) and Urumqi River basin (URB) (86°80′E–87°29′E; 43°02′N–43°38′N) originating from the northern slope of the Tianshan Mountains, and Kashgar River basin (KaRB) (83°00′E–85°00′E; 43°21′N–44°06′N), Toxkan River basin (TRB) (75°54′E–78°62′E; 40°28′N–41°50′N), Kumaric River basin (KuRB) (78°07′E–80°33′E; 41°39′N–42°48′N), and Kaidu River basin (KRB) (82°57′E–86°06′E; 42°06′N–43°21′N), originating from the southern slope of the Tianshan Mountains (Figure 1). There is only one meteorological station in the KRB and TRB, and no meteorological station in the KuRB, KaRB, MRB, or URB (Figure 1). Therefore, the above basins are typical data-scarce mountainous areas.

The Kashgar River is the main tributary of the Yili River, with a mean elevation of 3100 m [39]. In KaRB, the terrain is open to the west, and the westerly airflow brings abundant water vapor [40]. Among the rivers on the northern slope of the Tianshan Mountains, the Manas River has the most abundant runoff [41]. The Aksu River and KRB provide more than 80% of runoff for the Tarim River in Northwestern China [42]. The Toshkan River and Kumarik River are the main sources of the Aksu River [43]. The mean elevations are 3737 m and 3550 m, respectively, and the basin areas in the KuRB and TRB are 13,557 km and 18,835 km^2, respectively. The Kaidu River originates in the Sarming Mountains, with an average elevation of 2990 m and an area of 18,727 km^2. Snow and glaciers are widely distributed above the mountainous basins, which are an important supply of runoff [44]. There is a hydrological station in the mountain pass of each basin, the observed data of which can reflect runoff changes.

2.2. Datasets

The data used in this study included observed data and reanalysis data. We first downscaled temperature and precipitation based on reanalysis products and verified the accuracy of the downscaled results by observed data from meteorological stations. Then, based on the downscaled climate data and observed runoff, we calculated the contributions of temperature and precipitation to runoff change.

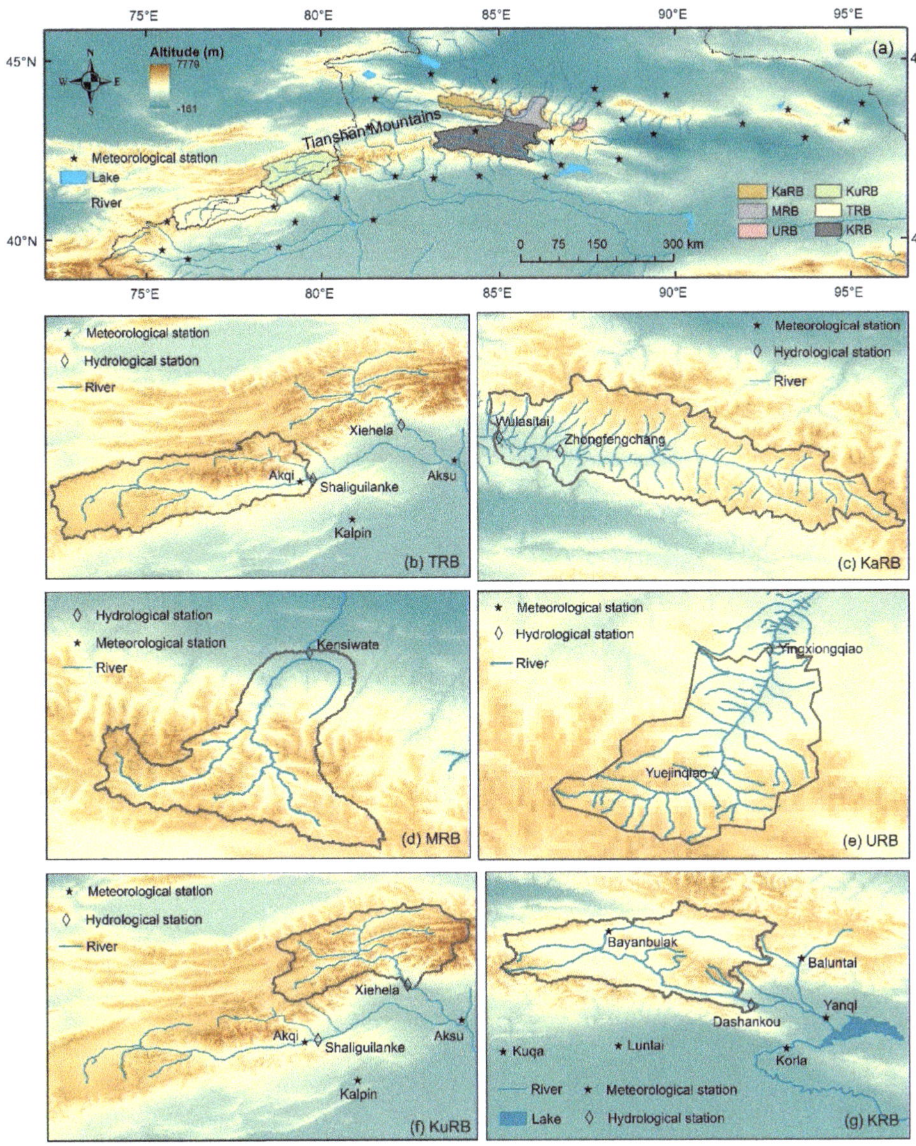

Figure 1. The study area. (**a**) Tianshan Mountains, (**b**) Toxkan River basin, (**c**) Kashgar River basin, (**d**) Manas River basin, (**e**) Urumqi River basin, (**f**) Kumaric River basin, and (**g**) Kaidu River basin.

The reanalysis products include ERAP, ERAT, and the Digital Elevation Model (DEM). The United States Geological Survey provided the DEM (http://srtm.csi.cgiar.org, accessed on 4 July 2022), with a resolution of 90 m × 90 m. The ERAT and ERAP are the third- and fifth-generation products of ECMWF, respectively. The spatial resolution of ERAT is 0.125° × 0.125°, and that of ERAP is 0.25° × 0.25° (https://apps.ecmwf.int/datasets/data/, accessed on 4 July 2022). The period of ERAT is from January 1979 to August 2019, and that of ERAP is from January 1979 to December 2020. To analyze the mechanism of climate change driving runoff, this study used a long-time series data set of snow depth in China, with a time resolution of one day, and a spatial resolution of 25 km. The data were provided by the National Glacier-Permafrost Desert Science Data Center (http://www.ncdc.ac.cn/portal/, accessed on 10 July 2022), from 1979 to 2016 [45].

To verify the accuracy of downscaled results, we used the observed monthly temperature and precipitation from 17 meteorological stations located near the six basins (http://data.cma.cn/, accessed on 17 June 2022). The data were downloaded from the National Meteorological Information Center, from 1979 to 2020. The monthly runoff was provided by the Hydrological Bureau of Xinjiang Uygur Autonomous Region. The runoff data for URB, MRB, and KaRB are from 1980 to 1987 and 2006 to 2011, and for KRB, TRB, and KuRB are from 1979 to 2015.

2.3. Methods

To calculate the contribution of climate change to runoff in data-scarce mountainous rivers, we integrated climate downscaling, the Mann–Kendall test, ICEEMDAN, and BGDLR. We first downscaled temperature and precipitation using machine learning algorithms, then developed a BGDLR model to calculate the contributions of temperature and precipitation to runoff change.

2.3.1. Climate Downscaling

Topography and geographical location are the main factors affecting precipitation and temperature distribution in mountainous areas [46,47]. Introducing terrain and geographic location to the downscaling of precipitation and temperature is useful [48]. This research fitted nonlinear models for temperature and precipitation, then trained the models using a gradient descent algorithm [49]. The downscaling models can be expressed as follows:

$$T = F_1\,(A,\ B,\ C,\ D,\ E) + \Delta T, \tag{1}$$

$$P = F_2\,(A,\ B,\ C,\ D,\ E) + \Delta P, \tag{2}$$

where ΔT and ΔP are residuals; A, B, C, D, and E are elevation, latitude, longitude, aspect, and slope, respectively. The steps build on Fan et al. [50,51].

2.3.2. Climate and Hydrological Process Analysis

The Mann–Kendall test [52,53] is an effective tool for analyzing trends in time series [22,54]. We applied this method to explore the changes in climate and runoff. In addition, we used Sen's slope [55] to verify trends in the Mann–Kendall test. For the steps of the Mann–Kendall test and Sen's slope, please see Wang et al. [15].

Runoff and climate have non-linear changes [56,57]. ICEEMDAN can effectively decompose signals [58–60]. In this research, ICEEMDAN was used to extract the multi-scale changes in runoff and climate. For the steps of ICEEMDAN, please see Ali and Prasad [61].

2.3.3. Contributions of Climate Change to Runoff

Precipitation and temperature are the main climatic variables affecting runoff changes in mountainous basins. Affected by factors such as geographic environment and altitude, the contribution of climate change to runoff varies in different basins and seasons. Linear regressions can clarify the relationship between dependent and independent variables [62,63]; the regression may be applied to calculate the contributions of independent variables to a dependent variable [64]. In this study, we developed a multi-linear model of runoff with temperature and precipitation and solved the model using a batch gradient descent (BGD) algorithm [49]. Before applying this method, normalization of temperature, precipitation, and runoff was performed with the equation as follows:

$$X_{is} = (x_i - x_{min})/(x_{max} - x_{min}), \tag{3}$$

where x_i is the sequential data and X_{is} is the normalized variable.

To avoid collinearity in the model, multicollinearity among the explanatory variables was evaluated, using the tolerance and variance inflation factor [65]. We also used partial correlation analysis to remove the effect between explanatory variables. According to Hair et al. [65], when the tolerance of independent variables is >0.1, the variance inflation coefficient is <10, indicating that there is no collinearity between independent variables. The resulting beta coefficients (partial regression coefficients) for the explanatory variables represent the independent contributions of each explanatory variable [66]. The regression model was as follows:

$$R = c_1T + c_2P + a, \tag{4}$$

where R, T, and P, are normalized runoff, temperature, and precipitation, respectively; c_1 is the regression coefficient of T; c_2 is the regression coefficient of P; and a is the regression constant.

In the model, the cost function for regression is:

$$J(\theta) = \frac{1}{2m}\sum\nolimits_{i=1}^{m}\left(h_\theta\left(x^{(i)}\right) - y^{(i)}\right)^2, \tag{5}$$

where m is the number of samples; $x^{(i)}$ represents the characteristics of the sample; and $y^{(i)}$ is the target.

Then, we used the BGD algorithm to train the dataset to minimize $J(\theta)$ and find the optimal solution of θ, which are as follows:

$$\frac{\partial}{\partial\theta_j}J(\theta) = \frac{1}{m}\sum\nolimits_{i=1}^{m}\left(h_\theta\left(x^{(i)}\right) - y^{(i)}\right)x_j^{(i)}, \tag{6}$$

The correction function for θ is:

$$\theta_j = \theta_j - \alpha\frac{1}{m}\sum\nolimits_{i=1}^{m}\left(h_\theta\left(x^{(i)}\right) - y^{(i)}\right)x_j^{(i)}, \tag{7}$$

where α is the learning rate.

Based on the regression coefficients, the contributions of precipitation and temperature to runoff are calculated as:

$$\eta_1 = |c_1|/(|c_1| + |c_2|), \tag{8}$$

$$\eta_2 = |c_2|/(|c_1| + |c_2|), \tag{9}$$

where η_1 is the contribution of temperature to runoff and η_2 is the contribution of precipitation to runoff.

3. Results

3.1. Accuracy of Downscaled Climate Data

After climate downscaling, we spatially distributed precipitation and temperature in six basins. The resolution of downscaled data is 90 m × 90 m. The observations from 17 meteorological stations distributed in and near the basins were used to verify the accuracy of the downscaling results. Tables S1 and S2 indicated the robust performance of the developed method. At 17 meteorological stations, the slope between the downscaled and observed data was close to 1, and the NSE was higher than 0.5. At most stations, the MAE and RMSE between downscaled temperature and observations were <3 °C (Table S1), and that between downscaled precipitation and observations were <10 mm (Table S2). At 11 stations, the NSE of downscaled temperature and the observed temperature was >0.9 (Table S1). The downscaled data accounted for scarce observations and reveal climate change in mountainous basins.

3.2. Climate Change

The mean and slope of downscaled grid data showed the characteristics of temperature and precipitation change. According to Figure 2, the temperature is <0 °C and annual precipitation is <600 mm in six basins. In valleys and plains, the annual precipitation is <300 mm, and the temperature is >0 °C. In mountainous areas, the annual precipitation is >300 mm, and the temperature is <0 °C.

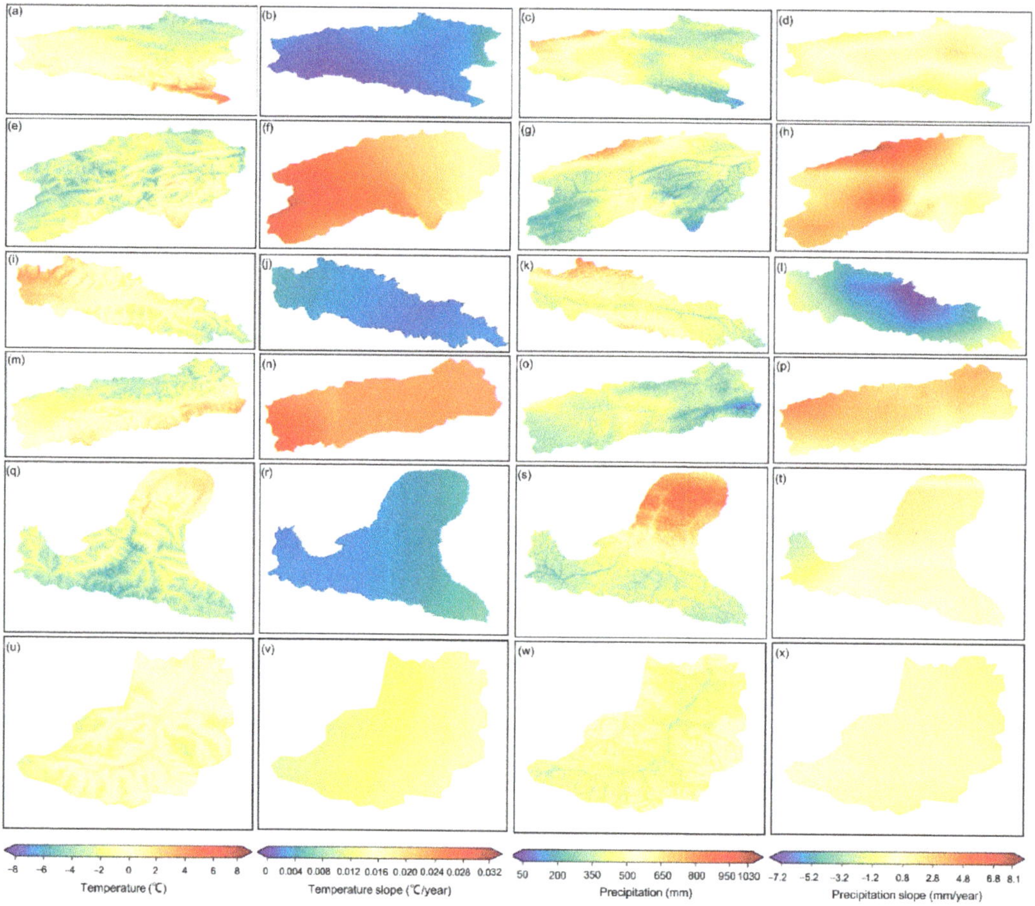

Figure 2. (**a**–**d**), spatial patterns and trends of precipitation and temperature from 1979 to 2020 in KRB; (**e**–**h**) KuRB; (**i**–**l**) KaRB; (**m**–**p**) TRB; (**q**–**t**) MRB; and (**u**–**x**) URB; (**a**,**e**,**i**,**m**,**q**,**u**) are annual average temperature; (**b**,**f**,**j**,**n**,**r**,**v**) are temperature trend; (**c**,**g**,**k**,**o**,**s**,**w**) are annual precipitation; (**d**,**h**,**l**,**p**,**t**,**x**) are precipitation trend.

There are significant differences in the temperature and precipitation changes in different basins (Figure 2). In the past 40 years, KRB has experienced wetting, with a humidification rate of 9 mm/10a (Figure 2b,d). At the same time, KaRB became increasingly dry at 40 mm/10a (Figure 2l). Different from the above basins, TRB, KuRB, MRB, and URB have experienced warming and wetting. In TRB and KuRB, the rate of warming and humidification gradually slowed from west to east. In TRB, the temperature increased 0.29 °C/10a in the west and 0.20 °C/10a in the east (Figure 2n); the precipitation increased 50 mm/10a in the west and 7.42 mm/10a in the east (Figure 2p). In KuRB, the temperature increased 0.21 °C/10a in the west and 0.18 °C/10a in the east (Figure 2f); the precipitation increased 30 mm/10a in the west and 27 mm/10a in the east (Figure 2h). On the whole, compared with valleys and plains, the high mountains have more dramatic changes in temperature and precipitation.

3.3. *Impact of Climatic Variables on Runoff*

3.3.1. Correlation of Runoff with Temperature and Precipitation

Figure 3 displays the correlation coefficients of runoff with climate on a monthly scale. In KaRB, the runoff has a stronger correlation with temperature, whereas in MRB, URB, KuRB, and KRB, the runoff has a stronger correlation with precipitation (Figure 3). Studies have shown that the correlations between runoff and climate varies in different seasons. Therefore, on a monthly scale, temperature and precipitation are the main driving factors for the runoff changes in mountainous watersheds. The difference is that, in KaRB, the temperature has a greater effect on runoff than precipitation, whereas, in MRB, URB, KuRB, and KRB, precipitation dominates the monthly runoff changes.

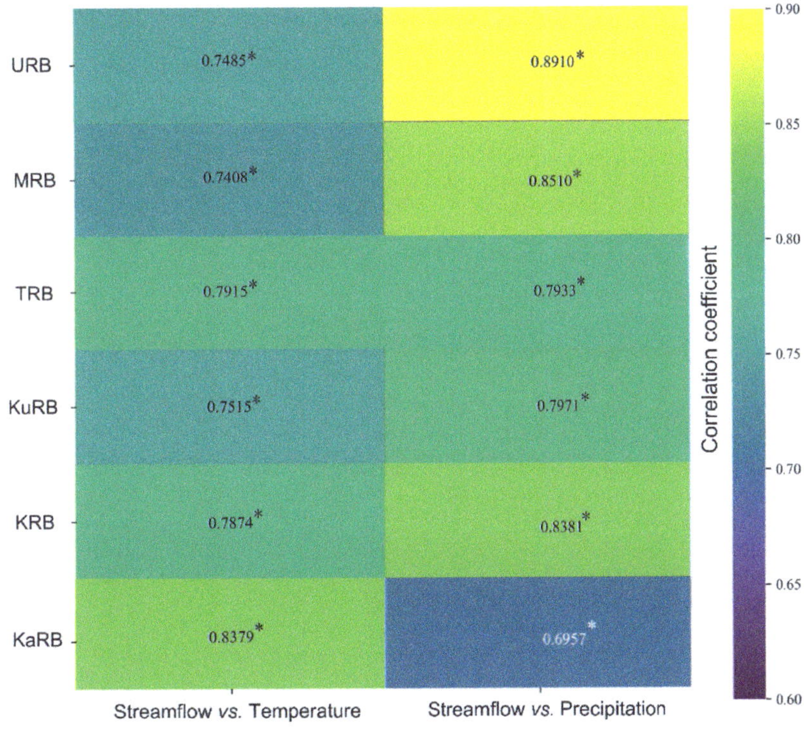

Figure 3. Correlation coefficients between climate and runoff on a monthly scale. * indicates the significance of a < 0.01.

The seasonal correlation coefficients between runoff and climate are shown in Table 1. In URB, a positive correlation between runoff and precipitation is only shown in summer, which indicates that the runoff in URB is mainly supplied by summer precipitation. In MRB, the runoff has significant positive correlations with both temperature and precipitation in summer, indicating that the runoff is mainly replenished by summer precipitation and glacier meltwater. In KaRB and KRB, a significant positive correlation is shown between runoff and precipitation in spring and summer, so the runoff in KaRB and KRB is mainly replenished by spring and summer precipitation. In TRB, the correlation is positive between precipitation and runoff in spring and autumn, indicating that the runoff in TRB is mainly replenished by spring and autumn precipitation. In KuRB, the correlation between runoff and temperature is significant and positive in spring, summer, and autumn, which indicates that the snowmelt water in spring and autumn, and melted ice water in summer, are the main replenishments for the runoff in KuRB (Table 1). There are obvious differences in the correlations between runoff and climate in different seasons. Overall, the changes in spring and autumn temperature, and summer precipitation, have an important impact on runoff.

Table 1. Seasonal correlation coefficients between climate and runoff.

Basin	Season	Streamflow vs. Temperature	Streamflow vs. Precipitation
MRB	Spring	0.05	0.37 *
	Summer	0.69 *	0.54 *
	Autumn	0.32	0.21
	Winter	0.22	0.37 *
URB	Spring	−0.29 *	0.08
	Summer	−0.13	0.65 **
	Autumn	−0.12	−0.09
	Winter	0.14	0.12
KaRB	Spring	0.09	0.54 *
	Summer	−0.02	0.74 *
	Autumn	−0.02	0.07
	Winter	0.28	0.12
TRB	Spring	−0.17	0.46 *
	Summer	0.02	0.24
	Autumn	0.30	0.44 *
	Winter	0.13	0.09
KuRB	Spring	0.38 *	0.05
	Summer	0.67 *	0.09
	Autumn	0.58 *	0.12
	Winter	−0.01	0.02
KRB	Spring	−0.02	0.32 *
	Summer	0.06	0.55 *
	Autumn	−0.01	0.24
	Winter	−0.06	0.10

Table note: * indicates the significance of a < 0.05 and ** indicates the significance of a ≤ 0.01.

ICEEMDAN was used to explore the multi-scale variations of climate and runoff. Table S3 shows that the runoff has similar cycles with temperature and precipitation. In MRB, URB, and KaRB, five intrinsic mode functions (IMF) and one residual component (RES) were obtained after the decomposition of runoff, temperature, and precipitation. While in other basins, six IMF and one RES were obtained after the decompositions. We reconstructed the precipitation, temperature, and runoff on inter-seasonal, inter-annual, and inter-decadal scales [67], then calculated the correlation coefficients between runoff and climate on different scales (Table 2). In MRB, URB, and KaRB, the inter-decadal variations of runoff are not shown because of the short period of runoff data. On the seasonal scale, the runoff has a stronger correlation with temperature in the KaRB and KuRB, and has a stronger correlation with precipitation in other basins; on the inter-annual scale, the correlation between runoff and temperature is stronger than that of precipitation; on the inter-decadal scale, the runoff has a stronger correlation with temperature in TRB and KRB, and has a stronger correlation with precipitation in KuRB (Table 2). It is worth noting that the runoff and temperature in KRB are significantly negatively correlated on the inter-decadal scale, and the specific mechanism(s) needs to be further studied.

Table 2. Multi-scale correlation coefficients between runoff and climate.

Basin	Scale	Streamflow vs. Temperature	Streamflow vs. Precipitation
MRB	Seasonal	0.7548 *	0.1586 *
	Inter-annual	0.4392 *	0.7284 *
URB	Seasonal	0.7915 *	0.1678 *
	Inter-annual	0.1214 *	0.6550 *
KaRB	Seasonal	0.0156	0.1681 *
	Inter-annual	0.2283 *	0.5309 *
TRB	Seasonal	0.7467 *	0.3494 *
	Inter-annual	0.2812 *	0.3866 *
	Inter-decadal	0.1837 *	0.9174 *
KuRB	Seasonal	0.1431 *	0.2772 *
	Inter-annual	0.6746 *	0.7185 *
	Inter-decadal	0.3832 *	0.2023 *
KRB	Seasonal	0.1458 *	−0.0136
	Inter-annual	0.2524 *	0.8536 *
	Inter-decadal	0.2008 *	−0.2041 *

Table note: * indicates the significance of a < 0.05.

3.3.2. Contributions of Climate Change to Runoff

Table S4 shows the collinearity tests for regression models. In the six basins, the tolerances were all > 0.8, and VIF were all < 2, indicating that the models did not have collinearity. According to partial regression coefficients, the contributions of temperature and precipitation to runoff can be calculated (Figure 4). Statistical results indicate significant differences in the contribution ratios among different basins. In KuRB, the temperature dominates annual runoff changes, whereas precipitation contributes more than temperature to runoff changes in other basins. Among the six watersheds, the KuRB has the highest elevation, and glaciers and permanent snow are widely distributed. Therefore, in high mountainous areas, the temperature contributed more than precipitation to runoff changes.

Across the seasons, there are obvious differences in the contribution of temperature and precipitation change to runoff (Figure 4). Generally, summer runoff accounts for the highest proportion of annual runoff in alpine basins [68]. Therefore, the relative contribution of temperature and precipitation to summer runoff changes can reflect the replenishment of runoff. In MRB, the contribution of precipitation change to runoff is 97% in spring, whereas the temperature change has a higher contribution than precipitation to runoff in summer and autumn, indicating that the glacier and snow melt water in summer is the main replenishment for runoff in MRB, followed by spring and summer precipitation. In URB, in spring, the contribution of temperature changes to runoff is higher than precipitation, indicating that glacier and snowmelt water are important replenishments for runoff; in summer, the runoff is mainly affected by precipitation change, and in autumn and winter, temperature and precipitation contribute equally to runoff. In KaRB, the relative contribution of precipitation changes to runoff is 72%, 88%, and 99% in spring, summer, and winter, respectively. Located in the Ili River Valley, KaRB is affected by warm and humid water vapor from the Atlantic Ocean; therefore, the runoff changes in KaRB are mainly affected by precipitation.

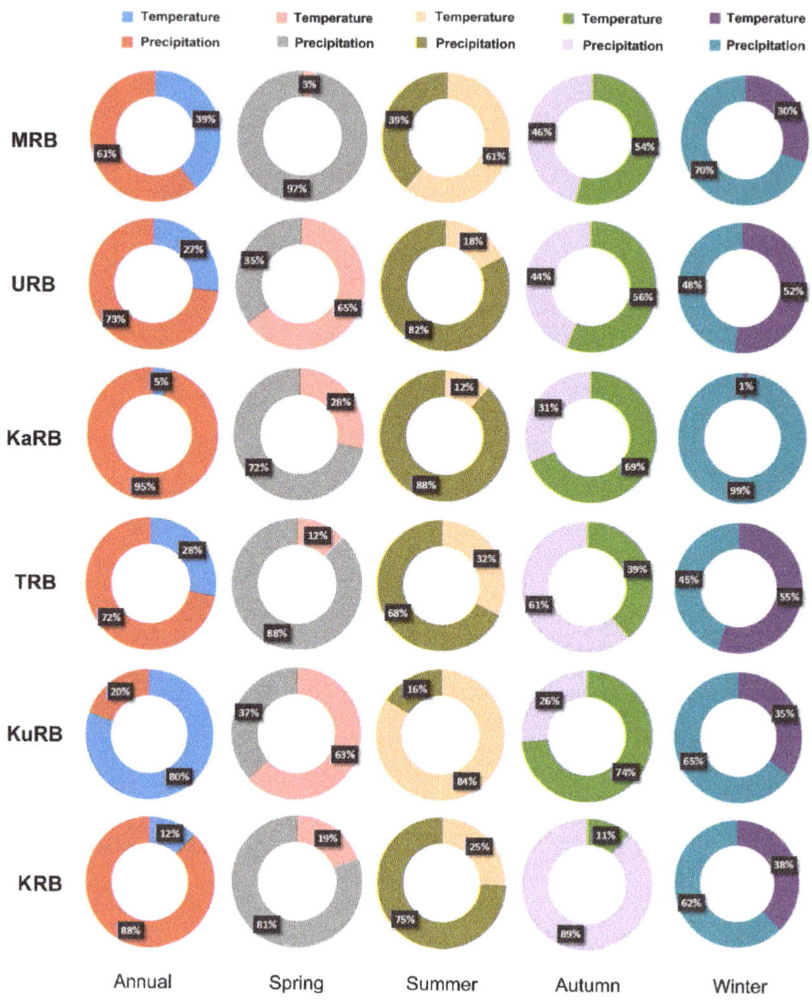

Figure 4. Contributions of precipitation and temperature change to runoff.

In TRB, precipitation contributed more than temperature to runoff in spring, summer, and autumn, whereas temperature has a higher contribution than precipitation to runoff in winter. This indicates that summer precipitation is the main replenishment of runoff in TRB, followed by spring and autumn precipitation. Different from other basins, in KuRB the temperature changes have a higher contribution than precipitation to runoff; therefore, glacier and snow melt water in summer is the main recharge of the runoff. In KRB, the contribution of precipitation changes to runoff is higher than temperature in each season, indicating that summer precipitation is the most important replenishment for runoff in KRB. In MRB and KuRB, glacier and snow melt water in summer is the main recharge of runoff, and runoff is mainly affected by temperature changes, whereas in other watersheds, summer precipitation is the main recharge of runoff, and runoff is mainly affected by precipitation changes.

4. Discussion

4.1. Climate Downscaling

Almost all previous studies have calculated the contribution of climate change to runoff using observed precipitation and temperature [11,16]. Generally, observed data only reflect climate change at immediate locations [51]. For data scarce areas, reanalysis products spatially distributed temperature and precipitation [31,32], whereas the low resolution eliminates the climate heterogeneity in basins [69]. Based on climate downscaling, this research obtained high-resolution temperature and precipitation data. We extracted the downscaled temperature (September 2017) and compared it with ERAT and observations at the corresponding time (Figure 5). There is only one meteorological station in KRB and TRB, and there is no meteorological station in KuRB, KaRB, MRB, and URB. Compared with observations and ERAT, downscaled data more accurately showed temperature changes in data-scarce mountain basins (Figure 5).

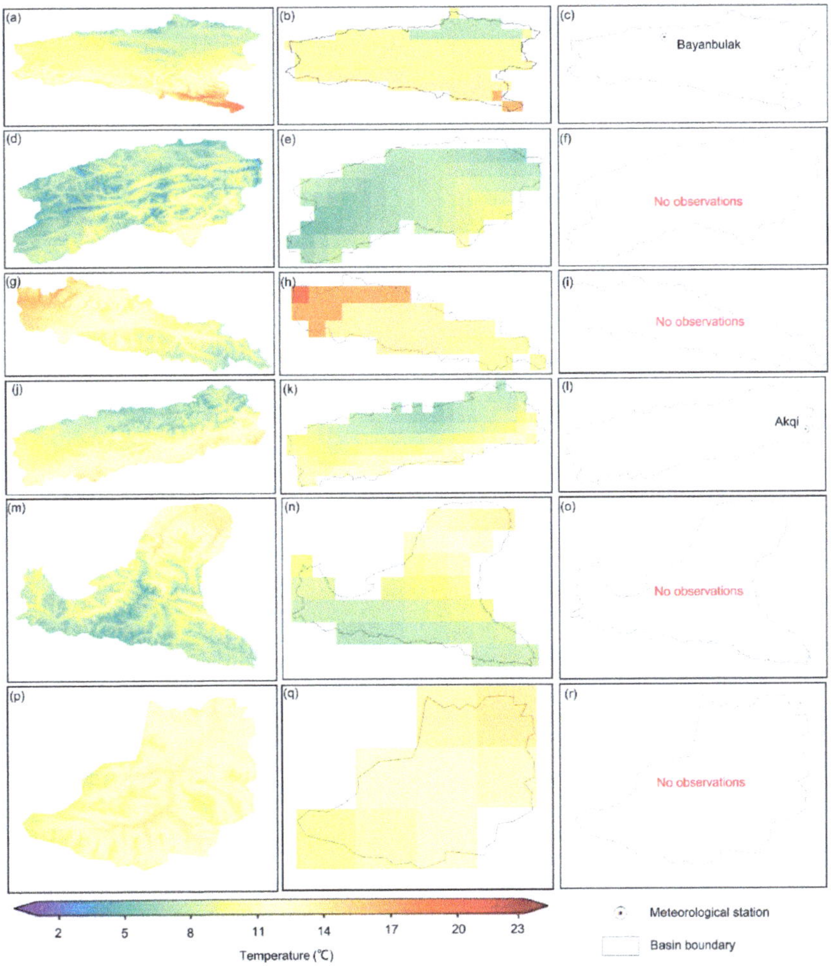

Figure 5. (**a–c**), comparison of downscaled temperature (90 m × 90 m), ERAT (0.125° × 0.125°), and observations in the KRB; (**d–f**) KuRB; (**g–i**) KaRB; (**j–l**) TRB; (**m–o**) MRB; and (**p–r**) URB; (**a,d,g,j,m,p**) are downscaled temperature; (**b,e,h,k,n,q**) are ERAT; (**c,f,i,l,o,r**) are observations.

4.2. Climate and Runoff Processes in Mountainous Rivers

In high mountainous watersheds, runoff is mainly supplied by precipitation and glacial snowmelt water, among which precipitation mainly supplies runoff in the form of rainfall and snowfall [44,70]. Above the snow line, snow transforms into glaciers and snow, which melts as temperatures increase, thereby replenishing runoff [71]. Studies have shown that glacier meltwater accounts for about 20% to 40% of total runoff in the Tianshan Mountains [72]. In the context of climate warming, the precipitation form in mountainous areas has changed, which has caused changes in the runoff process [73,74]. In rivers dominated by snowmelt runoff, a decrease in snowfall rates will lead to a shift to precipitation [75], thereby altering the seasonal distribution of runoff and leading to earlier flood peaks [76]. Studies have shown that the contributions of precipitation and temperature to runoff were 36% and 64% in KuRB, 52% and 48% in TRB [70], and 56% and 44% in KRB [15], which is consistent with our results.

To further understand the mechanisms of climate change driving runoff in the Tianshan Mountains, selecting KRB as an example, this section first compared the changes of temperature, precipitation, and runoff, then analyzed glacier and snow changes. Figure 6a shows that runoff has an annual distribution consistent with temperature and precipitation. In summer, temperature is the highest, precipitation is abundant, and runoff is the most abundant. Interannual variation shows that runoff is consistent with the changes in temperature and precipitation, and runoff increases (decreases) when the temperature and precipitation increase (decrease) (Figure 6).

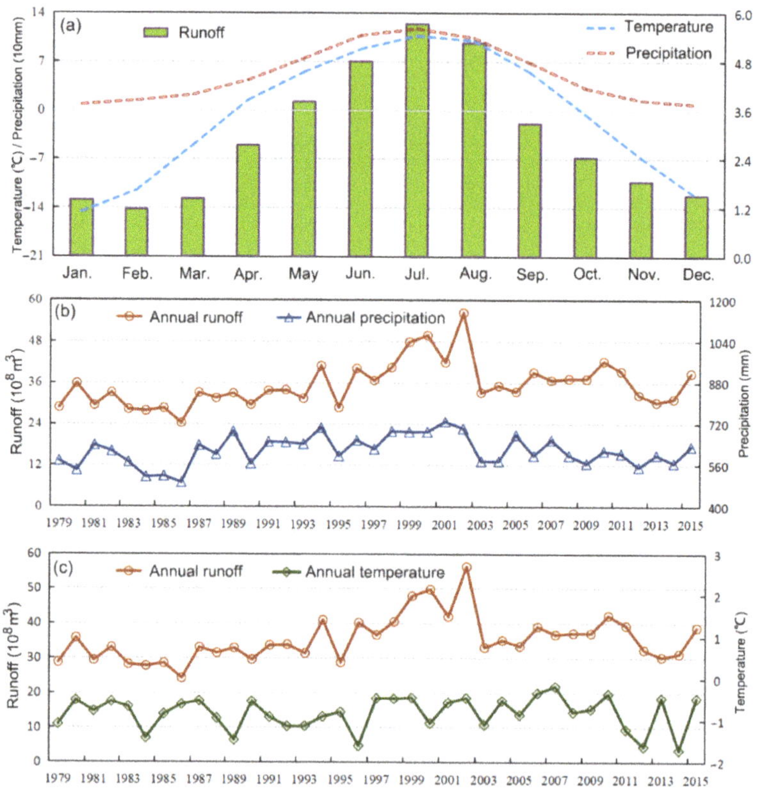

Figure 6. Changes in the temperature, precipitation, and runoff in KRB from 1979 to 2015: (**a**) annual distribution of temperature, precipitation, and runoff; (**b**) changes in precipitation and runoff; and (**c**) changes in temperature and runoff.

In the context of global warming, glacier and snow areas in KRB have changed. Bai [77] extracted the glacier area of KRB based on the cataloging data of two glaciers periods. Compared with the first statistical period (1956–1983), the glacier area in the second statistical period (2005–2010) decreased by 45.27% (Figure 7a). Warming is not only causing glaciers to retreat, but also accelerating snow melt. In the past few decades, the snow depth of KRB decreased; downward trends were most obvious in central and eastern regions. Relative to January 1980, snow depth in the northeastern mountains decreased by 10 cm (Figure 7b). Previous studies have shown that the correlation coefficients between temperature and snow cover in KRB are −0.81, −0.48, −0.80, and −0.82 in spring, summer, autumn, and winter, respectively. The above results confirmed that temperature has an important impact on the runoff change in high mountainous areas.

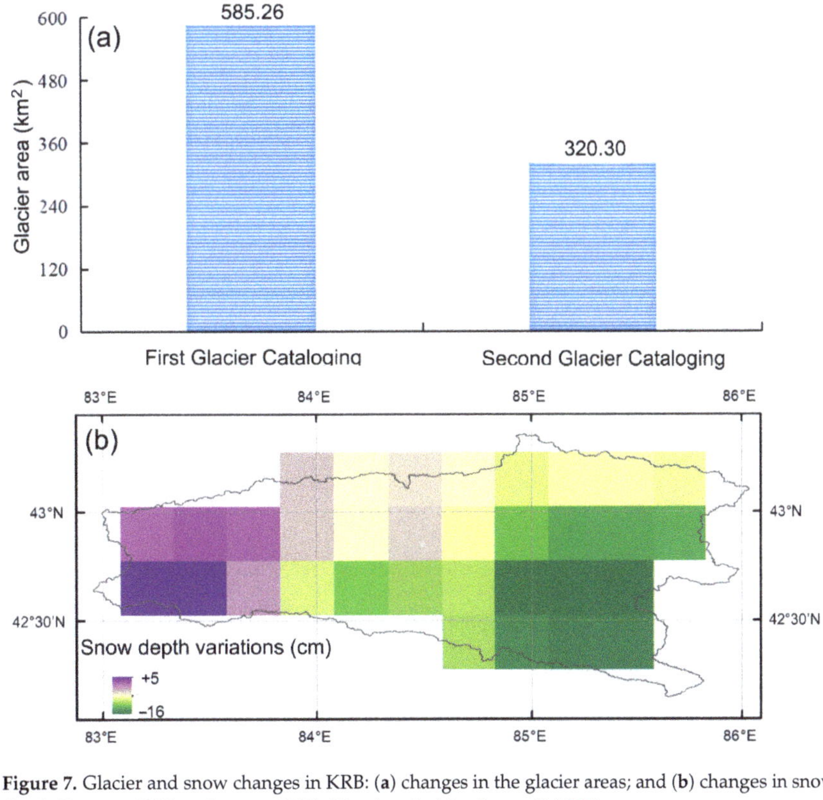

Figure 7. Glacier and snow changes in KRB: (**a**) changes in the glacier areas; and (**b**) changes in snow depth (January 2015 vs. January 1980). The data in (**a**) refers to Bai [75].

This research indicates that high mountains have more sensitive responses than plains to global climate changes. In mountainous areas, glaciers and snow are widely distributed, and plant diversity is abundant [78]. In valleys and plains, topography, latitude, and ecosystem stability may cause a buffering effect [79], with slower increases in temperature and precipitation than in mountains. The increase of precipitation is faster in mountainous areas than in plains, which is consistent with results from other studies [80]. In mountainous areas, increasing temperature accelerated the melting of snow and glaciers, leading to increased water vapor. Moreover, low saturated water vapor pressure in mountainous areas is conducive to the formation of precipitation [81].

Driven by climate change, runoff decreased in the KaRB in the past 40 years, which was mainly caused by decreasing precipitation. The KaRB is located near the Ili River Valley, and the decrease in precipitation may be related to the North Atlantic Drift [82].

At the same time, runoff increased in MRB, URB, KuRB, TRB, and KRB, but there are significant differences in the increase rate among different basins, and previous studies [70] have supported this.

4.3. Limitations

This study simulated high-resolution temperature and precipitation by downscaling reanalysis products. For alpine mountain areas, reanalysis products spatially distributed climate data. However, there is a deviation between the simulations of reanalyzed products and observed data, and the accuracy of products needs to be further improved. In addition, vegetation impacts climate change in mountainous areas, especially precipitation [83]. Due to the short time series and low spatial resolution of existing vegetation data, this study only introduced geographical and terrain factors in climate downscaling. In future research, the accuracy of downscaling results can be further improved if vegetation data with high temporal and spatial resolution are obtained.

This study mainly analyzed the impact on runoff of two main climate variables, temperature, and precipitation. In high mountainous basins, glaciers and snow melt are the main supplies of runoff. In future research, it is necessary to quantify the contributions of glaciers and snow to runoff change, to more clearly understand the mechanisms of climate change driving runoff. In addition, this study did not analyze the impact of alpine permafrost thawing on runoff change.

This study used beta coefficients (partial regression coefficients) of regression models to calculate the contributions of temperature and precipitation to runoff change. The bias of precipitation and temperature will lead to uncertainty in the calculation results of contribution ratio. In addition, as a component of the regression model, the contribution of regression constants to runoff were not calculated in this study. In future research, we will make improvements to the model to enhance its applicability.

5. Conclusions

Based on reanalysis products, this study developed an integrated method to calculate the contributions of climate change on runoff, in data scarce high mountains. Applying this method to six mountainous basins originating from the high Tianshan Mountains, this study found that, in high mountains, runoff changes are mainly affected by temperature, whereas in lowlands, precipitation contributes more than temperature to runoff changes. The contributions of precipitation and temperature change to runoff were 20% and 80%, respectively, in the Kumarik River. This study also found that climate changes are more significant in high mountains than in lowlands.

The present study lays the groundwork for future research using runoff simulations, which is of great significance to hydrological forecasting and water resource management in mountainous basins. This study highlights the impact of glaciers and snow on mountainous runoff. Thus, introducing the distribution of glaciers and snow into a hydrological modeling framework could better characterize runoff. Based on future climate scenarios, water management planning should be oriented to generate new strategies to cope with possible future changes in the strength of seasonality, and other variables.

Supplementary Materials: The following supporting information can be downloaded at: https://www.mdpi.com/article/10.3390/rs14164015/s1, Table S1: accuracy of the downscaled temperature at 17 meteorological stations; Table S2: accuracy of the downscaled precipitation at 17 meteorological stations; Table S3: cycles (month) of the runoff, temperature, and precipitation in six basins; Table S4: collinearity tests for regression models.

Author Contributions: Conceptualization, M.F. (Mengtian Fan) and J.X.; methodology, software, validation, formal analysis, investigation, and data curation, M.F. (Mengtian Fan); writing—original draft preparation, J.X. and M.F. (Meihui Fan); writing—review and editing, Y.C., W.Y. and M.F. (Meihui Fan); visualization and supervision, Y.C. and W.L. All authors have read and agreed to the published version of the manuscript.

Funding: This research is supported by the National Natural Science Foundation of China (Grant No. 42130512, 41871025, U20A2098) and the special fund for the introduction of talents in Nanjing University of Information Science and Technology (Grant No. 1521582201013).

Data Availability Statement: The data used in this study are available from Mengtian Fan (mtfan@nuist.edu.cn).

Acknowledgments: The authors gratefully acknowledge the Youth Innovation Promotion Association of the Chinese Academy of Sciences (2019431) and the State Key Laboratory of Desert and Oasis Ecology, Xinjiang Institute of Ecology and Geography, Chinese Academy of Sciences.

Conflicts of Interest: The authors declare no conflict of interest.

References

1. Pacheco, F.A. Regional groundwater flow in hard rocks. *Sci. Total Environ.* **2015**, *506*, 182–195. [CrossRef] [PubMed]
2. Wang, S.; Yang, Y.; Gong, W.; Che, Y.; Ma, X.; Xie, J. Reason Analysis of the Jiwenco Glacial Lake Outburst Flood (GLOF) and Potential Hazard on the Qinghai-Tibetan Plateau. *Remote Sens.* **2021**, *13*, 3114. [CrossRef]
3. Allen, M.R.; Frame, D.J.; Huntingford, C.; Jones, C.D.; Lowe, J.A.; Meinshausen, M.; Meinshausen, N. Warming caused by cumulative carbon emissions towards the trillionth tonne. *Nature* **2009**, *458*, 1163–1166. [CrossRef] [PubMed]
4. Arunrat, N.; Sereenonchai, S.; Chaowiwat, W.; Wang, C. Climate change impact on major crop yield and water footprint under CMIP6 climate projections in repeated drought and flood areas in Thailand. *Sci. Total Environ.* **2022**, *807*, 150741. [CrossRef]
5. Gardner, L.R. Assessing the effect of climate change on mean annual runoff. *J. Hydrol.* **2009**, *379*, 351–359. [CrossRef]
6. Fernandes, A.C.P.; de Oliveira Martins, L.M.; Pacheco, F.A.L.; Fernandes, L.F.S. The consequences for stream water quality of long-term changes in landscape patterns: Implications for land use management and policies. *Land Use Policy* **2021**, *109*, 105679. [CrossRef]
7. Pacheco, F.A.; Van der Weijden, C.H. Modeling rock weathering in small water-sheds. *J. Hydrol.* **2014**, *513*, 13–27. [CrossRef]
8. Chen, Y.; Takeuchi, K.; Xu, C.; Chen, Y.; Xu, Z. Regional climate change and its effects on river runoff in the Tarim Basin, China. *Hydrol. Process.* **2006**, *20*, 2207–2216. [CrossRef]
9. Dierauer, J.R.; Whitfield, P.H.; Allen, D.M. Climate Controls on Runoff and Low Flows in Mountain Catchments of Western North America. *Water Resour. Res.* **2018**, *54*, 7495–7510. [CrossRef]
10. Chawla, I.; Mujumdar, P.P. Isolating the impacts of land use and climate change on streamflow. *Hydrol. Earth Syst. Sci.* **2015**, *19*, 3633–3651. [CrossRef]
11. Bai, P.; Liu, W.H.; Guo, M.J. Impacts of climate variability and human activities on decrease in streamflow in the Qinhe River, China. *Theor. Appl. Climatol.* **2014**, *117*, 293–301. [CrossRef]
12. Meng, D.; Mo, X. Assessing the effect of climate change on mean annual runoff in the Songhua River basin, China. *Hydrol. Process.* **2011**, *26*, 1050–1061. [CrossRef]
13. Yang, H.B.; Yang, D.W. Derivation of climate elasticity of runoff to assess the effects of climate change on annual runoff. *Water Resour. Res.* **2011**, *47*, W07526. [CrossRef]
14. Fu, G.; Charles, S.P.; Chiew, F.H.S. A two-parameter climate elasticity of streamflow index to assess climate change effects on annual streamflow. *Water Resour. Res.* **2007**, *43*, W11419. [CrossRef]
15. Wang, C.; Xu, J.; Chen, Y.; Li, W. An approach to simulate the climate-driven streamflow in the data-scarce mountain basins of Northwest China. *J. Earth Syst. Sci.* **2019**, *128*, 95. [CrossRef]
16. Han, H.; Hou, J.; Huang, M.; Li, Z.; Xu, K.; Zhang, D.; Bai, G.; Wang, C. Impact of soil and water conservation measures and precipitation on streamflow in the middle and lower reaches of the Hulu River Basin, China. *Catena* **2020**, *195*, 104792. [CrossRef]
17. Patterson, L.A.; Lutz, B.; Doyle, M.W. Climate and direct human contributions to changes in mean annual streamflow in the South Atlantic, USA. *Water Resour. Res.* **2013**, *49*, 7278–7291. [CrossRef]
18. Pham, Q.B.; Afan, H.A.; Mohammadi, B.; Ahmed, A.N.; Linh, N.T.T.; Vo, N.D.; Moazenzadeh, R.; Yu, P.-S.; El-Shafie, A. Hybrid model to improve the river streamflow forecasting utilizing multi-layer perceptron-based intelligent water drop optimization algorithm. *Soft. Comput.* **2020**, *24*, 18039–18056. [CrossRef]
19. Bao, Z.; Zhang, J.; Wang, G.; Fu, G.; He, R.; Yan, X.; Jin, J.; Liu, Y.; Zhang, A. Attribution for decreasing streamflow of the Haihe River basin, northern China: Climate variability or human activities? *J. Hydrol.* **2012**, *460*, 117–129. [CrossRef]
20. Chien, H.; Yeh, P.J.-F.; Knouft, J.H. Modeling the potential impacts of climate change on streamflow in agricultural watersheds of the Midwestern United States. *J. Hydrol.* **2013**, *491*, 73–88. [CrossRef]
21. Tan, M.L.; Ibrahim, A.L.; Yusop, Z.; Duan, Z.; Ling, L. Impacts of land-use and climate variability on hydrological components in the Johor River basin, Malaysia. *Hydrol. Sci. J.* **2014**, *60*, 873–889. [CrossRef]
22. Xu, J.; Li, W.; Ji, M.; Lu, F.; Dong, S. A comprehensive approach to characterization of the nonlinearity of runoff in the headwaters of the Tarim River, Western China. *Hydrol. Process.* **2010**, *24*, 136–146. [CrossRef]
23. Bouhalais, M.L.; Nouioua, M. The analysis of tool vibration signals by spectral kurtosis and ICEEMDAN modes energy for insert wear monitoring in turning operation. *Int. J. Adv. Manuf. Technol.* **2021**, *115*, 2989–3001. [CrossRef]

24. Wang, D.; Hejazi, M. Quantifying the relative contribution of the climate and direct human impacts on mean annual streamflow in the contiguous United States. *Water Resour. Res.* **2011**, *47*, W00J12. [CrossRef]
25. Li, B.; Chen, Y.; Chen, Z.; Li, W. Trends in runoff versus climate change in typical rivers in the arid region of Northwest China. *Quat. Int.* **2012**, *282*, 87–95. [CrossRef]
26. Xiang, Y.; Wang, Y.; Chen, Y.; Zhang, Q. Impact of Climate Change on the Hydrological Regime of the Yarkant River Basin, China: An Assessment Using Three SSP Scenarios of CMIP6 GCMs. *Remote Sens.* **2022**, *14*, 115. [CrossRef]
27. Jena, P.P.; Chatterjee, C.; Pradhan, G.; Mishra, A. Are recent frequent high floods in Mahanadi basin in eastern India due to increase in extreme rainfalls? *J. Hydrol.* **2014**, *517*, 847–862. [CrossRef]
28. Yu, J.; Fu, G.; Cai, W.; Cowan, T. Impacts of precipitation and temperature changes on annual streamflow in the Murray–Darling Basin. *Water Int.* **2010**, *35*, 313–323. [CrossRef]
29. Brown, M.E.; Racoviteanu, A.E.; Tarboton, D.G.; Sen Gupta, A.; Nigro, J.; Policelli, F.; Habib, S.; Tokay, M.; Shrestha, M.S.; Bajracharya, S.; et al. An integrated modeling system for estimating glacier and snow melt driven streamflow from remote sensing and earth system data products in the Himalayas. *J. Hydrol.* **2014**, *519*, 1859–1869. [CrossRef]
30. O'Neel, S.; Hood, E.; Arendt, A.; Sass, L. Assessing streamflow sensitivity to variations in glacier mass balance. *Clim. Change* **2014**, *123*, 329–341. [CrossRef]
31. Hu, X.; Yuan, W.; Yu, R.; Zhang, M. The evolution process of warm season intense regional rainfall events in Yaan. *Clim. Dyn.* **2020**, *54*, 3245–3258. [CrossRef]
32. Decker, M.; Brunke, M.A.; Wang, Z.; Sakaguchi, K.; Zeng, X.; Bosilovich, M.G. Evaluation of the reanalysis products from GSFC, NCEP, and ECMWF using flux tower observations. *J. Clim.* **2012**, *25*, 1916–1944. [CrossRef]
33. Hamm, A.; Arndt, A.; Kolbe, C.; Wang, X.; Thies, B.; Boyko, O.; Reggiani, P.; Scherer, D.; Bendix, J.; Schneider, C. Intercomparison of Gridded Precipitation Datasets over a Sub-Region of the Central Himalaya and the Southwestern Tibetan Plateau. *Water* **2020**, *12*, 3271. [CrossRef]
34. Huai, B.; Wang, J.; Sun, W.; Wang, Y.; Zhang, W. Evaluation of the near-surface climate of the recent global atmospheric reanalysis for Qilian Mountains, Qinghai-Tibet Plateau. *Atmos. Res.* **2021**, *250*, 105401. [CrossRef]
35. Huang, D.; Gao, S. Impact of different reanalysis data on WRF dynamical downscaling over China. *Atmos. Res.* **2018**, *200*, 25–35. [CrossRef]
36. Wang, S.; Zhang, M.; Sun, M.; Wang, B.; Huang, X.; Wang, Q.; Feng, F. Comparison of surface air temperature derived from NCEP/DOE R2, ERA-Interim, and observations in the arid northwestern China: A consideration of altitude errors. *Theor. Appl. Climatol.* **2015**, *119*, 99–111. [CrossRef]
37. Gaur, A.; Simonovic, S.P. Extension of physical scaling method and its application towards downscaling climate model based near surface air temperature. *Int. J. Climatol.* **2017**, *37*, 3353–3366. [CrossRef]
38. Liu, Z.; Liu, Y.; Wang, S.; Yang, X.; Wang, L.; Baig, M.H.A.; Chi, W.; Wang, Z. Evaluation of Spatial and Temporal Performances of ERA-Interim Precipitation and Temperature in Mainland China. *J. Clim.* **2018**, *31*, 4347–4365. [CrossRef]
39. Yu, Y.M.; Mu, Z.X. The impact of DEM resolution on the accuracy of runoff simulation in the Kashgar River Basin. *Hydropower Energy Sci.* **2016**, *34*, 19–23. (In Chinese)
40. Song, Q.; Mu, Z.X.; Jiang, H.F. Application of SWAT model in the western mountainous area of Tianshan Mountains. *J. Water Resour. Water Eng.* **2013**, *24*, 46–49. (In Chinese)
41. Ran, S.H.; Wang, X.L.; Luo, Y. Multi-model prediction of climate change and its impact on runoff in the snow and ice watershed. *Arid. Land Geogr.* **2021**, *44*, 12. (In Chinese)
42. Chen, Y.; Xu, C.; Hao, X.; Li, W.; Chen, Y.; Zhu, C.; Ye, Z. Fifty-year climate change and its effect on annual runoff in the Tarim River Basin, China. *Quat. Int.* **2009**, *208*, 53–61. [CrossRef]
43. Ji, H.; Chen, Y.; Fang, G.; Li, Z.; Duan, W.; Zhang, Q. Adaptability of machine learning methods and hydrological models to discharge simulations in data-sparse glaciated watersheds. *J. Arid Land* **2021**, *13*, 549–567. [CrossRef]
44. Chen, Y.; Cui, W.; Li, W.; Zhang, Y.; Chen, Y. Utilization of water resources and ecological protection in the Tarim River. *Acta Geograph. Sin.* **2003**, *58*, 215–222. (In Chinese) [CrossRef]
45. Che, T.; Li, X.; Jin, R.; Armstrong, R.; Zhang, T. Snow depth derived from passive microwave remote-sensing data in China. *Ann. Glaciol.* **2008**, *49*, 145–154. [CrossRef]
46. Guan, H.; Wilson, J.L.; Xie, H. A cluster-optimizing regression-based approach for precipitation spatial downscaling in mountainous terrain. *J. Hydrol.* **2009**, *375*, 578–588. [CrossRef]
47. Revadekar, J.V.; Hameed, S.; Collins, D.; Manton, M.; Sheikh, M.; Borgaonkar, H.P.; Kothawale, D.R.; Adnan, M.; Ahmed, A.U.; Ashraf, J.; et al. Impact of altitude and latitude on changes in temperature extremes over South Asia during 1971–2000. *Int. J. Climatol.* **2012**, *33*, 199–209. [CrossRef]
48. Wacker, S.; Gröbner, J.; Vuilleumier, L. A method to calculate cloud-free long-wave irradiance at the surface based on radiative transfer modeling and temperature lapse rate estimates. *Theor. Appl. Climatol.* **2014**, *115*, 551–561. [CrossRef]
49. Zuo, J.; Xu, J.; Chen, Y.; Wang, C. Downscaling precipitation in the data-scarce inland river basin of Northwest China based on Earth system data products. *Atmosphere* **2019**, *10*, 613. [CrossRef]
50. Fan, M.; Xu, J.; Chen, Y.; Li, W. Reconstructing high-resolution temperature for the past 40 years in the Tianshan Mountains, China based on the Earth system data products. *Atmos. Res.* **2021**, *253*, 105493. [CrossRef]

51. Fan, M.; Xu, J.; Li, D.; Chen, Y. Response of Precipitation in Tianshan to Global Climate Change Based on the Berkeley Earth and ERA5 Reanalysis Products. *Remote Sens.* **2022**, *14*, 519. [CrossRef]
52. Kendall, M.G. *Rank Correlation Methods*; Springer: Oxford, UK, 1948; Available online: https://psycnet.apa.org/record/1948-15040-000 (accessed on 17 May 2022).
53. Mann, H.B. Nonparametric tests against trend. *Econometrica* **1945**, *13*, 245–259. [CrossRef]
54. Xu, J.; Chen, Y.; Li, W.; Nie, Q.; Hong, Y.; Yang, Y. The nonlinear hydroclimatic process in the Yarkand River, northwestern China. *Stoch. Env. Res. Risk Assess.* **2013**, *27*, 389–399. [CrossRef]
55. Sen, P.K. Estimates of the regression coefficient based on Kendall's Tau. *J. Am. Stat. Assoc.* **1968**, *63*, 1379–1389. [CrossRef]
56. Xu, J.; Chen, Y.; Lu, F.; Li, W.; Zhang, L.; Hong, Y. The nonlinear trend of runoff and its response to climate change in the Aksu River, western China. *Int. J. Climatol.* **2011**, *31*, 687–695. [CrossRef]
57. Xu, J.; Chen, Y.; Li, W.; Liu, Z.; Wei, C.; Tang, J. Understanding the Complexity of Temperature Dynamics in Xinjiang, China, from Multitemporal Scale and Spatial Perspectives. *Sci. World J.* **2013**, *2013*, 259248. [CrossRef]
58. Ghodratigohar, M.; Ghanadian, H.; Osman, H. A remote respiration rate measurement method for non-stationary subjects using CEEMDAN and machine learning. *IEEE Sensors J.* **2020**, *20*, 1400–1410. [CrossRef]
59. Wu, Z.; Huang, N.E. Ensemble empirical mode decomposition: A noise-assisted data analysis method. *Adv. Adapt. Data Anal.* **2009**, *1*, 1–41. [CrossRef]
60. Wu, Z.; Huang, N.E.; Wallace, J.M.; Smoliak, B.V.; Chen, X. On the time-varying trend in global-mean surface temperature. *Clim. Dyn.* **2011**, *37*, 759–773. [CrossRef]
61. Ali, M.; Prasad, R. Significant wave height forecasting via an extreme learning machine model integrated with improved complete ensemble empirical mode decomposition. *Renew. Sust. Energ. Rev.* **2019**, *104*, 281–295. [CrossRef]
62. Clow, D.W. Changes in the timing of snowmelt and streamflow in Colorado: A response to recent warming. *J. Clim.* **2010**, *23*, 2293–2306. [CrossRef]
63. Wang, Z.; Xie, P.; Lai, C.; Chen, X.; Wu, X.; Zeng, Z.; Li, J. Spatiotemporal variability of reference evapotranspiration and contributing climatic factors in China during 1961–2013. *J. Hydrol.* **2017**, *544*, 97–108. [CrossRef]
64. Haan, C.T. *Statistical Methods in Hydrology*; Iowa State University Press: Ames, IA, USA, 1977.
65. Hair, J.F.; Black, W.C.; Babin, B.J.; Anderson, R.E.; Tatham, R.L. *Multivariate Data Analysis*, 6th ed.; Prentice Hall: Hoboken, NJ, USA, 2005; 928p.
66. Kachigan, S.K. *Statistical Analysis: An Interdisciplinary Approach to Univariate and Multivariate Methods*; Radius Press: New York, NY, USA, 1986; 589p.
67. Bai, L.; Chen, Z.; Xu, J.; Li, W. Multi-scale response of runoff to climate fluctuation in the headwater region of Kaidu River in Xinjiang of China. *Theor. Appl. Climatol.* **2016**, *125*, 703–712. [CrossRef]
68. Wang, C.; Xu, J.; Chen, Y.; Bai, L.; Chen, Z. A hybrid model to assess the impact of climate variability on streamflow for an ungauged mountainous basin. *Clim. Dyn.* **2018**, *50*, 2829–2844. [CrossRef]
69. Praskievica, S. Downscaling climate-model output in mountainous terrain using local topographic lapse rates for hydrologic modeling of climate-change impacts. *Phys. Geogr.* **2018**, *39*, 99–117. [CrossRef]
70. Li, C.; Fang, H. Assessment of climate change impacts on the streamflow for the Mun River in the Mekong Basin, Southeast Asia: Using SWAT model. *Catena* **2021**, *201*, 105199. [CrossRef]
71. Bravo, C.; Loriaus, T.; Rivera, A.; Brock, B.W. Assessing glacier melt contribution to streamflow at Universidad Glacier, central Andes of Chile. *Hydrol. Earth Syst. Sci.* **2017**, *21*, 3249–3266. [CrossRef]
72. Chen, Y.N.; Li, Z.; Fang, G.H. Changes of key hydrological elements and research progress of water cycle in the Tianshan Mountains, Central Asia. *Arid. Land Geogr.* **2022**, *45*, 1–8. (In Chinese) [CrossRef]
73. Fan, M.; Xu, J.; Chen, Y.; Li, W. Modeling streamflow driven by climate change in data-scarce mountainous basins. *Sci. Total Environ.* **2021**, *790*, 148256. [CrossRef]
74. Fan, M.; Xu, J.; Chen, Y.; Li, W. Simulating the climate driven runoff in data-scarce mountains by machine learning and downscaling reanalysis data. *Stoch. Env. Res. Risk Assess.* **2022**. [CrossRef]
75. Barnett, T.P.; Adam, J.C.; Lettenmaier, D.P. Potential impacts of a warming climate on water availability in snow-dominated regions. *Nature* **2005**, *438*, 303–309. [CrossRef]
76. Liu, T.; Willems, P.; Pan, X. Climate change impact on water resource extremes in a headwater region of the Tarim basin in China. *Hydrol. Earth Syst. Sci.* **2011**, *15*, 3511–3527. [CrossRef]
77. Bai, L. *Climate Change and Its Impacts on Runoff Process for Typical River Basin in the Southern Slopes of Tianshan Mountains*; East China Normal University: Shanghai, China, 2016. (In Chinese)
78. Li, B.; Chen, Y.; Shi, X.; Chen, Z.; Li, W. Temperature and precipitation changes in different environments in the arid region of northwest China. *Theor. Appl. Climatol.* **2013**, *112*, 589–596. [CrossRef]
79. Chen, Y.; Li, Z.; Fan, Y.; Wang, H.; Deng, H. Progress and prospects of climate change impacts on hydrology in the arid region of northwest China. *Environ. Res.* **2015**, *139*, 11–19. [CrossRef]
80. Fu, A.; Chen, Y.; Li, W.; Li, B.; Yang, Y.; Zhang, S. Spatial and temporal patterns of climate variations in the Kaidu River Basin of Xinjiang, Northwest China. *Quat. Int.* **2013**, *311*, 117–122. [CrossRef]

81. Yang, Y.; Li, W.; Wei, W.; Hao, X.; Wang, M.; Li, H. Discrepancy analysis of the climate changes among mountain, plain, oasis and desert in an inland river basin in the northern slopes of the Tianshan Mountains-a case study in the Sangong river basin. *J. Glaciol. Geocryol.* **2009**, *31*, 1094–1100. [CrossRef]
82. Alpuim, T.; El-Shaarawi, A. Modeling monthly temperature data in Lisbon and Prague. *Environmetrics* **2009**, *20*, 835–852. [CrossRef]
83. Kawabata, A.; Ichii, K.; Yamaguchi, Y. Global monitoring of interannual changes in vegetation activities using NDVI and its relationships to temperature and precipitation. *Int. J. Remote Sens.* **2001**, *22*, 1377–1382. [CrossRef]

remote sensing

Article

Quantifying the Trends and Variations in the Frost-Free Period and the Number of Frost Days across China under Climate Change Using ERA5-Land Reanalysis Dataset

Hongyuan Li [1,2], Guohua Liu [3], Chuntan Han [1,2], Yong Yang [1] and Rensheng Chen [1,4,*]

1 Qilian Alpine Ecology and Hydrology Research Station, Northwest Institute of Eco-Environment and Resources, Chinese Academy of Sciences, Lanzhou 730000, China; lihongyuan@nieer.ac.cn (H.L.); hancht@lzb.ac.cn (C.H.); yy177@lzb.ac.cn (Y.Y.)
2 University of Chinese Academy of Sciences, Beijing 100049, China
3 College of Geography and Tourism, Hengyang Normal University, Hengyang 421000, China; lgh1990@lzb.ac.cn
4 College of Urban and Environmental Sciences, Northwest University, Xi'an 710000, China
* Correspondence: crs2008@lzb.ac.cn

Abstract: Understanding the spatio-temporal variations in the frost-free period (FFP) and the number of frost days (FD) is beneficial to reduce the harmful effects of climate change on agricultural production and enhancing agricultural adaptation. However, the spatio-temporal variations in FFP and FD and their response to climate change remain unclear across China. To investigate the impact of climate change on FFP and FD, the trends and variations in FFP and FD across China from 1950 to 2020 were quantified using ERA5-Land, a reanalysis dataset with high spatial and temporal resolution. The results showed that ERA5-Land has good applicability in quantifying the trends and variations in FFP and FD across China under climate change. The spatial distribution of multi-year average FFP and FD across China showed significant latitudinal zonality and altitude dependence, i.e., FFP decreased with increasing latitude and altitude, while FD increased with increasing latitude and altitude. As a result of climate warming across China, the FFP showed an increasing trend with an increase rate of 1.25 d/10a and the maximum increasing rate of FFP in the individual region was 6.2 d/10a, while the FD showed a decreasing trend with a decrease rate of 1.41 d/10a and the maximum decreasing rate of FD in the individual region was −6.7 d/10a. Among the five major climate zones in China, the subtropical monsoon climate zone (SUMZ) with the greatest increasing rate of 1.73 d/10a in FFP, while the temperate monsoon climate zone (TEMZ) with the greatest decreasing rate of −1.72 d/10a in FD. In addition, the coefficient of variation (C_v) of FFP showed greater variability at higher altitudes, while the C_v of FD showed greater variability at lower latitudes in southern China. Without considering the adaptation to temperature of crops, a general increase in FFP and a general decrease in FD were both beneficial to agricultural production in terms of FFP and FD promoting a longer growing period and reducing frost damage on crops. This study provides a comprehensive understanding of the trends and variations in FFP and FD under climate change, which is of great scientific significance for the adjustment of the agricultural production layout to adapt to climate change in China.

Keywords: frost-free period; frost days; ERA5-Land; climate change; agricultural production

Citation: Li, H.; Liu, G.; Han, C.; Yang, Y.; Chen, R. Quantifying the Trends and Variations in the Frost-Free Period and the Number of Frost Days across China under Climate Change Using ERA5-Land Reanalysis Dataset. *Remote Sens.* **2022**, *14*, 2400. https://doi.org/10.3390/rs14102400

Academic Editor: Xander Wang

Received: 13 April 2022
Accepted: 15 May 2022
Published: 17 May 2022

1. Introduction

The frost-free period (FFP) is the period of the year after the last frost day (LFD) and before the first frost day (FFD), during which no hoarfrost occurs [1–3]. The FFP is closely related to the growing period of crops, as a long FFP is associated with a long growing period of crops, and the length of the FFP is a very important heat index in agriculture [1,2]. The number of frost days (FD) is the total number of days in a year that frost occurs [4,5].

Frost is a relatively common extreme agro-meteorological disaster that occurs in winter and spring [6–8], primarily caused by the sharp drop in temperature below 0 °C in a short time due to the southward movement of the cold wave, or by strong radiation cooling at ground level on the night of the day when the weather changes from cloudy to clear sky condition after being affected by a cold wave [6,9–12].

Climate change has become a common concern for scientists and the public, with the global surface temperature rising by 1.09 °C from 2011 to 2020 compared to the industrial revolution [13]. The trend of surface temperature in China is basically the same as the global one in the last hundred years, but the warming in the last 50 years is more significant than the global average [14]. Under climate warming, the frequency of cold events and the dates of frost have changed significantly in many regions of the world [15–17]. Climate warming has resulted in an earlier LFD, a later FFD, an increase in FFP and a decrease in FD [7,18–23], and variations in FFP and FD have had a significant positive effect on agricultural production [16,17,24–29]. This not only affected seasonal agricultural planting planning, but also the growth and development of crops [2,18,30–33]. Consequently, determining the trends and variations in FFP and FD under climate change is important for the planning and management of agricultural production, such as agricultural zoning planning, agricultural production restructuring, optimization of agricultural production structures and regional layout [2,18,30,31,34].

Research on the trends and variations in FFP and FD are generally based on surface and air temperatures observed by ground-based meteorological stations [1,22,24,35]. Although the research on trends and variations in FFP and FD based on ground-based meteorological stations is relatively effective, there are certain disadvantages to the quantification of trends and variations in FFP and FD [22,29]. This is due to the uneven distribution of meteorological stations in China and the limited number of meteorological stations, especially in the vast regions of northwest China and the Qinghai–Tibet Plateau, where the meteorological stations are very sparse [36–38]. Even the above studies extrapolated the results to space using algorithms such as inverse distance interpolation based on station observed data, which creates large spatial uncertainties in regions with sparse station distribution and complex topography [39]. The uneven distribution, complex topography and limited number of meteorological stations results in the less accurate assessment of many climatic and meteorological conditions [38,40–42], including FFP and FD.

With the development of satellite remote sensing and data assimilation techniques as well as land surface and atmospheric models, the spatial and temporal resolution of reanalysis dataset has gradually increased and is widely used for the analysis of meteorological and hydrological processes applications [43–48]. Reanalysis dataset is the result of reprocessing and analysis of historical meteorological observations to reproduce past atmospheric conditions using well-established numerical prediction models and assimilation analysis [49,50]. Reanalysis dataset is generally gridded to improve the estimation of the meteorological elements with higher spatial and temporal resolution relative to meteorological stations, and is particularly useful in the study of meteorological and hydrological processes in regions where meteorological stations are sparse and unavailable [43,51,52].

Although many studies have investigated the FFP and FD in single or multiple provinces [9,18,22,29,35,53–55], the spatial and temporal variations in FFP and FD, as well as their response to climate change remain unclear across China. China covers a vast region and a variety of climate zones [56,57], and the geographical distribution of FFP and FD is zonal with evident spatial differences [6,10,22]. The layout of agricultural production is closely related to climatic conditions [58], and FFP and FD are also influenced by climatic conditions. Therefore, based on the quantification of FFP and FD at a national level, it is more beneficial to understand the relationship between FFP and FD and agricultural production by further quantifying the trends and variations in FFP and FD in different climate zones.

In the current study, the daily minimum skin temperature (T_s) and daily minimum 2 m temperature (T_a) from the ERA5-Land reanalysis dataset [50,59,60] were selected as data sources for quantifying the trends and variations in FFP and FD across China, and the data sources were validated using observed daily minimum ground surface temperature (GST) and daily minimum air temperature (TEM) from the Daily Meteorological Dataset of basic meteorological elements of China National Surface Weather Stations (DMDC) (V3.0) [61]. The main objectives of this study are (1) to analyze the spatial pattern of FFP and FD across China, (2) to quantify the trends and variations in FFP and FD across China during 1950~2020 and (3) to explore the effect of variations in FFP and FD on agricultural production in China.

2. Materials and Methods

2.1. Study Area

The vast size of China, the long distances between east and west, north and south, and the very different topographical and climatic characteristics of the various provinces and regions have resulted in a great difference in FFP and FD across China. In general, China can be divided into five main climate zones: the alpine climate zone (ALCZ), the temperate continental climate zone (TECZ), the temperate monsoon climate zone (TMCZ), the subtropical monsoon climate zone (SUMZ) and the tropical monsoon climate zone (TRMZ) [57] (Figure 1). There is consistency in the trends and changes in FFP and FD for each climate zone, while there is wide variability in the trends and variations in FFP and FD between climate zones. In addition, China contains 34 provinces, and the trends and variations in FFP and FD for each province are highly variable. Therefore, an integrated quantitative analysis of trends and variations in FFP and FD from a national perspective and in different climatic zones is of great scientific significance to adjust agricultural production to climate change. In this study, observed daily minimum surface and air temperatures from 822 national surface weather stations (including 620 national basic synoptic stations and 202 national reference climatological stations) (Figure 1) with longer time series were selected to evaluate the applicability of ERA5-Land reanalysis dataset.

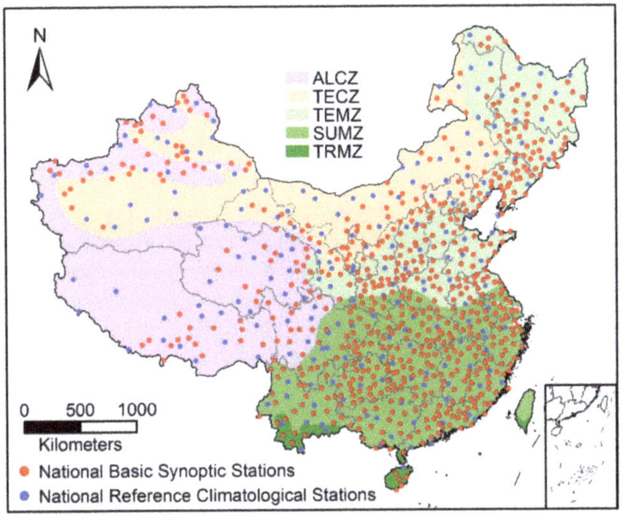

Figure 1. China's five main climate zones: the alpine climate zone (ALCZ), the temperate continental climate zone (TECZ), the temperate monsoon climate zone (TMCZ), the subtropical monsoon climate zone (SUMZ) and the tropical monsoon climate zone (TRMZ), and the distribution of 822 national surface weather stations in China.

2.2. *Data Collection*

2.2.1. Reanalysis Dataset of ERA5-Land

In this study, the daily minimum skin temperature (T_s) and daily minimum 2 m temperature (T_a) from the ERA5-Land reanalysis dataset were selected as the data source for quantifying the trends and variations of FFP and FD across China. ERA5-Land is a reanalysis dataset providing a consistent view of the evolution of land variables over several decades at an enhanced resolution, and ERA5-Land has been produced by replaying the land component of the ECMWF ERA5 climate reanalysis [50,59,60]. ERA5-Land has a high spatial resolution of 0.1 degree (approximately 10 km), a temporal resolution of one hour and a time of 1950 to the present. Because the temporal resolution of skin and air temperatures from ERA5-Land is one hour, the daily minimum T_s and daily minimum T_a selected in this study are the minimum values of 24 skin temperatures and 24 air temperatures in a day, respectively. Although the original dataset covers the 1950 to the present, we selected 1950 to 2020 as the study period. The high spatial and temporal resolution of ERA5-Land makes this dataset very useful for all kinds of land surface applications [50], such as FFP and FD quantification.

2.2.2. Dataset for ERA5-Land Evaluation

For the evaluation of daily minimum T_s and T_a from ERA5-Land, the observed daily minimum ground surface temperature (GST) and daily minimum air temperature (TEM) used to validate these land variables were collected from 822 national surface weather stations in China (Figure 1) during the 1951–2015 period and were provided by the China Meteorological Administration. Data from these observations are stored in the dataset of Daily Meteorological Dataset of basic meteorological elements of China National Surface Weather Stations (DMDC)(V3.0), which includes the daily minimum GST and daily minimum TEM [61]. Due to the early and late construction of each weather station, the time series of the observations is not consistent across weather stations. Only a few weather station observations had time series from 1951 to 2015, and most weather station observations had time series between 1951 and 2015. In particular, the time series of some weather station observations in the Qinghai–Tibet Plateau do not exceed 30 years. Therefore, the daily minimum T_s and T_a from ERA5-Land were validated by the corresponding daily minimum GST and TEM, respectively.

2.3. *Methods*

2.3.1. Quantitative Criteria

The FFP is the period between the LFD and the FFD, but there is no uniform definition of the LFD and FFD [22,53]. Generally, the last time the minimum surface temperature lower than 0 °C in the first half of the year is defined as the LFD, the first time the minimum surface temperature lower than 0 °C in the second half of the year is usually defined as the FFD and the period between the LFD and the FFD is defined as the FFP [53,62]. Therefore, the daily minimum T_s from ERA5-Land was used to calculate the LFD and FFD, and thus the FFP in this study.

The FD is different from the FFP, which is a period of continuity, whereas the FD is the total number of days in a year that hoarfrost occurs. The FD is one of the 27 core extreme climate change indices revisited by the joint CCl/CLIVAR/JCOMM Expert Team (ET) on Climate Change Detection and Indices (ETCCDI) [4,5], and the definition of FD is the annual count of days when the daily minimum temperature is lower than 0 °C. In view of the above definition, the daily minimum T_a from ERA5-Land was used to calculate the FD in this study.

2.3.2. Data Evaluation

Statistical indices were used for quantitative analysis of the performance of ERA5-Land in estimating the daily minimum T_s and daily minimum T_a, and four statistical indices were used in this study as follows:

$$R^2 = \left[\frac{\text{cov}(R,O)}{\sigma R \sigma O}\right]^2 \quad (1)$$

$$NSE = 1 - \frac{\sum_{i=1}^{n}(R_i - O_i)^2}{\sum_{i=1}^{n}(O_i - \overline{O})^2} \quad (2)$$

$$MBE = \frac{1}{n}\sum_{i=1}^{n}(R_i - O_i) \quad (3)$$

$$RMSE = \sqrt{\frac{1}{n}\sum_{i=1}^{n}(R_i - O_i)^2} \quad (4)$$

where R^2, NSE, MBE and $RMSE$ are the coefficient of determination, Nash–Sutcliffe efficiency coefficient, mean bias error and root mean square error, respectively. R_i and O_i represent the estimated (reanalysis) and observed data at time i, respectively. n is the total number of time steps. cov(R,O) is the covariance of estimated (reanalysis) and observed data and σR and σO are the standard deviations of estimated (reanalysis) and observed data, respectively.

2.3.3. Trend Analysis

In this study, the non-parametric Mann–Kendall (MK) test [63–66] was used to analyze the trend and significance level of the T_s, T_a, FFP and FD.

$$S = \sum_{i=1}^{n-1}\sum_{j=i+1}^{n} \text{sgn}(x_j - x_i) \quad (5)$$

$$\text{sgn}(x_j - x_i) = \begin{cases} 1 & (x_j - x_i) > 0 \\ 0 & (x_j - x_i) = 0 \\ -1 & (x_j - x_i) < 0 \end{cases} \quad (6)$$

$$\text{var}(S) = \frac{n(n-1)(2n+5) - \sum_{k=1}^{m} t_k(t_k - 1)(2t_k + 5)}{18} \quad (7)$$

$$Z_c = \begin{cases} (S-1)/\sqrt{\text{var}(S)} & S > 0 \\ 0 & S = 0 \\ (S+1)/\sqrt{\text{var}(S)} & S < 0 \end{cases} \quad (8)$$

where S is the statistic of the dataset, n is the length of the dataset, x_i and x_j are the sequential data values in time series i and j, m is the number of tied groups and t_k denotes the number of ties of extent k and a tied group is a set of sample data having the same value. Z_c is the standardized statistics of the dataset and the positive and negative values of Z_c indicate increasing and decreasing trends, respectively. If $|Z_c| > Z_{1-\alpha/2}$, the trend is statistically significant, otherwise, the trend is not statistically significant. Testing trends was carried out at a specific α significance level, and the significance level of α = 0.05 (95% confidence level) was applied in this study.

For the measurement of the trend in variation, the Sen's slope method [65–67] was used to analyze the slope of the variation, and the slope is expressed as follows:

$$\beta = \text{Median}\left(\frac{x_j - x_i}{j - i}\right) 1 < i < j < n \quad (9)$$

where β is the slope of the data variation, a positive β denotes an increasing trend, while a negative β means a decreasing trend.

3. Results

3.1. Performance of ERA5-Land

The daily minimum GST from DMDC (V3.0) was used to validate the daily minimum T_s from ERA5-Land, and the daily minimum TEM from DMDC (V3.0) was used to validate the daily minimum T_a from ERA5-Land. Figures 2 and 3 show the statistical indices between the daily minimum T_s and T_a from ERA5-Land and the daily minimum GST and TEM from DMDC (V3.0) at 822 national surface weather stations in China, respectively.

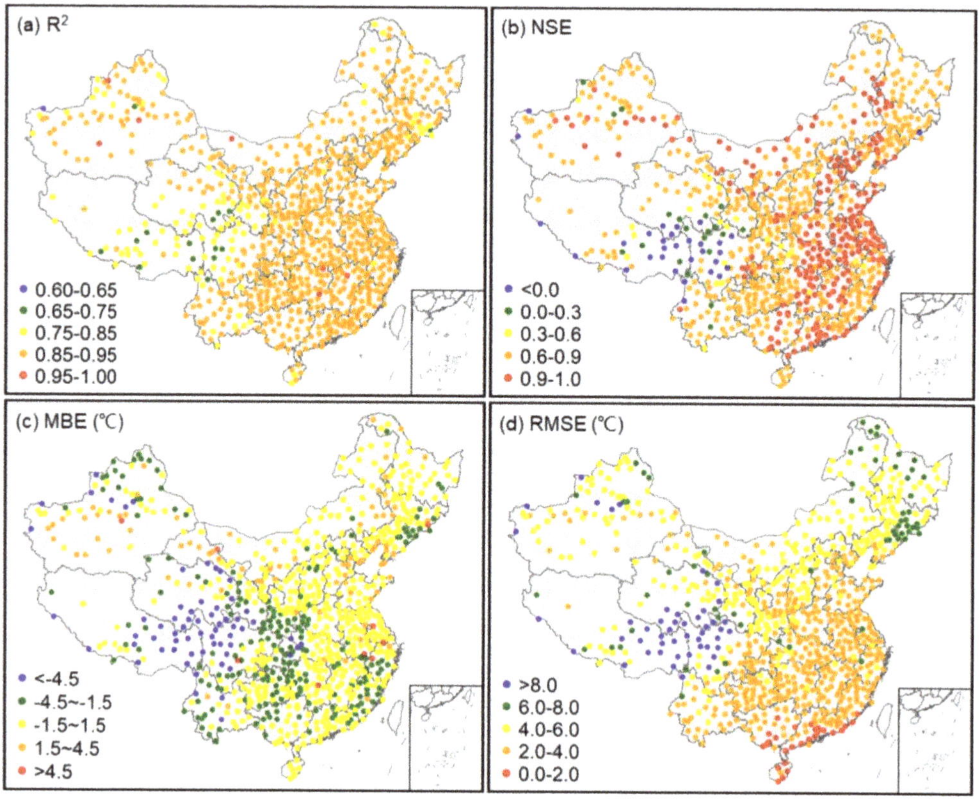

Figure 2. Statistical indices of the daily minimum T_s from ERA5-Land against observed daily minimum GST from CMCD (V3.0) at 822 national surface weather stations in China. (**a**) R^2, coefficient of determination; (**b**) *NSE*, Nash–Sutcliffe efficiency coefficient; (**c**) *MBE*, mean bias error; (**d**) *RMSE*, root mean square error.

As can be seen from Figures 2a and 3a, the correlation between ERA5-Land simulated and meteorological stations observed daily minimum surface temperature and air temperatures is pretty good, with an R^2 greater than 0.6 at all 822 national surface weather stations across China. Overall, the performance of ERA5-Land for daily minimum T_s and T_a simulation is better in East China than in West China, and better in flat regions than in high-altitude regions, and the performance of ERA5-Land for daily minimum T_s simulations is better than that for daily minimum T_a simulations. Meanwhile, it can be seen from Figures 2 and 3 that there is a significant underestimation of daily minimum T_s and T_a simulated by ERA5-Land, and this underestimation shows a regional pattern, with the underestimation of daily minimum T_s and T_a mainly concentrated in regions with high altitude and complex terrain, such as the southeastern Qinghai–Tibet Plateau. The regional underestimation of daily minimum T_s and T_a from ERA5-Land may have a negative impact on the quantification of local trends and variations in FFP and FD.

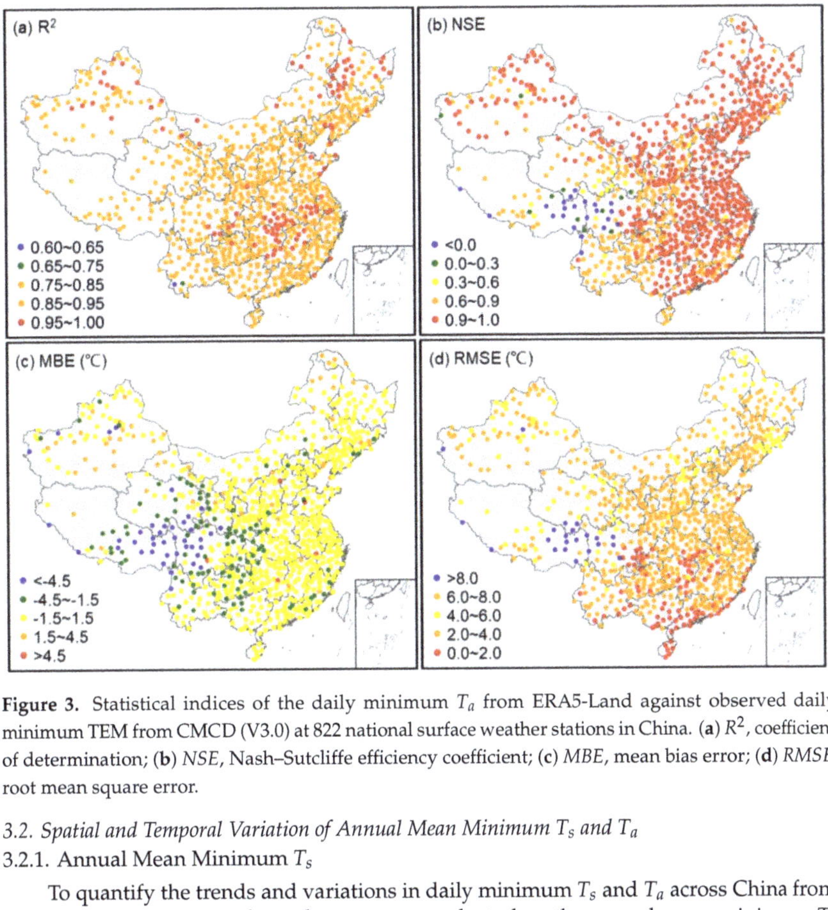

Figure 3. Statistical indices of the daily minimum T_a from ERA5-Land against observed daily minimum TEM from CMCD (V3.0) at 822 national surface weather stations in China. (**a**) R^2, coefficient of determination; (**b**) *NSE*, Nash–Sutcliffe efficiency coefficient; (**c**) *MBE*, mean bias error; (**d**) *RMSE*, root mean square error.

3.2. *Spatial and Temporal Variation of Annual Mean Minimum T_s and T_a*

3.2.1. Annual Mean Minimum T_s

To quantify the trends and variations in daily minimum T_s and T_a across China from 1950 to 2020, statistics and trend tests were conducted on the annual mean minimum T_s and T_a, as shown in Figure 4. From Figure 4a, the spatial distribution of the annual mean minimum T_s has significant latitudinal zonality and altitudinal dependence. The annual mean minimum T_s decreases with increasing altitude in high-altitude regions such as the Qinghai–Tibet Plateau, the Tianshan Mountains and the Altai Mountains, while the annual minimum T_s decreases with increasing latitude in East China and other regions.

Based on the MK trend test (Figure 4b), it shows that the annual mean minimum T_s generally shows a significant increasing trend in other regions of China, except for the borderlands in West China, the eastern Qinghai–Tibet Plateau and the Yunnan–Kweichow Plateau, where there is no significant increasing or decreasing trend of annual mean minimum T_s, but the degree of increase in annual mean minimum T_s varies greatly between regions. As can be seen from Figure 4c, Northeast China, northern Xinjiang, North China and the central and southeastern Qinghai–Tibet Plateau were the regions with the largest rates of increase in annual mean minimum T_s in China. In particular, the rate of increase in annual mean minimum T_s in Northeast China exceeded 0.2 °C/10a, with most regions experiencing increases of between 0.2 °C/10a and 0.53 °C/10a, indicating that Northeast

China was the region with the most significant increase in annual mean minimum T_s in China. In addition, as can also be seen from Figure 4c, individual regions, such as the eastern and southern margins of the Tianshan Mountains, parts of the eastern Qinghai–Tibet Plateau and parts of northwest Yunnan Province showed decreasing trends in annual mean minimum T_s, but for the whole country, the annual mean minimum T_s was generally increasing rapidly.

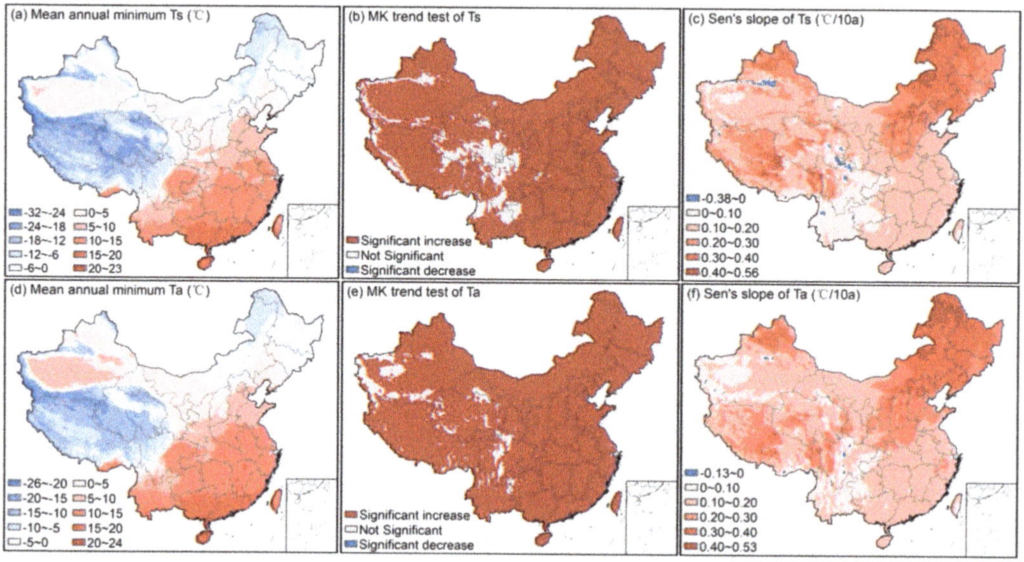

Figure 4. Spatial distribution of annual mean minimum T_s and T_a. (**a**) Trends of annual mean minimum T_s based on the Mann–Kendall trend test (95% confidence level), (**b**) Sen's slope of the annual mean minimum T_s, (**c**) annual mean minimum T_a, (**d**) trends of annual mean minimum T_a based on the Mann–Kendall trend test (95% confidence level), (**e**) Sen's slope of the annual mean minimum T_a (**f**) across China during the 1950~2020.

3.2.2. Annual Mean Minimum T_a

Similar to the spatial distribution pattern of the multi-year average annual mean minimum T_s, the spatial distribution of the annual mean minimum T_a also has obvious latitudinal zonality and altitudinal dependence (Figure 4d). Except for high-altitude regions such as the Qinghai–Tibet Plateau, the Tianshan Mountains and the Altai Mountains, the annual mean minimum T_a in East China decreases with increasing latitude, and the Tarim Basin is the region with the highest annual mean minimum T_a in Northwest China due to its relatively closed geographical environment and geological conditions.

Except for the periphery of the Qinghai–Tibet Plateau and parts of the central Tarim Basin, where there is no significant trend in annual mean minimum T_a, the rest of China shows a significant increasing trend (Figure 4e). Most regions of China show a significant increasing trend in annual mean minimum T_a, but the degree of increase in annual mean minimum T_a varies considerably from place to place. As can be seen from Figure 4f, Northeast China, northern Xinjiang, North China and the eastern and central Qinghai–Tibet Plateau were the regions with the greatest rate of increase in annual mean minimum T_a across China, especially Northeast China, where the rate of increase in annual mean minimum T_a greater than 0.2 °C/10a, with most regions experiencing increases ranging from 0.2 °C/10a to 0.53 °C/10a, indicating that Northeast China was the region with the most significant increase in annual mean minimum T_a across China. Although very few regions in the Sichuan Province and Turpan Basins showed a decreasing trend in annual mean minimum T_a, nationally, the annual mean minimum T_a was generally increasing rapidly.

In addition, as can be seen from Figure 4c,f, there was considerable spatial consistency in the variability of the annual mean minimum T_s and T_a across China.

3.3. Spatial and Temporal Variation of FFP

3.3.1. Spatial Distribution

Figure 5a shows the spatial distribution pattern of the multi-year average FFP across China from 1950 to 2020, from which the spatial distribution of FFP across China was characterized by obvious latitudinal zonality and altitude dependence. It is obvious that the high-altitude mountainous regions of the Qinghai–Tibet Plateau, the Qilian Mountains, the Tianshan Mountains and the Altai Mountains were the regions with the smallest FFP, especially in the northern Tibetan Plateau and the Kunlun Mountains in the northwestern part of the Qinghai–Tibet Plateau, where FFP was even less than 10 days and there was almost no FFP throughout the year. Despite the high altitude of the Tsaidam Basin in the northeastern Qinghai–Tibet Plateau, its relatively enclosed geography has resulted in a much higher FFP in the basin than in the rest of the Qinghai–Tibet Plateau, making it an extremely special region on the Qinghai–Tibet Plateau. Except for the high-altitude regions, FFP in East and Northwest China largely showed a trend of decreasing with increasing latitude. Provinces in the tropical monsoon climate zone, such as Yunnan, Guangxi, Guangdong, Taiwan and Hainan, have very high FFP, all greater than 350 days. The Sichuan Basin was also a region with very high FFP, with most regions exceeding 300 days. Overall, the spatial distribution of FFP was consistent with the spatial variation of multi-year average surface temperatures in China.

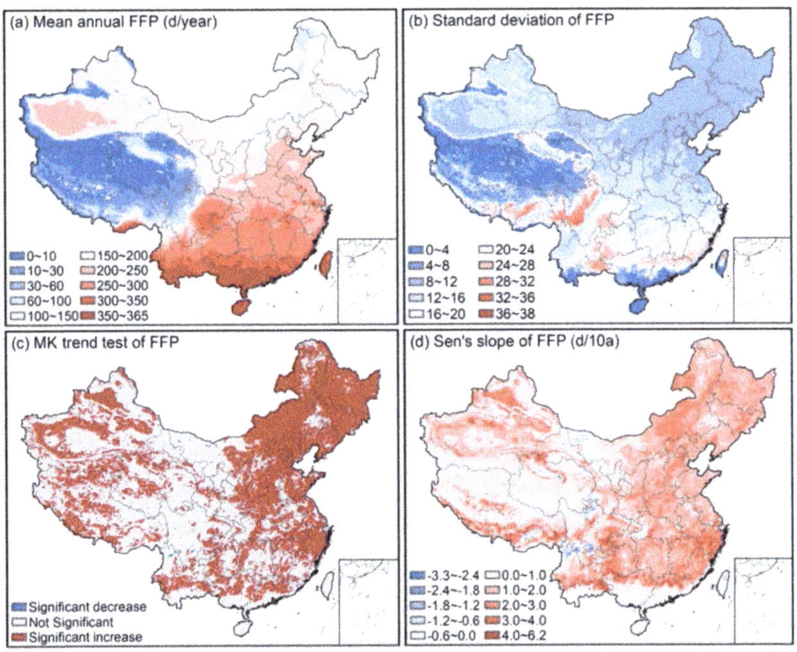

Figure 5. Spatial distribution of multi-year average FFP, (**a**) standard deviation of FFP, (**b**) trends of FFP based on the Mann–Kendall trend test (**c**) and Sen's slope of FFP (**d**) across China during the 1950~2020.

The standard deviation of FFP from 1950 to 2020 in Figure 5b shows that the spatial distribution of FFP varies significantly across China, with the eastern and southern parts of the Qinghai–Tibet Plateau, the regions around the Tsaidam Basin and the Sichuan Basin, as well as the southern edge of the subtropical monsoon climate zone, being regions of high interannual variability in FFP in China. Especially in the eastern and southern parts of the Qinghai–Tibet Plateau, where the standard deviation of FFP was greater than 20 days, indicating that the annual mean surface temperature in these regions was more variable indirectly. The standard deviation of FFP in other regions of China was not very variable and did not show a clear latitudinal zonality.

3.3.2. Temporal Variations

Interannual trends in FFP across China from 1950 to 2020 were quantified at the 95% confidence level based on the MK trend test and Sen's slope method, as shown in Figure 5c,d. As can be seen from Figure 5c, there were significant regional differences in the interannual trends of FFP across China. Overall, North China, Northeast China and Zhejiang Province were the regions where the significant increase in FFP was more concentrated. The outer Tarim Basin, the Junggar Basin and the Turpan Basin were also regions with a relatively high concentration of significant increases in FFP. In addition, the FFP in some regions of the Qinghai–Tibet Plateau and South China also showed a significant increasing trend, but did not show a clear regional pattern. Only a very few regions showed a significant decreasing trend in FFP, such as parts of the southern Sichuan Province and parts of the northern Yunnan Province. A comparison of Figure 5c,b shows that the interannual trends in FFP and the annual mean minimum T_s across China were not entirely consistent.

From the decadal rates of variation in FFP in Figure 5d, there were large differences in the degree of interannual variability in FFP across China. Most regions of China showed increasing trends in FFP, with rates of variation in FFP ranging from 0 to 6.2 d/10a, except for parts of western Inner Mongolia, northern Gansu Province, the Kunlun Mountains, the central and eastern Qinghai–Tibet Plateau and northern Yunnan Province. Overall, the rapid increase in FFP was the main feature of the variations in FFP across China during 1950~2020.

3.4. Spatial and Temporal Variation of FD

3.4.1. Spatial Distribution

Similar to the spatial distribution of FFP, the spatial distribution of FD also showed a clear latitudinal zonality and altitudinal dependence (Figure 6a). Figure 6a shows the spatial pattern of the multi-year average FD across China from 1950 to 2020. The FD was greatest in the high-altitude mountainous regions of the Qinghai–Tibet Plateau, the Qilian Mountains, the Tianshan Mountains and the Altai Mountains, especially in the Kunlun Mountains in the northwestern part of the Qinghai–Tibet Plateau, where the FD even greater than 350 days. Despite its high altitude, the relatively enclosed geography of the Tsaidam Basin results in much lower FD in the basin than in other parts of the Tibetan Plateau, making the Tsaidam Basin a relatively special region on the Qinghai–Tibet Plateau. Apart from the high-altitude mountains, the FD in East China and Northwest China basically showed an increasing trend with increasing latitude. Provinces in the tropical monsoon climate zone, such as Yunnan, Guangxi, Guangdong, Taiwan and Hainan, have very low FD, all less than 10 days. The Sichuan Basin was also a region with very few FD, with most regions having less than 10 days. Overall, the spatial distribution of FD was consistent with the spatial variation of the annual mean minimum air temperature and was the opposite of the spatial distribution of FFP in China.

The standard deviation of FD from 1950 to 2020 (Figure 6b) shows that there were significant differences in the spatial variation of FD across China, with the central Qinghai–Tibet Plateau, the region around the Tsaidam Basin and the region around the Sichuan Basin being regions with large interannual variability in FD in China, indicating greater variability in air temperature in these regions indirectly. The standard deviation of FD in the rest of China varied little and was similar to the spatial variation of FPP, showing no clear latitudinal zonality.

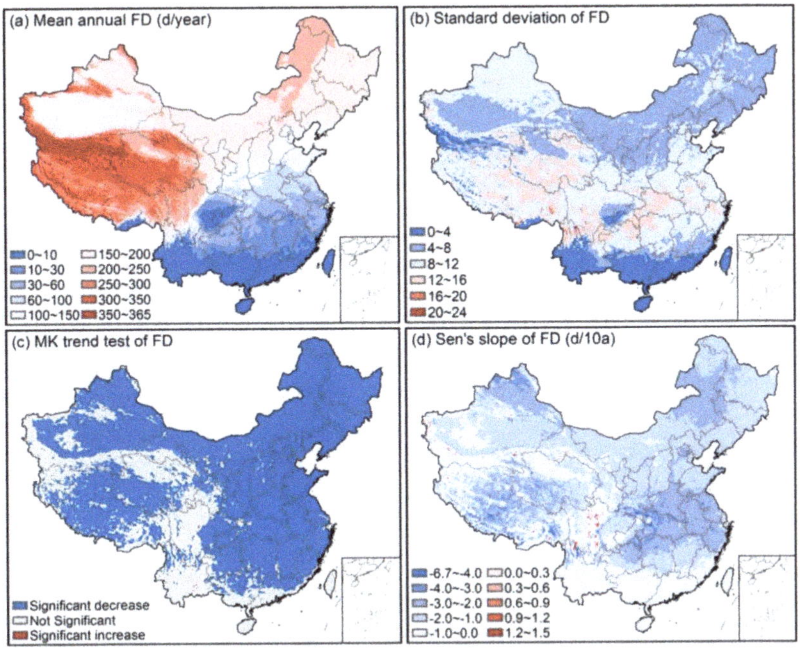

Figure 6. Spatial distribution of multi-year average annual FD, (**a**) standard deviation of FD, (**b**) trends of FD based on the Mann–Kendall trend test (**c**) and Sen's slope of the FD (**d**) across China during the 1950~2020.

3.4.2. Temporal Variations

Interannual trends in FD across China from 1950 to 2020 were quantified at the 95% confidence level based on the MK trend test and Sen's slope method, as shown in Figure 6c,d. As can be seen from Figure 6c, the trends of FD across China showed obvious regional differences. Overall, the FD in the peripheral alpine regions of the Qinghai–Tibet Plateau, Sichuan Basin, most of Yunnan Province, southern Guangxi Province, southern Guangdong Province, Taiwan Province and Hainan Province did not show a significant decreasing or increasing trend from 1950 to 2020. Except for the above-mentioned regions, all other regions in China showed a significant decreasing trend in FD, and only a very few regions showed a significant increasing trend in FD, indicating that the trends in FD across China were dominated by significant increasing trends, which was consistent with the trends in air temperature across China in the past decades.

Despite the clear regionalization of FD trends, the degree of increase or decrease in FD varied widely across China. Except for some alpine regions on the periphery of the Qinghai–Tibet Plateau, where an increasing trend in FD is seen, most regions in China showed a decreasing trend in FD, with rates of variation in FD ranging from 0 to −6.7/10a. The central and western parts of the Qinghai–Tibet Plateau, the middle and lower reaches of the Yangtze River, the Huaihe River Basin, northern Inner Mongolia and northern Xinjiang were the regions with the largest FD reduction rates.

3.5. Regional Interannual Trends and Variations in FFP and FD

The interannual trends and variations in FFP and FD for different climate zones were calculated by averaging the data for all gridded annual FFP and annual FD within each climate zone. Figure 7 shows the interannual trends and variations in FFP and FD across China and different climate zones from 1950 to 2020. As can be from Figure 7a, the annual FFP of China and different climate type zones showed a significant increasing trend, among which, the increase in FFP of SUMZ was the largest and the increase in FFP of TRMZ was the smallest. The national average rate of increase in FFP was 1.25 d/10a, and the rate of increase in FFP in SUMZ was 1.73 d/10a. The increase in annual FFP of China and different climate type zones was SUMZ > TEMZ > China > TECZ > ALCZ > TRMZ in order. Although the annual FFP showed an increasing trend across China and different climate zones, the process of increase was simultaneously phased. It was clear from Figure 7a that the FFP experienced a decreasing process for both China and the different climate zones from 1950 to the 1970s and 1980s, and then maintained a continuous fluctuating increasing trend from the 1970s and 1980s to the present. The phased variations in annual FFP across China were consistent with the phased variations in surface temperature across China, which was also consistent with the apparent warming of China from the 1970s and 1980s onwards [68,69].

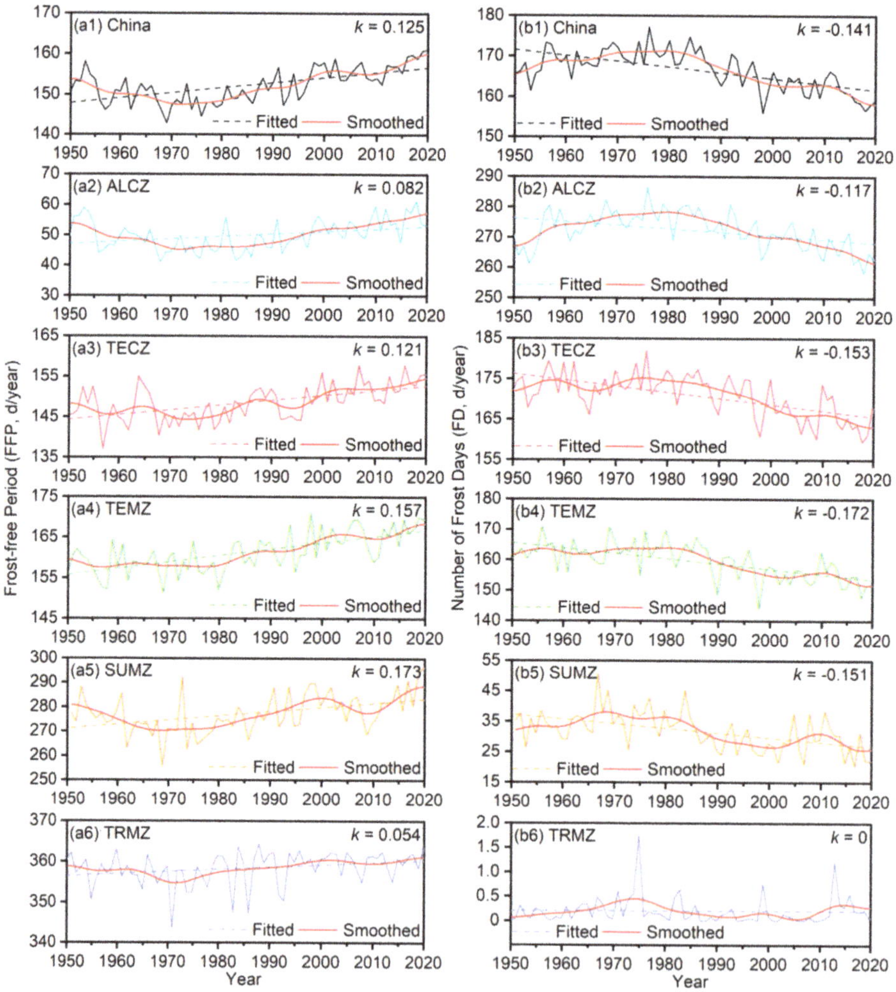

Figure 7. Interannual trends and variations of FFP and FD for China and different climate type zones in China during the 1950~2020. k represents the degree of variation in annual FFP and FD, respectively, i.e., the slope of the data variation. **(a1–a6)** Annual FFP; **(b1–b6)** Annual FD.

Contrary to the trends and variations of annual FFP across China and different climate zones, except for the TRMZ, the annual FD showed a clear decreasing trend across China and other climate zones, where the TEMZ was the climate zone with the greatest decrease in FD. Despite the large interannual fluctuations in the FD of the TRMZ, the FD of the TRMZ from 1950 to 2020 did not show a significant decreasing trend. The national average rate of decrease in FD was −1.41 d/10a, and the rate of decrease in FD in SUMZ was −1.72 d/10a. The degree of decrease in annual FD across China and different climate zones was TEMZ > TECZ > SUMZ > China > ALCZ > TRMZ in order. Similarly, while there was a general decreasing trend in FD across China and in different climate zones, the decrease has been phased. It is clear from Figure 7b that the FD experienced an increasing process for both China and the different climate zones from 1950 to the 1970s and 1980s, and then maintained a continuous, fluctuating, decreasing trend from the 1970s and 1980s to the present. The phased variations in FD across China were consistent with the phased variations in air temperature across China, which was also consistent with the marked warming of China from the 1970s and 1980s onwards.

Among the different climate zones in China, the SUMZ was the zone with the most significant increase in FFP, while the TEMZ was the zone with the most significant decrease in FD, indicating that the increase in FFP and the decrease in FD was not consistent across China.

3.6. Regional Variation Differences in FFP and FD

In order to explore the degree of difference in the interannual variations of FFP and FD across China, the coefficient of variation (C_V) of FFP and FD across China was calculated. As can be seen from Figure 8, there were significant regional differences in the C_V of FFP and FD across China. It was obvious that, except for the Tarim Basin, high-altitude regions such as the Qinghai–Tibet Plateau, the Tianshan Mountains and the Altai Mountains were the regions with the largest C_V in FFP, with C_V ranging from 0.3 to 8.4. In particular, the C_V of FFP was even greater than 1 in the Kunlun Mountains in the northwestern part of the Qinghai–Tibet Plateau, while the C_V of FFP in the rest of China was not very variable, with C_V ranging from 0 to 0.3. Similarly, the C_V of FD across China also has obvious regional differences, with the Sichuan Basin, South China and southern Tibet being the regions with the largest C_V of FD, with a C_V of FD ranging from 0.3 to 8.4. In particular, the regions near the Tropic of Cancer were the regions with the largest C_V of FD, with C_V of FD greater than 1.

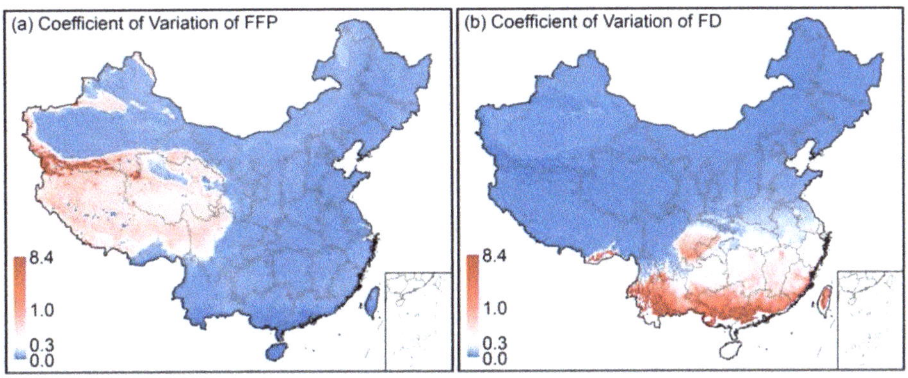

Figure 8. Comparison of the coefficient of variation (C_V) of FFP and FD across China from 1950~2020. (**a**) C_V of FFP; (**b**) C_V of FD.

The obvious regional differences in the C_V of FFP and FD across China indicating that the higher the altitude and lower the temperature, the greater the interannual variability of FFP, while the higher the temperature, the greater the interannual variability of FD. In addition, since there was little FD in the tropical monsoon climate zone in China, the C_v of FD in this region can be ignored.

Although the large regional variability exhibited by FFP and FD is related to regional climatic characteristics, the variability in FFP and FD differs from the latitudinal zonality and altitudinal dependence of climate. In particular, the spatial characteristics of FD variability are hardly latitudinally zoned, and the large regional variability of FFP and FD may have important implications for regional agricultural production, e.g., in the subtropical monsoon region of southern China, an increase in FD may cause damage to spring crops, which in turn may lead to reduced crop yields. Therefore, the large regional variability of FFP and FD may indicate the instability of regional climatic characteristics and regions where agrometeorological disasters occur more frequently to a certain extent.

4. Discussion

4.1. Uncertainty of ERA5-Land on the Quantification of FFP and FD

As can be seen from Figures 2a and 3a, the daily minimum T_s and T_a from ERA5-Land correlate reasonably well with the daily minimum GST and TEM from DMDC (V3.0). However, as can also be seen from Figures 2b–d and 3b–d, there is a clear regional underestimation of the daily minimum T_s and T_a from ERA5-Land. Based on the four evaluation indices, including the R^2, NSE, MBE and $RMSE$, ERA5-Land is generally more applicable to surface and air temperatures simulation in East China than in West China and the Qinghai–Tibet Plateau, more applicable in flatter regions than in regions with high altitude and complex terrain, and more applicable in regions with dense weather stations than in regions with sparse weather stations. In particular, the southeast Qinghai–Tibet Plateau is the region where the ERA5-Land underestimates daily minimum T_s and T_a most significantly (Figures 2 and 3).

Studies have shown that the applicability of reanalysis dataset is poor in regions with sparse meteorological stations and complex terrain, such as Northwest China and the Qinghai–Tibet Plateau [43,70–73]. Reanalysis dataset is an important product of numerical modelling, obtained by assimilating quality-controlled observations (including ground, sounding, satellite, radar, buoy, aircraft, ships and other observations) into global or regional numerical model calculations [74,75]. Therefore, the accuracy of the reanalysis dataset for the simulation of land variables is reduced when ground-based or airborne observations are scarce. In addition, considering the complex terrain of the southeastern Qinghai–Tibet Plateau, the terrain height of the reanalysis dataset may differ from the actual altitude, leading to a deviation in the surface and air temperatures between the reanalysis and the in situ observed data [43]. In regions with complex terrain, the meteorological stations are generally constructed at lower elevations in flatter river valleys [36], and the terrain around the stations is highly undulating, with average elevations much higher than the elevations of the meteorological stations, making the average surface and air temperatures over a larger region much lower than those at lower elevations due to the temperature lapse rate [43].

The underestimation of daily minimum surface and air temperatures in a reanalysis dataset such as ERA5-Land is inevitable due to the scarcity of meteorological stations at high altitudes on the southeastern Qinghai–Tibet Plateau. Consequently, it is inevitable that the lack of ground-based meteorological stations on the southeastern Qinghai–Tibet Plateau leads to a reduction in the simulation accuracy of the ERA5-Land reanalysis dataset, which in turn leads to a reduction in the quantification accuracy of FFP and FD in this region. In fact, high-altitude mountainous regions such as the southeastern Qinghai–Tibet Plateau are not major crop production regions in China, so the uncertainty in the trends and variations in FFP and FD in this region based on the ERA5-Land does not affect the accuracy of quantifying the trends and variations in FFP and FD in China as a whole.

4.2. Effect of Variations in FFP and FD on Agriculture Production

As an important characteristic quantity of agro-climatic heat conditions, the FFP plays a crucial role in agricultural production, especially in North China [31]. Generally, the longer the FFP, the longer the growth period of crops, and the more conducive to the development of agricultural production [30,31]. Under climate warming, the increase in FFP leads to an increase in heat resources, and the increase in heat resources can prolong the growth period of crops or increase agricultural biological yields per unit region [22,31,32,76]. Frost is a meteorological disaster that can cause serious damage to crop growth [34,77]. The decrease in FD reduces the risk of frost damage to crops and is extremely beneficial to agricultural production [34,55]. The decrease in FD in North China has also led to a further expansion of wheat cultivation northwards [78,79]. In summary, the increase in FFP and the decrease in FD have had three main effects on agricultural production: firstly, the increase in FFP has contributed to a longer crop growing period; secondly, the decrease in FD has reduced the risk of extreme frost damage to crops; and thirdly, the increase in FFP and the decrease in FD have contributed to an adjustment in the layout of agricultural production to adapt to climate change.

However, the increase in FFP and the decrease in FD across China are the result of climate warming [16] which is not entirely positive for agricultural production. Generally, crops are divided into thermophilic crops such as rice and maize and chimonophilous crops such as wheat and barley [80–82], where climate warming is beneficial to the growth of thermophilic crops but detrimental to the growth of chimonophilous crops [81–84]. Studies have shown that climate warming leads to a reduction in the yields of chimonophilous crops such as barley on the Qinghai–Tibet Plateau, as it leads to a greater difference in saturated water vapor pressure, which in turn reduces the stomatal conductivity of plant leaves, reduces photosynthesis and ultimately leads to lower crop yields [85]. Climate warming has led to a longer growing season for thermophilic crops, but also to a shorter growing season for chimonophilous crops [81,82]. Therefore, when planning the layout of agricultural production, it is necessary to expand the cultivation of thermophilic crops in accordance with the increase in FFP and the decrease in FD, while shifting the cultivation of chimonophilous crops to regions with lower temperatures, thus adjusting the layout of agricultural production more scientifically in order to better adapt to climate change.

5. Conclusions

To investigate the trends and variations in FFP and FD across China under climate change, this study conducted a quantitative statistics and trend test of the FFP and FD across China from 1950 to 2020 based on the ERA5-Land reanalysis dataset. Although the ERA5-Land simulations of daily minimum T_s and T_a are significantly underestimated regionally in regions such as the southeastern Qinghai–Tibet Plateau with high altitude and complex terrain, from a national perspective, the ERA5-Land has good applicability for quantifying the trends and variations in FFP and FD.

The study revealed that both the annual mean minimum T_s and T_a showed a significant increasing trend from 1950 to 2020, with particularly pronounced warming in Northeast China, North China and northern Xinjiang, which has a lasting impact on the variations in FFP and FD. Quantitative statistics and trend test of FFP and FD across China from 1950 to 2020 found that both FFP and FD showed significant latitudinal zonality and altitude dependence, i.e., FFP decreased with increasing latitude and altitude, while FD increased with increasing latitude and altitude; at the 95% confidence level, FFP generally showed an increasing trend, while FD generally showed a decreasing trend across China, but there were large spatial differences in the increase in FFP and decrease in FD between regions, suggesting that the increase in FFP and decrease in FD was not entirely consistent in space.

For China as a whole, FFP showed a significant increasing trend with average increasing rate of 1.25 d/10a, the maximum increasing rate of FFP in individual regions was 6.2 d/10a, while FD showed a significant decreasing trend with an average decreasing rate of −1.41 d/10a, the maximum decreasing rate of FD in individual regions was −6.7 d/10a.

Among the five major climate zones in China, the SUMZ was the region with the largest increase in FFP, with a rate of 1.73/10a, while the TEMZ was the region with the largest decrease in FD, with a rate of −1.72 d/10a. The variations in FFP and FD across China and in different climate zones were phased, which was consistent with the phased variations of surface and air temperatures across China over the past 70 years.

There is no denying that the variations in FFP and FD across China have a positive impact on agricultural production, but the adaptability of different crop species to climate change should also be considered when adjusting the layout of agricultural production to make it better and more scientifically adapted to climate change.

Author Contributions: Conceptualization, H.L. and C.H.; methodology, H.L. and Y.Y.; software, H.L. and G.L.; investigation, H.L., G.L. and C.H.; data curation, C.H. and R.C.; writing—original draft preparation, H.L.; writing—review and editing, G.L., C.H., Y.Y. and R.C.; funding acquisition, C.H., R.C. and Y.Y. All authors have read and agreed to the published version of the manuscript.

Funding: This research was funded by the National Natural Sciences Foundation of China (41971041), the Joint Research Project of Three-River Headwaters National Park, Chinese Academy of Sciences and the People's Government of Qinghai Province (LHZX-2020-11), the Sciences and Technology Plan Project of Gansu Province (21JR7RA056) and the Open Research Fund of the National Cryosphere Desert Data Center (2021kf09).

Institutional Review Board Statement: Not applicable.

Informed Consent Statement: Not applicable.

Data Availability Statement: The ERA5-Land reanalysis dataset from the ECMWF used in this study can be accessed online (https://cds.climate.copernicus.eu/cdsapp#!/dataset/reanalysis-era5-land?tab=overview/ (accessed on 1 March 2022)). The Daily Meteorological Dataset of basic meteorological elements of China National Surface Weather Stations provided by the China Meteorological Administration can be accessed online (http://data.cma.cn/ (accessed on 1 March 2022)).

Acknowledgments: The authors would like to thank the European Centre for Medium-Range Weather Forecasts (ECMWF) (https://cds.climate.copernicus.eu/ (accessed on 1 March 2022)) for providing the ERA5-Land data, and the authors would also like to thank all team members in Qilian Alpine Ecology and Hydrology Research Station, Northwest Institute of Eco-Environment and Resources, Chinese Academy of Sciences.

Conflicts of Interest: The authors declare no conflict of interest.

References

1. Fan, X.; Wang, Q.; Wang, M. Variations in frost-free period in Shanxi province, China: 1959–2008. *Ecol. Environ. Sci.* **2010**, *19*, 2393–2397.
2. Yu, W.; Yang, Q. Change Laws of Frost-free Period and Its Impact on Agriculture in Linjiang. *J. Anhui Agric. Sci.* **2012**, *40*, 8248–8249.
3. Gao, M.; Jia, H.; Pei, Z.; Yang, X.; Wang, H. The Changing Trends of the First and Last Frost Dates, Frost-free Periods in Recent 58 Years in Linfen City. *Chin. Agric. Sci. Bull.* **2014**, *30*, 272–275.
4. Karl, T.R.; Ghazi, N.A. Clivar/GCOS/WMO Workshop on Indices and Indicators for Climate Extremes Workshop Summary. In *Weather and Climate Extremes*; Springer: Dordrecht, The Netherlands, 1999.
5. Peterson, T.; Folland, C.; Gruza, G.; Hogg, W.; Mokssit, A.; Plummer, N. *Report on the Activities of the Working Group on Climate Change Detection and Related Rapporteurs 1998–2001*; World Meteorological Organization: Geneva, Switzerland, 2001.
6. Ma, Z. Variation of Frost Days and Its Relationship to Regional Warming in Northern China. *Acta Geogr. Sin.* **2003**, *58*, 31–37.
7. Kong, X.; Feng, C.; Cao, D.; Liu, Y.; Han, X. Analysis on the Changes of the First and Last Frost Date and the Frost-free Period in Linqing City in Recent 30 Years. *J. Hebei Agric. Sci.* **2014**, *18*, 97–99, 103.
8. Erlata, E.; Türke, M. Analysis of observed variability and trends in numbers of frost days in Turkey for the period 1950–2010. *Int. J. Climatol.* **2012**, *32*, 1889–1898. [CrossRef]
9. Gao, H.; Cai, G.; Lu, Y.; Zhang, X. Spatial and temporal variations of frost days in Guizhou province from 1960 to 2013. *J. Agric. Sci.* **2017**, *38*, 16–19, 42.
10. Wang, G.; Xu, Y.; Zhu, Y.; Zhang, H.; Liao, J. The Spatial and Temporal Characteristics and Trend of Frost Days over China for the Past 50 Years. *Meteorol. Mon.* **2009**, *35*, 61–67.
11. Wypych, A.; Ustrnul, Z.; Sulikowska, A.; Chmielewski, F.-M.; Bochenek, B. Spatial and temporal variability of the frost-free season in Central Europe and its circulation background. *Int. J. Climatol.* **2016**, *37*, 3340–3352. [CrossRef]

12. Chervenkov, H.; Slavov, K. Inter-annual variability and trends of the frost-free season characteristics over Central and Southeast Europe in 1950–2019. *J. Cent. Eur. Agric.* **2022**, *23*, 154–164. [CrossRef]
13. IPCC. *Climate Change 2021: The Physical Science Basis*; IPCC: Geneva, Switzerland, 2021.
14. Chao, Q.; Yan, Z.; Sun, Y.; Jiang, Z.; Liao, H.; Jia, S.; Cai, R. A recent scientific understanding of climate change in China. *China Popul. Resour. Environ.* **2020**, *30*, 1–9.
15. Li, Y.; Xie, L. Current situation and cause analysis of global climate change. In Proceedings of the 2010 Annual academic Meeting of Chinese Crop Society, Shenyang, China, 13–14 September 2010.
16. Popov, T.; Gnjato, S.; Trbic, G. Trends in frost days in Bosnia and Herzegovina. *Glas. Srp. Geogr. Drus.* **2017**, *97*, 35–55. [CrossRef]
17. Easterling, D. Recent Changes in Frost Days and the Frost-Free Season in the United States. *Bull. Am. Meteorol. Soc.* **2002**, *83*, 1327–1332. [CrossRef]
18. Xu, F.; Zhang, S.; Wang, Q.; Chang, C.; Qin, Y. First, Last Frost Dates and Frost-free Period During 1961–2015 in Shouguang:Variation Characteristics and Its Effect on Agriculture. *Chin. Agric. Sci. Bull.* **2019**, *35*, 112–117.
19. Zhou, X.; Zhao, C.; Cui, Y.; Liu, M.; Ao, X.; Yi, X.; Hou, Y.; Zhang, X. Variation Characteristics of the Date of First Frost and Last Frost and the Frost-free Period in the Northeast China during 1961–2013. *J. Nat. Resour.* **2017**, *32*, 494–506.
20. Yang, X.; Hu, J.; Zhang, Y. Change Characteristics of First and Last Frost Date and Frost-Free Period in Eastern Hexi Corridor. *Chin. Agric. Ence Bull.* **2016**, *32*, 149–155.
21. Bao, Y. Analysis on Variations of First Frost Date, Latest Frost Date and Frost-free Period in Ewenki Autonomous Banner of Inner Mongolia. *Anim. Husb. Feed Sci.* **2016**, *37*, 54–56.
22. Ning, X.; Liu, G.; Zhang, L.; Qin, X.; Zhou, S.; Qin, Y. The spatio-temporal variations of frost-free period in China from 1951 to 2012. *J. Geogr. Sci.* **2017**, *27*, 23–42. [CrossRef]
23. Rahimi, M. Trends in frost days, late and early frost dates and the frost-free season over Iran during 1951–2005. In Proceedings of the WCRP OSC Climate Research in Service to Society, Denver, CO, USA, 24–28 October 2011.
24. Li, Z.; You, Q.; Liu, H.; Yin, Z.; Duan, L. Analysis of Climate Change and Circulation Features of Frost Days in Hunan Province, China in Recent 67 Years. *J. Geosci. Environ. Prot.* **2019**, *7*, 124–137. [CrossRef]
25. Malinovic-Milicevic, S.; Stanojevic, G.; Radovanovic, M. Recent changes in first and last frost dates and frost-free period in Serbia. *Geogr. Ann. Ser. A Phys. Geogr.* **2018**, *100*, 44–58. [CrossRef]
26. McCabe, G.; Betancourt, J.; Feng, S. Variability in the Length of Frost-Free Periods in the Conterminous US during the Past Century. *Int. J. Climatol.* **2015**, *35*, 4673–4680. [CrossRef]
27. Longley, R.W. The frost-free period in alberta. *Can. J. Plant Sci.* **1967**, *47*, 239–249. [CrossRef]
28. Meehl, G.A.; Tebaldi, C.; Nychka, D. Changes in frost days in simulations of twentyfirst century climate. *Clim. Dyn.* **2004**, *23*, 495–511. [CrossRef]
29. Liu, B.; Henderson, M.; Xu, M. Spatiotemporal change in China's frost days and frost-free season, 1955–2000. *J. Geophys. Res. Atmos.* **2008**, *113*, D12104. [CrossRef]
30. Li, L.; Li, X.; Xu, X.; Wang, X.; Yang, S. Climatic characteristics of frost-free period in Datong and its impact on agricultural production. *Agric. Technol.* **2020**, *40*, 114–115.
31. Pan, Z.; Guan, X.; Chen, X. Analysis of frost free-period and its impact on agricultural production. *Sci. Tech. Inf. Gansu* **2018**, *47*, 22–26.
32. Yu, W.; Cao, D.; Gao, S. Effects of Frost-free Date Change on the Agricultural Production in Xinmin City from 1960 to 2009. *J. Anhui Agric. Sci.* **2010**, *38*, 11929–11930, 11932.
33. Xu, B.; Zhang, L. Research on the characteristics of castor growth in an area with short frost-free periods. *Chin. J. Oil Crop Sci.* **1988**, *9*, 47–49.
34. Du, J.; Shi, L.; Yuan, L. Responses of Climatic Change on the Frost Days in Main Agricultural Area of Tibet from 1961 to 2010. *Chin. J. Agrometeorol.* **2013**, *34*, 264–271.
35. Du, L.; Chen, X.; Yang, D.; Shang, K.; Wang, S. Variation Characteristics of the Large-scale Frost in East Region of the Yellow River of Gansu in Recent 40 Years. *Meteorol. Environ. Res.* **2011**, *2*, 42–45, 59.
36. Zou, H.; Zhu, J. Applicability of reanalysis data in the study of surface temperature in Qinghai–Tibet Plateau. In Proceedings of the 28th Annual Meeting of China Meteorological Society, Xiamen, China, 1–3 November 2011.
37. Zhang, X. *Dynamics Changes of Typical Inland Lakes on Tibetan Plateau Using Multi-Sensor Remote Sensing Data*; Northwest A&F University: Xi'an, China, 2015.
38. Liao, R.; Cao, L.; Zhang, D.; Li, Y. Validation of Gridded Precipitation and Temperature Data over China. *Meteorol. Sci. Technol.* **2017**, *45*, 364–374.
39. Bai, L.; Zhang, F.; Wen, Y.; Shi, C.; Wu, J.; Shang, M.; Zhu, Z.; Meng, J. Evolution of the Frost Hazards Based on Gridded Meteorological Data across China in 1961−2018. *Chin. J. Agrometeorol.* **2021**, *42*, 761–774.
40. Liu, T.; Kun, Y.; Jun, Q.; Fuqiang, T. Construction and applications of time series of monthly precipitation at weather stations in the central and eastern Qinghai-Tibetan Plateau. *Plateau Meteorol.* **2018**, *37*, 1449–1457.
41. Liang, Z.; Shang, Y.; Wang, X.; Fang, Y.; Zhang, T.; Ren, Y.; Yang, X. Analysis on applicability of meteorological rainfall data in hydrological simulation of ungaged watershed. *Water Resour. Hydropower Eng.* **2021**, *52*, 1–12.
42. Li, H. *DEM Dividing Regions Optimization of China Land's Accumulated Temperature Spatial Interpolation and Precision Analysis*; Northwest Normal University: Lanzhou, China, 2007.

43. Yang, J.; Huang, M.; Zhai, P. Performance of the CRA-40/Land, CMFD, and ERA-Interim Datasets in Reflecting Changes in Surface Air Temperature over the Tibetan Plateau. *J. Meteorol. Res.* **2021**, *35*, 663–672. [CrossRef]
44. Yue, S.; Yan, Y.; Zhang, S.; Yang, J.; Wang, W. Spatiotemporal variations of soil freeze-thaw state in Northeast China based on the ERA5-LAND dataset. *Acta Geogr. Sin.* **2021**, *76*, 2765–2779.
45. Chen, X.; Su, B.; Ma, Y.; Trigo, I.F.; Gentine, P. Remote Sensing of Global Daily Evapotranspiration based on a Surface Energy Balance Method and Reanalysis Data. *J. Geophys. Res. Atmos.* **2021**, *126*, e2020JD032873. [CrossRef]
46. Chen, H.; Gao, H.; Wang, T.; Zhang, W.; Chen, X.; Nie, N.; Liu, H. Combining sparse observations and reanalysis data for refining spatiotemporal variability in near-surface air temperature lapse rates over China. *Int. J. Climatol.* **2021**, *41*, 6768–6784. [CrossRef]
47. Gao, Z.; Huang, B.; Ma, Z.; Chen, X.; Qiu, J.; Liu, D. Comprehensive Comparisons of State-of-the-Art Gridded Precipitation Estimates for Hydrological Applications over Southern China. *Remote Sens.* **2020**, *12*, 3997. [CrossRef]
48. Xiang, Y.; Chen, J.; Li, L.; Peng, T.; Yin, Z. Evaluation of Eight Global Precipitation Datasets in Hydrological Modeling. *Remote Sens.* **2021**, *13*, 2831. [CrossRef]
49. Wang, M.; Yao, S.; Jiang, L.; Liu, Z.; Shi, C.; Hu, K.; Zhang, T.; Zhang, Z.; Liu, J. Collection and Pre-Processing of Satellite RemoteSensing Data in CRA-40 (CMA's Global Atmospheric ReAnalysis). *Adv. Meteorol. Sci. Technol.* **2018**, *8*, 158–163.
50. Muñoz-Sabater, J.; Dutra, E.; Agustí-Panareda, A.; Albergel, C.; Arduini, G.; Balsamo, G.; Boussetta, S.; Choulga, M.; Harrigan, S.; Hersbach, H.; et al. ERA5-Land: A state-of-the-art global reanalysis dataset for land applications. *Earth Syst. Sci. Data* **2021**, *13*, 4349–4383. [CrossRef]
51. Zhang, J.; Zhao, T.; Li, Z.; Li, C.; Li, Z.; Ying, K.; Shi, C.; Jiang, L.; Zhang, W. Evaluation of Surface Relative Humidity in China from the CRA-40 and Current Reanalyses. *Adv. Atmos. Sci.* **2021**, *38*, 1958–1976. [CrossRef]
52. Ye, M.; Yao, X.; Zhang, T.; Xu, X.; Wang, S. Intercomparison of CRA-Interim Precipitation Products with ERA5 and JRA-55. *J. Trop. Meteorol.* **2020**, *26*, 136–147.
53. Han, R.; Li, W.; Ai, W.; Song, Y.; Ye, D.; Hou, W. The Climatic Variability and Influence of First Frost Dates in Northern China. *Acta Geogr. Sin.* **2010**, *65*, 525–532.
54. Zhang, L.; Yang, Y.; Zhang, X.; Yuan, H.; Jin, F.; Du, H. Characteristics and Change Trend of Frost Days in Ning Xia in Recent 50 Years. *Chin. Agric. Sci. Bull.* **2015**, *31*, 214–219.
55. Labaciren; Suolangjiacuo; Baima. Spatial and temporal distribution of frost days over Tibet from 1981 to 2010. *Acta Geogr. Sin.* **2014**, *69*, 690–696.
56. Yang, Y.; Chen, R.; Liu, G.; Liu, Z.; Wang, X. Trends and variability in snowmelt in China under climate change. *Hydrol. Earth Syst. Sci.* **2022**, *26*, 305–329. [CrossRef]
57. Yang, Y.; Chen, R.; Han, C.; Liu, Z. Evaluation of 18 models for calculating potential evapotranspiration in different climatic zones of China. *Agric. Water Manag.* **2021**, *244*, 106545. [CrossRef]
58. Tang, J.; Li, W.; Zhao, S. Progress and directions in studying the impact of climate change on agriculture and grain production in China. *Chin. J. Agric. Resour. Reg. Plan.* **2013**, *34*, 1–7.
59. Muñoz-Sabater, J. ERA5-Land hourly data from 1950 to 1980. Copernicus Climate Change Service (C3S) Climate Data Store (CDS). 2021. Available online: https://cds.climate.copernicus.eu/cdsapp#!/dataset/reanalysis-era5-land?tab=overview/ (accessed on 10 March 2022).
60. Muñoz-Sabater, J. ERA5-Land Hourly Data from 1981 to Present. Copernicus Climate Change Service (C3S) Climate Data Store (CDS). 2019. Available online: https://cds.climate.copernicus.eu/cdsapp#!/dataset/reanalysis-era5-land?tab=overview/ (accessed on 10 March 2022).
61. Center, N.M.I. Daily Meteorological Dataset of Basic Meteorological Elements of China National Surface Weather Station (V3.0) (1951–2010). National Tibetan Plateau Data Center. 2019. Available online: https://data.tpdc.ac.cn/en/ (accessed on 10 March 2022).
62. Liu, Z.; Tao, L. Discussion on Statistical Methods of Frost-free Period in Strict Meaning. *Mod. Agric. Sci. Technol.* **2019**, *47*, 160–163.
63. Mann, H. Non-Parametric Test Against Trend. *Econometrica* **1945**, *13*, 245–259. [CrossRef]
64. Kendall, M.G. *Rank Correlation Methods*, 4th ed.; Charles Griffin: London, UK, 1975.
65. Agarwal, S.; Suchithra, A.S.; Singh, S.P. Analysis and Interpretation of Rainfall Trend using Mann-Kendall's and Sen's Slope Method. *Indian J. Ecol.* **2021**, *48*, 453–457.
66. Jagadeesh, P.; Agrawal, S. Investigation of trends and its magnitude by non-parameteric Mann-Kendall and Sen's slope methods. *Int. J. Hydrol. Sci. Technol.* **2015**, *5*, 83–94. [CrossRef]
67. Sen, P.K. Estimates of the Regression Coefficient Based on Kendall's Tau. *J. Am. Stat. Assoc.* **1968**, *63*, 1379–1389. [CrossRef]
68. Zhang, H.; Wang, E.; Zhou, D.; Luo, Z.; Zhang, Z. Rising soil temperature in China and its potential ecological impact. *Sci. Rep.* **2016**, *6*, 35530. [CrossRef]
69. Liu, B.; Xu, M.; Henderson, M.; Qi, Y.; Li, Y. Taking China's Temperature: Daily Range, Warming Trends, and Regional Variations, 1955–2000. *J. Clim.* **2004**, *17*, 4453–4462. [CrossRef]
70. Li, S.; Wu, W.; Huang, X. Wind Speed Evaluation of CLDAS and ERA5-Land in Sichuan. *Plateau Mt. Meteorol. Res.* **2020**, *40*, 70–76.
71. Namaiti, H.; Rusuli, Y.; Dilixiati, M.; Aikemu, R. Adaptability Analysis of ERA-Interim and GHCN-CAM Reanalyzed Data Temperature Values in Tianshan Mountains Area, China. *Mt. Res.* **2019**, *37*, 613–621.
72. Zhao, J.; Guo, Z.; He, X.; Xu, T.; Liu, S.; Xu, Z. Uncertainty Assessment of Temperature and PrecipitationReanalysis Data in Heihe River Basin. *J. Arid Meteorol.* **2019**, *37*, 529–539.

73. Gao, R.; Mu, Z.; Peng, L.; Zhou, Y.; Yi, Z.; Tang, R. Application of CFSR and ERA-Interim Reanalysis Data in Runoff Simulation in High Cold Alpine Areas. *Water Resour. Power* **2017**, *35*, 8–12.
74. Deng, M.; Meng, X.; Ma, Y.; An, Y. Analysis on Soil Moisture Characteristics of Tibetan Plateau Based on GLDAS. *J. Arid Meteorol.* **2018**, *36*, 595–602.
75. Qing, S.; Sun, X.; Li, Y. Evaluation of ERA5 reanalysis soil moisture over inner mongolia. *Sci. Technol. Eng.* **2020**, *20*, 2161–2168.
76. Zheng, Y.; Bai, C.; Hu, X. Variation of frost-free period and its impact on agricultural production. *Liaoning Meteorol. Q.* **2004**, *20*, 23–24.
77. Shen, H.; Sun, X.; Lin, X. Study on Risk Assessment of Winter Wheat Frost Damage in Huanghuai Wheat Production Zone. *J. Inst. Disaster Prev.* **2011**, *13*, 71–77.
78. Zhang, H. *Characteristics of Climatic Resources and Extreme Climate Events in Gansu Province and Their Effects on Winter Wheat Productivity*; Nanjing University of Information Science and Technology: Nanjing, China, 2018.
79. Wang, L.; Liu, C.; Li, Q.; Wu, D.; Wang, Q.; Cheng, W. The Northern Boundary Variation of Winter Wheat in Beijing-Tianjin-Hebei under Climate Warming. *Crops* **2017**, *32*, 61–67.
80. Deng, Z.; Zhang, Q.; Wang, Q.; Qing, J.; Huang, L.; Liu, W.; Xu, J. Impact of climate warming and drying on food crops in northern China and the countermeasures. *Acta Ecol. Sin.* **2010**, *30*, 6278–6288.
81. Si, W.; Zhang, M.; Liu, P. Spatial and temporal response differences to global warming hiatus of the ther-mophilic crops climate growth period in oases of China. *Arid Land Geogr.* **2020**, *43*, 899–908.
82. Zhang, W.; Zhang, M.; Liu, P.; Wang, T. Spatiotemporal responses of the growth period of chimonophilous crop to global warming hiatus—Take the example of Chinese oases. *China Environ. Sci.* **2020**, *40*, 2254–2261.
83. Wang, L.; Xie, X.; Su, W.; Guo, X. Changes of maximum and minimum temperature and thier impacts in northern China over second half of the 20th century. *J. Nat. Resour.* **2004**, *19*, 337–343.
84. Liu, Z.; Yang, X.; Wang, W. Changes of China agricultural climate resources under the background of climate change. IV. Spatiotemporal change characteristics of agricultural climate resources in sub-humid warm-temperate irrigated wheat-maize agricultural area of Huang-Huai-Hai Plain. *Chin. J. Appl. Ecol.* **2011**, *22*, 905–912.
85. Dorji, T.; Piao, S.; Wang, X.; Zhao, C.; Liu, B.; Chen, A.; Wang, S.; Wang, T. Emerging Negative Warming Impacts on Tibetan Crop Yield. *Engineering*, 2021; *in press*. [CrossRef]

remote sensing

MDPI

Article

Flood Management, Characterization and Vulnerability Analysis Using an Integrated RS-GIS and 2D Hydrodynamic Modelling Approach: The Case of Deg Nullah, Pakistan

Ijaz Ahmad [1], Xiuquan Wang [2,*], Muhammad Waseem [1], Muhammad Zaman [3], Farhan Aziz [2], Rana Zain Nabi Khan [1] and Muhammad Ashraf [4,5]

1 Centre of Excellence in Water Resources Engineering, University of Engineering and Technology, Lahore 54890, Pakistan; dr.ijaz@uet.edu.pk (I.A.); waseem.jatoi@cewre.edu.pk (M.W.); znk@cewre.edu.pk (R.Z.N.K.)
2 School of Climate Change and Adaptation, University of Prince Edward Island, 550 University Ave., Charlottetown, PE C1A 4P3, Canada; faziz@upei.ca
3 Department of Irrigation and Drainage, University of Agriculture, Faisalabad 38000, Pakistan; muhammad.zaman@uaf.edu.pk
4 Department of Agricultural Engineering, Khwaja Fareed University of Engineering and Information Technology, Rahim Yar Khan 64200, Pakistan; dr.ashraf@kfueit.edu.pk
5 Commonwealth Scientific and Industrial Research, Organization Land and Water, Canberra 2601, Australia
* Correspondence: xxwang@upei.ca

Abstract: One-dimensional (1D) hydraulic models have been extensively used to conduct flood simulations for investigating flood depth and extent maps. However, the 1D models cannot simulate many other flood characteristics, such as flood velocity, duration, arrival time and recession time when the flow is not restricted within the channel. These flood characteristics cannot be disregarded as they play an important role in developing flood mitigation and evacuation strategies. This study formulates a two-dimensional (2D) hydrodynamic model combined with remote sensing (RS) and geographic information system (GIS) approach to generate additional flood characteristic maps that cannot be produced with 1D models. The model was applied to a transboundary river of Deg Nullah in Pakistan to simulate an extreme flood event experience in 2014. The flood extent images from the moderate resolution imaging spectroradiometer (MODIS) and observed flood extents were used to evaluate the model performance. Moreover, an entropy distance-based approach was proposed to facilitate the integrated multivariate flood vulnerability classification. The simulated 2D flood modeling results showed a good agreement with the flood extents registered by MODIS and the observed ones. The northwest parts of Deg Nullah near Seowal, Dullam Kahalwan and Zafarwal were the most vulnerable areas due to high flood depths and prolonged flooding duration. Whereas high flood velocities, short flood arrival time, prolonged flood duration and recession times were observed in the upper reach of Deg Nullah thereby making it the most susceptible, critical and vulnerable region to flooding events.

Keywords: HEC-RAS 2D model; flood characterization; flood vulnerability; flood hazard maps; Deg Nullah

Citation: Ahmad, I.; Wang, X.; Waseem, M.; Zaman, M.; Aziz, F.; Khan, R.Z.N.; Ashraf, M. Flood Management, Characterization and Vulnerability Analysis Using an Integrated RS-GIS and 2D Hydrodynamic Modelling Approach: The Case of Deg Nullah, Pakistan. *Remote Sens.* **2022**, *14*, 2138. https://doi.org/10.3390/rs14092138

Academic Editor: Luca Brocca

Received: 24 February 2022
Accepted: 27 April 2022
Published: 29 April 2022

1. Introduction

Floods are considered the most devastating hazard all over the world. Floods claim more loss of life than any other hydrometeorological disasters [1], and negatively affect the socioeconomic development of many countries [2]. Moreover, the frequency and intensity of these flood events are increasing in many regions around the world, making it a problem of major concern in the sustainable development of countries vulnerable to flooding events [3]. The main reason behind these unprecedented flood events is the rise in temperature and precipitation since the 1950s due to anthropogenic activities and increased rates of greenhouse gas emission [4]. Moreover, the increase in population growth rate and urbanization led to increased impervious areas and increased runoff due to the reduced infiltration [5]. Recent flood events are becoming more frequent and severe in Pakistan due to the population growth and increasing anthropogenic activities in the floodplains [6].

Pakistan faces almost one major flood event every three years [7] and drought-like conditions every six years [8], which is one of the main challenges to the country's sustainable economic development. Pakistan faced the worst flood event of its history in 2010 [9]. Scientists from the World Climate Research Program and World Meteorological Organization (WMO) stated that climate change was one of the main reasons behind this unprecedented sequence of weather events in Pakistan [10]. Moreover, a United Nations (UN) based scientific body stated that the hot extremes, heat waves and heavy precipitation events are expected to be more frequent and intense in the future in this country [11]. WMO also qualified the assessments that the flood event of 2010 fit the sequence as predicted by climate experts and stated that these events match the projections of more frequent and intense climatic events due to global warming [12].

In recent years, remote sensing (RS) and geographic information systems (GIS) played an important role in presenting and validating the flood risk and hazard mapping results [13,14]. RS and GIS are considered effective geospatial tools assisting in preparing the flood characteristic maps and their impacts on the environmental, social and economic aspects [15]. These geospatial tools have been extensively used to map the results of simulation models of past flood events such as flood inundation extents, flood levels and to identify the critical facilities at risk [16,17]. These tools also support developing the flood forecasting and early warning systems in flood-prone areas, which are helpful in flood hazard management and preparedness planning [18,19]. RS is an important tool in acquiring the data required to formulate and validate the flood model result, flood risk assessment and to identify the infrastructure at risk (e.g., health and education facilities, transportation infrastructure, urban and commercial areas, etc.). Whereas, use of GIS helps in developing automated flood models and flood risk investigations (identification of urban areas, settlements and agricultural lands to different stages of floods), to generate flood inundation maps and to conduct other statistical analyses that may be useful in developing the flood disaster management strategies at various levels of flood events [20]. For instance, the analytical hierarchy process (AHP) method and a geographical information system (GIS) were utilized to create the flood hazard assessment map in the burned and urban areas of central Greece [21].

Assessment of flood risk and management practices is important in identifying the flood hazards, flood-prone areas and future mitigation strategies [22]. Understanding the characteristics of floods is essential in developing flood management action plans and analyzing the impact of various proposed management measures. In situ observations of flood events are the simplest ways to interpret the flood risk [23]. However, in situ observations of the floods are not available in many cases in developing countries. Moreover, flood analysis based on these observations is limited to specific flood events. Therefore, one of the other ways to conduct flood risk modeling studies is to use the RS data [24,25]. However, flood events usually occur under cloudy conditions, limiting the application of remotely sensed data in the visible and infrared spectrum due to visibility issues [26]. Therefore, these observation maps cannot be used directly to predict future flood events and to analyze the efficacy of various structural measures in developing flood mitigation strategies. The numerical models are considered more suitable for the simulation and prediction of flood events. Moreover, in many previous studies, numerical models have been efficiently applied to simulate flood events and investigate the flood hazards to develop management strategies [27–29].

Both one-dimensional (1D) and two-dimensional (2D) numerical hydrodynamic models combined with RS and GIS approaches have been extensively used in numerous studies for simulating flood events and risk assessment [4,25,30]. Although 1D flood models can be used for simulating the water levels and discharges when the flow is restricted within the channels [31,32], these models present many limitations when applied for simulating the overflow conditions. Moreover, methodologies used by 1D flood models are limited to generating flood hazard and depth maps. Flood depth maps are considered the primary input in evaluating flood impacts. However, additional flood characteristic maps, i.e., flood velocities, flood arrival times, flood duration and flood recession times, cannot be disregarded as they play an important role in the detailed design and analysis stages of flood mitigation studies [20]. For instance, high flood velocities with a longer flood duration can damage the flood mitigation structures and result in erosion and water pollution, which may cause environmental degradation, economic and life losses [33]. Awadallah et al. [34] used the LiDAR datasets in the HEC-RAS 2D model for comparing the flood inundations maps using the bathymetric and topographic terrain models in Norway. Therefore, the use of 2D models becomes indispensable to address the above-mentioned issues.

In previous studies, several researchers have used the 2D hydrodynamic models to assess flood risks and develop flood mitigation strategies. For instance, Santillan et al. [20] applied an integrated RS, GIS and 2D hydrodynamic modeling approach to generate flood characteristic maps, i.e., flood velocity, arrival time, flood duration and recession time, for improved flood disaster management in the Philippines. Quiroga et al. [31] used a 2D numerical model to perform the unsteady flow analysis and compared the model results with the satellite imagery in the floodplains of Bolivian Amazonia, Brazil. In another study, an integrated 1D/2D hydrodynamic model was applied to generate the flood hazard maps and velocities in the floodplains of the Musi river basin of Indonesia [35]. Bhandari et al. [4] employed the 2D hydrodynamic modeling approach to perform the unsteady flow analysis and flood inundation mapping to outline the flood-prone areas in the lower reaches of the Brazos river basin, USA. Elkhrachy [36] used the remote sensing images as independent variables in the HEC-RAS 2D simulation model to estimate the post-flash flood event depths of 24–26 April 2018 in New Cairo, Egypt. Moreover, 2D hydrodynamic modeling approaches combined with RS and GIS have been applied since the release of the HEC-RAS 2D model in different regions of the world to investigate the flood events and to develop mitigation strategies [37–42].

Even with the extensive flood simulations using the hydrodynamic models to generate the flood hazard and depth maps, it is almost impossible for the decision-makers to develop flood mitigation and management strategies without analyzing the vulnerability of different elements at risk in the floodplains [43]. The concept of vulnerability has changed drastically in various disaster management studies based on their objectives. Consequently, several attempts were made to define the term vulnerability. For instance, Cutter [44] defined vulnerability as a hazard that includes natural risks and human response to these risks. However, Mitchell [45] stated that vulnerability is the degree of loss to a given element at risk (or set of elements) resulting from a given hazard at a given severity level. Moreover, Merz et al. [46] and Salami et al. [47] stated that it is a function of the vulnerability definition to elements at risk, exposure and susceptibility. The vulnerability of different elements (e.g., road networks, bridges, urban areas/settlements, etc.) at risk in the floodplains largely depends on water depth, flood velocities and duration of flooding [48]. Santillan et al. [20] categorized the levels of the flood hazard based on maximum depths into three classes, i.e., low (<0.5 m depth), medium (0.5–1.5 m depth) and high (>1.5 m depth) for the residential buildings. However, in most of the previous studies, flood extents and water depth were considered; whereas the flood variables such as flow velocities, flood duration and recession times were discussed only in a few studies [31,49]. For instance, Balijepalli and Oppong [50] suggested that the flood velocities of less than 2–3 m/s do not affect the physical infrastructure (e.g., road networks, bridges and urban areas/settlements) in floodplains. However, to the best of the author's knowledge, no study has been conducted that performs the multivariate flood vulnerability analysis by using a 2D hydrodynamic flood modeling approach. Moreover, no flood research has been carried out by using an integrated RS-GIS and 2D hydrodynamic modeling approach together with vulnerability analysis in Pakistan to generate flood characteristics other than depth for developing flood management strategies.

Therefore, this study aims to simulate an extreme flood event with a return period of 200 years in Deg Nullah of Pakistan using an integrated RS-GIS and a 2D hydrodynamic modeling approach to generate the additional flood characteristic maps, i.e., flood velocities, arrival times, duration and flood recession times. The study area is subjected to frequent floods and result in disruption of socio-economic activities. Moreover, flood vulnerability analysis was performed to investigate the sensitivity of roads, railway tracks and urban areas/settlements at risk to different flood variables such as flood depths, velocities, arrival times, duration and recession times in the study area. Therefore, it is believed that the results of the present study would help develop improved flood management strategies.

2. Study Area

Deg Nullah, a natural drain in Rachna Doab of Punjab province, was analyzed to simulate the 2014 major flood event. Two streams, Divak and Basenter join together in the Indian Administered Kashmir to form Deg Nullah before diverting into Pakistan near Lehri check post, located in the northeast of Narowal district. Deg Nullah is a straight twisted channel with wide and shallow cross-sections and uneven slopes. Flows of Deg Nullah often overspill during the monsoon season as its banks are not high enough to pass the large discharges. There is a reach of 75 km from Kingra-more to the Bambanwala-Ravi-Bedian (BRB) canal, as shown in Figure 1. Deg Nullah mostly remains dry, but carries a large volume of flows during monsoon season. The catchment of Deg Nullah receives most of the precipitation in summer (July to September) [51]. This heavy precipitation resulted from the combined effects of the monsoon precipitation regime from the Arabian Sea and Bay of Bengal, making it a meteorological complex region. Precipitation over the catchment of Deg Nullah causes high flow peaks during the monsoon season and the deteriorated condition of shallow banks intensifies the flooding problem and affects the road networks, urban areas, settlements, agricultural lands, etc.

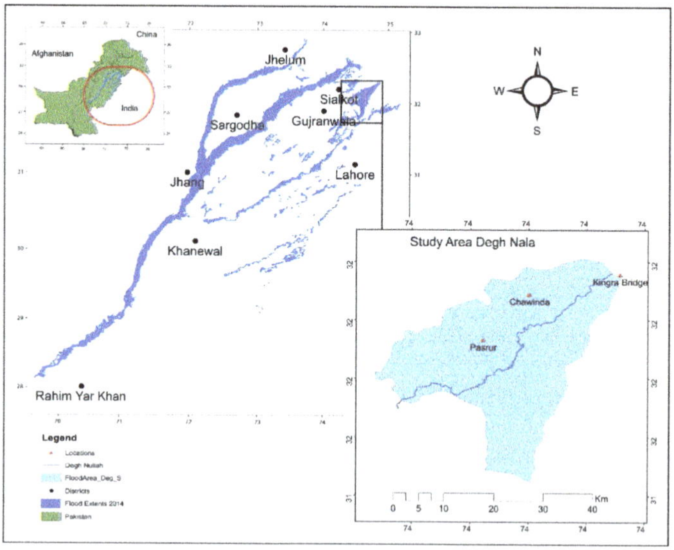

Figure 1. Location of the study area.

In 2014, Deg Nullah experienced one of the worst flood events of its history and caused severe damage to its surrounding areas. A peak flood of 2050 m^3/s with a return level of 200 years was observed in Deg Nullah on 6 September 2014 [52]. Therefore, water overflowed from the banks of the Nullah and caused submergence in Zafarwal city, Qila Ahmedabad town and 55 villages around Deg Nullah. This flood event caused the displacement of a vast population from urban areas/settlements, road networks were damaged due to submergence and negatively affected the standing crops over a large area. Moreover, more than 25 health centers and 104 government schools were submerged [53]. Residents of this area blamed the Deg Nullah authorities for not providing accurate information regarding the flood events. Therefore, accurate evaluations of the flood extents, depth, velocity, arrival and recession times are essential for developing improved flood management strategies in this region.

3. Materials and Methods

3.1. Numerical Model

This study used a 2D hydrodynamic model (HEC-RASv5) due to its various advantages over 1D models. For instance, 1D flood models are considered suitable when the flow is restricted within the channel. However, when the overflow conditions prevail, it limits the applicability of 1D flood models in simulating flood characteristics and defining the flood flow directions. It is not always realistic to define the flow paths or directions in the floodplains, e.g., in flat areas with large variations in water levels, 1D models do not provide any information regarding the velocity distribution when the water leaves the main channel to the floodplains. However, these limitations can be efficiently addressed by 2D hydrodynamic flood modeling. These flood models can simulate additional flood characteristics such as flood arrival times, flood velocities, flood duration and flood recession times [54]. Moreover, remotely sensed datasets have been extensively used with the HEC-RAS 2D model to simulate, validate and forecast flood events in different regions of the world [16,20,34,36].

Therefore, a 2D hydrodynamic model (HEC-RAS 2D) developed by the USACE, was used to simulate the selected flood events in the Deg Nullah catchment area. This model solved either the full 2D Saint-Venant equations or 2D diffusive equation as given below:

$$\frac{\partial \xi}{\partial t} + \frac{\partial p}{\partial x} + \frac{\partial q}{\partial y} = 0 \tag{1}$$

$$\frac{\partial p}{\partial t} + \frac{\partial}{\partial x}\left(\frac{p^2}{h}\right) + \frac{\partial}{\partial y}\left(\frac{pq}{h}\right) = -\frac{n^2 pg\sqrt{p^2+q^2}}{h^2} - gh\frac{\partial \xi}{\partial x} + pf + \frac{\partial}{\rho \partial x}(h\tau_{xx}) + \frac{\partial}{\rho \partial y}(h\tau_{yy}) \tag{2}$$

$$\frac{\partial q}{\partial t} + \frac{\partial}{\partial y}\left(\frac{q^2}{h}\right) + \frac{\partial}{\partial x}\left(\frac{pq}{h}\right) = -\frac{n^2 pg\sqrt{p^2+q^2}}{h^2} - gh\frac{\partial \xi}{\partial y} + qf + \frac{\partial}{\rho \partial y}(h\tau_{yy}) + \frac{\partial}{\rho \partial x}(h\tau_{xx}) \tag{3}$$

where represents the surface elevation (m); p and q represent the specific flows in the x and y directions (m^2/s); h depicts the depth of water (m); g is the acceleration due to gravity (m/s^2); n is the Manning roughness coefficient, q represents the density of water (kg/m^3); s_{xx}, s_{yy} and s_{xy} are the components of the effective shear stress and f is the Coriolis (s^{-1}) [31].

3.2. Overall View of the Research Approach

Numerical simulations of physical processes are considered efficient through an iterative development of the model. The accuracy of the numerical models depends upon the quality of the available data and hydrologic understanding of the problem being investigated. A brief description of the study approach being adopted to achieve the objectives is presented in the flowchart (Figure 2) and described in the subsequent paragraphs.

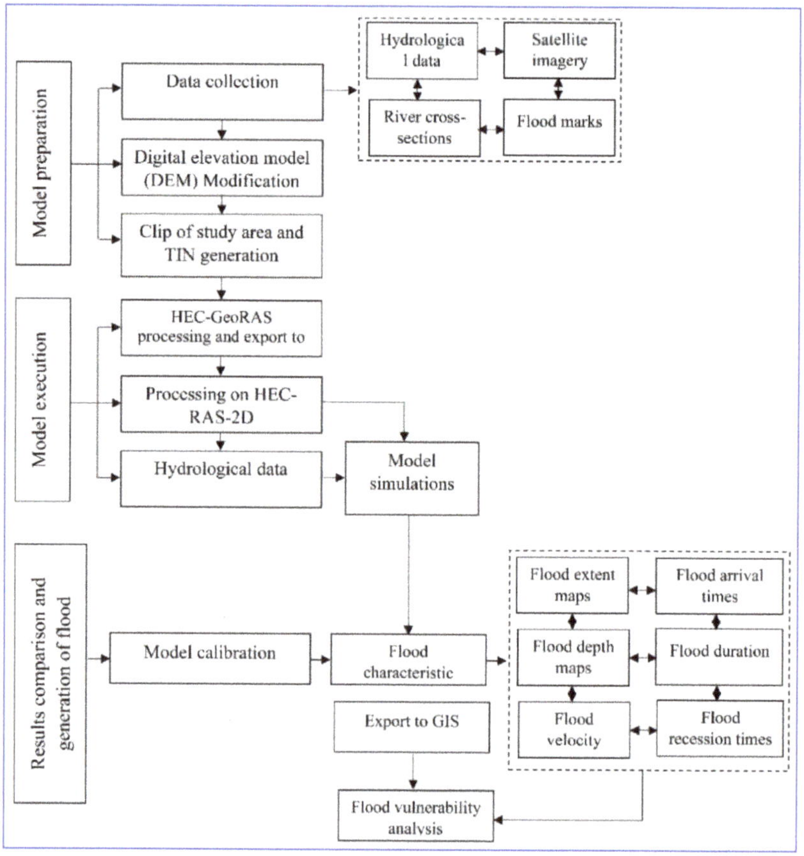

Figure 2. Overall summary of steps of methodology in the form of a flowchart.

The flow hydrographs of Deg Nullah at Kingra bridge station were collected from the Punjab Irrigation Department for the monsoon periods from 2014–2017 (Figure 3). The hydrograph data comprises an interval of one day. Among the collected hydrographs, the 2014 flood was the severest and was considered for flood analysis. The flow hydrograph of the Kingra bridge was used as the upstream boundary condition in the model.

The Shuttle Radar Topography Mission (SRTM) data are considered efficient in modeling the hydrological and hydraulic processes because of their free availability, homogeneity and consistency [55]. Therefore, the present study explores the hydrological modeling with the SRTM digital elevation model (DEM). The DEM with a resolution of 12.5 × 12.5 m was used. River cross-sections were considered the major input in both the 1D and 2D hydraulic modeling. The DEM was integrated with the observed cross-sections collected from the Punjab Irrigation Department at 53 locations across the Deg Nullah (Figure 4). The accuracy of the 2D hydrodynamic flood model, to analyze the water flow within the channel and floodplains (overflow conditions) was improved by integrating the DEM with the field observed cross-sections [20] and after that the study area was clipped from the DEM for further analysis. The DEM was then converted to a triangular irregular network (TIN) surface model whose surface does not deviate from the input raster. The DEM of the study area, along with the river cross-sections, is presented in Figure 4.

HEC-GeoRAS is a GIS extension designed to process geospatial data, allowing users to create an HEC-RAS import file containing geometric data. The results of the water surface profiles can be interpreted to examine the flood depths and extents. In this study, preprocessing was conducted in HEC-GeoRAS to generate the model files. Afterward, the HEC-GeoRAS file consisting of geometric data was imported to HEC-RAS to perform the hydraulic computations.

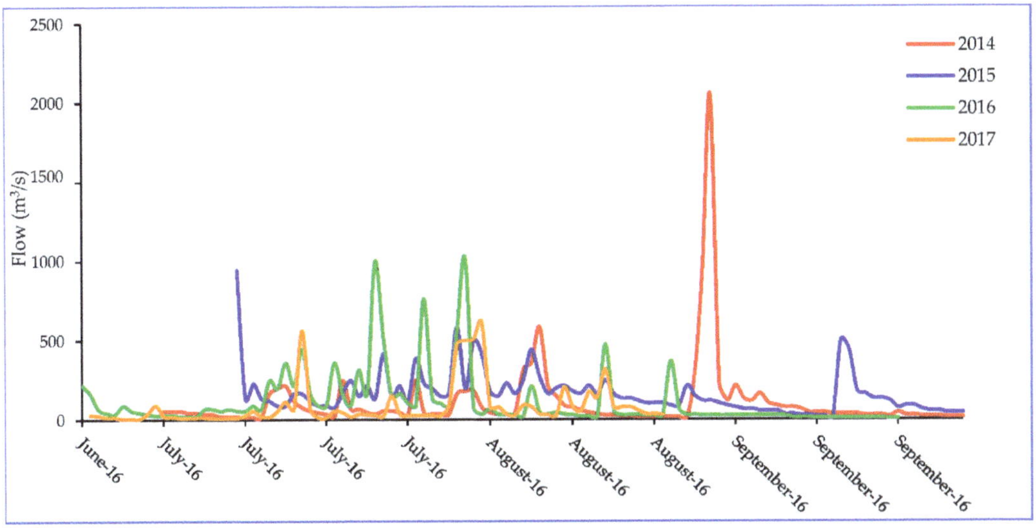

Figure 3. Observed flow hydrographs at Kingra bridge during monsoon periods of 2014–2017.

HEC-RAS model can either be used as an integrated 1D–2D model or as a fully 2D model to simulate the main rivers and floodplains, respectively. Although an integrated 1D–2D model might be faster compared to only the 2D modeling engine, it is necessary to define the connections of overflow locations between the 1D and 2D modules. In this study, a fully hydrodynamic 2D model was used, as the connections of overflow locations were unknown. The computational model domain (2D flow area) area was approximately 888 km^2. This area was discretized into 100 × 100 m size cells/grids. The model was run from 1 July 2014 to 7 October 2014 at a time step of 1 min, whereas output results were generated at 1-h intervals. Moreover, two types of boundary conditions were used in the 2D hydraulic model, i.e., inflow hydrograph as an upstream boundary condition at Kingra bridge of Deg Nullah and average riverbed slope was considered as the downstream boundary condition.

To compare the model-generated flood map with the observed flood extents (Table 1), the MODIS extent images were downloaded from https://modis.gsfc.nasa.gov/data/ (accessed on 28 March 2019) for the monsoon period of 2014. In addition, two field visits were also conducted on 30 April 2019 and 15 May 2019 to collect the flood extents data at a few locations to verify the model simulated flood maps, ground-truthing of settlements and riverbed elevations by using GPS, and interviewing the locals and personnel from the irrigation department. A total of 32 imageries were downloaded from MODIS and a map showing the maximum extent was selected for the comparison of results.

Table 1. Observed flood extent locations.

Location No.	Longitude (E)	Latitude (N)
1	74°53′12.97′′	32°18′39.65′′
2	74°51′55.39′′	32°14′15.79′′
3	74°48′1.75′′	32°22′22.1′′

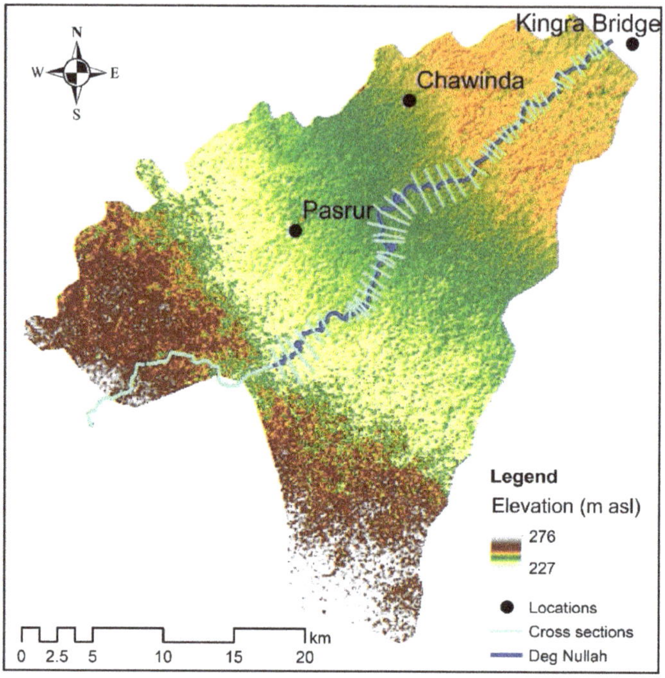

Figure 4. Observed river cross-sections, location of Deg Nullah and DEM of the study area.

3.3. Flood Vulnerability Analysis

In this study, first, we digitized all the road networks, railway tracks and urban areas/settlements within the study area by using the satellite imageries (Google Earth Pro) and further verified their locations during the field visits, as shown in Figure 5. Secondly, various flood characteristic maps, i.e., flood depths, velocities, arrival times, duration and recession times, were superimposed on the digitized maps of the road networks and urban areas/settlements to investigate the flood risks at different locations in the study area. The details regarding the flood hazard classification based on flood depth, velocity, arrival time and duration are presented in Table 2 [31,50,56,57]. Moreover, the details regarding important urban areas (cities, villages, etc.) and road networks in the study area are presented in Table 3.

Table 2. Flood hazard classification based on maximum depth, velocity, duration and arrival time [26,44,51,52].

Hazard Level	Depth (m)	Velocity (m/s)	Duration (h)	Arrival Time (h)
Very low	<0.50	<1	<20	<14
Low	0.50–1.0	1–2	20–40	14–28
Medium	1.0–2.0	2–3.8	40–60	28–42
High	2.0–5.0	3.8–5.8	60–80	42–56
Extreme	>5.0	>5.8	>80	56–70

Table 3. Detaild of important locations (cities, villages, etc.) and road networks in the study area.

Sr. No.	Important Locations		Road Networks		
	ID on Map	Name of Location	ID on Map	Name of Road	Length of Road (km)
1	1	Kamalpur Bajwa	A	Pasrur road	29.0
2	2	Seowal	B	Chawinda road	28.5
3	3	Dullam Kahalwan	C	Gujranwala-Pasrur	19.1
4	4	Shahzada	D	Pasrur-Zafarwal	16.6
5	5	Kotli Haji Pur	E	Narowal road	20.3
6	6	Ahmad Abad	F	Kalasawala road	16.3
7	7	Dodha	G	Muridke road	12.1
8	8	Talwandi Bhindran	H	MirakPur road	16.2
9	9	Satrah Sandhuan	I	Nonar road	6.5
10	10	Pasrur	J	Daska road	10.2
11	11	Chawinda			
12	12	Zafarwal			

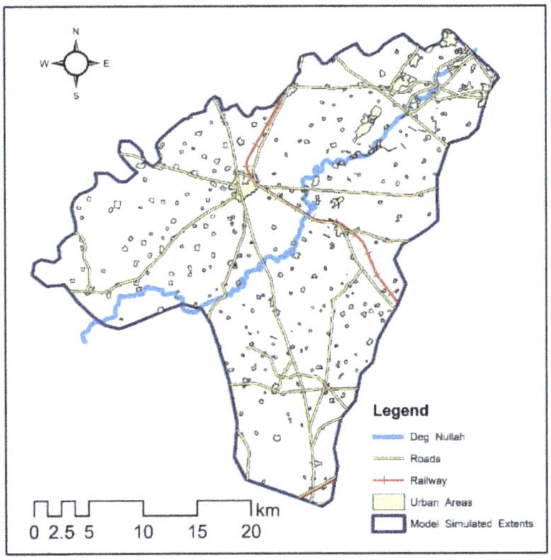

Figure 5. Digitized locations of the road networks and urban areas/settlements.

3.4. Multivariate Flood Vulnerability Classification

Once the standard values of the flood hazard level, i.e., very low, low, medium, high and extreme, of different flood variables, i.e., depth, velocity, arrival time and duration, were selected along with all the related information (Table 2), the entropy distance-based multivariate flood vulnerability classification was performed by using the following approach.

Let S_{ij} be the standard value of hazard level i of the jth flood variable, and P_{jk} is the value of the jth flood variable at location k.

Where i = ith hazard level = 1, 2, . . . , 5; j = jth flood variable = 1, 2, . . . , 4 and k = kth location = 1, 2, . . . , 383.

Normalization of S_{ij} and P_{jk} was performed to avoid any dimension-related issues by using Equations (4) and (5), respectively.

$$N_{ij} = \frac{abs(S_{ij})}{\sum_{i=1}^{n=5} abs(S_{ij})} \tag{4}$$

$$M_{jk} = \frac{abs\left(P_{jk}\right)}{\sum_{k=1}^{n=383} abs\left(P_{jk}\right)} \tag{5}$$

After that, difference of the normalized values and entropy-based weights were calculated by using the following equations

$$d_{jk} = M_{jk} - N_{ij} \tag{6}$$

$$w_j = \frac{e_j}{\sum_{j=1}^{4} e_j} \tag{7}$$

in which

$$e_j = 1 - \left[\frac{-1}{\ln(N)} \sum_{k=1}^{n=383} M_{jk} \ln\left(M_{jk}\right) \right] \tag{8}$$

Finally, the entropy-based distance is calculated by using the following equation.

$$D_k = \sqrt{\sum_{k=1}^{383} w_j \cdot \left(d_{jk}\right)^2} \tag{9}$$

where w_j is the entropy-based weight assigned to individual flood variables. D_k is the weighted Euclidean distance between the standard value of flood hazard level and the value obtained at different locations. A minimum value of the entropy-based distance of flood hazard level at a selected location represents the hazard level.

4. Results

4.1. Model Performance Evaluation

The performance of the HEC-RAS 2D model was evaluated by comparing the model simulated flood extents with the satellite image registered by MODIS on 6 September 2014. Moreover, two field visits were also conducted to identify the flood extents at three locations in the study area as shown in Figure 6 and summarized in Table 1. The image from 6 September 2014 was selected based on its maximum flood extents for calibration. Flood extents registered by MODIS and the simulated flood extents by the HEC-RASv5 model, along with the observed flood extents, are presented in Figure 6. It is clear from Figure 6 that the model simulated flood extents were consistent with the MODIS registered flood extents and the observed ones. The simulated flood extent was found 6% less compared to the flood extents obtained from the satellite image, which is considered acceptable due to the complexity of the system.

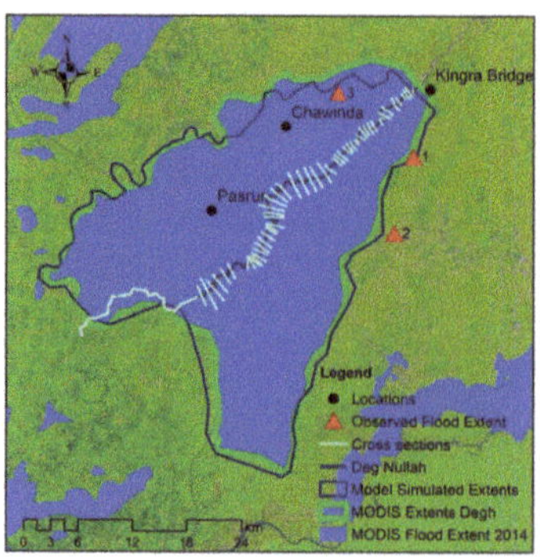

Figure 6. Comparison of model results with MODIS extents and observed flood marks.

4.2. Model Performance Evaluation

After successfully comparing the model results with the satellite imageries and observed flood extents, the HEC-RAS model was simulated to generate the flood characteristic maps of flood depths, velocities, arrival times, duration and recessions times. These flood characteristic maps are briefly discussed in the subsequent sections.

4.2.1. Maximum Flood Depths

The maximum flood depth map was generated using the 2D hydraulic model, as shown in Figure 7. This map was generated by considering the maximum depth of water in each cell during the simulation period regardless of the time when this depth was achieved. It was observed that the majority of the flooded areas north of the Deg Nullah have water depths of less than 1 m. Whereas the lower parts of the Deg Nullah have depths ranging from 0.5 m to 2 m. The most extreme flood depths were observed in the northeast part of the study area. However, these cells represent a small portion of the area compared to the rest of the wet cells. This map of flood depths can be useful in flood preparedness strategies and in developing evacuation plans when a flood of the same magnitude is expected to occur. This map can also help identify the areas required to be alerted or need emergency evacuation before future flood events. Moreover, the map of the maximum flood depths (Figure 7) can also be evaluated based on the various flood hazard levels. For instance, upstream reaches of the Deg Nullah experienced extreme flood hazards compared to the lower reaches as the flooded water overflowing from the channel spread over vast areas.

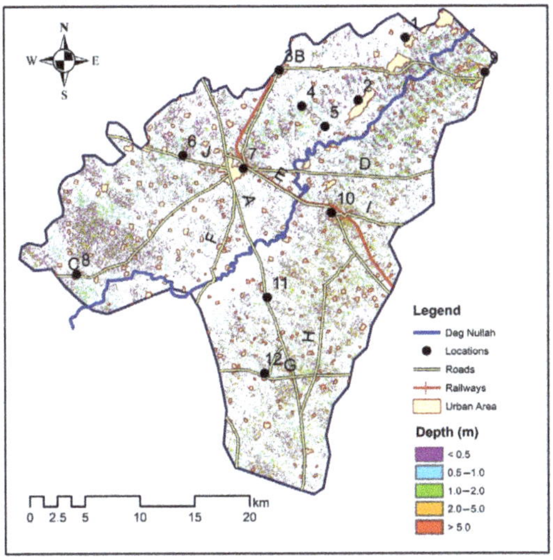

Figure 7. Map of the flood based on the maximum flood depths.

4.2.2. Maximum Flood Velocities

In addition to flood extent and depth maps, the HEC-RAS 2D model has the capability to generate velocity maps to depict the flood velocity at various locations in the study area as shown in Figure 8. The map of flood velocities would explain how fast the flood water will reach a particular location and thereby helps in developing flood mitigation strategies and evacuation planning. In most of the areas above the Deg Nullah, flood velocities vary between 0 to 1 m/s; whereas, in the western parts of the study area, the flood velocities were less than 0.5 m/s. Therefore, these areas are not expected to face severe flood risks due to flood velocities. However, flood velocities in the south of Deg Nullah vary from 0.5 to 1.5 m/s. At the upper reaches of Deg Nullah near Kingra bridge, flood velocities of more than 1.5 m/s were also observed. Such high flood velocities cause more water overflows from this point. Moreover, the areas with flood velocities of about 1 m/s or higher may pose additional hazards and evacuation will be more problematic due to the flash nature of flooding. These maps of flood velocities indicate the level of harm to which a community is exposed; therefore, they play an important role in developing flood risk management strategies.

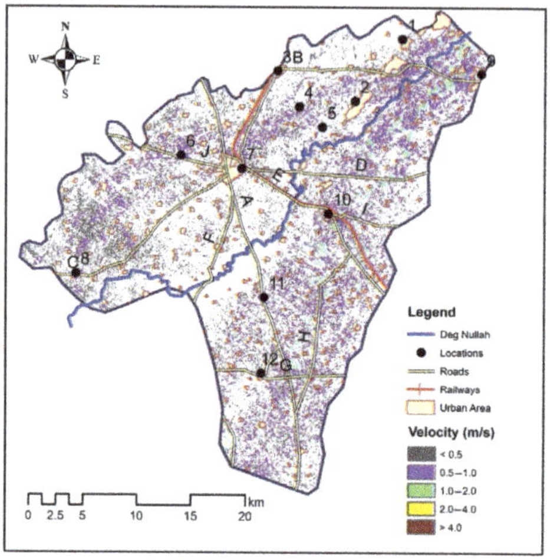

Figure 8. Map of the maximum flood velocities.

4.2.3. Flood Arrival Times

The maps of flood arrival times represent the model computed time from a specified simulation time when the water depth at a location reaches a specified inundation depth [54]. In this study, a flood arrival time map was generated in hours when a location is inundated at maximum depth (Figure 9). Most of the areas in the upper reaches of Deg Nullah achieved the maximum inundation depths within the first fourteen (14) hours. Whereas the lower reaches of the study area become flooded after 28 h from initiation of the flood event and so on. This implies that the areas in the upper reaches of the Deg Nullah are more vulnerable to floods and require a swifter evacuation response to flooding. These maps of flood arrival times are helpful in developing flood management and evacuation strategies.

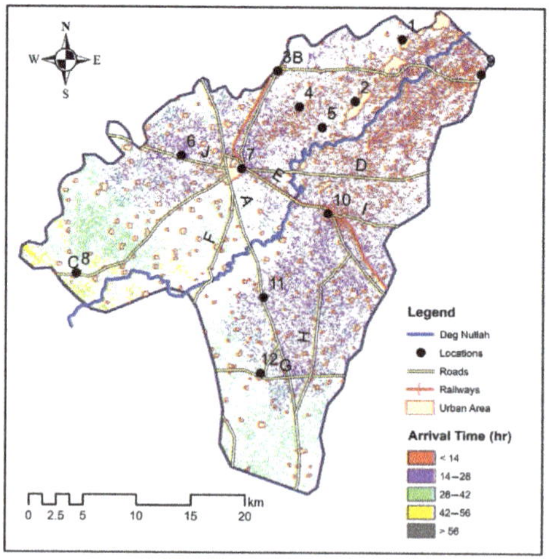

Figure 9. Map showing flood arrival times from the start of the simulation period at different locations of the study area.

4.2.4. Flood Duration

Figure 10 showed the flood duration map for the study. This map helps to understand the flood propagation in the areas that remain flooded for an extended duration. These maps are considered very useful in estimating the time required for a flood-affected community to return to their respective areas and in assessing the flood damages to crops, road networks and other critical facilities located in the area. Figure 10 showed that most of the areas in the upper reaches remain flooded for more than 80 h. However, flood durations decreased in the areas of lower reaches of Deg Nullah.

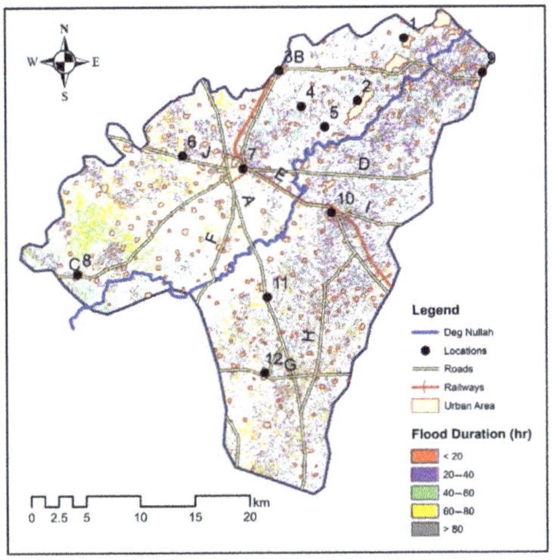

Figure 10. Map of flood duration at different locations of the study area.

4.2.5. Flood Recession Times

The map of flood recession times is presented in Figure 11. This map represents the number of hours that floodwater required to recede from different locations in the study area. The map of flood recession time generated showed that the flood-affected area required about 99 h to completely recede the inundation area generated due to the flood event of 2014. However, this duration decreased at locations far from the mainstream of Deg Nullah. This map has the same importance as flood duration in developing flood mitigation and management strategies.

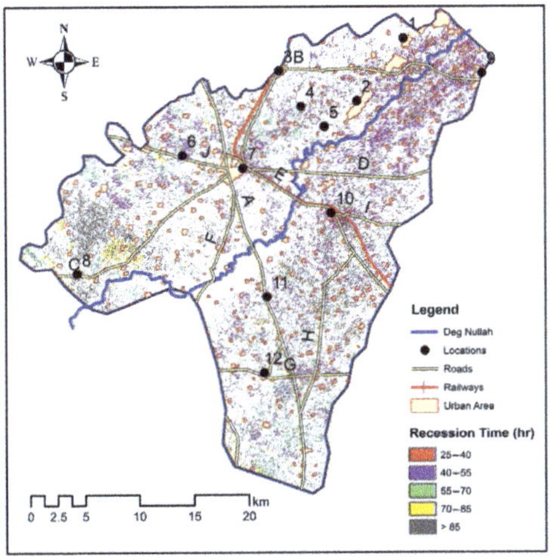

Figure 11. Map of flood recession times at different locations in the study area.

4.3. Flood Vulnerability Analysis

Physical infrastructure (e.g., road networks, railway tracks and urban areas/settlements) in the floodplains was considered to determine their vulnerability to flooding. The road networks and urban areas/settlements were digitized using the satellite imageries (Google Earth Pro) and later verified during the field visits. A total of 381 polygons of urban areas/settlements were marked in the study area. These polygons cover a total urban area/settlements of 47 km^2, which is about 6% of the entire study area. Whereas the total length of road networks in the study area was about 175 km connecting these urban areas/settlements and other major cities. The spatial flood characteristic maps, such as flood depths, velocities, arrival time, duration and recession time generated from the model for the flood event of September 2014, were superimposed on the maps of urban areas/settlements and road networks for vulnerability analysis. The results of the flood vulnerabilities to urban areas/settlements and road networks for various flood characteristic maps are presented in Figure 12. The flood vulnerability map is self-explanatory; however, flood hazard levels to maximum water depths and velocities are explained in the subsequent paragraph based on the water depths and velocities around the urban areas/settlements and road networks. Different flood hazard levels based on maximum depths are presented in Table 2.

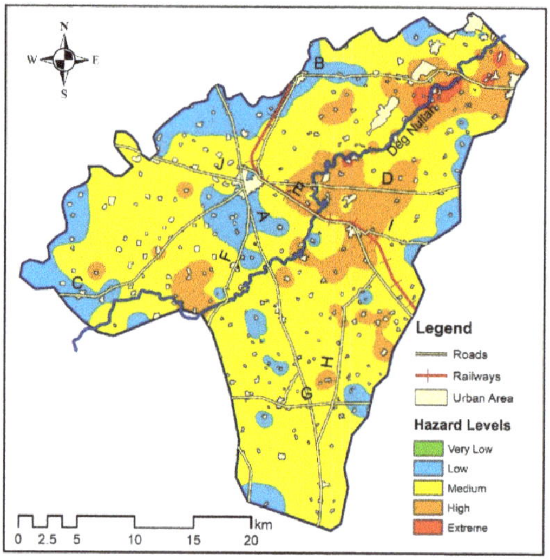

Figure 12. Multivariate flood hazard classification map along with the urban areas/settlements and road networks.

5. Discussion

Flooding events can significantly impact economic conditions and even threaten human life. These events are considered as the most disastrous among the hydro-meteorological hazards [58]. Hence, investigation of these events is highly important in developing mitigation strategies to reduce their negative impacts on human life and infrastructure. Flood simulation through hydraulic models has been extensively used for this purpose. The present research used the HEC-RAS 2D model for simulating an extreme flood event observed in a transboundary Deg Nullah which often experiences floods during the monsoon season and results in economic and human life loss. However, before the application of the model to generate flood characteristic maps, such as flood depth, velocity, arrival time, duration and recession time, the model simulated flood extents were compared with the extents registered by MODIS and later verified during the field visits. The model simulated flood showed good performance when compared with a satellite image of the flood event [31]. The flood extent area simulated by the model was about 6% less than that as registered by MODIS extents. However, during the field visits, maximum flooding extents were verified by observations and discussions with locals at three different locations (Figure 6). Moreover, previous studies [27,59,60] confirmed the acceptability of the present study results in which the model simulated flood areas were found to be less compared to that of the satellite imageries. This reduction in the model simulated areas compared to satellite imageries may be due to the courser DEM resolution [61] and due to the limitations of the HEC-RAS sub-grid configuration to produce a continuous inundation pattern that restricts the calculation of flood extent and the distribution of local water depth values [61]. Therefore, further studies may be carried out using the high-resolution DEM for improved model performance.

The maps of various flood characteristics such as flood depths, velocities, arrival times, duration and recession times were prepared to evaluate their distribution in the study area for an extreme flood event of 2050 m^3/s [52]. The flood depth allows identifying areas exposed to different hazard levels. In most of the areas located in the northern part of Deg Nullah, the flood depths are less than 1 m whereas in its southern parts depth varies between 0.5–2 m, thereby posing low and low medium threats, respectively [56]. However, results have demonstrated that most of the settlements are constructed in the flood prone areas, which necessitates proactive planning and selection of the proper construction rules for the prevention and mitigation of the consequences of flood hazards. Over most of the flooded area, the water velocity was less than 0.1 m/s. However, water velocity in the south of Deg Nullah varies between 1–1.5 m/s, whereas at Kingra bridge it exceeds 1.5 m/s (Figure 8). The evacuation process may become difficult in areas with floodwater velocities of more than 1 m/s [57]. Flood arrival times at particular locations are one of the major variables along with velocity, depth, safe locations, etc., used to prepare the evacuation strategies [62]. The floodwater in the upper reaches of Deg Nullah attains the maximum depth within the first 14 h after the overflow condition, whereas in lower parts of the study area, it took more than 28 h and thereby, hazard levels are classified as low to medium [47]. The north part of the study area is the most exposed to the flood; it showed a larger flood extent, longer flood duration and deeper water depth. The flood that threatens the Seowal and Dullam-Kahalwan towns originates from the Kingra bridge, where the safe carrying capacity of the channel is only 280 m^3/s [53]. The flood from the north gets close to Seowal hours after it begins to overflow, while it takes one day for the flood to reach the southern parts of the study area when the channel begins to overflow. Moreover, the flood in the north region is deeper than the flood in the south, the flood from the north begins to flood before the south. Therefore, flood hazard levels are higher in the northern region compared to the south and required efficient flood mitigation strategies. The study showed the applicability and the value of the 2D capabilities of the new HEC-RAS for flood studies.

Flood vulnerability of settlements and road networks to maximum water depths is presented in Figure 12. It is recorded that the vast majority of road networks are vulnerable to medium to high flood risks and may reduce the accessibility to certain parts of the areas or in the extreme, they may remain completely cut off from the rest until the affected links are revived [21]. Therefore, the vulnerability of road networks along with the urban areas and settlements to flood events was investigated by superimposing their digitized maps over the flood characteristic maps. The settlements in the upper reaches experienced medium to extreme flood hazard levels based on the water depths. However, these hazards decreased gradually as the flood water arrived in the lower reaches of Deg Nullah. Moreover, Balijepalli and Oppong [50] stated that the physical infrastructure (roads, bridges and buildings) is not subjected to severe damage for velocities below 2–3 m/s. In most areas, flood velocities are lower than 3 m/s and only a few locations in the upper reaches of Deg Nullah are experiencing flood velocities of more than 3 m/s as shown in Figure 8. However, the vulnerability of different infrastructures needs to be evaluated by considering the flood depths and velocities together. Therefore, multivariate flood vulnerability classification was performed to assess the combined effect of different flood variables (i.e., flood depth, velocity, arrival time and duration and recession times) at various locations in the study area. Most of the study area lies within the medium hazard levels; however, upper and middle reaches near the channel pose high to extreme threats (Figure 12). Therefore, detailed studies related to the level of extreme events and the vulnerability of the exposed elements at risk (i.e., urban areas/settlements) will help propose the proper protection measures [63]. Moreover, the appropriate mitigation or hazard reduction approaches can be more effectively designed and applied. Furthermore, the awareness of the non-safe regions related to flood hazards might be helpful in emergency preparedness planning.

Therefore, scientists, engineers, stakeholders, planners and decision-makers may utilize the proposed approach in forthcoming spatial planning projects [64,65]. Additionally, the local authorities may use the produced map to guide the adoption of measures and strategies aiming toward flood hazard mitigation and post-fire management. The present study was limited to simulating the flood event of 2014 with return levels of 200 years. Further studies may be carried out for the floods of different return periods to develop flood mitigation strategies. Moreover, sediment and riverbed morphology may be included in the simulation for detailed investigation in future studies.

6. Conclusions

In this study, the extreme flood event of 6 September 2014 of Deg Nullah was simulated using a 2D HEC-RAS hydrodynamic model. The model simulated results showed a good agreement with flood extents registered by satellite images (MODIS) and observed flood extents. Moreover, flood characteristics, including depth, velocity, arrival time, duration and recession time, were generated by the HEC-RAS 2D model. Model simulated flood extents were found consistent with the flood extents registered by MODIS along with the field observed flood extents. Model simulated flood extent area was found 6% less than the flood extents obtained from the satellite image, which is considered acceptable.

The results of this study showed that the northwest parts of Deg Nullah near Seowal (ID-2), Dullam Kahalwan (ID-3) and Zafarwal (ID-12) were found most hazardous due to high flood depths and longer flood duration. Therefore, these areas are considered the most critical for evacuation. The flood velocities were found to be less than 1.5 m/s at most of the study area locations, thereby posing no serious threats due to the low velocity of the floodwater. However, in the upper reaches of Deg Nullah, floodwater velocities of greater than 2 m/s were observed. Moreover, flood duration in this region was also the longest among the rest of the areas. Therefore, high flood depths, flood velocities, longer flood duration and recession times and less flood arrival times make it the most critical and sensitive region to future flood events.

The present study analyzes the flood vulnerability of residential areas and road networks in the floodplains of Deg Nullah. Based on the hydrodynamic flood modeling results, it is noticeable that most of the residential areas and transportation infrastructure currently exists in medium to high flood risk zones and are likely to be exposed to future flood hazards of the same or more magnitudes. However, it was observed that the areas lying in the upper reaches of Deg Nullah were exposed to severe flood hazards due to high flood depths and velocities, faster flood arrival times, longer flood duration and recession times. Moreover, most of the urban areas/settlements in the middle and lower reaches of Deg Nullah were categorized in the range of low to medium hazard levels and the population may be considered safe inside their homes. Similarly, the road networks in the upper reaches were subjected to high flood depths and velocities which may influence the evacuation. The study showed that the HEC-RAS 2D model is an effective tool for studying and assessing the risk of flood events by coupling it with MODIS and other RS data in areas with little or no post-flood event information available. Future applications of the HEC-RAS 2D model with high-resolution DEM and other observed high floodwater marks may help analyze possible flood management strategies.

Author Contributions: Conceptualization, I.A. and R.Z.N.K.; data curation, M.W.; formal analysis, M.A.; funding acquisition, I.A.; investigation, M.W.; methodology, I.A., M.Z. and F.A.; project administration, I.A.; software, M.A., X.W., F.A. and R.Z.N.K.; supervision, X.W.; validation, M.Z.; visualization, M.Z.; writing—original draft, I.A. and F.A.; writing—review and editing, X.W. and M.A. All authors have read and agreed to the published version of the manuscript.

Funding: This research was funded funded by the Higher Education Commission of Pakistan (grant: SRGP-1593).

Data Availability Statement: Some or all data, models or code that support the findings of this study are available from the corresponding author upon reasonable request.

Acknowledgments: This research was funded funded by the Higher Education Commission of Pakistan (grant: SRGP-1593). The authors also thank the Punjab Irrigation Department for providing the data required for this research without any cost.

Conflicts of Interest: The authors declare that they have no conflict of interest in this work.

References

1. Shadmehri, T.A.; Doulabian, S.; Ghasemi, T.E.; Calbimonte, G.H.; Alaghmand, S. Large-scale flood hazard assessment under climate change: A case study. *Ecol. Eng.* **2020**, *147*, 105765. [CrossRef]
2. Smith, D.I. *Floods: Physical Processes and Human Impacts*; Smith, K., Ward, R., Eds.; John Wiley & Sons: Chichester, UK, 1999.
3. Dewan, T.H. Societal impacts and vulnerability to floods in Bangladesh and Nepal. *Weather Clim. Extrem.* **2015**, *7*, 36–42. [CrossRef]
4. Bhandari, M.; Nyaupane, N.; Mote, S.R.; Kalra, A.; Ahmad, S. 2D Unsteady Flow Routing and Flood Inundation Mapping for Lower Region of Brazos River Watershed. In Proceedings of the World Environmental and Water Resources Congress 2017, Sacramento, CA, USA, 21–25 May 2017; American Society of Civil Engineers: Reston, VA, USA, 2017; pp. 292–303.
5. Thakur, B.; Parajuli, R.; Kalra, A.; Ahmad, S.; Gupta, R. Coupling HEC-RAS and HEC-HMS in Precipitation Runoff Modelling and Evaluating Flood Plain Inundation Map. In Proceedings of the World Environmental and Water Resources Congress 2017, Sacramento, CA, USA, 21–25 May 2017.
6. FFC. *Annual Flood Report 2017*; Federal Flood Commission (FFC), Ministry of Water Resources, Government of Pakistan: Islamabad, Pakistan, 2017.
7. Hashmi, H.N.; Siddiqui, Q.T.M.; Ghumman, A.R.; Kamal, M.A.; Mughal, H.R. A critical analysis of 2010 floods in Pakistan. *Afr. J. Agric. Res.* **2012**, *7*, 1054–1067.
8. Ohlsson, L. Pakistan: IRIN Special Report on Water Crisis, Integrated Regional Information Network (IRIN). Available online: http://www.padrigu.gu.se/EDCNews (accessed on 31 May 2019).
9. FFC. *Annual Flood Report 2010*; Federal Flood Commission (FFC), Ministry of Water & Power, Government of Pakistan: Islamabad, Pakistan, 2011.
10. Coumou, D.; Rahmstorf, S. A decade of weather extremes. *Nat. Clim. Chang.* **2012**, *2*, 491–496. [CrossRef]
11. IPCC, Climate Change. *The Physical Science Basis. Contribution of Working Group I to the Fourth Assessment Report of the Intergovernmental Panel on Climate Change*; Jansen, E., Overpeck, J., Briffa, K.R., Duplessy, J.C., Joos, F., Masson-Delmotte, V., Olago, D., Otto-Bliesner, B., Peltier, W.R., Rahmstorf, S., et al., Eds.; Cambridge University Press: Cambridge, UK, 2007.
12. Gronewold, N. Is the Flooding in Pakistan a Climate Change Disaster? Available online: http://www.scientificamerican.com/article/is-the-flooding-in-pakist/ (accessed on 31 May 2019).
13. Sajjad, A.; Lu, J.; Chen, X.; Chisenga, C.; Saleem, N.; Hassan, H. Operational Monitoring and Damage Assessment of Riverine Flood-2014 in the Lower Chenab Plain, Punjab, Pakistan, Using Remote Sensing and GIS Techniques. *Remote Sens.* **2020**, *12*, 714. [CrossRef]
14. Tayyab, M.; Zhang, J.; Hussain, M.; Ullah, S.; Liu, X.; Khan, S.N.; Baig, M.A.; Hassan, W.; Al-Shaibah, B. GIS-Based Urban Flood Resilience Assessment Using Urban Flood Resilience Model: A Case Study of Peshawar City, Khyber Pakhtunkhwa, Pakistan. *Remote Sens.* **2021**, *13*, 1864. [CrossRef]
15. Sidek, L.M.; Chua, L.H.C.; Azizi, A.S.M.; Basri, H.; Jaafar, A.S.; Moon, W.C. Application of PCSWMM for the 1-D and 1-D–2-D Modeling of Urban Flooding in Damansara Catchment, Malaysia. *Appl. Sci.* **2021**, *11*, 9300. [CrossRef]
16. Santillan, J.R.; Amora, A.M.; Makinano-Santillan, M.; Marqueso, J.T.; Cutamora, L.C.; Serviano, J.L.; Makinano, R.M. Assessing the impacts of flooding caused by extreme rainfall events through a combined geospatial and numerical modeling approach. *ISPRS—Int. Arch. Photogramm. Remote Sens. Spat. Inf. Sci.* **2016**, *XLI-B8*, 1271–1278. [CrossRef]
17. Dasallas, L.; Lee, S.; Dasallas, L.; Lee, S. Topographical Analysis of the 2013 Typhoon Haiyan Storm Surge Flooding by Combining the JMA Storm Surge Model and the FLO-2D Flood Inundation Model. *Water* **2019**, *11*, 144. [CrossRef]
18. Asare-Kyei, D.; Forkuor, G.; Venus, V. Modeling Flood Hazard Zones at the Sub-District Level with the Rational Model Integrated with GIS and Remote Sensing Approaches. *Water* **2015**, *7*, 3531–3564. [CrossRef]
19. Al Baky, M.A.; Islam, M.; Paul, S. Flood Hazard, Vulnerability and Risk Assessment for Different Land Use Classes Using a Flow Model. *Earth Syst. Environ.* **2020**, *4*, 225–244. [CrossRef]

20. Santillan, J.R.; Marqueso, J.T.; Makinano-Santillan, M.; Serviano, J.L. Beyond Flood Hazard Maps: Detailed Flood Characterization with Remote Sensing, GIS and 2D Modelling. *ISPRS—Int. Arch. Photogramm. Remote Sens. Spat. Inf. Sci.* **2016**, *XLII-4/W1*, 315–323. [CrossRef]
21. Skilodimou, H.D.; Bathrellos, G.D.; Alexakis, D.E. Flood Hazard Assessment Mapping in Burned and Urban Areas. *Sustainability* **2021**, *13*, 4455. [CrossRef]
22. Su, X.; Shao, W.; Liu, J.; Jiang, Y.; Wang, K. Dynamic Assessment of the Impact of Flood Disaster on Economy and Population under Extreme Rainstorm Events. *Remote Sens.* **2021**, *13*, 3924. [CrossRef]
23. Hagen, E.; Lu, X.X. Let us create flood hazard maps for developing countries. *Nat. Hazards* **2011**, *58*, 841–843. [CrossRef]
24. Psomiadis, E.; Diakakis, M.; Soulis, K.X. Combining SAR and Optical Earth Observation with Hydraulic Simulation for Flood Mapping and Impact Assessment. *Remote Sens.* **2020**, *12*, 3980. [CrossRef]
25. Albu, L.M.; Enea, A.; Iosub, M.; Breaban, I.G. Dam Breach Size Comparison for Flood Simulations. A HEC-RAS Based, GIS Approach for Drăcșani Lake, Sitna River, Romania. *Water* **2020**, *12*, 1090. [CrossRef]
26. Tahsin, S.; Medeiros, S.C.; Hooshyar, M.; Singh, A. Optical Cloud Pixel Recovery via Machine Learning. *Remote Sens.* **2017**, *9*, 527. [CrossRef]
27. Di Baldassarre, G.; Schumann, G.; Bates, P.D. A technique for the calibration of hydraulic models using uncertain satellite observations of flood extent. *J. Hydrol.* **2009**, *367*, 276–282. [CrossRef]
28. Schober, B.; Hauer, C.; Habersack, H. A novel assessment of the role of Danube floodplains in flood hazard reduction (FEM method). *Nat. Hazards* **2015**, *75*, 33–50. [CrossRef]
29. Ruiz-Bellet, J.L.; Balasch, J.C.; Tuset, J.; Barriendos, M.; Mazon, J.; Pino, D. Historical, hydraulic, hydrological and meteorological reconstruction of 1874 Santa Tecla flash floods in Catalonia (NE Iberian Peninsula). *J. Hydrol.* **2015**, *524*, 279–295. [CrossRef]
30. Bhandari, S.; Jobe, A.; Thakur, B.; Kalra, A.; Ahmad, S. Flood Damage Reduction in Urban Areas with Use of Low Impact Development Designs. In Proceedings of the World Environmental and Water Resources Congress 2018, Minneapolis, MN, USA, 3–7 June 2018; American Society of Civil Engineers: Reston, VA, USA, 2018; pp. 52–61.
31. Quiroga, V.M.; Kure, S.; Udo, K.; Mano, A. Application of 2D numerical simulation for the analysis of the February 2014 Bolivian Amazonia flood: Application of the new HEC-RAS version 5. *RIBAGUA—Rev. Iberoam. Agua* **2016**, *3*, 25–33. [CrossRef]
32. Gilles, D.; Young, N.; Schroeder, H.; Piotrowski, J.; Chang, Y.-J.; Gilles, D.; Young, N.; Schroeder, H.; Piotrowski, J.; Chang, Y.-J. Inundation Mapping Initiatives of the Iowa Flood Center: Statewide Coverage and Detailed Urban Flooding Analysis. *Water* **2012**, *4*, 85–106. [CrossRef]
33. Ahmadisharaf, E.; Kalyanapu, A.J.; Chung, E.-S. Evaluating the Effects of Inundation Duration and Velocity on Selection of Flood Management Alternatives Using Multi-Criteria Decision Making. *Water Resour. Manag.* **2015**, *29*, 2543–2561. [CrossRef]
34. Awadallah, M.O.M.; Juárez, A.; Alfredsen, K. Comparison between Topographic and Bathymetric LiDAR Terrain Models in Flood Inundation Estimations. *Remote Sens.* **2022**, *14*, 227. [CrossRef]
35. Al Amin, M.B.; Sarino; Haki, H. Floodplain simulation for Musi River using integrated 1D/2D hydrodynamic model. *MATEC Web Conf.* **2017**, *101*, 05023. [CrossRef]
36. Elkhrachy, I. Flash Flood Water Depth Estimation Using SAR Images, Digital Elevation Models, and Machine Learning Algorithms. *Remote Sens.* **2022**, *14*, 440. [CrossRef]
37. Liu, Z.; Merwade, V.; Jafarzadegan, K. Investigating the role of model structure and surface roughness in generating flood inundation extents using one- and two-dimensional hydraulic models. *J. Flood Risk Manag.* **2019**, *12*, e12347. [CrossRef]
38. Yalcin, E. Two-dimensional hydrodynamic modelling for urban flood risk assessment using unmanned aerial vehicle imagery: A case study of Kirsehir, Turkey. *J. Flood Risk Manag.* **2018**, *12*, e12499. [CrossRef]
39. Alivio, M.B.T.; Puno, G.R.; Talisay, B.A.M. Flood hazard zones using 2d hydrodynamic modeling and remote sensing approaches. *Glob. J. Environ. Sci. Manag.* **2019**, *5*, 1–16. [CrossRef]
40. Naeem, B.; Azmat, M.; Tao, H.; Ahmad, S.; Khattak, M.U.; Haider, S.; Ahmad, S.; Khero, Z.; Goodell, C.R. Flood Hazard Assessment for the Tori Levee Breach of the Indus River Basin, Pakistan. *Water* **2021**, *13*, 604. [CrossRef]
41. Papaioannou, G.; Varlas, G.; Terti, G.; Papadopoulos, A.; Loukas, A.; Panagopoulos, Y.; Dimitriou, E. Flood Inundation Mapping at Ungauged Basins Using Coupled Hydrometeorological–Hydraulic Modelling: The Catastrophic Case of the 2006 Flash Flood in Volos City, Greece. *Water* **2019**, *11*, 2328. [CrossRef]
42. Garcia, M.; Juan, A.; Bedient, P. Integrating Reservoir Operations and Flood Modeling with HEC-RAS 2D. *Water* **2020**, *12*, 2259. [CrossRef]
43. Tariq, M.A.U.R.; Farooq, R.; van de Giesen, N. A Critical Review of Flood Risk Management and the Selection of Suitable Measures. *Appl. Sci.* **2020**, *10*, 8752. [CrossRef]
44. Cutter, S. Vulnerability to Environmental Hazards. *Progress Human Geogr.* **1996**, *20*, 529–532. [CrossRef]
45. Mitchell, J. Urban disasters as indicators of global environmental change: Assessing functional varieties of vulnerability. In Proceedings of the Symposium on Disaster Reduction and Global Environmental Change, Federal Foreign Office, Berlin, Germany, 20–21 June 2002.
46. Merz, B.; Thieken, A.H.; Gocht, M. *Flood Risk Mapping at the Local Scale: Concepts and Challenges*; Springer: Dordrecht, The Netherlands, 2007; pp. 231–251.
47. Salami, R.O.; von Meding, J.K.; Giggins, H. Vulnerability of human settlements to flood risk in the core area of Ibadan metropolis, Nigeria. *Jamba* **2017**, *9*, 371. [CrossRef] [PubMed]

48. Pistrika, A.; Tsakiris, G.; Nalbantis, I. Flood Depth-Damage Functions for Built Environment. *Environ. Process.* **2014**, *1*, 553–572. [CrossRef]
49. Citeau, J.M. A New Flood Control Concept in the Oise Catchment Area: Definition and Assessment of Flood Compatible Agricultural Activities. In Proceedings of the FIG Working Week 2003, Paris, France, 13–17 April 2003.
50. Balijepalli, C.; Oppong, O. Measuring vulnerability of road network considering the extent of serviceability of critical road links in urban areas. *J. Transp. Geogr.* **2014**, *39*, 145–155. [CrossRef]
51. Hussain, M.S.; Lee, S. The regional and the seasonal variability of extreme precipitation trends in Pakistan. *Asia-Pacific J. Atmos. Sci.* **2013**, *49*, 421–441. [CrossRef]
52. Faisal, M.; Muzammil, M.; Azam, M.I.; Yaseen, M.; Abbas, T.; Nabi, G. Flood Hazard Mapping and Risk Zoning of the Nullah Deg, Pakistan Using Hydraulic Simulation Model (a Case Study). *Sci. Int.* **2015**, *27*, 6459–6464.
53. Faisal, M. *Assessment of Flood Inundation Using Hydraulic Simulation Model*; University of Engineering and Technology: Lahore, Pakistan, 2015.
54. USACE-HEC. *HEC RAS River Analysis System 2D Modelling User's Manual Version 5.0*; Hydrologic Engineering Center, United States Corps of Engineer: Davis, CA, USA, 2016.
55. Yang, L.; Meng, X.; Zhang, X. SRTM DEM and its application advances. *Int. J. Remote Sens.* **2011**, *32*, 3875–3896. [CrossRef]
56. Dang, N.M.; Babel, M.S.; Luong, H.T. Evaluation of food risk parameters in the Day River Flood Diversion Area, Red River Delta, Vietnam. *Nat. Hazards* **2011**, *56*, 169–194. [CrossRef]
57. Kreibich, H.; Piroth, K.; Seifert, I.; Maiwald, H.; Kunert, U.; Schwarz, J.; Merz, B.; Thieken, A.H. Is flow velocity a significant parameter in flood damage modelling? *Nat. Hazards Earth Syst. Sci.* **2009**, *9*, 1679–1692. [CrossRef]
58. Lea, D.; Yeonsu, K.; Hyunuk, A. Case Study of HEC-RAS 1D–2D Coupling Simulation: 2002 Baeksan Flood Event in Korea. *Water* **2019**, *11*, 2048. [CrossRef]
59. Horritt, M.S.; Di Baldassarre, G.; Bates, P.D.; Brath, A. Comparing the performance of a 2-D finite element and a 2-D finite volume model of floodplain inundation using airborne SAR imagery. *Hydrol. Process.* **2007**, *21*, 2745–2759. [CrossRef]
60. Horritt, M.S.; Bates, P.D. Evaluation of 1D and 2D numerical models for predicting river flood inundation. *J. Hydrol.* **2002**, *268*, 87–99. [CrossRef]
61. Shustikova, I.; Domeneghetti, A.; Neal, J.C.; Bates, P.; Castellarin, A. Comparing 2D capabilities of HEC-RAS and LISFLOOD-FP on complex topography. *Hydrol. Sci. J.* **2019**, *64*, 1769–1782. [CrossRef]
62. Musolino, G.; Ahmadian, R.; Xia, J. Enhancing pedestrian evacuation routes during flood events. *Nat. Hazards* **2022**, *2022*, 1–25. [CrossRef]
63. Bathrellos, G.D.; Skilodimou, H.D.; Chousianitis, K.; Youssef, A.M.; Pradhan, B. Suitability estimation for urban development using multi-hazard assessment map. *Sci. Total Environ.* **2017**, *575*, 119–134. [CrossRef]
64. Saunders, W.S.A.; Kilvington, M. Innovative land use planning for natural hazard risk reduction: A consequence-driven approach from New Zealand. *Int. J. Disaster Risk Reduct.* **2016**, *18*, 244–255. [CrossRef]
65. Panagopoulos, G.P.; Bathrellos, G.D.; Skilodimou, H.D.; Martsouka, F.A. Mapping Urban Water Demands Using Multi-Criteria Analysis and GIS. *Water Resour. Manag.* **2012**, *26*, 1347–1363. [CrossRef]

remote sensing

MDPI

Article

Resolution-Sensitive Added Value Analysis of CORDEX-CORE RegCM4-7 Past Seasonal Precipitation Simulations over Africa Using Satellite-Based Observational Products

Gnim Tchalim Gnitou [1], Guirong Tan [2,*], Yan Hongming [3], Isaac Kwesi Nooni [1,4] and Kenny Thiam Choy Lim Kam Sian [1]

[1] Binjiang College, Nanjing University of Information Science & Technology, Wuxi 214105, China; patrickgnitou@yahoo.fr (G.T.G.); nooni25593@alumni.itc.nl (I.K.N.); kennlks@gmail.com (K.T.C.L.K.S.)
[2] Collaborative Innovation Center on Forecast and Evaluation of Meteorological Disasters, Key Laboratory of Meteorological Disaster, Ministry of Education, Nanjing University of Information Science and Technology, Nanjing 210044, China
[3] Yunnan Climate Center, Kunming 650034, China; y-hm@netease.com
[4] School of Geographical Sciences, Nanjing University of Information Science & Technology, Nanjing 210044, China
* Correspondence: tanguirong@nuist.edu.cn

Abstract: This study adopts a two-way approach to CORDEX-CORE RegCM4-7 seasonal precipitation simulations' Added Value (AV) analysis over Africa, which aims to quantify potential improvements introduced by the downscaling approach at high and low resolution, using satellite-based observational products. The results show that RegCM4-7 does add value to its driving Global Climate Models (GCMs) with a positive Added Value Coverage (AVC) ranging between 20 and 60% at high resolution, depending on the season and the boundary conditions. At low resolution, the results indicate an increase in the positive AVC by up to 20% compared to the high-resolution results, with an up to 8% decrease for instances where an increase is not observed. Typical climate zones such as West Africa, Central Africa, and Southern East Africa, where improvements by Regional Climate Models (RCMs) are expected due to strong dependence on mesoscale and fine-scale features, show positive AVC greater than 20%, regardless of the season and the driving GCM. These findings provide more evidence for confirming the hypothesis that the RCMs AV is influenced by their internal physics rather than being the product of a mere disaggregation of large-scale features provided by GCMs. Although the results show some dependencies to the driving GCMs relating to their equilibrium climate sensitivity nature, the findings at low resolutions similar to the native GCM resolutions make the influence of internal physics more important. The findings also feature the CORDEX-CORE RegCM4-7 precipitation simulations' potential in bridging the quality and resolution gap between coarse GCMs and high-resolution remote sensing datasets. Even if further post-processing activities, such as bias correction, may still be needed to remove persistent biases at high resolution, using upscaled RCMs as an alternative to GCMs for large-scale precipitation studies over Africa can be insightful if the AV and other performance statistics are satisfactory for the intended application.

Keywords: regional climate models; global climate models; precipitation; Africa; added value; satellite-based observations

Citation: Gnitou, G.T.; Tan, G.; Hongming, Y.; Nooni, I.K.; Lim Kam Sian, K.T.C. Resolution-Sensitive Added Value Analysis of CORDEX-CORE RegCM4-7 Past Seasonal Precipitation Simulations over Africa Using Satellite-Based Observational Products. *Remote Sens.* **2022**, *14*, 2102. https://doi.org/10.3390/rs14092102

Academic Editors: Xander Wang and Beatriz M. Funatsu

Received: 2 March 2022
Accepted: 25 April 2022
Published: 27 April 2022

1. Introduction

The distillation of regional climate information is crucial for anticipating the potential threats of regional-to-local climate change and formulating actionable adaptation and mitigation plans [1]. Carrying out these activities with primary climate models, known as Global Climate Models (GCMs), has been prohibitive due to the relatively coarse resolutions at which they are produced and the subsequent computational burden that could arise from increasing such resolutions. Moreover, the regional nature of decision-making expected from these activities demands the accountability of fine-scale atmospheric phenomena and heterogeneity of surface properties not explicitly resolved by GCMs [2].

To address this central issue, limited-area nested Regional Climate Models (RCMs) are used as a tool to downscale large-scale boundary conditions from GCMs or reanalysis data. The expectation from the RCMs is to serve as a "magnifying glass" to reveal fine-scale details that are hampered by the impossibility of running the GCMs at the desired high resolution [3]. This one-way nesting approach to dynamical downscaling has been the backbone of the three decade's worth of researching and developing RCMs, with substantial applications and use cases for various scientific problems worldwide [4,5]. Although limited-area two-way nesting models have been introduced for understanding feedback from a regional to a global scale, which one-way models do not consider, their results remained inconclusive and sometimes difficult to obtain due to their computationally demanding nature [4].

The large adoption of the one-way nesting methods for the production of RCMs also contributed to a paradigm shift in climate model evaluation and validation methods. Unlike traditional validation methods where the skill of the climate model output is directly assessed by comparison with observations, the one-way nesting method has triggered the need to quantify the improvement by the downscaling process, known as Added Value (AV) methods [6]. Since then, the AV concept has undergone many refinements and precondition relaxation to accommodate various validation scenarios. These scenarios range from a conjectural and subjective expectation of an AV to a more objective quantification based on observations [7].

Sticking with the principal aim of RCMs, which is to produce reliable high-resolution data for regional to local decision-making, AV studies remained mostly observational, with a one-way perspective of quantifying improvements by downscaling models from the coarse resolution GCMs to the high-resolution RCM outputs. Consequently, less interest was shown in a second type of observational AV, which could complementarily quantify in a two-way manner the AV feedback from a regional to a global scale [6]. Such a secondary metric could constitute a reliable way of quantifying the resolution-sensitivity of the AV results and disentangling results due to the RCMs' internal physics from the ones due to a mere disaggregation of the boundary condition [8].

The availability of RCM data in the public domain has served as a good playground for exploring the AV by one-way nested high-resolution RCMs [9]. This was made possible thanks to the World Research Climate Program (WRCP) under the Coordinated Regional Downscaling Experiment (CORDEX) [10–12], which made available a series of reanalysis and GCM-driven RCM outputs at approximately 50 km resolution in its first phase, while giving the highest priority to the African continent. Recently, the second phase, which aims at guaranteeing the availability of a homogeneous set of simulations over all the regions of the world, called the Common Regional Experiment Framework (CORDEX-CORE) [13–16], was launched by making available for its first experiments a set of simulations at an unprecedented resolution of 25 km and, thus, reaching common satellite-based observational products scale.

For Africa, which is a priority region in the context of CORDEX, the production of highly resolved simulations has not been followed by an increase in the resolution of local or global ground-based observational datasets. This situation is detrimental to observational AV studies. In practice, one has to rely on satellite-based observational datasets produced at similar resolutions by keeping in mind that they might also have some biases. Moreover, RCMs AV studies over Africa [17–21] were conducted under the one-way paradigm, except for the recent work by Dosio et al. [19], where a two-way approach is adopted in the context of precipitation climate change projections. These studies show that the evaluated RCMs add value to their boundary conditions over Africa. Still, the extent to which such an AV could be sensitive to the resolution is usually not analyzed.

Beyond the valuable information that could be obtained using a two-way approach to AV analysis over Africa, the use of satellite-based observational products as a reference for such analysis offers the opportunity to explore the potential of dynamical downscaling to bridge the resolution and quality gap between GCMs and observational products based on remote sensing technology. These newly emerging needs are highly relevant for both the climate modelling and the remote sensing communities, especially for the precipitation variable over Africa, where a consensual and unified characterization is needed [22–24].

In this context, the AV metrics can represent reliable quantitative metrics for estimating how the downscaling approach improves the driving GCMs towards reproducing observed features by remotely sensed datasets. Such applications of the AV analysis could be valuable for climate data distillation [1] and for exploring the possibility of using highly resolved RCMs as proxies for satellite-based observations in climate projections, where observations are not available. Additionally, the AV analysis could be instrumental in choosing post-processing methods such as bias correction if the RCMs' performances are inadequate for the intended application [21,25].

In this study, we propose an analysis of the CORDEX-CORE RegCM4-7 past precipitation simulations over Africa, with a perspective on the contribution of the resolution to AV results. The AV by the RCM simulations over the driving GCMs is computed and analyzed at fine- and large-scale resolutions to represent an improvement at a regional and global scale and to further understand AV sensitivity to resolution. The study results are also used to distinguish the role of the resolution from the role of the physics parameterizations used for the downscaling experiment of CORDEX-CORE RegCM4-7 over Africa and quantify its contribution to bridging the gap between GCMs and satellite-based observational precipitation products.

2. Study Area, Data, and Methods

2.1. Study Area

The African continent was designated as the highest priority region in the context of the CORDEX framework, owing to its vulnerability to global warming and its deficit of infrastructural resources needed to carry out climate projection modeling activities [13]. Africa represents a key domain of the 9 out of 14 continental CORDEX domains considered for the resolution doubling CORDEX-CORE simulations [16], and an undeniable testbed region for typical improvement expected from highly resolved RCMs, especially for fine scale-dependent variables such as precipitation [7,9,13]. This is particularly true concerning Africa's complex topographic structure, and its unique and homogeneous climate zones such as the Sahara (SAH), West Africa (WAF), Central Africa (CAF), Northern East Africa (NEAF), Southern East Africa, Eastern South Africa (ESAF) and Western South Africa (WSAF) (see Figure 1).

Moreover, Africa is a hotspot for large-scale precipitation patterns, which are still not adequately reproduced by state-of-the-art GCMs [26,27]. Although the AV by RCMs is usually expected at a finer scale, the extent to which such improvements can cumulatively enhance large-scale precipitation patterns and represent a large-scale AV at GCMs resolutions is an open question [6,7,28]. Addressing such a question for Africa is critical, especially from the model users' perspective, given the increasing availability of various climate models over the continent and the possible risk of data misuse [1]. Additionally, such a distinction between AVs at large- and fine-scale can be resourceful in disentangling the role of resolution from the RCMs internal physics and reveal how sensitive the AV is to resolution.

Last but not least, the production of precipitation estimates complementing existing ones, such as satellite-based products, reanalysis data and traditional climate model data, has been recommended as a prerequisite toward unifying and further understanding rainfall over Africa [29]. This alternative to rainfall stations observation data is becoming unavoidable given the serious decline in the very few stations that have been operating for the past few decades [22]. The presence of an AV by CORDEX-CORE RegCM4-7 could be instrumental, not only to address the need for a consensus on rainfall over Africa, but also for a process-based understanding of decades of satellite-based climate data records available over the continent [24,30]. Another opportunity for the African modeling community could be the potential collaborations with the remote sensing community in order to leverage the gigantesque amount of literature available on data processing [31] and use it to enhance RCMs quality.

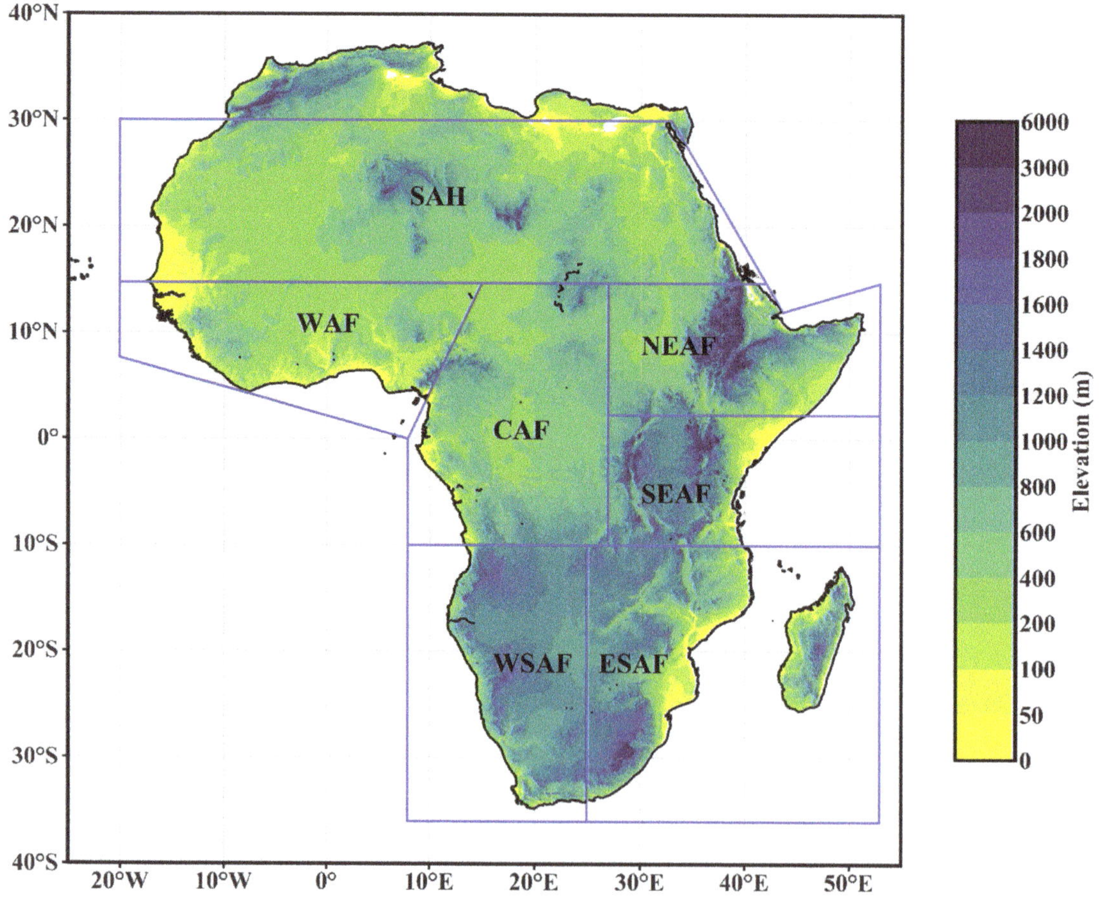

Figure 1. Topographical features of the African domain and its key climate zones.

2.2. Data and Method

The resolution-sensitive analysis of the AV by RCMs output, proposed in the present study, is mainly based on the CORDEX-CORE RegCM4-7 past precipitation simulations over the African domain. The RegCM4-7 RCM [32] is developed by the Abdus Salam International Center for Theoretical Physics (ITCP), located in Trieste, Italy. As part of the CORDEX-CORE's first experiment, the RegCM4-7 was used to downscale the ERA-Interim (ERA-INT) reanalysis data [33] for evaluation runs and 3 GCMs of the Coupled Model Intercomparison Project, phase 5 (CMIP5) [34] for the historical runs. HadGEM2-ES [35], MPI-ESM-MR [36], and NorESM1-M [37] GCMs, respectively, corresponding to high, medium and low equilibrium climate sensitivity, were used for the historical runs to capture the sensitivity range of the CMIP5 ensemble given the computational limitations related to downscaling a larger ensemble of simulations [38]. The AV sensitivity analysis is carried out using the 25 km resolution satellite-based product of the Climate Hazard Group InfraRed Precipitation with Station data version 2 (CHIRPS-2.0) [39] to represent fine-scale observations and its upscaled version at 250 km resolution to represent large-scale observation. The 250 km resolution used for the upscaling process is obtained from the Global Precipitation Climatology Project monthly data version 2.3(GPCPv2.3) [40,41], which is widely used for large-scale climate analysis. Although not directly presented in the results, GPCPv2.3 is used as a supplement to show potential uncertainty implications that may prevail at low resolutions in the context of resolution-sensitive AV studies. Further details about the different datasets used are provided in Table 1.

Table 1. List of multi-source datasets used in this study.

Data Type	Name	Institution	Resolution
Ground and satellite-blended observations	CHIRPSv2.0	Climate Hazard Group of the University of California, Sancta Barbara	0.25° × 0.25°
Ground, satellite, and sounding-blended observations	GPCPv2.3	University of Maryland	2.5° × 2.5°
Global Climate Model (GCM)	HadGEM2-ES	Met Office Hadley Centre (MOHC)	1.25° × 1.85°
Global Climate Model (GCM)	MPI-ESM-MR	Max Plank Institute for Meteorology (MPI-M)	1.8653° × 1.875°
Global Climate Model (GCM)	NorESM1-M	Norwegian Climate Centre (NCC)	1.8947° × 2.5°
Reanalysis	ERA-Interim	European Centre for Medium-Range Weather Forecasts (ECMWF)	0.75° × 0.75°
Regional Climate Model (RCM)	RegCM4-7	Abdus Salam International Center for Theoretical Physics (ITCP)	0.22° × 0.22°

Both the climate simulations and observations datasets are acquired from 1981 to 2005. The driving GCMs datasets and the RegCM4-7 outputs are first interpolated at CHIRPS' 25 km grid resolution for high-resolution analysis and then upscaled to 250 km resolution to match CHIRPS' upscaled grids for large-scale analysis. The analysis focuses on the seasonal mean bias pattern of the GCM-driven RegCM4-7 precipitation outputs and their consistency with observations and structural biases from ERA-INT driven simulations, considering both high resolution (25 km) and coarse resolution (250 km). The analysis is carried out for the December–January–February (DJF), the March–April–May (MAM), the June–July–August (JJA), and the September–October–November (SON) seasons. The potential improvement of the RegCM4-7 outputs as compared to the driving GCMs at higher and coarser resolution is also analyzed, using the *AV* metric proposed by Dosio et al., [17], for which the formula is given as follows:

$$AV = \frac{(X_{GCM} - X_{OBS})^2 - (X_{RCM} - X_{OBS})^2}{Max\left((X_{GCM} - X_{OBS})^2, (X_{RCM} - X_{OBS})^2\right)} \quad (1)$$

where $X_{GCM}, X_{RCM}, X_{OBS}$ represent, respectively, the GCM, the RCM and the observation's statistics for which the *AV* is evaluated. The *AV* values vary between -1 and 1 to capture possible improvement or degradation of the RCM over the GCM.

The seasonal precipitation bias patterns of the GCMs and RegCM4-7 outputs and the subsequent *AV* outcomes are aggregated using the different climate zones over Africa (see Figure 1). Specifically, the Added Value Coverage (*AVC*), representing the percentage of grid cells showing a positive, negative, or non-significant *AV* is computed for each climate zone. The *AVC* provides a general figure of the improvement by the RegCM4-7 RCM, and allows adequate comparison among different zones, seasons and resolutions. The *AVC* formula is given as follows:

$$AVC_{pos/neg/ns} = \frac{N_{pos/neg/ns}}{N_{tot}} \times 100 \quad (2)$$

where $AVC_{pos/neg/ns}$ represents the positive, negative, or non-significant *AV* coverage, $N_{pos/neg/ns}$, the number of pixels with a positive, negative, or non-significant *AV* and N_{tot}, the total number of pixels over the region considered.

Following previous use of the *AVC* [20,21], we use a threshold of 0.1 for significant positive *AV* and -0.1 for significant negative *AV*. This means that any pixel with an $AV > 0.1$ is considered a pixel with significant positive *AV*, and any pixel with an $AV < -0.1$ is considered a pixel with significant negative *AV*. Any pixel with an *AV* between -0.1 and 0.1 is considered non-significant.

3. Results

3.1. Evaluation Results for DJF Season

Figures 2 and S1 depict the DJF season's mean bias results by the driving GCMs and their RegCM4-7 dynamically downscaled outputs at a high resolution (25 km) and coarse resolution (250 km), using CHIRPS observations. Rain-abundant areas such as the southern CAF, SEAF, northern WSAF, and ESAF show similar patterns at high and coarse resolutions (Figures 2a and S1a). The driving GCMs (Figures 2c–f and S1c–f) tend to show wet biases of 1 to 10 mm/day over WSAF, ESAF, and parts of CAF with pronounced quantities of more than 6 mm/day in NorESM1-M (Figures 2d and S1d). Over northern CAF and NEAF, a dominance of dry biases ranging from 0 to -4 mm/day is observed with HadGEM2-ES (Figures 2c and S1c) and MPI-ESM-MR (Figures 2d and S1d), while wet biases of 0 to 8 mm/day are depicted by NorESM1-M (Figures 2d and S1d).

In accordance with RegCM4-7 evaluation runs by ERA-INT (Figures 2b and 3b), the dynamically downscaled outputs (Figures 2g–j and S1g–j) show persisting wet biases of about 1 to 8 mm/day, both at high and coarse resolution over Southern Africa (SAF), resulting in degraded AV for most simulations such as HadGEM2-ES (Figures 2k and S1k) and MPI-ESM-MR (Figures 2l and S1l) driven RegCM4-7 outputs. The presence of generally wet biases in the dynamically downscaled outputs compared to mainly dry biases in the driving GCMs is confirmed by the spatially averaged bias results reported in Table 2.

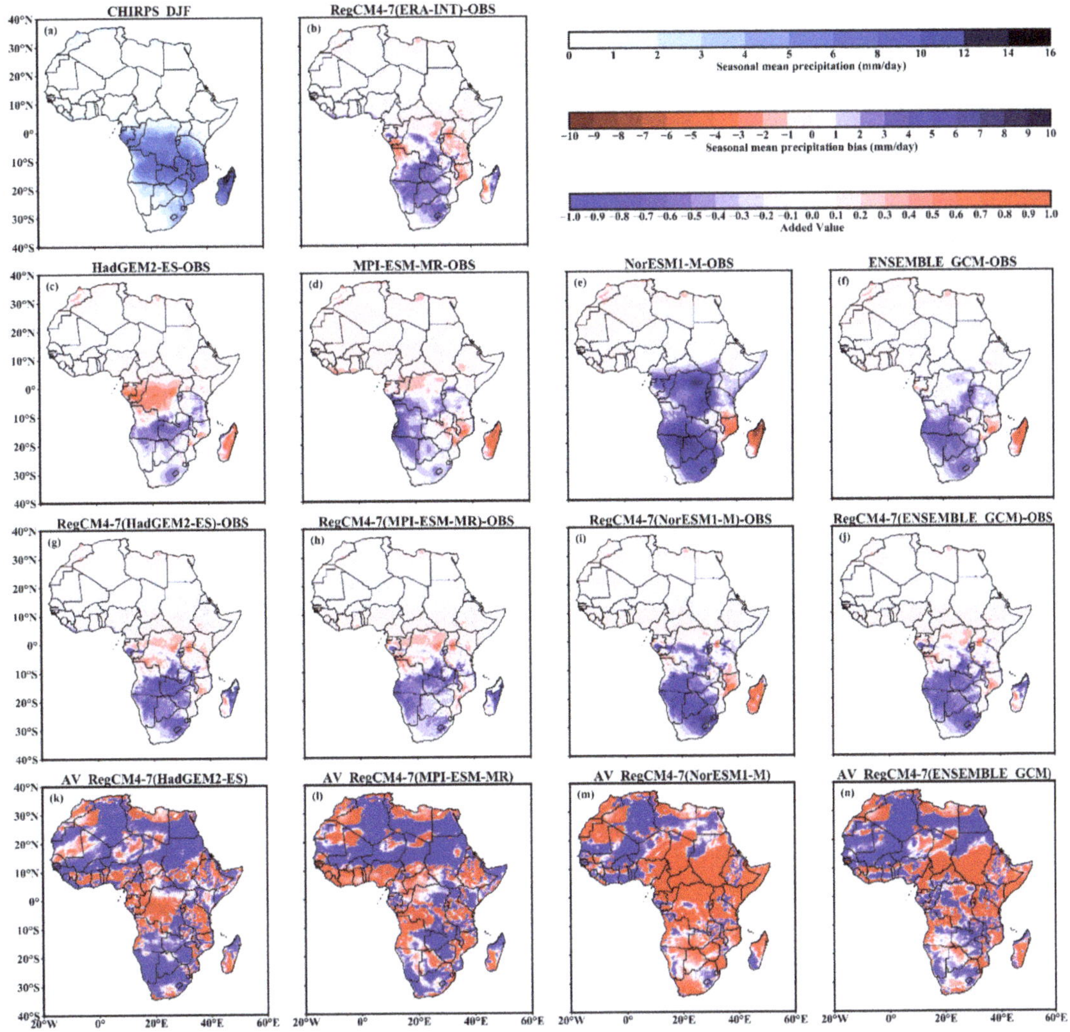

Figure 2. Performance of African precipitation in DJF season at high resolution compared to (**a**) CHIRPS, for (**b**) RegCM4-7 evaluation run driven by ERA-INT, (**c**) driving GCM HadGEM2-ES, (**d**) driving GCM MPI-ESM-MR, (**e**) driving GCM NorESM1-M, (**f**) ensemble mean of the driving GCMs, (**g**) RegCM4-7 historical run driven by HadGEM2-ES, (**h**) RegCM4-7 historical run driven by MPI-ESM-MR, (**i**) RegCM4-7 historical run driven by NorESM1-M, (**j**) RegCM4-7 historical runs' ensemble mean, (**k**) Added Value by RegCM4-7 to HadGEM2-ES, (**l**) Added Value by RegCM4-7 to MPI-ESM-MR, (**m**) Added Value by RegCM4-7 to NorESM1-M, (**n**) Added Value by RegCM4-7 to the ensemble mean of the driving GCMs.

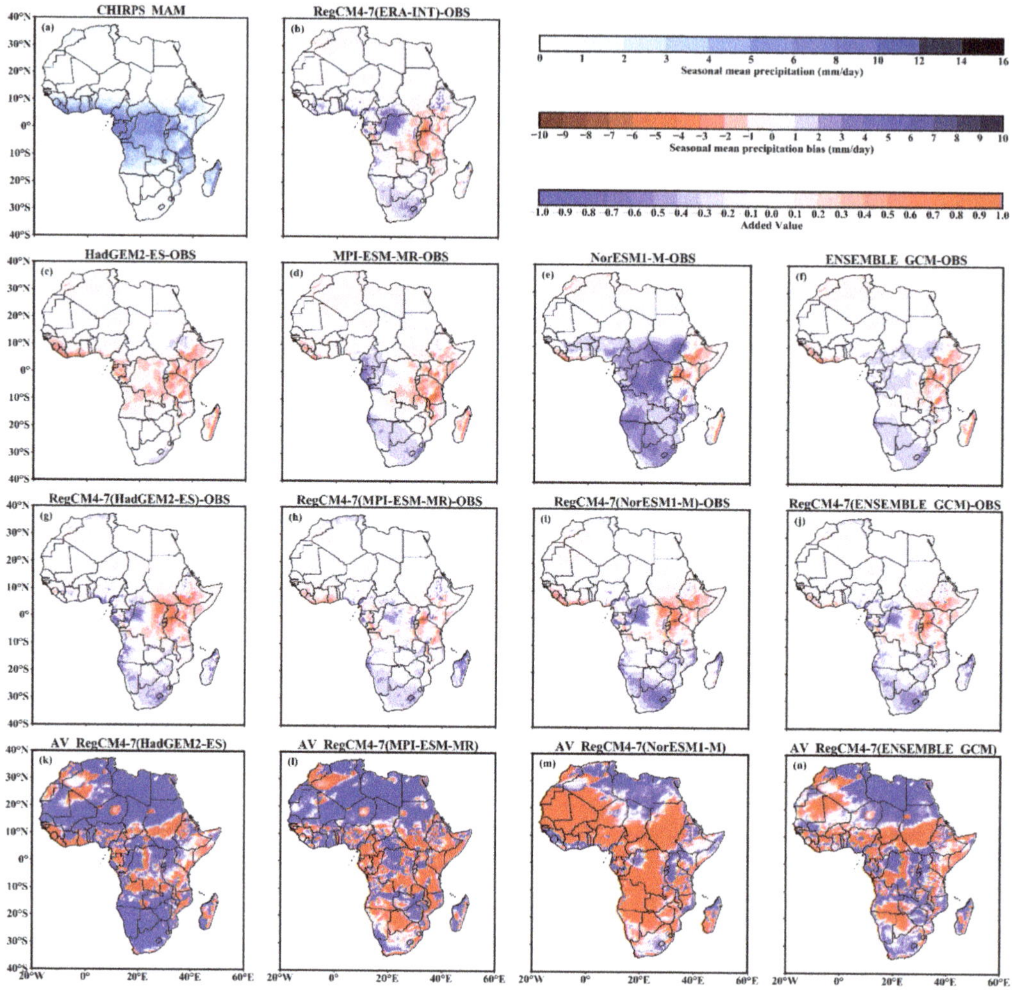

Figure 3. Performance of African precipitation in MAM season at high resolution compared to (**a**) CHIRPS, for (**b**) RegCM4-7 evaluation run driven by ERA-INT, (**c**) driving GCM HadGEM2-ES, (**d**) driving GCM MPI-ESM-MR, (**e**) driving GCM NorESM1-M, (**f**) ensemble mean of the driving GCMs, (**g**) RegCM4-7 historical run driven by HadGEM2-ES, (**h**) RegCM4-7 historical run driven by MPI-ESM-MR, (**i**) RegCM4-7 historical run driven by NorESM1-M, (**j**) RegCM4-7 historical runs' ensemble mean, (**k**) Added Value by RegCM4-7 to HadGEM2-ES, (**l**) Added Value by RegCM4-7 to MPI-ESM-MR, (**m**) Added Value by RegCM4-7 to NorESM1-M, (**n**) Added Value by RegCM4-7 to the ensemble mean of the driving GCMs.

Over CAF, SEAF, and NEAF, reduced intensity of the dry biases by HadGEM2-ES and MPI-ESM-MR is observed in the downscaled outputs, with a sign shift from wet biases to slightly dry biases in NorESM1-M downscaled output. Consequently, a positive AV is observed for all simulations and their ensemble mean (Figures 2k–n and S1k–n) over CAF. The positive AV is extended to other dry areas such as WAF, NEAF and SAH. The relatively similar AV patterns at high and low resolutions are further confirmed by the AVC results (see Table 3), which report an increase/decrease of roughly 8%. Other differences are also observed in the error amplitude, which tends to show a systematic reduction from high- to low-resolution (see Table 2).

Table 2. Spatially averaged statistics over continental Africa at high resolution (HR) and low resolution (LR).

	HadGEM2-ES		RegCM4-7 (HadGEM2-ES)		MPI-ESM-MR		RegCM4-7 (MPI-ESM-MR)		NorESM1-M		RegCM4-7 (NorESM1-M)		ENSEMBLE_GCM		RegCM4-7 (ENSEMBLE-GCM)	
	HR	**LR**	**HR**	**LR**	**HR**	**LR**	**HR**	**LR**	**HR**	**LR**	**HR**	**LR**	**HR**	**LR**	**HR**	**LR**
DJF																
BIAS	−0.09	−0.43	1.72	1.88	0.18	−0.4	1.77	2.01	−0.42	−0.62	−0.01	0.03	−0.11	−0.49	1.16	1.31
RMSE	6.17	1.52	9.71	3.63	6.81	1.72	10.44	3.51	5.19	2.15	7.16	2.49	5.73	1.54	8.24	2.47
MAM																
BIAS	−0.38	−0.52	0.69	0.39	0.04	−0.51	0.36	0.42	0	−0.01	0.59	0.38	−0.11	−0.35	0.55	0.4
RMSE	3.67	1.39	5.69	2.08	4.04	1.49	6.36	2.37	3.88	1.56	6.58	2.2	3.56	1.24	5.37	1.72
JJA																
BIAS	−1.82	−1.56	0.77	−0.26	−0.8	−0.66	0.54	0.3	−1.47	−0.87	−086	−0.58	−1.36	−1.03	0.15	−0.18
RMSE	2.35	2.22	3.62	2.38	2.08	1.51	2.66	2.14	2.16	2.21	2.75	2.14	2.08	1.78	2.69	1.67
SON																
BIAS	−0.78	−0.64	2.83	1.24	0.14	0.08	0.96	0.99	−0.71	−0.48	−0.27	0.07	−0.45	−0.35	1.17	0.76
RMSE	2.78	1.97	4.25	2.69	2.62	1.62	3.63	2.52	2.18	1.89	2.45	1.92	2.3	1.44	2.83	1.71

Table 3. Added Value Coverage results of RegCM4-7 simulations over continental Africa at high resolution (HR) and low resolution (LR).

		Negative AVC		Non-Significant AVC		Positive AVC	
		HR	LR	HR	LR	HR	LR
DJF	RegCM4-7(HadGEM2-ES)	65.22%	52.08%	5.67%	26.36%	29.1%	21.56%
	RegCM4-7(MPI-ESM-MR)	52.26%	53.73%	4.24%	7.45%	43.5%	38.82%
	RegCM4-7(NorESM1-M)	46.41%	43.4%	5.05%	0%	48.55%	56.6%
	RegCM4-7(ENSEMBLE_GCM)	62.05%	60.55%	2.32%	2.77%	35.63%	35.68%
MAM	RegCM4-7(HadGEM2-ES)	68.16%	75.43%	3.66%	0%	28.18%	24.57%
	RegCM4-7(MPI-ESM-MR)	63.86%	56.28%	5.26%	5%	30.89%	38.72%
	RegCM4-7(NorESM1-M)	52.4%	41.18%	3.96%	0.37%	43.65%	58.45%
	RegCM4-7(ENSEMBLE_GCM)	54.7%	51.6%	5.05%	1.23%	40.25%	47.17%
JJA	RegCM4-7(HadGEM2-ES)	64.47%	53.52%	1.6%	1.16%	33.94%	45.32%
	RegCM4-7(MPI-ESM-MR)	48.32%	32.3%	4.35%	0%	47.33%	67.7%
	RegCM4-7(NorESM1-M)	30.37%	26.66%	9.65%	14.25%	59.98%	59.09%
	RegCM4-7(ENSEMBLE_GCM)	49.65%	41.04%	4.36%	0.42%	45.99%	58.53%
SON	RegCM4-7(HadGEM2-ES)	61.52%	49.14%	2.52%	4.3%	35.96%	46.56%
	RegCM4-7(MPI-ESM-MR)	73.86%	77.47%	3.44%	0%	22.71%	22.53%
	RegCM4-7(NorESM1-M)	37.98%	37.2%	3.09%	0.13%	58.93%	62.67%
	RegCM4-7(ENSEMBLE_GCM)	57.48%	55.54%	2.32%	0.09%	40.19%	44.37%

3.2. Evaluation Results for MAM Season

In the MAM season, the performances of the RegCM4-7 outputs and their boundary forcing at downscaled and upscaled resolutions are shown in Figures 3 and S2. At high and low resolution, CHIRPS (Figures 3a and S2a) show similar rain belt expansion patterns over both WAF and NEAF in addition to rain-abundant areas such as CAF.

An underestimation of rainfall quantities of 0 to −4 mm/day is mostly found over NEAF and SEAF in all the driving GCMs (Figures 3c–f and S2c–f) and their dynamically downscaled outputs (Figures 3g–j and S2g–j). These biases tend to be identical except for MPI-ESM-MR (Figures 3h and S2h) driven RegCM4-7 output, which shows a reduction and a subsequent positive AV (Figures 3m and S2m). Although the biases in the driving GCMs (Figures 3c–f and S2c–f) over WAF and CAF show unique patterns based on the model used, their dynamically downscaled simulations (Figures 3g–j and S2g–j) share unique features that are highly similar to the evaluation runs driven by ERA-INT (Figures 3b and S2b).

An intensification of HadGEM2-ES's (Figures 3c and S2c) slightly wet biases (0–1 mm/day) in its downscaled output (Figures 3g and S2g) is observed over SAF, leading to negative AV (Figures 3k and S2k). A general dominance of dry biases for the driving GCMs and slightly wet biases for RegCM4-7 outputs are observed as reported in Table 2, which is similar to the DJF season results. The positive AVC results in Table 3 show an increase from high- to low-resolution results for MPI-ESM-MR, NorESM1-M and the ensemble mean downscaled outputs. At the same time, the HadGEM2-ES-based RegCM4-7 simulations report a decrease. Compared to the DJF season results, the results during MAM season are still within the 8% increase/decrease range, except for the downscaled NorESM1-M results, which show an increase of nearly 15%. The systematic reduction in the averaged error from high to low resolution remains the same as in DJF, even if the error amplitudes are lower (see Table 2).

3.3. Evaluation Results for JJA Season

In the JJA season, the mean biases of RegCM4-7 simulations and their driving GCMs at high and low resolutions are summarized in Figures 4 and S3. The results show dry and wet bias signals in all the climate simulations along the rain belt depicted over WAF, northern CAF and NEAF by CHIRPS (Figures 4a and S3a) at both resolutions. Dry biases of about 0–4 mm/day are observed over CAF and southern SAH, and wet biases of about 1–8 mm/day are observed along the remaining part of the rain belt over NEAF, SEAF and ESAF, in both the driving GCMs (Figures 4c–f and S3c–f) and the RegCM4-7 downscaled simulations (Figures 4g–j and S3g–j).

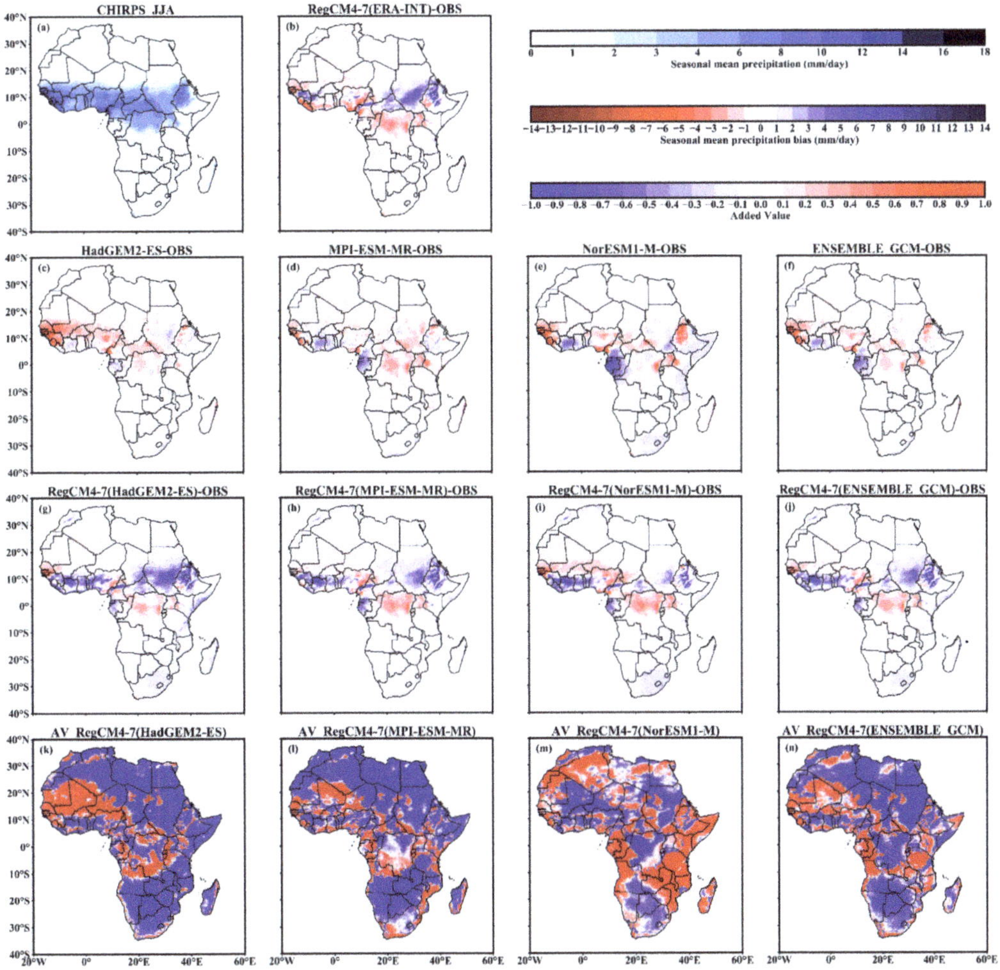

Figure 4. Performance of African precipitation in JJA season at high resolution compared to (**a**) CHIRPS, for (**b**) RegCM4-7 evaluation run driven by ERA-INT, (**c**) driving GCM HadGEM2-ES, (**d**) driving GCM MPI-ESM-MR, (**e**) driving GCM NorESM1-M, (**f**) ensemble mean of the driving GCMs, (**g**) RegCM4-7 historical run driven by HadGEM2-ES, (**h**) RegCM4-7 historical run driven by MPI-ESM-MR, (**i**) RegCM4-7 historical run driven by NorESM1-M, (**j**) RegCM4-7 historical runs' ensemble mean, (**k**) Added Value by RegCM4-7 to HadGEM2-ES, (**l**) Added Value by RegCM4-7 to MPI-ESM-MR, (**m**) Added Value by RegCM4-7 to NorESM1-M, (**n**) Added Value by RegCM4-7 to the ensemble mean of the driving GCMs.

The spatially averaged bias results (Table 2) show a systematic reduction in the RegCM4-7 averaged bias, regardless of the driving GCM, with the ensemble mean reporting improvement over some parts of CAF for all the GCM-based downscaled outputs (Figures 4g–j and S3g–j), compared to individual simulations. A positive AV is observed over parts of WAF for RegCM4-7 simulations driven by HadGEM2-ES and MPI-ESM-MR, while NorESM1-M-driven RegCM4-7 simulation depicts a positive AV over SAH, NEAF, SEAF and parts of ESAF and WSAF.

At a high resolution, RegCM4-7 reports a positive AVC of 33.94% for the simulation driven by HadGEM2-ES, 47.33% for the simulation driven by MPI-ESM-MR and 59.98% for NorESM1-M-driven simulation. The change from high to low resolution ranges between roughly 12 and 20% for all RegCM4-7 simulations, except the one driven by NorESM1-M, which decreases by 0.89% (Table 3). The error amplitude reduction from high to low resolution is also observed in the JJA season, as shown in Table 2.

3.4. Evaluation Results for SON Season

The evaluation and historical runs of RegCM4-7 and their driving GCMs results for SON season over Africa at high and low resolution are given in Figures 5 and S4. High resolution and upscaled CHIRPS observations (Figures 5a and S4a) show a retreat of the monsoonal belt toward coastal WAF and CAF with lightweight rain quantities over NEAF.

The driving GCMs (Figures 5c–f and S4c–f) exhibit wet biases ranging from 1 to 8 mm/day, mostly over CAF, with an extension to NEAF, SEAF and SAF in NorESM1-M results. Dry biases of −4 to 0 mm/day are observed, especially over WAF in the HadGEM2-ES results. The historical runs of RegCM4-7 (Figures 5g–j and S4g–j) depict wet biases, which tend to represent substantial reduction compared to the driving GCM results (Figures 5c–f and S4c–f) over parts of CAF and some parts of NEAF and SEAF. However, the downscaled simulations also show a second type of wet biases over most parts of WAF, ESAF and WSAF, particularly in the HadGEM2-ES- and MPI-ESM-MR-driven RegCM4-7 results, which tend to degrade the driving GCMs' results.

The spatially averaged bias results from Table 2 show a dominance of wet and relatively high biases in the RegCM4-7 simulations compared to dry and relatively low biases for the driving GCMs, when HadGEM2-ES and MPI-ESM-MR are considered. For NorESM1-M-based results, a systematic reduction in the driving GCM dry biases in the downscaled outputs is observed. These findings are reflected in the AV results (Figures 5k–n and S4k–n), with positive AV pixels observed mostly over CAF and SEAF for all RegCM4-7 simulations and their ensemble mean, and a higher positive AVC for NorESM1-M at both high and low resolution. The positive AVC of the simulations (see Table 3) indicates an increasing tendency from high- to low-resolution results of, at most, 11%, except MPI-ESM-MR, which demonstrates a dynamically downscaled output with a 22.71% positive AVC at high resolution and a 22.53% positive AVC at low resolution. Similar to the previous seasons' results, the SON season spatially averaged results (see Table 2) indicate a decrease in the error amplitude from high to low resolution.

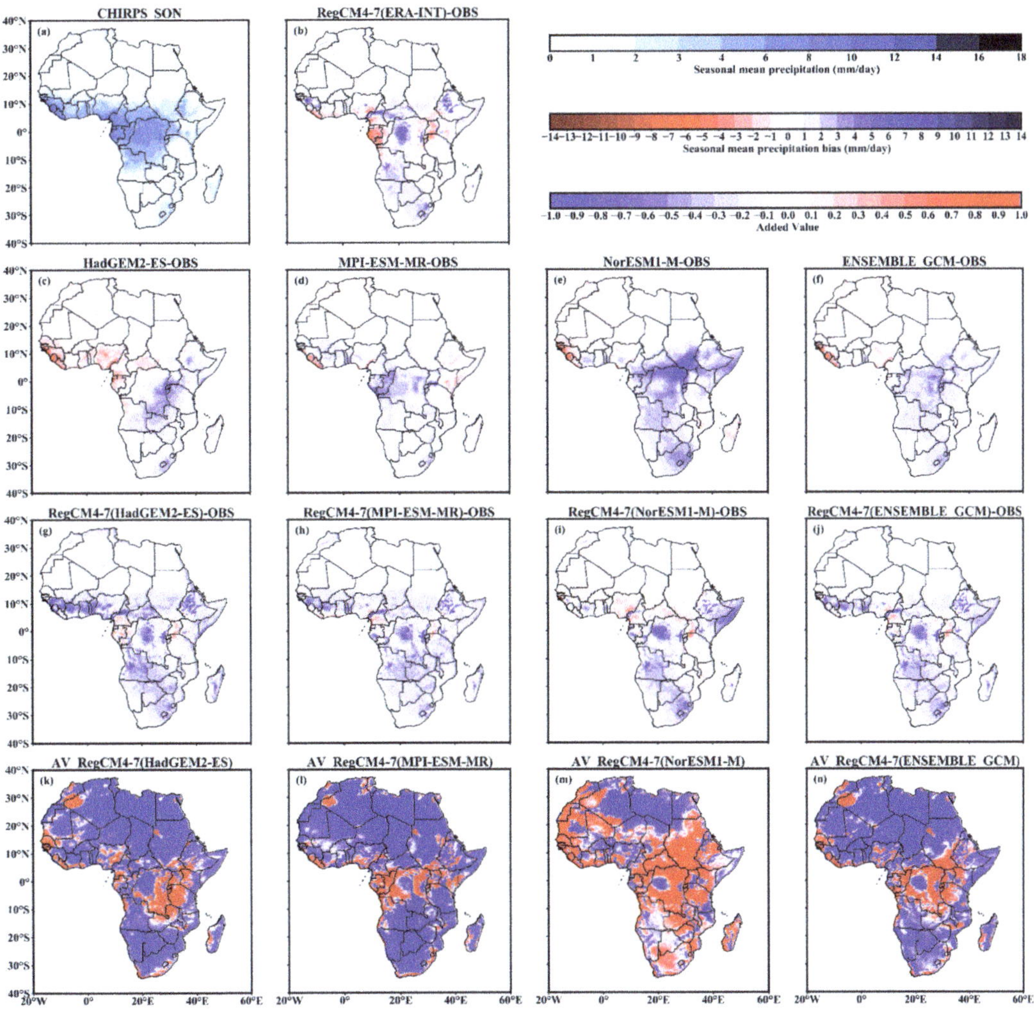

Figure 5. Performance of African precipitation in SON season at high resolution compared to (**a**) CHIRPS, for (**b**) RegCM4-7 evaluation run driven by ERA-INT, (**c**) driving GCM HadGEM2-ES, (**d**) driving GCM MPI-ESM-MR, (**e**) driving GCM NorESM1-M, (**f**) ensemble mean of the driving GCMs, (**g**) RegCM4-7 historical run driven by HadGEM2-ES, (**h**) RegCM4-7 historical run driven by MPI-ESM-MR, (**i**) RegCM4-7 historical run driven by NorESM1-M, (**j**) RegCM4-7 historical runs' ensemble mean, (**k**) Added Value by RegCM4-7 to HadGEM2-ES, (**l**) Added Value by RegCM4-7 to MPI-ESM-MR, (**m**) Added Value by RegCM4-7 to NorESM1-M, (**n**) Added Value by RegCM4-7 to the ensemble mean of the driving GCMs.

3.5. Unified Season, Sub-Area and Resolution-Based Results

The seasonal performances of the RegCM4-7 historical runs over continental Africa indicate a clear similarity pattern between the results at low and high resolution, even if a wide range of differences in terms of AVC and error amplitude are reported in Tables 2 and 3. Due to the heterogeneous nature of the results over continental Africa, climate zones and seasons-based partition of the overall AVC findings are further presented in Figure 6. Overall, the AVC results show various outcomes based on the season, sub-area, and driving

GCM. NorESM1-M-dynamically downscaled RegCM4-7 output tends to show the highest positive AVC for all sub-regions and seasons, with few exceptions such as WAF in DJF (Figure 6e) and JJA (Figure 6g), and NEAF in MAM (Figure 6n).

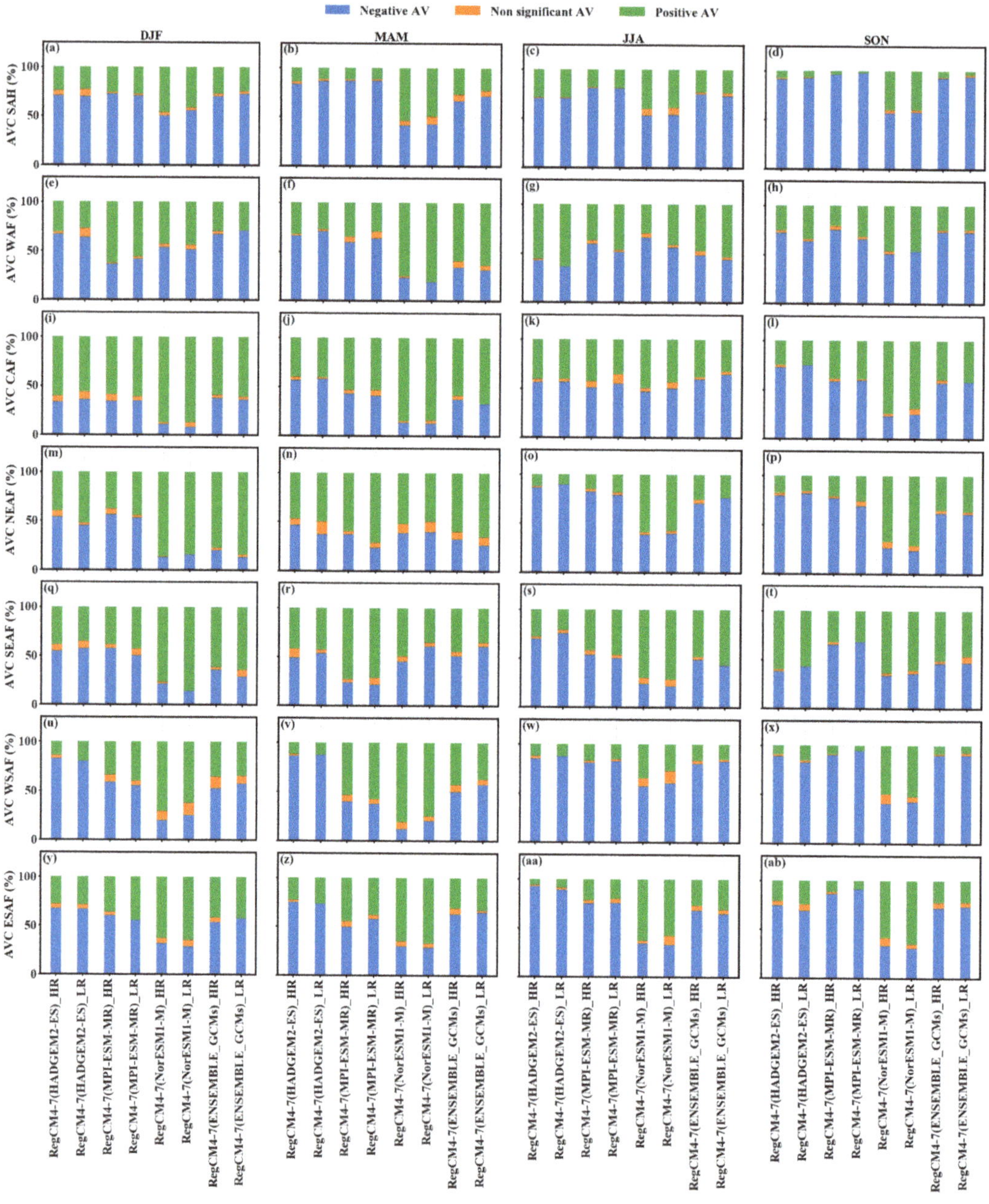

Figure 6. Added Value Coverage results for all seasons over (**a–d**) SAH region, (**e–h**) WAF region, (**i–l**) CAF region, (**m–p**) NEAF region, (**q–t**) SEAF region, (**u–x**) WSAF region, and (**y-ab**) ESAF region. "_HR" and "_LR", respectively refer to high-resolution and low-resolution results using CHIRPS observational data.

The positive AVC changes from high to low resolution are mostly in line with the overall seasonal results from Table 3. CAF (Figure 6i–l), NEAF (Figure 6m–p) and SEAF (Figure 6q–t) represent areas with the most consistent positive AVC (mostly >50%), regardless of the driving GCMs, especially in DJF and MAM seasons. RegCM4-7 ensemble mean provides an acceptable performance tradeoff but can be less satisfactory in terms of positive AVC in some specific regions and seasons. Typical examples are the SAH and WSAF regions in SON season (Figure 6d,x), where the positive AVC is less than 10%. In general, sub-regions such as WAF (Figure 6e–h), CAF (Figure 6i–l), and SEAF (Figure 6q–t) show a positive AVC greater than 20%, regardless of the season and driving GCMs.

4. Discussion

The potential applicability of climate models in different climate studies is highly constrained by model resolution. Beyond the ability of RCMs to integrate fine-scale features, their relatively high resolution compared to GCMs often influences data user preferences [7,9]. The expectation of an AV by RCMs at high resolution owing to their fundamental design choices constitutes another incentive for RCM data use. AV issues have been central to the past few decades of research and development of RCMs [4,5]. Still, their discussions were tailored using the one-way paradigm, where the information flow from the driving GCMs to the RCM is prioritized.

Although sufficient to prove the presence of improvement in a statistical sense, this perspective of the AV gives fewer insights into the attribution of such improvements [7]. Moreover, GCMs are still useful for large-scale studies, but using RCMs as alternatives is still an active research question, especially for data-scarce parts of the world such as Africa. The methodological choices of the present study were mainly motivated by these mentioned issues and the need to provide valuable information towards a better understanding of the resolution-sensitivity of AV by RCMs over Africa, and their potential to bridge resolution and quality gaps between the GCMs and the high-resolution satellite-based products.

The results mainly indicate AV by RegCM4-7 simulations at their native resolution, with a typical dependence on the driving GCM, the season, and the sub-area. Similar results have been highlighted by Gnitou et al., [21] in the context of other CORDEX-CORE precipitation simulations. Moreover, other challenges related to CORDEX-CORE precipitation simulations over Africa, such as the persistence of dry and wet biases at noticeably high amplitudes along the seasonal rain belt, were also found in the RegCM4-7 results. The historical run biases have shown clear consistency with the evaluation run driven by ERA-INT, thus suggesting that the RCM internal model physics might have more influence than the driving GCMs, particularly along the seasonal rain belt. Reasons for such deficiencies may be similar to previously known ones, including missing or misrepresented processes and regional model transferability issues [42,43].

Although the influence of RegCM4-7 internal model physics appears to be consistent with the dynamically downscaled outputs, typical dependencies to the driving GCMs are found in terms of AV. For instance, the NorESM1-M dynamically downscaled output shows significantly higher positive AVC coverage as compared to other driving GCMs. These performances may be due to the high, medium and low equilibrium climate sensitivity criteria under which the driving GCMs were chosen [16,38]. The typically low equilibrium climate sensitivity nature of the NorESM1-M driving GCM may explain the higher positive AVC performances. Since the low sensitivity relates to relatively low performances and high error, the downscaled NorESM1-M results reinforce early conclusions by Diaconescu and Laprise [28] on the fact that RCMs can bring substantial error reduction when the driving lateral boundary condition contains errors.

Beyond considerations related to the AV by regional climate models seen from a one-way perspective, adopting a second type of AV based on upscaled results from high resolution yielded supplementary results and insights. At low resolution, a typical increase in the positive AVC is observed with some exceptions. These exceptions are related to cases with a decrease in the positive AVC from high to low resolution. In the worst case, however, these exceptions represent a reduction of 8% and therefore indicate that RegCM4-7 simulations could be used for large-scale precipitation applications over Africa when sufficient AV is observed or considered enough for the intended application. These results make the RCMs influential contribution hypothesis to the AV more plausible than the idea of an AV due to a mere disaggregation of the driving GCMs. Previous findings by Gnitou et al. [21] and Sørland et al. [8] led to similar conclusions.

The resolution-sensitivity results also revealed that the improvement of the positive AVC from high to low resolution occurs in a climate zone where consistent and uniformly distributed spatial positive AV patterns are observed at high resolution. This general tendency is also true for areas with large negative AVC. Therefore, the AV at fine scale probably represents a precondition to the expectation of improved large-scale features by RCM outputs. In the context of climate change projection over Africa, some studies [19,44] documented such major differences between GCMs and RCMs known under the concept of potential or conjectural AV [6,7] since observations are not available to confirm it. The findings from the present study bring more arguments to the possibility that at least part of such GCM-RCM differences is due to salient positive AV of the RCMs that could not, however, be measured in the context of future projections.

The influential impact of the RCMs internal physics as compared to the driving GCMs is further confirmed by the regional results, which feature the WAF, CAF and SEAF for all seasons, and NEAF for DJF and MAM seasons, as the best performing sub-region with a positive AVC greater than 20%. For instance, the WAF and CAF sub-regions are known for mesoscale activities and land-atmosphere interactions, and the SEAF and NEAF for their topographic influence on rainfall quantities [19]. The significant positive AVC over these sub-areas represents additional evidence for the hypothesis that the observed AV originates from mesoscale and fine-scale features resolved by RCMs [2,4].

Despite the encouraging results obtained at low resolution and their dependence on the AV at high resolution considering CHIRPS data, cautionary considerations of the present study results are still needed regarding observational uncertainties. Although the availability of CHIRPS data produced at a resolution similar to CORDEX-CORE simulations motivated the present resolution-sensitivity study, alternative datasets widely used at low resolutions may provide different estimates compared to CHIRPS upscaled results and therefore represent a source of uncertainty for AV results at such resolutions. For instance, as shown in Table S1, the AVC values at low resolution using GPCP data show some differences compared to upscaled CHIRPS estimates used in this study. The upscaling of CHIRPS at the GPCP grid for the present study is motivated by these reasons. This implies that more due diligence is needed from users in their data choices to avoid misuses, as suggested by Giorgi [1].

Overall, the AV results by the CORDEX-CORE RegCM4-7 precipitation simulations and their highly likely attribution to internal RCM physics opens a wide range of opportunities for both the climate modeling and the satellite remote sensing communities. For the former community, these results will be instrumental for improving RCMs simulations, and therefore, continuously bridging the resolution and quality gap between traditional climate models and high-resolution satellite-based precipitation products. For the latter, these results expand the remote sensing data applications spectrum, emphasizing on climate change applications and opening new avenues toward a process-based understanding of remotely sensed earth observations [30,31].

5. Conclusions

This study evaluates the AV by CORDEX-CORE RegCM4-7 seasonal precipitation simulations over Africa, using both high- and low-resolution satellite-based observations, with the aim of quantifying large- and fine-scale improvements by the downscaled outputs in comparison to their driving GCMs. The study yielded substantial results, which can be summarized as follows:

1. At high-resolution, pixels with at least 10% improvement by the CORDEX-CORE RegCM4-7 downscaling approach represent 20 to 60% of the overall African domain depending on the driving GCM and the season.
2. The results obtained at low resolution mostly show an increase in the improvement coverage, with at most 8% spatial coverage reduction for cases where a decrease is observed as compared to high-resolution findings.
3. A relatively high improvement coverage of NorESM1-M downscaled results is found for nearly all seasons and climate zones, with West Africa, Central Africa, and Southern East Africa showing the overall best performances for all seasons and driving GCMs.
4. The results appear to be mostly influenced by the RCMs internal physics, with a typical highlight over sub-regions where fine-scale features-driven AV is expected. A better or nearly similar AV pattern is also observed at low resolution, suggesting an enhancement of the driving GCMs by sub-grid processes resolved by the RCM physics.
5. Observational uncertainty is found to have a significant influence, as one should have expected, given observational data scarcity over Africa.
6. For the first attempt to produce RCM data at a resolution similar to common satellite-based products resolution, the AV results are relatively good. They feature the potential of dynamical downscaling tools in filling resolution and quality gaps between GCMs and remotely sensed technology-based datasets.

In comparison with the 60 years of research and development of earth observation satellites [31], it is fair to say that the present study's results represent a remarkable achievement of the RCM community, which is only 30 years old [4,5]. For instance, the AV results from this study bring new pieces of evidence on the ability of dynamical downscaling to bridge the resolution and quality gap between coarse GCMs and high-resolution satellite-based precipitation products. For Africa, the opportunities ahead are enormous in the context of unifying multi-source precipitation estimates and undertaking process-based climate assessments and projections, especially for applications where high-resolution data is needed [24].

Overall, RegCM4-7 outputs will still need further processing, such as bias correction, by leveraging available historical high-resolution satellite-based products for more plausible future projection analysis due to the persistent biases observed in the present study [21,25], until new developments and improvements become available. This is true for alternative tools such as Empirical Statistical Downscaling (ESD) [45] and convection-permitting simulations [46], which constitute the next step toward further understanding regional climates. The second component of the CORDEX phase II framework, named Flagship Pilot Studies (FPSs), will provide a coordinated setting for testing these emerging tools [13]. Upcoming studies looking at Vulnerability, Impacts and Adaptation (VIA) assessments are to be expected in the near future to explore other CORDEX-CORE data applications.

Supplementary Materials: The following supporting information can be downloaded at: https://www.mdpi.com/article/10.3390/rs14092102/s1, Figure S1 Performance of African precipitation in DJF season at low resolution compared to (a) CHIRPS, for (b) RegCM4-7 evaluation run driven by ERA-INT, (c) driving GCM HadGEM2-ES, (d) driving GCM MPI-ESM-MR, (e) driving GCM NorESM1-M, (f) ensemble mean of the driving GCMs, (g) RegCM4-7 historical run driven by HadGEM2-ES, (h) RegCM4-7 historical run driven by MPI-ESM-MR, (i) RegCM4-7 historical run driven by NorESM1-M, (j) RegCM4-7 historical runs' ensemble mean, (k) Added Value by RegCM4-7 to HadGEM2-ES, (l) Added Value by RegCM4-7 to MPI-ESM-MR, (m) Added Value by RegCM4-7 to NorESM1-M, (n) Added Value by RegCM4-7 to the ensemble mean of the driving GCMs. Figure S2 Performance of African precipitation in MAM season at low resolution compared to (a) CHIRPS, for (b) RegCM4-7 evaluation run driven by ERA-INT, (c) driving GCM HadGEM2-ES, (d) driving GCM MPI-ESM-MR, (e) driving GCM NorESM1-M, (f) ensemble mean of the driving GCMs, (g) RegCM4-7 historical run driven by HadGEM2-ES, (h) RegCM4-7 historical run driven by MPI-ESM-MR,

(i) RegCM4-7 historical run driven by NorESM1-M, (j) RegCM4-7 historical runs' ensemble mean, (k) Added Value by RegCM4-7 to HadGEM2-ES, (l) Added Value by RegCM4-7 to MPI-ESM-MR, (m) Added Value by RegCM4-7 to NorESM1-M, (n) Added Value by RegCM4-7 to ensemble mean of the driving GCMs. Figure S3 Performance of African precipitation in JJA season at low resolution compared to (a) CHIRPS, for (b) RegCM4-7 evaluation run driven by ERA-INT, (c) driving GCM HadGEM2-ES, (d) driving GCM MPI-ESM-MR, (e) driving GCM NorESM1-M, (f) ensemble mean of the driving GCMs, (g) RegCM4-7 historical run driven by HadGEM2-ES, (h) RegCM4-7 historical run driven by MPI-ESM-MR, (i) RegCM4-7 historical run driven by NorESM1-M, (j) RegCM4-7 historical runs' ensemble mean, (k) Added Value by RegCM4-7 to HadGEM2-ES, (l) Added Value by RegCM4-7 to MPI-ESM-MR, (m) Added Value by RegCM4-7 to NorESM1-M, (n) Added Value by RegCM4-7 to ensemble mean of the driving GCMs. Figure S4 Performance of African precipitation in SON season at low resolution compared to (a) CHIRPS, for (b) RegCM4-7 evaluation run driven by ERA-INT, (c) driving GCM HadGEM2-ES, (d) driving GCM MPI-ESM-MR, (e) driving GCM NorESM1-M, (f) ensemble mean of the driving GCMs, (g) RegCM4-7 historical run driven by HadGEM2-ES, (h) RegCM4-7 historical run driven by MPI-ESM-MR, (i) RegCM4-7 historical run driven by NorESM1-M, (j) RegCM4-7 historical runs' ensemble mean, (k) Added Value by RegCM4-7 to HadGEM2-ES, (l) Added Value by RegCM4-7 to MPI-ESM-MR, (m) Added Value by RegCM4-7 to NorESM1-M, (n) Added Value by RegCM4-7 to ensemble mean of the driving GCMs. Table S1 Added Value Coverage results of RegCM4-7 simulations over continental Africa at low resolution for CHIRPS data (LR_C) and GPCP data (LR_G).

Author Contributions: Conceptualization, G.T.G., Y.H. and G.T.; methodology, G.T.G., Y.H. and G.T.; software, G.T.G., K.T.C.L.K.S., Y.H. and G.T.; validation, Y.H. and G.T.; formal analysis, G.T.G. and K.T.C.L.K.S.; investigation, G.T.G. and Y.H.; resources, Y.H. and G.T.; data curation, G.T.G. and K.T.C.L.K.S.; writing—original draft preparation, G.T.G.; writing—review and editing, I.K.N., K.T.C.L.K.S. and Y.H; visualization, G.T.G., K.T.C.L.K.S. and I.K.N.; supervision, Y.H. and G.T.; project administration, Y.H. and G.T.; funding acquisition, Y.H. and G.T. All authors have read and agreed to the published version of the manuscript.

Funding: The National Natural Science Foundation of China, Grant Nos: U1902209 and the National Natural Science Foundation of Yunnan Province (202201AS070069) supported this work, through the Climate Center of Yunnan Meteorological Bureau, with GST registration number 12530000431270053A.

Institutional Review Board Statement: Not applicable.

Informed Consent Statement: Not applicable.

Data Availability Statement: Data will be provided upon request from the corresponding author.

Acknowledgments: We acknowledge the producers of the CMIP5 and CORDEX-CORE precipitation datasets from the Earth System Grid Federation (ESGF), the GPCP dataset from the University of Maryland, and the CHIRPS dataset from the University of California Sancta Barbara Climate Hazard Group. These datasets were used in accordance with their respective terms and conditions.

Conflicts of Interest: The authors declare no conflict of interest.

References

1. Giorgi, F. Producing actionable climate change information for regions: The distillation paradigm and the 3R framework. *Eur. Phys. J. Plus* **2020**, *135*, 435. [CrossRef]
2. Gutowski, W.J.; Ullrich, P.A.; Hall, A.; Leung, L.R.; O'Brien, T.A.; Patricola, C.M.; Arritt, R.W.; Bukovsky, M.S.; Calvin, K.V.; Feng, Z.; et al. The Ongoing Need for High-Resolution Regional Climate Models: Process Understanding and Stakeholder Information. *Bull. Am. Meteorol. Soc.* **2020**, *101*, E664–E683. [CrossRef]
3. Laprise, R.; de Elía, R.; Caya, D.; Biner, S.; Lucas-Picher, P.; Diaconescu, E.; Leduc, M.; Alexandru, A.; Separovic, L. Challenging some tenets of Regional Climate Modelling. *Meteorol. Atmos. Phys.* **2008**, *100*, 3–22. [CrossRef]
4. Giorgi, F. Thirty Years of Regional Climate Modeling: Where Are We and Where Are We Going next? *J. Geophys. Res. Atmos.* **2019**, *124*, 5696–5723. [CrossRef]
5. Tapiador, F.J.; Navarro, A.; Moreno, R.; Sánchez, J.L.; García-Ortega, E. Regional climate models: 30 years of dynamical downscaling. *Atmos. Res.* **2020**, *235*, 104785. [CrossRef]
6. Di Luca, A.; de Elía, R.; Laprise, R. Potential for added value in precipitation simulated by high-resolution nested Regional Climate Models and observations. *Clim. Dyn.* **2012**, *38*, 1229–1247. [CrossRef]

7. Di Luca, A.; de Elía, R.; Laprise, R. Challenges in the Quest for Added Value of Regional Climate Dynamical Downscaling. *Curr. Clim. Chang. Rep.* **2015**, *1*, 10–21. [CrossRef]
8. Sørland, S.L.; Schär, C.; Lüthi, D.; Kjellström, E. Bias patterns and climate change signals in GCM-RCM model chains. *Environ. Res. Lett.* **2018**, *13*, 074017. [CrossRef]
9. Rummukainen, M. Added value in regional climate modeling. *Wiley Interdiscip. Rev. Clim. Chang.* **2016**, *7*, 145–159. [CrossRef]
10. Giorgi, F.; Jones, C.; Asrar, G. Addressing climate information needs at the regional level: The CORDEX framework. *World Meteorol. Organ. Bull.* **2009**, *58*, 175.
11. Jones, C.; Giorgi, F.; Asrar, G. The Coordinated Regional Downscaling Experiment: CORDEX, an International Downscaling Link to CMIP5. *CLIVAR Exch.* **2011**, *16*, 34–39.
12. Gutowski, J.W.; Giorgi, F.; Timbal, B.; Frigon, A.; Jacob, D.; Kang, H.S.; Raghavan, K.; Lee, B.; Lennard, C.; Nikulin, G.; et al. WCRP COordinated Regional Downscaling EXperiment (CORDEX): A diagnostic MIP for CMIP6. *Geosci. Model Dev.* **2016**, *9*, 4087–4095. [CrossRef]
13. Giorgi, F.; Gutowski, W.J. Coordinated Experiments for Projections of Regional Climate Change. *Curr. Clim. Chang. Rep.* **2016**, *2*, 202–210. [CrossRef]
14. Teichmann, C.; Jacob, D.; Remedio, A.R.; Remke, T.; Buntemeyer, L.; Hoffmann, P.; Kriegsmann, A.; Lierhammer, L.; Bülow, K.; Weber, T.; et al. Assessing mean climate change signals in the global CORDEX-CORE ensemble. *Clim. Dyn.* **2020**, *57*, 1269–1292. [CrossRef]
15. Ciarlo, J.M.; Coppola, E.; Fantini, A.; Giorgi, F.; Gao, X.J.; Tong, Y.; Glazer, R.H.; Torres Alavez, J.A.; Sines, T.; Pichelli, E.; et al. A new spatially distributed added value index for regional climate models: The EURO-CORDEX and the CORDEX-CORE highest resolution ensembles. *Clim. Dyn.* **2020**, *57*, 1403–1424. [CrossRef]
16. Giorgi, F.; Coppola, E.; Jacob, D.; Teichmann, C.; Abba Omar, S.; Ashfaq, M.; Ban, N.; Bülow, K.; Bukovsky, M.; Buntemeyer, L.; et al. The CORDEX-CORE EXP-I initiative: Description and highlight results from the initial analysis. *Bull. Am. Meteorol. Soc.* **2021**, *103*, E293–E310. [CrossRef]
17. Dosio, A.; Panitz, H.J.; Schubert-Frisius, M.; Lüthi, D. Dynamical downscaling of CMIP5 global circulation models over CORDEX-Africa with COSMO-CLM: Evaluation over the present climate and analysis of the added value. *Clim. Dyn.* **2015**, *44*, 2637–2661. [CrossRef]
18. Gibba, P.; Sylla, M.B.; Okogbue, E.C.; Gaye, A.T.; Nikiema, M.; Kebe, I. State-of-the-art climate modeling of extreme precipitation over Africa: Analysis of CORDEX added-value over CMIP5. *Theor. Appl. Climatol.* **2019**, *137*, 1041–1057. [CrossRef]
19. Dosio, A.; Jones, R.G.; Jack, C.; Lennard, C.; Nikulin, G.; Hewitson, B. What can we know about future precipitation in Africa? Robustness, significance and added value of projections from a large ensemble of regional climate models. *Clim. Dyn.* **2019**, *53*, 5833–5858. [CrossRef]
20. Gnitou, G.T.; Tan, G.; Ma, T.; Akinola, E.O.; Nooni, I.K.; Babaousmail, H.; Al-Nabhan, N. Added value in dynamically downscaling seasonal mean temperature simulations over West Africa. *Atmos. Res.* **2021**, *260*, 105694. [CrossRef]
21. Gnitou, G.T.; Tan, G.; Niu, R.; Nooni, I.K. Assessing past climate biases and the added value of cordex-core precipitation simulations over Africa. *Remote Sens.* **2021**, *13*, 2058. [CrossRef]
22. Nicholson, S.E.; Klotter, D.A. Assessing the reliability of satellite and reanalysis estimates of rainfall in equatorial Africa. *Remote Sens.* **2021**, *13*, 3609. [CrossRef]
23. Nicholson, S.E.; Klotter, D.; Zhou, L.; Hua, W. Validation of satellite precipitation estimates over the Congo Basin. *J. Hydrometeorol.* **2019**, *20*, 631–656. [CrossRef]
24. Dosio, A.; Pinto, I.; Lennard, C.; Sylla, M.B.; Jack, C.; Nikulin, G. What Can We Know About Recent Past Precipitation Over Africa? Daily Characteristics of African Precipitation From a Large Ensemble of Observational Products for Model Evaluation. *Earth Space Sci.* **2021**, *8*, 8. [CrossRef]
25. Dieng, D.; Cannon, A.J.; Laux, P.; Hald, C.; Adeyeri, O.; Rahimi, J.; Srivastava, A.K.; Mbaye, M.L.; Kunstmann, H. Multivariate Bias-Correction of High-Resolution Regional Climate Change Simulations for West Africa: Performance and Climate Change Implications. *J. Geophys. Res. Atmos.* **2022**, *127*, 127. [CrossRef]
26. Dike, V.N.; Shimizu, M.H.; Diallo, M.; Lin, Z.; Nwofor, O.K.; Chineke, T.C. Modelling present and future African climate using CMIP5 scenarios in HadGEM2-ES. *Int. J. Climatol.* **2015**, *35*, 1784–1799. [CrossRef]
27. Zebaze, S.; Jain, S.; Salunke, P.; Shafiq, S.; Mishra, S.K. Assessment of CMIP5 multimodel mean for the historical climate of Africa. *Atmos. Sci. Lett.* **2019**, *20*, 20. [CrossRef]
28. Diaconescu, E.P.; Laprise, R. Can added value be expected in RCM-simulated large scales? *Clim. Dyn.* **2013**, *41*, 1769–1800. [CrossRef]
29. Washington, R.; Harrison, M.; Conway, D.; Black, E.; Challinor, A.; Grimes, D.; Jones, R.; Morse, A.; Kay, G.; Todd, M. African climate change: Taking the shorter route. *Bull. Am. Meteorol. Soc.* **2006**, *87*, 1355–1365. [CrossRef]
30. Yang, J.; Gong, P.; Fu, R.; Zhang, M.; Chen, J.; Liang, S.; Xu, B.; Shi, J.; Dickinson, R. The role of satellite remote sensing in climate change studies. *Nat. Clim. Chang.* **2013**, *3*, 875–883. [CrossRef]
31. Zhao, Q.; Yu, L.; Du, Z.; Peng, D.; Hao, P.; Zhang, Y. An Overview of the Applications of Earth Observation Satellite Data: Impacts and Future Trends. *Remote Sens.* **2022**, *14*, 1863. [CrossRef]
32. Giorgi, F.; Coppola, E.; Solmon, F.; Mariotti, L.; Sylla, M.B.; Bi, X.; Elguindi, N.; Diro, G.T.; Nair, V.; Giuliani, G.; et al. RegCM4: Model description and preliminary tests over multiple CORDEX domains. *Clim. Res.* **2012**, *52*, 7–29. [CrossRef]

33. Dee, D.P.; Uppala, S.M.; Simmons, A.J.; Berrisford, P.; Poli, P.; Kobayashi, S.; Andrae, U.; Balmaseda, M.A.; Balsamo, G.; Bauer, P.; et al. The ERA-Interim reanalysis: Configuration and performance of the data assimilation system. *Q. J. R. Meteorol. Soc.* **2011**, *137*, 553–597. [CrossRef]
34. Taylor, K.E.; Stouffer, R.J.; Meehl, G.A. An overview of CMIP5 and the experiment design. *Bull. Am. Meteorol. Soc.* **2012**, *93*, 485–498. [CrossRef]
35. Collins, W.J.; Bellouin, N.; Doutriaux-Boucher, M.; Gedney, N.; Halloran, P.; Hinton, T.; Hughes, J.; Jones, C.D.; Joshi, M.; Liddicoat, S.; et al. Development and evaluation of an Earth-System model—HadGEM2. *Geosci. Model Dev.* **2011**, *4*, 485–498. [CrossRef]
36. Zanchettin, D.; Rubino, A.; Matei, D.; Bothe, O.; Jungclaus, J.H. Multidecadal-to-centennial SST variability in the MPI-ESM simulation ensemble for the last millennium. *Clim. Dyn.* **2013**, *40*, 1301–1318. [CrossRef]
37. Zhang, Z.S.; Nisancioglu, K.; Bentsen, M.; Tjiputra, J.; Bethke, I.; Yan, Q.; Risebrobakken, B.; Andersson, C.; Jansen, E. Pre-industrial and mid-Pliocene simulations with NorESM-L. *Geosci. Model Dev.* **2012**, *5*, 523–533. [CrossRef]
38. Elguindi, N.; Giorgi, F.; Turuncoglu, U. Assessment of CMIP5 global model simulations over the subset of CORDEX domains used in the Phase I CREMA. *Clim. Chang.* **2014**, *125*, 7–21. [CrossRef]
39. Funk, C.C.; Peterson, P.J.; Landsfeld, M.F.; Pedreros, D.H.; Verdin, J.P.; Rowland, J.D.; Romero, B.E.; Husak, G.J.; Michaelsen, J.C.; Verdin, A.P. A Quasi-Global Precipitation Time Series for Drought Monitoring. *U.S. Geol. Surv. Data Ser.* **2014**, *832*, 4. [CrossRef]
40. Adler, R.F.; Huffman, G.J.; Chang, A.; Ferraro, R.; Xie, P.-P.; Janowiak, J.; Rudolf, B.; Schneider, U.; Curtis, S.; Bolvin, D.; et al. The Version-2 Global Precipitation Climatology Project (GPCP) Monthly Precipitation Analysis (1979–Present). *J. Hydrometeorol.* **2003**, *4*, 1147–1167. [CrossRef]
41. Adler, R.F.; Sapiano, M.R.P.; Huffman, G.J.; Wang, J.J.; Gu, G.; Bolvin, D.; Chiu, L.; Schneider, U.; Becker, A.; Nelkin, E.; et al. The Global Precipitation Climatology Project (GPCP) monthly analysis (New Version 2.3) and a review of 2017 global precipitation. *Atmosphere* **2018**, *9*, 138. [CrossRef] [PubMed]
42. Jacob, D.; Elizalde, A.; Haensler, A.; Hagemann, S.; Kumar, P.; Podzun, R.; Rechid, D.; Remedio, A.R.; Saeed, F.; Sieck, K.; et al. Assessing the Transferability of the Regional Climate Model. *Atmosphere* **2012**, *3*, 181–199. [CrossRef]
43. Remedio, A.R.; Teichmann, C.; Buntemeyer, L.; Sieck, K.; Weber, T.; Rechid, D.; Hoffmann, P.; Nam, C.; Kotova, L.; Jacob, D. Evaluation of new CORDEX simulations using an updated köppen-trewartha climate classification. *Atmosphere* **2019**, *10*, 726. [CrossRef]
44. Dosio, A.; Jury, M.W.; Almazroui, M.; Ashfaq, M.; Diallo, I.; Engelbrecht, F.A.; Klutse, N.A.B.; Lennard, C.; Pinto, I.; Sylla, M.B.; et al. Projected future daily characteristics of African precipitation based on global (CMIP5, CMIP6) and regional (CORDEX, CORDEX-CORE) climate models. *Clim. Dyn.* **2021**, *57*, 3135–3158. [CrossRef]
45. Kreienkamp, F.; Paxian, A.; Früh, B.; Lorenz, P.; Matulla, C. Evaluation of the empirical–statistical downscaling method EPISODES. *Clim. Dyn.* **2019**, *52*, 991–1026. [CrossRef]
46. Lucas-Picher, P.; Argüeso, D.; Brisson, E.; Tramblay, Y.; Berg, P.; Lemonsu, A.; Kotlarski, S.; Caillaud, C. Convection-permitting modeling with regional climate models: Latest developments and next steps. *Wiley Interdiscip. Rev. Clim. Chang.* **2021**, *12*, e731. [CrossRef]

remote sensing

MDPI

Article

Dynamical Downscaling of Temperature Variations over the Canadian Prairie Provinces under Climate Change

Xiong Zhou [1], Guohe Huang [1,2,*], Yongping Li [1], Qianguo Lin [3], Denghua Yan [4] and Xiaojia He [5]

1 China-Canada Center for Energy, Environment and Ecology Research, State Key Joint Laboratory of Environmental Simulation and Pollution Control, UR-BNU, School of Environment, Beijing Normal University, Beijing 100875, China; xzhou@bnu.edu.cn (X.Z.); yongping.li@iseis.org (Y.L.)
2 Faculty of Engineering and Applied Science, University of Regina, Regina, SK S4S 0A2, Canada
3 College of Environmental Science and Engineering, North China Electric Power University, Beijing 102206, China; qianguo.lin@gmail.com
4 China Institute of Water Resources and Hydropower Research, A-1 Fuxing Road, Beijing 100038, China; yandh@iwhr.com
5 The Administrative Center for China's Agenda 21, Beijing 100038, China; hexiaojia@acca21.org.cn
* Correspondence: huang@iseis.org; Tel.: +1-306-585-4095

Abstract: In this study, variations of daily mean, maximum, and minimum temperature (expressed as T_{mean}, T_{max}, and T_{min}) over the Canadian Prairie Provinces were dynamically downscaled through regional climate simulations. How the regional climate would increase in response to global warming was subsequently revealed. Specifically, the Regional Climatic Model (RegCM) was undertaken to downscale the boundary conditions of Geophysical Fluid Dynamics Laboratory Earth System Model Version 2M (GFDL-ESM2M) over the Prairie Provinces. Daily temperatures (i.e., T_{mean}, T_{max}, and T_{min}) were subsequently extracted from the historical and future climate simulations. Temperature variations in the two future periods (i.e., 2036 to 2065 and 2065 to 2095) are then investigated relative to the baseline period (i.e., 1985 to 2004). The spatial distributions of temperatures were analyzed to reveal the regional impacts of global warming on the provinces. The results indicated that the projected changes in the annual averages of daily temperatures would be amplified from the southwest in the Rocky Mountain area to the northeast in the prairie region. It was also suggested that the projected temperature averages would be significantly intensified under RCP8.5. The projected temperature variations could provide scientific bases for adaptation and mitigation initiatives on multiple sectors, such as agriculture and economic sectors over the Canadian Prairies.

Keywords: dynamical downscaling; projected variations; Canadian Prairies; global warming

Citation: Zhou, X.; Huang, G.; Li, Y.; Lin, Q.; Yan, D.; He, X. Dynamical Downscaling of Temperature Variations over the Canadian Prairie Provinces under Climate Change. *Remote Sens.* **2021**, *13*, 4350. https://doi.org/10.3390/rs13214350

Academic Editor: Jorge Vazquez

Received: 8 September 2021
Accepted: 26 October 2021
Published: 29 October 2021

1. Introduction

Climate warming is one of the most significant challenges currently facing the globe and humanity. Consequently, assessment of climate change impacts is needed to support adaptation and mitigation strategies [1–9]. Such impacts are commonly investigated based on future climate projections simulated by global climate models (GCMs) under multiple scenarios [10–13]. However, the mechanisms and processes of climate change at regional scales cannot be comprehensively reflected by these coarse-resolution projections [14–18]. Therefore, the development of fine-resolution climate projections is needed to explore the climate change impacts within a regional context.

Climate projections of GCMs have been primarily employed in previous studies to examine climate variations over the Canadian region in response to global warming [19–30]. More recently, several studies [31–34] have attempted to dynamically downscale GCMs over the Canadian prairies based on regional climate models (RCMs). For example, PaiMazumder et al. [33] projected changes in short- and long-term drought characteristics over the Canadian Prairies by using an ensemble of ten Canadian RCM (CRCM) simulations. These previous studies have demonstrated that RCMs have advantages in reflecting the climatology processes at local scales as compared to GCMs [9,35]. For instance, Zhou et al. [9] developed a dynamical-coupled downscaling approach and demonstrated its advantages in capturing the regional details.

RCMs can resolve more detailed features such as mountain ranges, coastal zones, and soil properties, which are consistent with physical mechanisms in GCMs. As a consequence, dynamical downscaling can provide climate variables at a fine spatio-temporal resolution in order to support a better understanding of climate change impacts within the global warming context [36]. However, there have been few reports of dynamic downscaling of daily temperatures over the Canadian Prairie Provinces at a high spatial resolution (e.g., 25 km) under the representative concentration pathways (RCPs).

In addition, the Canadian Prairie Provinces are affected by the Rocky Mountains and undulating prairies. Such complex topography will have a significant impact on the local climate conditions in this region. The observations in the Rocky Mountain region are sparse due to the complex topography. Initiatives of regional climate downscaling can provide gridded climatic information with a fine spatio-temporal resolution, supporting mitigation and adaptation of the severe impacts of climatic changes. The development of a fine-resolution climate fluctuations is thus essential to comprehensively reveal the potential impacts of climate change on the Prairie Provinces.

Therefore, the objective of this study is to examine how the regional climate over the Canadian Prairie Provinces will increase in response to global warming through the development of climate projections based on RCM. In detail, the Regional Climatic Model (RegCM) will be undertaken to downscale the boundary conditions of Geophysical Fluid Dynamics Laboratory Earth System Model Version 2M (GFDL-ESM2M). Daily mean, maximum, and minimum temperature are subsequently extracted from the historical and future climate simulations. Temperature variations in the two future periods (i.e., 2036 to 2065 and 2065 to 2095) are then investigated relative to the baseline period (i.e., 1985 to 2004). The spatial temperature distributions will be analyzed to reveal the impacts of global warming on the Canadian Prairie Provinces. It is expected that the projected temperature variations would provide scientific bases for adaptation and mitigation initiatives on multiple sectors, such as agriculture and economic sectors.

2. Model Setup, Study Area, and Data Sets

In this study, physically-based climate downscaling is based on the 4.6.0 version of RegCM, which is developed by the International Center for Theoretical Physics [37–40]. The RegCM4.6 model is employed to dynamically downscale the GFDL-ESM2M simulations [41–44], which are derived from the historical (1950–2005) and future (2006–2099) experiments under RCPs [35]. Specifically, the intermediate (i.e., RCP4.5) and heavy (i.e., RCP8.5) emission scenarios are chosen to reveal the range of possible future climate variations. Such emission scenarios are mainly distinctive from each other after the year 2050. More specific information for the scenarios is illustrated in the previous studies [9,35].

The study area is within the context of the Canadian Prairie Provinces, which include the Provinces of Manitoba, Saskatchewan, and Alberta. As shown in Figure 1, the topography in the study area is varied and complicated. The highest and lowest elevation is 3434 and 1 m, respectively. The Prairie Provinces have a total surface area of over 1,960,000 km^2, accounting for 19.6% of the entire area of Canada [9]. Moreover, the climatic patterns of the Arctic and the Rocky Mountain have significant impacts on the climate of the Canadian Prairie Provinces [35,45].

A spatial experiment domain of 108 longitude $\times$ 128 latitude grid points with a horizontal resolution of $0.22° \times 0.22°$ (i.e., approximately 25 km) is set up to cover the Prairie Provinces [9]. The land surface scheme is the 4.5 version of the Community Land Model (CLM4.5), while the multiple-phase cloud microphysics scheme is selected as the moisture scheme in RegCM4.6 [46]. The microphysics scheme is employed as the moisture scheme, while the lateral boundary condition scheme is the relaxation exponential technique [47,48]. The Holtslag PBL [49] is chosen as the boundary layer scheme and the cumulus convective scheme is the Emanuel scheme [50].

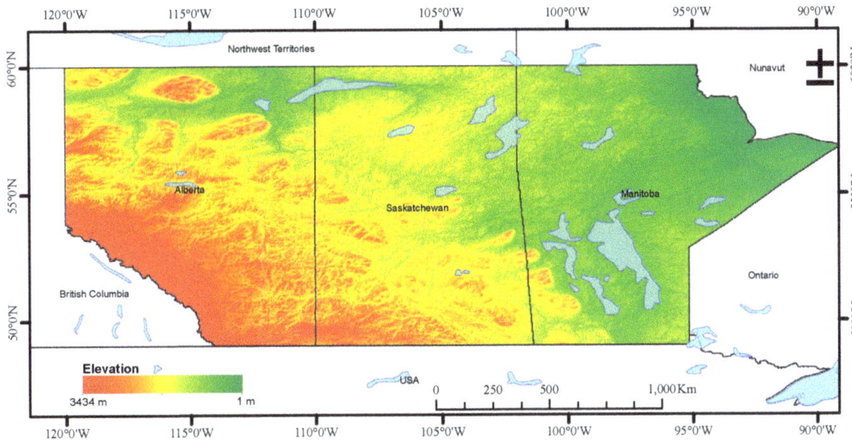

Figure 1. Topography of the study area.

The annual average of daily maximum temperature over the Canadian Prairies in the period of 1960 to 2005 ranged from −4.0 to 12.6 °C, whereas daily minimum temperature varied from −12.2 to 0.6 °C. From 1960 to 2005, the maximum and minimum temperature increased by 1.5 and 2.3 °C, respectively. The spatial average of daily mean temperature for the period of 1960 to 2005 fluctuated from 3.7 to 7.3 °C, which increased by 1.6 °C since 1895 [9].

Intensified greenhouse-gas impacts and aggravated land-cover changes have contributed to increases in annual averages of daily maximum and minimum temperature [51]. It is expected that such increases would be amplified in the future, which can significantly affect the agriculture sector in the Canadian Prairies. Therefore, a higher spatial resolution of climate projections is needed to reveal the impacts of climate changes on the Canadian Prairies, and thus facilitate proper mitigative and adaptive strategies.

The historical climate observations over the Canadian Prairies were derived from a 10 km gridded climate dataset, which was produced by the National Land and Water Information Service (NLWIS), Agriculture and Agri-Food, Canada [52]. In this study, daily maximum and minimum temperatures in the period of 1985 to 2004 were extracted to analyze the annual averages within the study area. The elevation of the Canadian Prairie Provinces with a horizontal resolution of 30 arc-seconds was acquired from the Canadian Digital Elevation Data, which is developed by the Canadian Forestry Service, Ontario region [53].

The RegCM4.6 model is undertaken to dynamically downscale daily mean, maximum, and minimum temperature (expressed as T_{mean}, T_{max}, and T_{min}) over the Canadian Prairies. Annual averages of the daily temperatures for future periods of both 2036 to 2065 and 2066 to 2095 are calculated. The changes in temperature relative to the baseline period (i.e., 1985 to 2004) are then analyzed. Particularly, the changes in temperature (i.e., V_{mean}, V_{max}, and V_{min}) can be calculated as follows:

$$V_{mean} = T^{f}_{mean} - T^{h}_{mean} \tag{1}$$

$$V_{max} = T^{f}_{max} - T^{h}_{max} \tag{2}$$

$$V_{min} = T^{f}_{min} - T^{h}_{min} \tag{3}$$

where T^f_{mean}, T^f_{max}, and T^f_{min} represent the annual average of daily mean, maximum, and minimum temperature in the future periods, respectively; whereas T^h_{mean}, T^h_{max}, and T^h_{min} indicate the annual average of daily mean, maximum, and minimum temperature for the historical period (i.e., 1985 to 2004). Detailed steps for projecting temperature variations through the RegCM4.6 model are summarized in Figure 2.

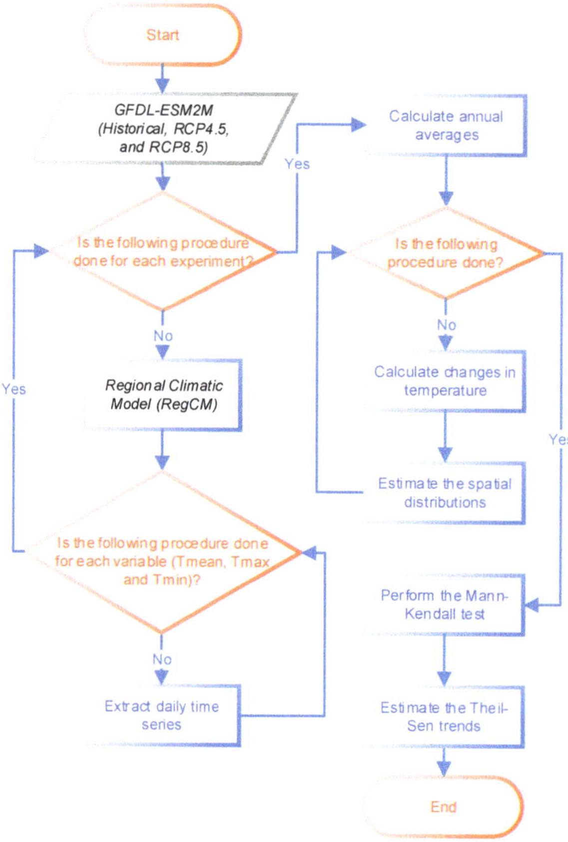

Figure 2. Methodological framework for projecting temperature variations.

3. Projected Variations of Temperature

Previous studies have illustrated that RCMs are capable of capturing regional terrestrial and atmospheric processes as compared with GCMs [9,35]. The performance of RCMs in reproducing the historical temperature observations has been evaluated in previous studies [9,35,45,54–60]. Therefore, the development of fine-resolution temperature projections is the focus of this study. To comprehensively reveal how future temperature would be changed over the Canadian Prairies, temperature variations (i.e., V_{mean}, V_{max}, and V_{min}) in two future periods of both 2036 to 2065 and 2066 to 2095 under RCPs are analyzed with respective to the baseline period of 1985 to 2004.

Figure 3 presents the projected variations of annual averages of T_{mean}, T_{max}, and T_{min} for the period of 2036 to 2065 under RCP45 relative to the baseline period. The figures in columns one and two show the spatial distribution of projected averages of daily temperatures for the historical and future period, respectively, whereas column three shows the calculated variations (i.e., V_{mean}, V_{max}, and V_{min}); rows one to three present T_{mean}, T_{max}, and T_{min}, respectively. For T_{mean}, the simulated spatial average and standard deviation over the entire area for the historical period are −0.4 and 2.1 °C, respectively (Figure 3a). It can be found that there is a decreasing pattern of the annual average from north to south. The spatial distribution from the future experiment in the period of 2036 to 2065 under RCP4.5 (Figure 3b) is closed to the historical one (Figure 3a), with an obviously increased magnitude. The result also indicates that the spatial average of T_{mean} over the provinces is projected to be 1.0 °C in the period of 2036 to 2065, where maximum and minimum values in the map reach −7.3 °C in the southwest and 5.8 °C in the southeast, respectively. However, a smaller standard deviation of 1.9 °C is projected for the period of 2036 to 2065, which indicates that the spatial variation would be reduced due to the increased magnitude of negative values under climate change. As shown in Figure 3c, the average of T_{mean} increases over the entire of the Prairie Provinces, with a northeastward-intensified pattern of projected changes. In addition, the RegCM model also simulates the largest increase at 56.60° N and 96.73° W in the northeast of Manitoba, and the smallest one at 51.47° N and 112.66° W in the southwest of Alberta.

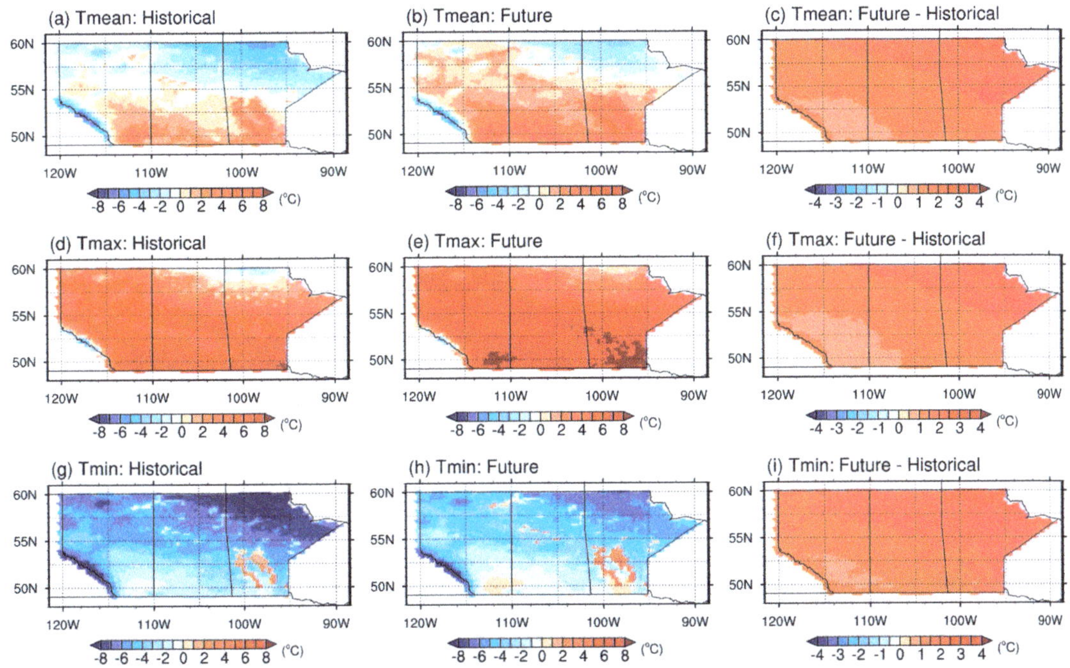

Figure 3. Projected variations of annual averages of T_{mean}, T_{max}, and T_{min} for the period of 2036 to 2065 under RCP45. (**a**,**d**,**g**) show the historical distributions of T_{mean}, T_{max}, and T_{min}, respectively; (**b**,**e**,**h**) show the future distributions of T_{mean}, T_{max}, and T_{min}, respectively; (**c**,**f**,**i**) show the projected variations of V_{mean}, V_{max}, and V_{min}, respectively.

Compared with T_{mean}, most of the spatial averages of T_{max} are generally simulated to be above zero degrees, with the magnitude varying from −5.7 to 8.2 °C. The results derived from the RegCM model also show that T_{max} has a spatial average of 4.3 °C and a standard deviation of 2.1 °C over the entire area for the historical period of 1985 to 2004 (Figure 3d). For the projected average of T_{max}, generally the future (Figure 3e) and the historical (Figure 3d) experiments have a similar gradient pattern, but amplified values are revealed in the future period of 2036 to 2065. In particular, a spatial average of 5.6 °C over the Canadian Prairies is projected, with a standard deviation of 2.1 °C in the future period under RCP4.5. Figure 3f presents the projected variations of annual averages of T_{max} for the future period relative to the baseline period. It can be observed that the entire Prairie Provinces would experience increases in T_{max}, with a gradient intensified pattern from southwest to northeast. Likewise, the spatial average of T_{max} would increase from 0.6 to 2.1 °C, while the most significant increase is found at the grid of 58.13° N and 94.91° W in the northeastern part.

In contrast, T_{min} over the Prairie Provinces is almost below zero degrees, with a spatial average of −4.6 °C and a standard deviation of 2.4 °C (Figure 3g). It also indicated that greater spatial variations are presented in T_{min} as compared with T_{max}. The spatial pattern of both simulated historical and projected future average of T_{min} matches each other quite well (Figure 3g,h), but with much greater values. A spatial average of −3.1 °C over the Canadian Prairies is projected, with a standard deviation of 2.2 °C in the period of 2036 to 2065 under RCP4.5. The greatest and smallest average are projected to be 3.4 °C in the southeast and −11.0 °C in the Rocky Mountain area, respectively. However, smaller spatial variations in the future might be caused by increased values of negative T_{min} under climate change. The projected variations of annual averages of T_{min} for the future period are presented in Figure 3i. Similar to T_{max} during the future period, the annual averages of T_{min} increase from 0.8 to 2.3 °C. It also can be observed that there is a gradient intensified pattern from southwest to northeast. Similarly, the most significant increase is located at 56.86° N and 88.97° W in the northeast, whereas the smallest one is at 51.47° N and 112.66° W in the southwest.

Moreover, Figure 4 depicts the annual averages of T_{mean}, T_{max}, and T_{min} for the period of 2066 to 2095 under RCP4.5, as well as their changes with respect to the baseline. Similarly, the spatial distributions of projected averages of daily temperatures for the historical and future period are shown in columns one and two, respectively. Column three depicts the computed changes in the projected values relative to the baseline period. In addition, the T_{mean}, T_{max}, and T_{min} respectively are presented in rows one to three. In general, there is a similar spatial distribution pattern of projected variations among T_{mean}, T_{max}, and T_{min} for the historical and future period. Similar to the revelations during the period of 2036 to 2065, the spatial average of T_{mean}, T_{max}, and T_{min} for the period of 2066 to 2095 increase by 1.7, 1.6, and 1.8 °C, respectively (Figure 4c,f,i). However, such increases are projected as slightly greater than those in the period 2036 to 2065 under RCP4.5. Moreover, the projected changes in the annual averages of T_{mean}, T_{max}, and T_{min} would be amplified from the southwest in the Rocky Mountain area to the northeast in the prairie region.

In order to further analyze the projected variations of T_{mean}, T_{max}, and T_{min}, the spatial distributions of their annual averages over the entire domain for the period of 2036 to 2065 and 2066 to 2095 under RCP4.5 are compared to those for the baseline period (Figure 5). In general, the shape of the distribution of T_{mean}, T_{max}, and T_{min} among three different periods matches each other. However, intensification of daily temperatures in the future is evident as shown from the changes of mean values, which is consistent with the previous revelations (Figures 3 and 4). It can be observed that there is no significant difference in the projected changes between the period of 2036 to 2065 and the period of 2066 to 2095 in terms of the mean and standard deviation. For example, the mean and standard deviation of projections for T_{min} in the period of 2036 to 2065 are projected to be −2.8 and 2.3 °C, respectively. Nevertheless, during the period of 2066 to 2095, the mean of projections is −3.1 °C with a standard deviation of 2.2 °C.

Figure 6 displays the projected annual averages of T_{mean}, T_{max}, and T_{min} for the period of 2036 to 2065 under RCP8.5. In the same way, the figures in columns one to three respectively present the simulated averages of daily temperatures for the historical and future periods, as well as their differences. The figures in rows one to three show T_{mean}, T_{max}, and T_{min}, respectively. Overall, the results show that the projections in the period of 2036 to 2065 under RCP8.5 are similar to those under RCP4.5. Moreover, there is an apparent warming pattern during the period of 2036 to 2065 under RCP8.5. For example, the RegCM model simulates the spatial averages of −4.6 and −2.5 °C for T_{min} during the period 1985 to 2004 and 2036 to 2065, respectively. The largest average over the map during the future period of 2036 to 2065 is located at 50.20° N and 98.04° W in the southeast, nevertheless, the smallest one is at 52.19° N and 117.24° W in the southwest (Figure 6b).

Moreover, significant increases under RCP8.5 are observed in the calculated differences between two periods. There is a similar intensified pattern from southwest to northeast, which is consistent with the results under RCP4.5.

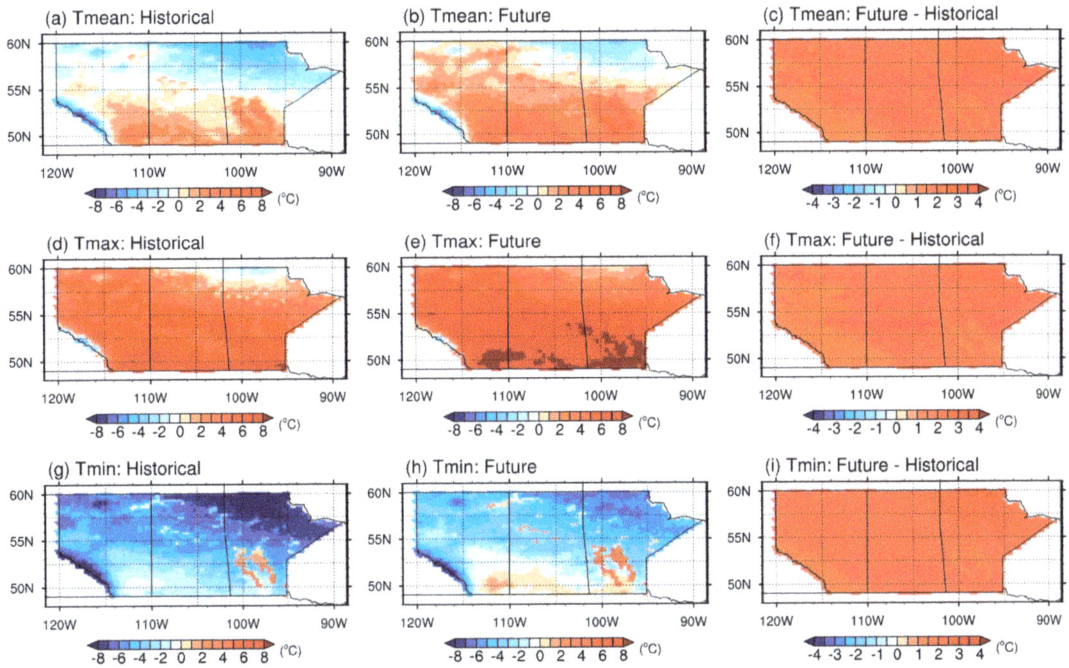

Figure 4. Projected variations of annual averages of T_{mean}, T_{max}, and T_{min} for the period of 2066 to 2095 under RCP45. (**a**,**d**,**g**) show the historical distributions of T_{mean}, T_{max}, and T_{min}, respectively; (**b**,**e**,**h**) show the future distributions of T_{mean}, T_{max}, and T_{min}, respectively; (**c**,**f**,**i**) show the projected variations of V_{mean}, V_{max}, and V_{min}, respectively.

To further analyze the temperature changes, Figure 7 presents the projected averages of T_{mean}, T_{max}, and T_{min} for the period of 2066 to 2095 under RCP8.5. Similar to the pattern during the period of 2036 to 2065 under RCP8.5, the increase in the averages would be reduced from the northeastern Prairie to the southeastern Rocky Mountain. However, it can be found that the projected increases during the period of 2066 to 2095 under RCP4.5 and RCP8.5 are significantly different from each other. For example, the maximum increases in annual averages of daily maximum temperature during the period of 2066 to 2095 under RCP4.5 and RCP8.5 are 2.5 and 4.5 °C, respectively. Such increases are amplified under RCP8.5 as caused by heavy GHG emissions.

The spatial distribution of annual averages over the entire Prairie Provinces for two future periods under RCP8.5 are in comparison with those for the baseline period (Figure 8). It is indicated that the distribution of T_{mean}, T_{max}, and T_{min} generally maintain the same shape among three different periods. Such revelations are also consistent with the results under RCP4.5. Nevertheless, the projected averages of daily temperatures are significantly intensified as seen from the shifted mean values. For example, the mean value of T_{max} in the period of 2066 to 2095 is projected to be 6.0 °C under RCP4.5, whereas it is enlarged to 7.8 °C under RCP8.5.

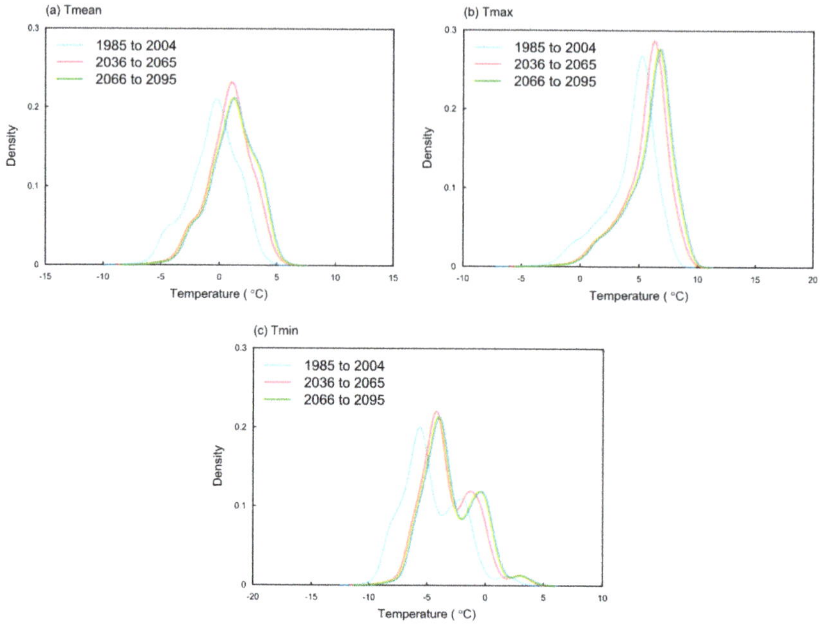

Figure 5. Spatial distribution of annual averages for the historical and future periods under RCP45. (**a**) T_{mean}, (**b**) T_{max}, and (**c**) T_{min}.

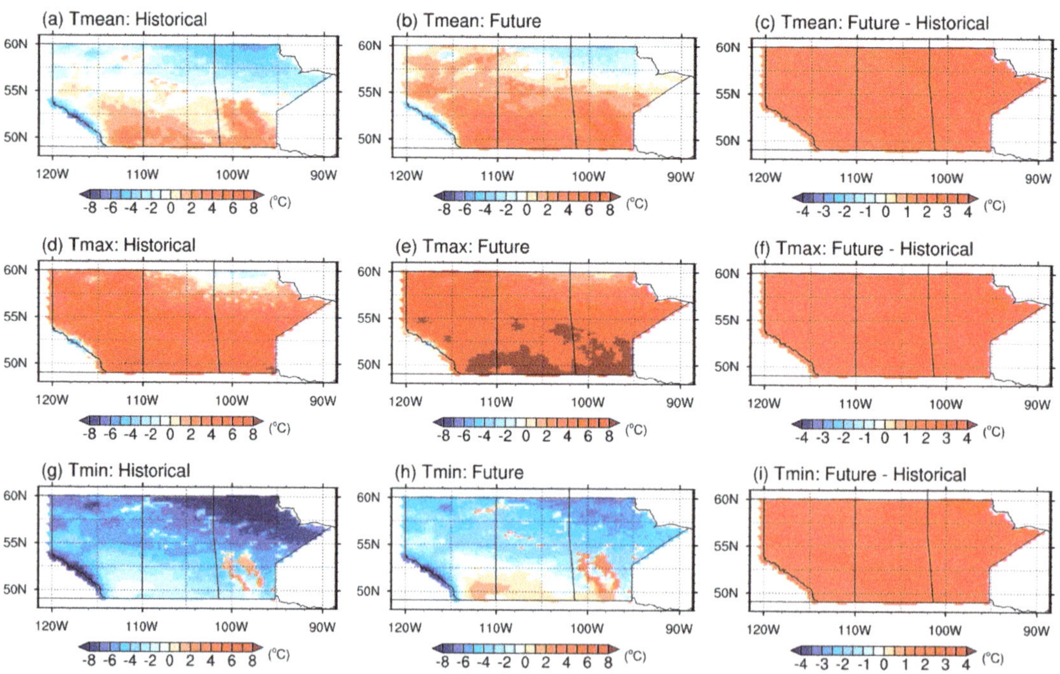

Figure 6. Projected variations of annual averages of T_{mean}, T_{max}, and T_{min} for the period of 2036 to 2065 under RCP85. (**a**,**d**,**g**) show the historical distributions of T_{mean}, T_{max}, and T_{min}, respectively; (**b**,**e**,**h**) show the future distributions of T_{mean}, T_{max}, and T_{min}, respectively; (**c**,**f**,**i**) show the projected variations of V_{mean}, V_{max}, and V_{min}, respectively.

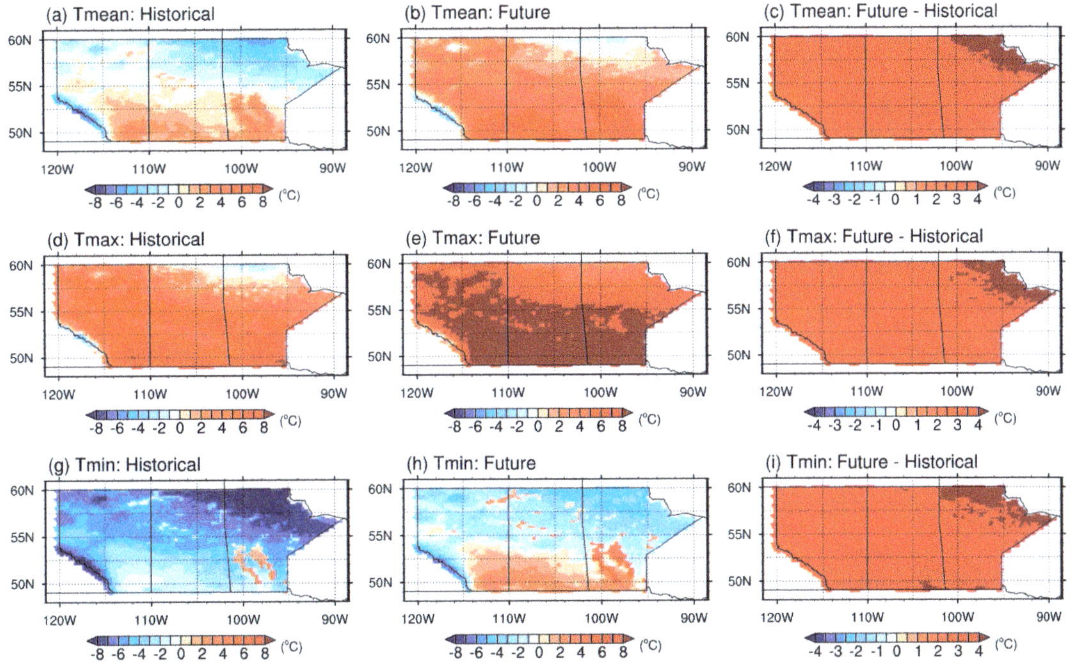

Figure 7. Projected variations of annual averages of T_{mean}, T_{max}, and T_{min} for the period of 2066 to 2095 under RCP85. (**a**,**d**,**g**) show the historical distributions of T_{mean}, T_{max}, and T_{min}, respectively; (**b**,**e**,**h**) show the future distributions of T_{mean}, T_{max}, and T_{min}, respectively; (**c**,**f**,**i**) show the projected variations of V_{mean}, V_{max}, and V_{min}, respectively.

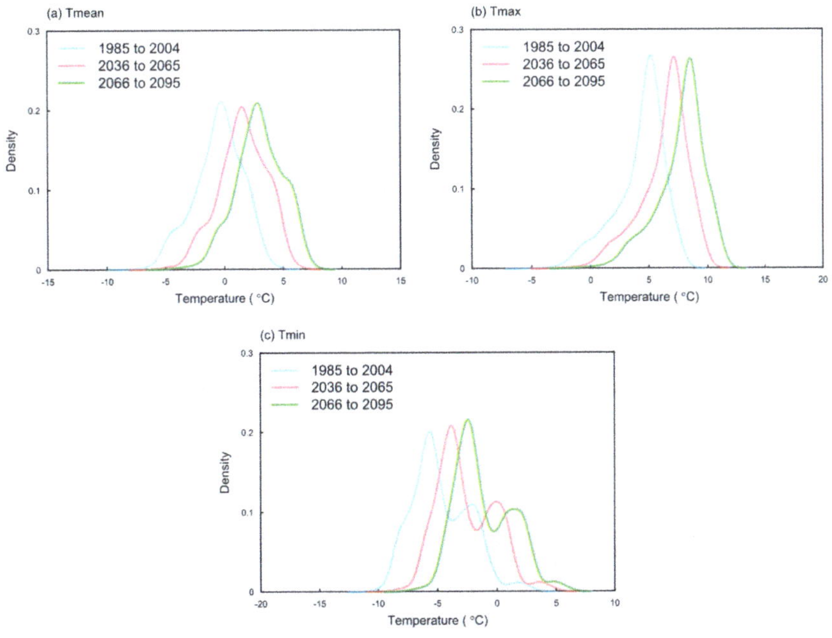

Figure 8. Spatial distribution of annual averages for the historical and future periods under RCP85. (**a**) T_{mean}, (**b**) T_{max}, and (**c**) T_{min}.

4. Projected Trends of Temperature

In order to further explore the impacts of climate change on the Prairie Provinces, the Mann–Kendall (MK) test [61,62] and the non-parametric Theil–Sen (TS) estimator [63,64] were performed to reveal the temporal trends in annual averages of T_{mean}, T_{max}, and T_{min} at multiple stations under RCPs. In this study, sixteen spatially distributed stations are chosen, while more details can be referred to in Zhou et al. [9]. The probability of the MK test statistics and the magnitude of the TS estimator are examined for demonstrating how the Prairie Provinces are affected within the context of global warming.

Figure 9 shows the time series and temporal trends of annual averages of T_{mean} at the 16 stations for two future periods under RCPs. It can be observed that the p-values of the MK test statistics at the majority of stations for future periods under RCP8.5 are less than the significance level ($\alpha = 0.05$), indicating that there are significant trends in the annual averages of T_{mean}. Particularly, T_{mean} at all the stations in the period of 2066 to 2095 under RCP4.5 shows insignificant trends, ranging from −0.029 to 0.020 °C/year. This might be attributable to the stabilization of the population growth, land-use conversions, and greenhouse gas (GHG) emissions under RCP4.5 after the year 2050.

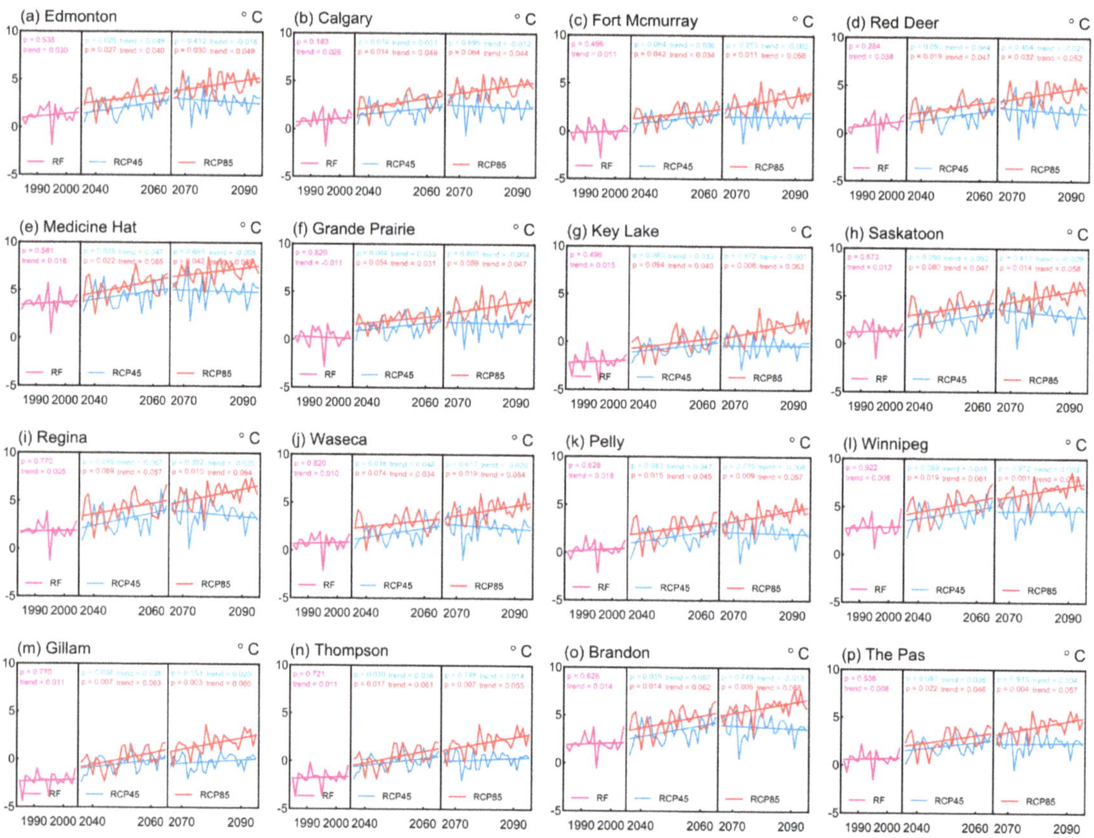

Figure 9. Temporal trends in annual averages of T_{mean} at the 16 stations. The value of p and trend is derived from the MK test and the Theil–Sen estimator, respectively. The straight line represents the linear trend derived from the Theil–Sen estimator. (**a**) Edmonton, (**b**) Calgary, (**c**) Fort Mcmurray, (**d**) Red Deer, (**e**) Medicine Hat, (**f**) Grande Prairie, (**g**) Key Lake, (**h**) Saskatoon, (**i**) Regina, (**j**) Waseca, (**k**) Pelly, (**l**) Winnipeg, (**m**) Gillam, (**n**) Thompson, (**o**) Brandon, and (**p**) The Pas.

In addition, temporal trends are investigated for annual averages of T_{max} and T_{min} for future periods under RCPs, which are presented in Figures 10 and 11, respectively. It can be noticed that similar warming trends are detected for the annual averages at all the stations under RCP8.5. However, the magnitude of positive trends in T_{mean} for most stations is generally smaller than T_{min}, while larger than T_{max}. This is also consistent with the results of projected variations, where larger increases are reported in T_{min}. It is also evident that the trends in T_{max} and T_{min} at all the stations in the period of 2066 to 2095 under RCP4.5 are insignificant, as seen from the p-values of the MK test statistics.

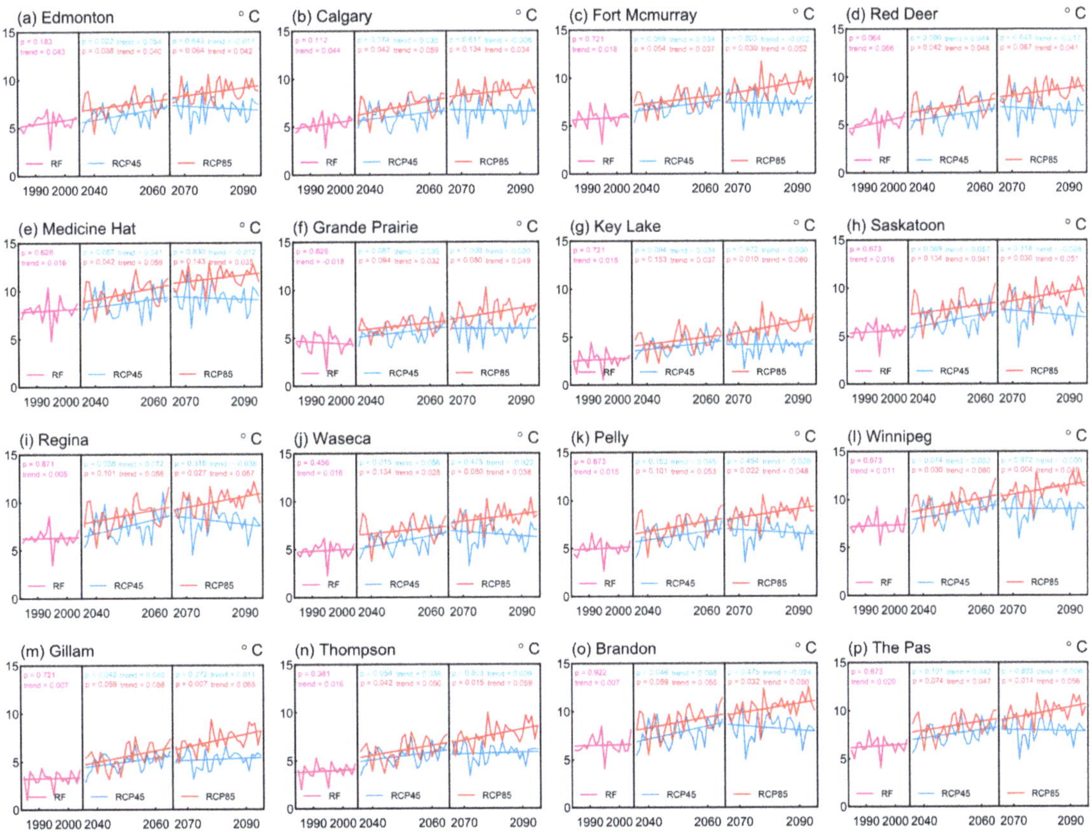

Figure 10. Temporal trends in annual averages of T_{max} at the 16 stations. The value of p and trend is derived from the MK test and the Theil–Sen estimator, respectively. The straight line represents the linear trend derived from the Theil–Sen estimator. (**a**) Edmonton, (**b**) Calgary, (**c**) Fort Mcmurray, (**d**) Red Deer, (**e**) Medicine Hat, (**f**) Grande Prairie, (**g**) Key Lake, (**h**) Saskatoon, (**i**) Regina, (**j**) Waseca, (**k**) Pelly, (**l**) Winnipeg, (**m**) Gillam, (**n**) Thompson, (**o**) Brandon, and (**p**) The Pas.

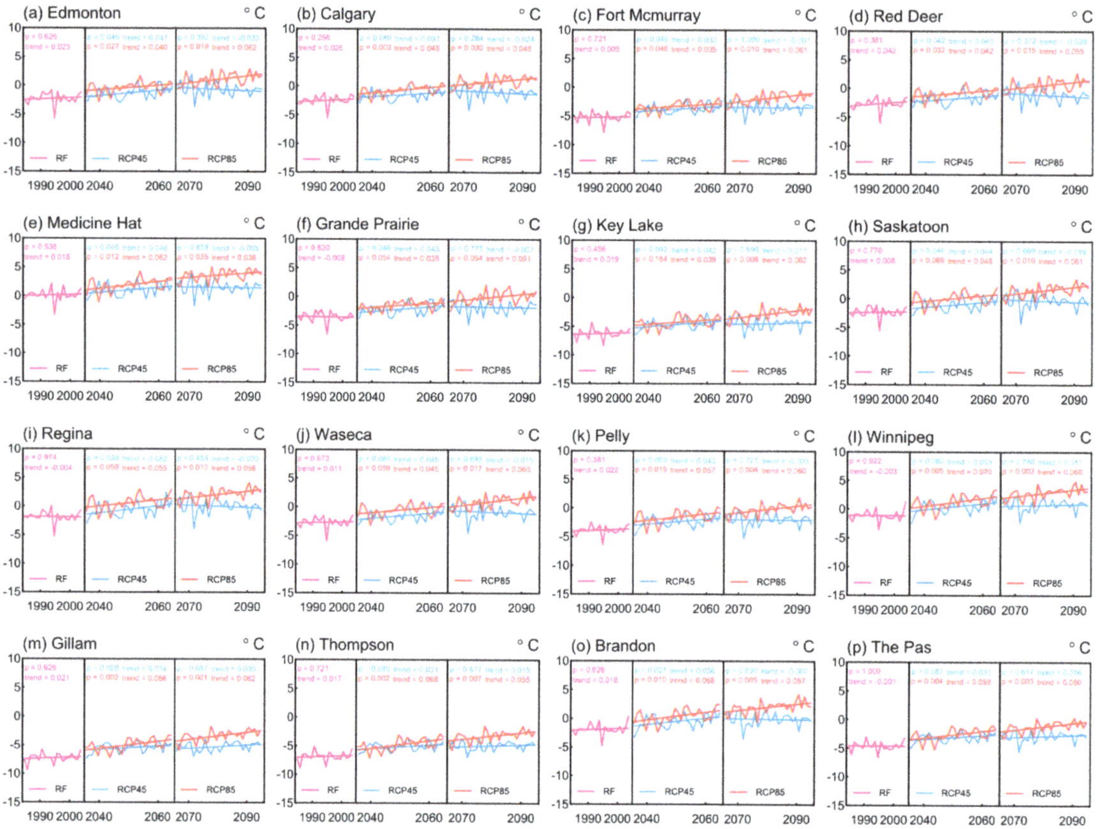

Figure 11. Temporal trends in annual averages of T_{min} at the 16 stations. The value of p and trend is derived from the MK test and the Theil–Sen estimator, respectively. The straight line represents the linear trend derived from the Theil–Sen estimator. (**a**) Edmonton, (**b**) Calgary, (**c**) Fort Mcmurray, (**d**) Red Deer, (**e**) Medicine Hat, (**f**) Grande Prairie, (**g**) Key Lake, (**h**) Saskatoon, (**i**) Regina, (**j**) Waseca, (**k**) Pelly, (**l**) Winnipeg, (**m**) Gillam, (**n**) Thompson, (**o**) Brandon, and (**p**) The Pas.

5. Discussion

In this study, the Canadian Prairies shows a larger increase in the annual averages of T_{min} than those of T_{max} (Figure 12). This also implies that projected increases in annual averages of T_{max} (Figure 12b) make a smaller contribution than T_{min} (Figure 12c) to the increase in annual averages of T_{mean} (Figure 12a). The downscaling experiments also reveal that the trends in T_{mean}, T_{max}, and T_{min} at all the stations in the 2080s under RCP4.5 are insignificant. This might be attributable to the stabilization of population growth, land-use conversions, and greenhouse gas (GHG) emissions under RCP4.5 after the year 2050. Moreover, we found that greater increases in annual averages of daily temperatures are projected under RCP8.5 than those under RCP4.5, indicating that a higher emission scenario will initiate a more significant increase in temperatures due to greenhouse gas effects.

The projected results of temperature variations reveal that increases in the annual averages of daily temperatures would be intensified from the southwest in the Rocky Mountain area to the northeast in the prairie region. This is most likely due to the dynamic effects of a strong positive ice-snow albedo feedback, amplifying climate change as a result of decreasing ice-snow cover in the prairie region. The contribution of ice-snow albedo feedback to larger increases in the annual averages of T_{min} in the northeast region would be more significant than T_{max} due to a similar amplification effect. This is also consistent with the previous studies, which indicate that the high-latitude ice-snow albedo feedback is a primary element in projected increases in temperatures under the global warming scenarios [65–67].

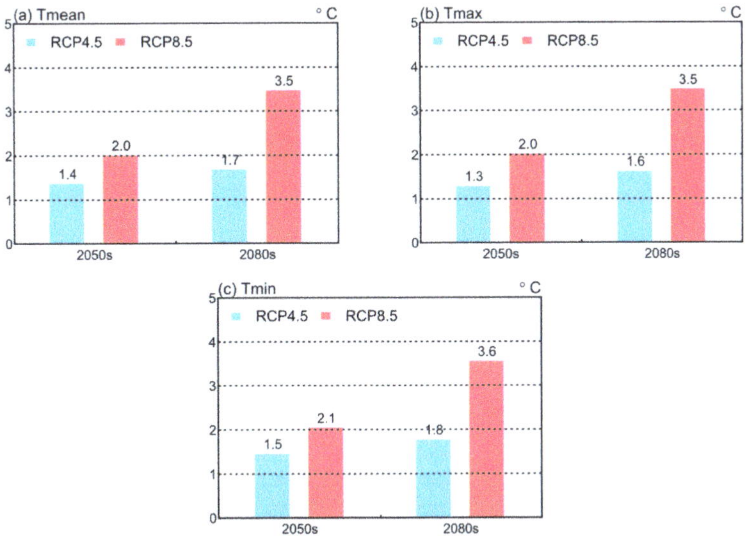

Figure 12. Spatial averages of projected changes for the future periods under RCPs. (**a**) T_{mean}, (**b**) T_{max}, and (**c**) T_{min}.

6. Conclusions

In this study, dynamically downscaled variations of daily mean, maximum, and minimum temperature over the Canadian Prairies were developed through regional climate simulations. How the regional climate will increase in response to global warming in the future was thus analyzed. Specifically, RegCM was undertaken to downscale the boundary conditions of GFDL-ESM2M. Daily temperatures were subsequently extracted from the historical and future simulations. Temperature variations in the future periods (i.e., 2036 to 2065 and 2065 to 2095) are then investigated relative to the baseline period (i.e., 1985 to 2004). The spatial temperature distributions were analyzed to reveal the regional impacts of global warming on the Prairie Provinces.

The results suggested that the spatial distribution from the future experiment in the period of 2036 to 2065 under RCP4.5 is closed to the historical one, with an obviously increased magnitude. Moreover, it could be observed that there is no significant difference in the projected changes in temperatures between the period of 2036 to 2065 and the period of 2066 to 2095 in terms of their mean and standard deviation. The results further indicated that the projected changes would be amplified from the southwest in the Rocky Mountain area to the northeast in the prairie region. It was also suggested that the projected changes would be significantly intensified under RCP8.5. The projected variations of daily mean, minimum, and maximum temperature could provide scientific bases for adaptation and mitigation initiatives on multiple sectors, such as agriculture and economic sectors.

Significant increases in temperatures over the Canadian Prairies have been revealed through a comprehensive analysis of the projected variations. Such revealed increases can provide scientific bases for identifying appropriate adaptation and mitigation initiatives on multiple sectors. This study is limited since the RegCM model is merely driven by one GCM, which has difficulty in reflecting the uncertainties of boundary conditions. However, such boundary data provided by GCM outputs could significantly affect the projected temperature variations. Therefore, future research efforts can be extended to investigate the full potential ranges of temperature increases through dynamical downscaling of multiple GCMs.

Author Contributions: Conceptualization, X.Z.; methodology, X.Z.; software, X.Z.; validation, X.Z.; formal analysis, X.Z.; investigation, X.Z.; resources, X.Z., G.H. and Y.L.; data curation, X.Z., Q.L., D.Y. and X.H.; writing—original draft preparation, X.Z.; writing—review and editing, X.Z; visualization, X.Z.; supervision, G.H. and Y.L.; funding acquisition, Y.L. All authors have read and agreed to the published version of the manuscript.

Funding: This research is supported by the Natural Science Foundation of China (51779008, U2040212) and the Fundamental Research Funds for the Central Universities.

Acknowledgments: We are very grateful for the helpful inputs and suggestions from the editor and anonymous reviewers.

Conflicts of Interest: The authors declare no conflict of interest.

References

1. Stern, P.C.; Perkins, J.H.; Sparks, R.E.; Knox, R.A. The challenge of climate-change neoskepticism. *Science* **2016**, *353*, 653–654. [CrossRef]
2. McCreath, M. Report of the Conference on Climate Change and the Law of the Sea: Adapting the Law of the Sea to Address the Challenges of Climate Change, Centre for International Law, National University of Singapore, Singapore, 13–14 March 2018. *Int. J. Mar. Coast. Law* **2018**, *33*, 836–846. [CrossRef]
3. Lewandowsky, S. Future Global Change and Cognition. *Top. Cogn. Sci.* **2016**, *8*, 7–18. [CrossRef] [PubMed]
4. Lawson, E. Debrief: Meeting the climate change challenge. *Br. J. Gen. Pract.* **2019**, *69*, 343. [CrossRef]
5. Kipling, R.; Topp, C.; Bannink, A.; Bartley, D.; Blanco-Penedo, I.; Cortignani, R.; del Prado, A.; Dono, G.; Faverdin, P.; Graux, A.-I.; et al. To what extent is climate change adaptation a novel challenge for agricultural modellers? *Environ. Model. Softw.* **2019**, *120*. [CrossRef]
6. Balasubramanian, M. Climate change, famine, and low-income communities challenge Sustainable Development Goals. *Lancet Planet. Health* **2018**, *2*, e421–e422. [CrossRef]
7. Argo, E.E. Migration Mismatch: Bird Migration and Phenological Mismatching. *Am. Biol. Teach.* **2018**, *80*, 540–543. [CrossRef]
8. Abdulai, A. The challenges and adaptation to climate change by farmers in Sub-Saharan Africa. *Agrekon* **2018**, *57*, 28–39. [CrossRef]
9. Zhou, X.; Huang, G.; Wang, X.; Fan, Y.; Cheng, G. A coupled dynamical-copula downscaling approach for temperature projections over the Canadian Prairies. *Clim. Dyn.* **2017**, *51*, 2413–2431. [CrossRef]
10. Peel, M.C.; Srikanthan, R.; McMahon, T.A.; Karoly, D.J. Approximating uncertainty of annual runoff and reservoir yield using stochastic replicates of global climate model data. *Hydrol. Earth Syst. Sci.* **2015**, *19*, 1615–1639. [CrossRef]
11. Nover, D.M.; Witt, J.W.; Butcher, J.; Johnson, T.E.; Weaver, C.P. The Effects of Downscaling Method on the Variability of Simulated Watershed Response to Climate Change in Five U.S. Basins. *Earth Interact.* **2016**, *20*, 1–27. [CrossRef] [PubMed]
12. Lee, S.; Wallace, C.W.; Sadeghi, A.M.; Mccarty, G.W.; Zhong, H.; Yeo, I.-Y. Impacts of Global Circulation Model (GCM) bias and WXGEN on Modeling Hydrologic Variables. *Water* **2018**, *10*, 764. [CrossRef]
13. Hosseinzadehtalaei, P.; Tabari, H.; Willems, P. Uncertainty assessment for climate change impact on intense precipitation: How many model runs do we need? *Int. J. Clim.* **2017**, *37*, 1105–1117. [CrossRef]
14. Zhang, W.; Zhang, X.; Li, W.; Hou, N.; Wei, Y.; Jia, K.; Yao, Y.; Cheng, J. Evaluation of Bayesian Multimodel Estimation in Surface Incident Shortwave Radiation Simulation over High Latitude Areas. *Remote Sens.* **2019**, *11*, 1776. [CrossRef]
15. Qiu, R.; Han, G.; Ma, X.; Xu, H.; Shi, T.; Zhang, M. A Comparison of OCO-2 SIF, MODIS GPP, and GOSIF Data from Gross Primary Production (GPP) Estimation and Seasonal Cycles in North America. *Remote Sens.* **2020**, *12*, 258. [CrossRef]
16. Fu, Y.; Lu, X.; Zhao, Y.; Zeng, X.; Xia, L. Assessment Impacts of Weather and Land Use/Land Cover (LULC) Change on Urban Vegetation Net Primary Productivity (NPP): A Case Study in Guangzhou, China. *Remote Sens.* **2013**, *5*, 4125–4144. [CrossRef]
17. Ahn, J.; Hong, S.; Cho, J.; Lee, Y.-W.; Lee, H. Statistical Modeling of Sea Ice Concentration Using Satellite Imagery and Climate Reanalysis Data in the Barents and Kara Seas, 1979–2012. *Remote Sens.* **2014**, *6*, 5520–5540. [CrossRef]
18. Addo, K.A.; Larbi, L.; Amisigo, B.; Ofori-Danson, P.K. Impacts of Coastal Inundation Due to Climate Change in a CLUSTER of Urban Coastal Communities in Ghana, West Africa. *Remote Sens.* **2011**, *3*, 2029–2050. [CrossRef]
19. Mika, A.M.; Weiss, R.M.; Olfert, O.; Hallett, R.H.; Newman, J. Will climate change be beneficial or detrimental to the invasive swede midge in North America? Contrasting predictions using climate projections from different general circulation models. *Glob. Chang. Biol.* **2008**, *14*, 1721–1733. [CrossRef]
20. McGinn, S.M.; Shepherd, A. Impact of climate change scenarios on the agroclimate of the Canadian prairies. *Can. J. Soil Sci.* **2003**, *83*, 623–630. [CrossRef]
21. Thorpe, J.; Wolfe, S.A.; Houston, B. Potential impacts of climate change on grazing capacity of native grasslands in the Canadian prairies. *Can. J. Soil Sci.* **2008**, *88*, 595–609. [CrossRef]

22. Bonsal, B.; Liu, Z.; Wheaton, E.; Stewart, R. Historical and Projected Changes to the Stages and Other Characteristics of Severe Canadian Prairie Droughts. *Water* **2020**, *12*, 3370. [CrossRef]
23. Wotton, B.M.; Martell, D.L.; Logan, K.A. Climate Change and People-Caused Forest Fire Occurrence in Ontario. *Clim. Chang.* **2003**, *60*, 275–295. [CrossRef]
24. Thorne, R. Uncertainty in the impacts of projected climate change on the hydrology of a subarctic environment: Liard River Basin. *Hydrol. Earth Syst. Sci.* **2011**, *15*, 1483–1492. [CrossRef]
25. Sultana, Z.; Coulibaly, P. Distributed modelling of future changes in hydrological processes of Spencer Creek watershed. *Hydrol. Process.* **2010**, *25*, 1254–1270. [CrossRef]
26. Shrestha, R.R.; Schnorbus, M.A.; Werner, A.T.; Zwiers, F.W. Evaluating Hydroclimatic Change Signals from Statistically and Dynamically Downscaled GCMs and Hydrologic Models. *J. Hydrometeorol.* **2014**, *15*, 844–860. [CrossRef]
27. Schertzer, W.M.; Croley, T.E. Climate change impact on hydrology and lake thermal structure. In Proceedings of the 27th Congress of the International-Association-for-Hydraulic-Research, San Francisco, CA, USA, 10–15 August 1997; pp. 919–924.
28. Mahat, V.; Anderson, A. Impacts of climate and catastrophic forest changes on streamflow and water balance in a mountainous headwater stream in Southern Alberta. *Hydrol. Earth Syst. Sci.* **2013**, *17*, 4941–4956. [CrossRef]
29. Cheng, C.S.; Lopes, E.; Fu, C.; Huang, Z. Possible Impacts of Climate Change on Wind Gusts under Downscaled Future Climate Conditions: Updated for Canada. *J. Clim.* **2014**, *27*, 1255–1270. [CrossRef]
30. Bonsal, B.R.; Prowse, T.D.; Pietroniro, A. An assessment of global climate model-simulated climate for the western cordillera of Canada (1961–90). *Hydrol. Process.* **2003**, *17*, 3703–3716. [CrossRef]
31. St-Jacques, J.-M.; Andreichuk, Y.; Sauchyn, D.J.; Barrow, E. Projecting Canadian Prairie Runoff for 2041-2070 with North American Regional Climate Change Assessment Program (NARCCAP) Data. *JAWRA J. Am. Water Resour. Assoc.* **2018**, *54*, 660–675. [CrossRef]
32. Qian, B.; Wang, H.; He, Y.; Liu, J.; De Jong, R. Projecting spring wheat yield changes on the Canadian Prairies: Effects of resolutions of a regional climate model and statistical processing. *Int. J. Clim.* **2015**, *36*, 3492–3506. [CrossRef]
33. PaiMazumder, D.; Sushama, L.; Laprise, R.; Khaliq, M.N.; Sauchyn, D. Canadian RCM projected changes to short- and long-term drought characteristics over the Canadian Prairies. *Int. J. Clim.* **2012**, *33*, 1409–1423. [CrossRef]
34. Kerr, S.A.; Andreichuk, Y.; Sauchyn, D.J. Re-Evaluating the Climate Factor in Agricultural Land Assessment in a Changing Climate—Saskatchewan, Canada. *Land* **2019**, *8*, 49. [CrossRef]
35. Zhou, X.; Huang, G.; Piwowar, J.; Fan, Y.; Wang, X.; Li, Z.; Cheng, G. Hydrologic Impacts of Ensemble-RCM-Projected Climate Changes in the Athabasca River Basin, Canada. *J. Hydrometeorol.* **2018**, *19*, 1953–1971. [CrossRef]
36. Schumacher, V.; Fernández, A.; Justino, F.; Comin, A. WRF High Resolution Dynamical Downscaling of Precipitation for the Central Andes of Chile and Argentina. *Front. Earth Sci.* **2020**, *8*, 328. [CrossRef]
37. Shalaby, A.; Zakey, A.S.; Tawfik, A.B.; Solmon, F.; Giorgi, F.; Stordal, F.; Sillman, S.; Zaveri, R.A.; Steiner, A.L. Implementation and evaluation of online gas-phase chemistry within a regional climate model (RegCM-CHEM4). *Geosci. Model Dev.* **2012**, *5*, 741–760. [CrossRef]
38. Sylla, M.; Giorgi, F.; Stordal, F. Large-Scale origins of rainfall and temperature bias in high-resolution simulations over southern Africa. *Clim. Res.* **2012**, *52*, 193–211. [CrossRef]
39. Zampieri, M.; Lionello, P. Anthropic land use causes summer cooling in Central Europe. *Clim. Res.* **2011**, *46*, 255–268. [CrossRef]
40. Maurya, R.K.S.; Mohanty, M.R.; Sinha, P.; Mohanty, U.C. Performance of hydrostatic and non-hydrostatic dynamical cores in RegCM4.6 for Indian summer monsoon simulation. *Meteorol. Appl.* **2020**, *27*, e1915. [CrossRef]
41. Balov, M.N.; Altunkaynak, A. Trend Analyses of Extreme Precipitation Indices Based on Downscaled Outputs of Global Circulation Models in Western Black Sea Basin, Turkey. *Iran. J. Sci. Technol. Trans. Civ. Eng.* **2019**, *43*, 821–834. [CrossRef]
42. Kohyama, T.; Hartmann, D.L.; Battisti, D.S. La Niña–like Mean-State Response to Global Warming and Potential Oceanic Roles. *J. Clim.* **2017**, *30*, 4207–4225. [CrossRef]
43. Matveeva, T.A.; Gushchina, D.Y.; Zolina, O.G. Large-Scale indicators of extreme precipitation in coastal natural-economic zones of the European part of Russia. *Russ. Meteorol. Hydrol.* **2015**, *40*, 722–730. [CrossRef]
44. Ng, B.; Cai, W.; Walsh, K. Nonlinear Feedbacks Associated with the Indian Ocean Dipole and Their Response to Global Warming in the GFDL-ESM2M Coupled Climate Model. *J. Clim.* **2014**, *27*, 3904–3919. [CrossRef]
45. Zhou, X.; Huang, G.; Wang, X.; Cheng, G. Dynamically-Downscaled temperature and precipitation changes over Saskatchewan using the PRECIS model. *Clim. Dyn.* **2017**, *50*, 1321–1334. [CrossRef]
46. Fisher, R.A.; Muszala, S.; Verteinstein, M.; Lawrence, P.; Xu, C.; McDowell, N.G.; Knox, R.G.; Koven, C.; Holm, J.; Rogers, B.M.; et al. Taking off the training wheels: The properties of a dynamic vegetation model without climate envelopes, CLM4.5(ED). *Geosci. Model Dev.* **2015**, *8*, 3593–3619. [CrossRef]
47. Adeniyi, M.O. Sensitivity of different convection schemes in RegCM4. 0 for simulation of precipitation during the Septembers of 1989 and 1998 over West Africa. *Theor. Appl. Climatol.* **2014**, *115*, 305–322. [CrossRef]
48. Davies, H.C. A lateral boundary formulation for multi-level prediction models. *Q. J. R. Meteorol. Soc.* **1976**, *102*, 405–418. [CrossRef]
49. Holtslag, A.; De Bruijn, E.; Pan, H.L. A high resolution air mass transformation model for short-range weather forecasting. *Mon. Weather Rev.* **1990**, *118*, 1561–1575. [CrossRef]

50. Emanuel, K.A.; Živković-Rothman, M. Development and evaluation of a convection scheme for use in climate models. *J. Atmos. Sci.* **1999**, *56*, 1766–1782. [CrossRef]
51. Skinner, W.; Majorowicz, J. Regional climatic warming and associated twentieth century land-cover changes in north-western North America. *Clim. Res.* **1999**, *12*, 39–52. [CrossRef]
52. NLWIS. A Daily 10 km Gridded Climate Dataset for Canada, 1961–2003. Available online: https://www.mcgill.ca/library/find/maps/dgcd10km (accessed on 14 February 2017).
53. Natural Resources Canada. Canada 3D—Digital Elevation Model of the Canadian Landmass. Available online: https://open.canada.ca/data/en/dataset/042f4628-94b2-40ac-9bc1-ca3ac2a27d82 (accessed on 14 July 2021).
54. Jeong, D.I.; Sushama, L.; Diro, G.T.; Khaliq, M.N. Projected changes to winter temperature characteristics over Canada based on an RCM ensemble. *Clim. Dyn.* **2015**, *47*, 1351–1366. [CrossRef]
55. Mladjic, B.; Sushama, L.; Khaliq, M.N.; Laprise, R.; Caya, D.; Roy, R. Canadian RCM Projected Changes to Extreme Precipitation Characteristics over Canada. *J. Clim.* **2011**, *24*, 2565–2584. [CrossRef]
56. Plummer, D.A.; Caya, D.; Frigon, A.; Côté, H.; Giguère, M.; Paquin, D.; Biner, S.; Harvey, R.; De Elia, R. Climate and Climate Change over North America as Simulated by the Canadian RCM. *J. Clim.* **2006**, *19*, 3112–3132. [CrossRef]
57. Sushama, L.; Khaliq, N.; Laprise, R. Dry spell characteristics over Canada in a changing climate as simulated by the Canadian RCM. *Glob. Planet. Chang.* **2010**, *74*, 1–14. [CrossRef]
58. Jeong, D.I.; Sushama, L.; Diro, G.T.; Khaliq, M.N.; Beltrami, H.; Caya, D. Projected changes to high temperature events for Canada based on a regional climate model ensemble. *Clim. Dyn.* **2015**, *46*, 3163–3180. [CrossRef]
59. Zhou, X.; Huang, G.; Baetz, B.W.; Wang, X.; Cheng, G. PRECIS-Projected increases in temperature and precipitation over Canada. *Q. J. R. Meteorol. Soc.* **2018**, *144*, 588–603. [CrossRef]
60. Zhou, X.; Huang, G.; Wang, X.; Cheng, G. Future Changes in Precipitation Extremes over Canada: Driving Factors and Inherent Mechanism. *J. Geophys. Res. Atmos.* **2018**, *123*, 5783–5803. [CrossRef]
61. Mann, H.B. Nonparametric tests against trend. *Econom. J. Econom. Soc.* **1945**, *13*, 245–259. [CrossRef]
62. Kendall, M.G. *Rank Correlation Methods*; Griffin: Oxford, UK, 1948.
63. Theil, H. A rank-invariant method of linear and polynomial regression analysis. In *Henri Theil's Contributions to Economics and Econometrics*; Springer: Berlin/Heidelberg, Germany, 1992; pp. 345–381.
64. Sen, P.K. Estimates of the regression coefficient based on Kendall's tau. *J. Am. Stat. Assoc.* **1968**, *63*, 1379–1389. [CrossRef]
65. Gorodetskaya, I.; Cane, M.A.; Tremblay, L.; Kaplan, A. The effects of sea-ice and land-snow concentrations on planetary albedo from the earth radiation budget experiment. *Atmos. Ocean* **2006**, *44*, 195–205. [CrossRef]
66. Pierrehumbert, R.T. Climate dynamics of a hard snowball Earth. *J. Geophys. Res. Space Phys.* **2005**, *110*, D01111. [CrossRef]
67. Sorteberg, A.; Kattsov, V.; Walsh, J.; Pavlova, T. The Arctic surface energy budget as simulated with the IPCC AR4 AOGCMs. *Clim. Dyn.* **2007**, *29*, 131–156. [CrossRef]

remote sensing

MDPI

Article

Long-Term Projection of Water Cycle Changes over China Using RegCM

Chen Lu [1], Guohe Huang [1,*], Guoqing Wang [2], Jianyun Zhang [2], Xiuquan Wang [3] and Tangnyu Song [1]

1 Faculty of Engineering and Applied Science, University of Regina, Regina, SK S4S 0A2, Canada; lu309@uregina.ca (C.L.); song203t@uregina.ca (T.S.)
2 State Key Laboratory of Hydrology-Water Resources and Hydraulic Engineering, Nanjing Hydraulic Research Institute, Nanjing 210029, China; gqwang@nhri.cn (G.W.); jyzhang@nhri.cn (J.Z.)
3 School of Climate Change and Adaptation, University of Prince Edward Island, 550 University Ave., Charlottetown, PE C1A 4P3, Canada; xxwang@upei.ca
* Correspondence: huangg@uregina.ca; Tel.: +1-306-585-4095; Fax: +1-306-585-5755

Abstract: The global water cycle is becoming more intense in a warming climate, leading to extreme rainstorms and floods. In addition, the delicate balance of precipitation, evapotranspiration, and runoff affects the variations in soil moisture, which is of vital importance to agriculture. A systematic examination of climate change impacts on these variables may help provide scientific foundations for the design of relevant adaptation and mitigation measures. In this study, long-term variations in the water cycle over China are explored using the Regional Climate Model system (RegCM) developed by the International Centre for Theoretical Physics. Model performance is validated through comparing the simulation results with remote sensing data and gridded observations. The results show that RegCM can reasonably capture the spatial and seasonal variations in three dominant variables for the water cycle (i.e., precipitation, evapotranspiration, and runoff). Long-term projections of these three variables are developed by driving RegCM with boundary conditions of the Geophysical Fluid Dynamics Laboratory Earth System Model under the Representative Concentration Pathways (RCPs). The results show that increased annual average precipitation and evapotranspiration can be found in most parts of the domain, while a smaller part of the domain is projected with increased runoff. Statistically significant increasing trends (at a significant level of 0.05) can be detected for annual precipitation and evapotranspiration, which are 0.02 and 0.01 mm/day per decade, respectively, under RCP4.5 and are both 0.03 mm/day per decade under RCP8.5. There is no significant trend in future annual runoff anomalies. The variations in the three variables mainly occur in the wet season, in which precipitation and evapotranspiration increase and runoff decreases. The projected changes in precipitation minus evapotranspiration are larger than those in runoff, implying a possible decrease in soil moisture.

Keywords: climate change; China; water cycle; precipitation; evapotranspiration; runoff; projection

Citation: Lu, C.; Huang, G.; Wang, G.; Zhang, J.; Wang, X.; Song, T. Long-Term Projection of Water Cycle Changes over China Using RegCM. *Remote Sens.* **2021**, *13*, 3832. https://doi.org/10.3390/rs13193832

Academic Editor: Magaly Koch

Received: 29 August 2021
Accepted: 23 September 2021
Published: 25 September 2021

Publisher's Note: MDPI stays neutral with regard to jurisdictional claims in published maps and institutional affiliations.

1. Introduction

The global water cycle is becoming more intense in a warming climate; increases in precipitation, evapotranspiration, and runoff can be widely observed over the world [1,2]. The resulting extreme rainstorms and heavy runoff can themselves lead to losses of life and damage to infrastructure (such as urban drainage systems), not to mention that they may also cause flood events that are of more devastating consequences [3]. On the other hand, the delicate balance of the three variables also deserves attention, as, according to the surface water budget equation $dS/dt = P - E - R$ (where S denotes the subsurface storage of water substances, P is precipitation, E is evapotranspiration, and R is runoff) [2,4], they are closely associated with variations in soil moisture. In cases where meteorological drought occurs (i.e., long-term rainfall deficit), the interactions among the water cycle components could affect how the meteorological drought is propagated to hydrological drought (i.e., long-term runoff deficit) and agriculture drought (i.e., long-term soil moisture deficit), which can have significant influences on regional water availability and agriculture [5]. In addition, variations in the water cycle can have profound long-term environmental [6–8] and ecological [9,10] implications. In China, disasters caused by variations in water cycle components have been recorded. For example, a series of flood events that occurred in the Yangtze River Basin in 2020, in which the inflow to the Three Gorges Dam once reached 75,000 m^3/s during the fifth flood, caused the death/disappearance of 219 people and an economic loss of 178.96 billion yuan by 13 August [11,12]. In order to avoid losses such as above, studies of future variations in the water cycle are needed to help provide scientific foundations for the design of relevant adaptation and mitigation measures.

Previous attempts have been made in projecting future changes in individual water cycle components. For example, Li et al. [13] applied a self-organizing map to study the future changes in summer precipitation in eastern China using global climate model (GCM) data of climate model intercomparison project phase 5 (CMIP5). Dong et al. [14] examined the climate change impacts on reference evapotranspiration in Xinjiang, China based on CMIP5 GCM projections. Yan et al. [15] projected future runoff in the Yellow River Basin using CMIP5 GCM data through global Bayesian model averaging. Fewer studies can be found focusing on more than one water-cycle-related variable. One example is the study of Gu et al. [16] on future precipitation and runoff changes and the resultant drought events over China using GCM and hydrological model ensembles. In another study conducted by Zhang et al. [2], future changes in the three water cycle components over global land monsoon regions were systematically examined using the GMIP5 GCM ensemble.

In general, most of the previous studies that simultaneously involve the projection of more than one water cycle component are based on GCM data. However, precipitation, evapotranspiration, and runoff are closely related to land surface processes that are largely dependent on topographic and land-use information, which is not adequately depicted in GCMs due to their coarse resolution [17,18]. Climate downscaling, including statistical and dynamical approaches, is commonly undertaken to derive regional- and/or local-scale information [18]. While both techniques have been demonstrated to be capable of correcting biases in GCM outputs, studies have shown that statistical simulations can inherit unphysical signals from the driving data, and thus fail to generate physically consistent regional climate projections as regional climate models (RCMs) used in the dynamical approach do [19]. The applicability of RCMs to the simulation of temperature and precipitation over China has been tested by various previous studies, which suggest that RCMs can refine large-scale climate information in complex terrains, correct certain biases in the boundary conditions, and provide results that correspond better with the observations [17,20]. Among all RCMs, RegCM has been noted to perform well over China [21,22]. For example, Pan et al. [23] conducted climate projections over Northwest China, and suggested that RegCM can reproduce the spatial patterns of temperature, precipitation, and climate extremes over this region. Gao [24] simulated climate over China using RegCM and WRF, and noticed that RegCM shows better skills in simulating the temperature and precipitation pattern and magnitude in dry regions. Jiang et al. [25] compared the performance of RegCM and PRECIS in modeling precipitation over China, and indicated that the former outperforms the latter in capturing annual precipitation and wet days in eastern China.

Therefore, as an extension of previous studies, the objective of this study is to examine, through the application of an RCM, climate change impacts on the water cycle components over China. Specifically, regional climate simulations over China are undertaken using RegCM. Model performance is evaluated via the comparison of the simulated climate against the remote sensing and gridded observation data. Then, high-resolution long-term projections of precipitation, evapotranspiration, and runoff under two emission scenarios are developed, and the changes and trends of these variables are subsequently evaluated. The results of this research could be beneficial to the forecast and control of flood and drought events in China.

2. Methodology and Data

Dynamical downscaling of water cycle components over China is developed using RegCM, which is a regional climate model developed by the International Center for Theoretical Physics [26]. The Community Land Model version 4.5 (CLM4.5) coupling is enabled in RegCM simulations to provide an improved description of land surface processes (e.g., carbon cycle, vegetation dynamics, and river routing) [27,28]. The detailed representation of water vapor fluxes for both non-vegetated and vegetated surfaces in CLM is expected to help with the simulation of evapotranspiration. In addition, CLM is embedded with SIMTOP (simple TOPMODEL-based runoff model) [29], which can take into account the influence of topological information in runoff generation. The simulated total runoff is then routed to active ocean or marginal seas through a river transport model [30]. More details about RegCM parameterization scheme configuration can be found in Lu et al. [22].

Two rounds of RegCM hindcast simulations are conducted for model validation purposes; one of them is driven by the ERA-Interim reanalysis data developed by the European Centre for Medium-Range Weather Forecasts [31], which provide the realistic historical climate over China; the other one is driven by the historical climate scenario of the Earth System Model developed by the Geophysical Fluid Dynamics Laboratory (GFDL) [32,33], which is used to provide the baseline for projections. The baseline period is 1986 to 2005.

The RegCM performance is validated through comparisons of the annual and seasonal averages of model results with those of the observations, remote sensing data, and reconstructed data. The months included in each season are as follows: December (of the previous year), January, and February for winter; March, April, and May for spring; June, July, and August for summer; September, October, and November for autumn. Spatial correlation is employed as a quantitative metric to reflect the similarity between the annual and seasonal averaged observation/remote sensing data/reconstructed data and the simulation results.

In this study, version 4 of the high-resolution gridded observation dataset from the Climate Research Unit (CRU) [34] is used for verifying the model-generated temperature and precipitation. This dataset is generated through the interpolation of extensive networks of gauge station observations into a 0.5° regular grid [34], and is widely applied for the calibration/validation of global and regional climate models [35].

For evapotranspiration, two sets of remote sensing data are employed, specifically, version 6 of the Resolution Imaging Spectroradiometer (MODIS) terrestrial evapotranspiration product (MOD16A2GF v006) [36], and the latest version of the Global Land Evapotranspiration Amsterdam Model (GLEAM, v3.5b) [37,38]. The MOD16A2GF dataset is created from remotely sensed data products of MODIS (e.g., land cover and albedo) based on the Penman–Monteith equation [39]. On the other hand, GLEAM assembles various satellite-based observations (e.g., radiation from GERES, precipitation from TMPA and MSWEP, air temperature from AIRS, soil moisture from SMOS and ERA CCI SM) and derives global evaporation variables with the Priestley and Taylor model [37]. Both datasets were demonstrated to be able to reasonably represent the actual evapotranspiration over China [40,41]. The evapotranspiration from MOD16A2GF and the actual evaporation from GLEAM are used for validating the RegCM-generated evapotranspiration. It is worth noting that the start years of these two remote sensing datasets are 2000 and 2003, respectively; therefore, RegCM results averaged over the same periods (i.e., 2000 to 2005 and 2003 to 2005) are used for comparison. To assist the validation for evapotranspiration over the entire baseline period, station-based observed tank evaporation from the National Meteorological Information Center of China (NIMC) is also used (data available at data.cma.cn; accessed on 9 April 2019). The locations of the stations are shown in Figure S1 in the Supplementary Materials.

For runoff, the validation was undertaken through the comparison of the model outputs with the Global Runoff Reconstruction (GRUN), which is constructed through machine learning techniques based on runoff and meteorological observations [42]. This dataset is widely used in weather and climate research and is shown to have a reasonable performance over China [43–45].

For future climate, GFDL projections of two representative concentration pathways (RCPs) are employed, which are RCP4.5 and RCP8.5, respectively, for intermediate and heavy emissions. Simulations are conducted for the entire twenty-first century. Three time-slices are considered in result analysis: 2020 to 2039 (or 2030s), 2040 to 2069 (or 2050s), and 2070 to 2099 (or 2080s); time averages and trends are calculated with respect to these periods. In addition, since the water cycle components are closely related to atmospheric moisture contents, and the saturation vapor pressure is related to the temperature following the Clausius–Clapeyron equation [46], different warming periods are also considered in this study. The warming periods are defined as twenty-year periods in which the domain average temperature increases by 1, 1.5, 2, 3, and 4 °C compared with the baseline average. Table 1 lists the respective periods for each warming level under the two emission scenarios. The numerical values of future trends are obtained based on Sen's slope estimator, and their statistical significance is examined by Mann–Kendall tests [47–50].

Table 1. Future periods in which warming over China reaches 1, 1.5, 2, 3, and 4 °C under RCP4.5 and RCP8.5.

Warming Level	RCP4.5	RCP8.5
1 °C	2021~2040	2008~2027
1.5 °C	2027~2046	2030~2049
2 °C	2067~2086	2046~2065
3 °C	-	2056~2075
4 °C	-	2069~2088

3. Results

3.1. Model Validation

Validation results for near-surface temperature, precipitation, evapotranspiration, and runoff are shown in Figures 1–4, respectively. The columns of each figure are for different datasets (i.e., CRU, MODIS, GLEAM, GRUN, RegCM driven by ERA-Interim, and RegCM driven by GFDL); the rows are for annual and seasonal averages. It can be observed that RegCM can reasonably reproduce the spatial distribution and seasonal variations of temperature over China. As shown by the CRU observations (Figure 1a), above-zero baseline-average temperature can be found in most parts of the domain, except for the Tibetan Plateau and a small part of northeastern China. This spatial feature is realistically generated in the two sets of RegCM results, although with underestimations of various degrees (Figure 1b,c). Such underestimations, as discussed by Lu et al. [22], are partly caused by the model setup and partly due to the driving GCM data. Temperature over China demonstrates clear seasonality, i.e., hot summer, cold winter, and mild spring and autumn. RegCM is able to capture the seasonal variations, although underestimations can still be observed. The spatial correlation between the observed and modeled temperature can be found in Table S1 in the Supplementary Materials. The high correlation (ranges between 0.94 and 0.98) indicates RegCM's good performance in temperature. In addition, the results of RegCM driven by GFDL have higher correlations with the observations than those of the raw GCM data (please refer to Table S2 in the Supplementary Materials), which suggests that RegCM is able to correct some biases in GFDL's temperature results.

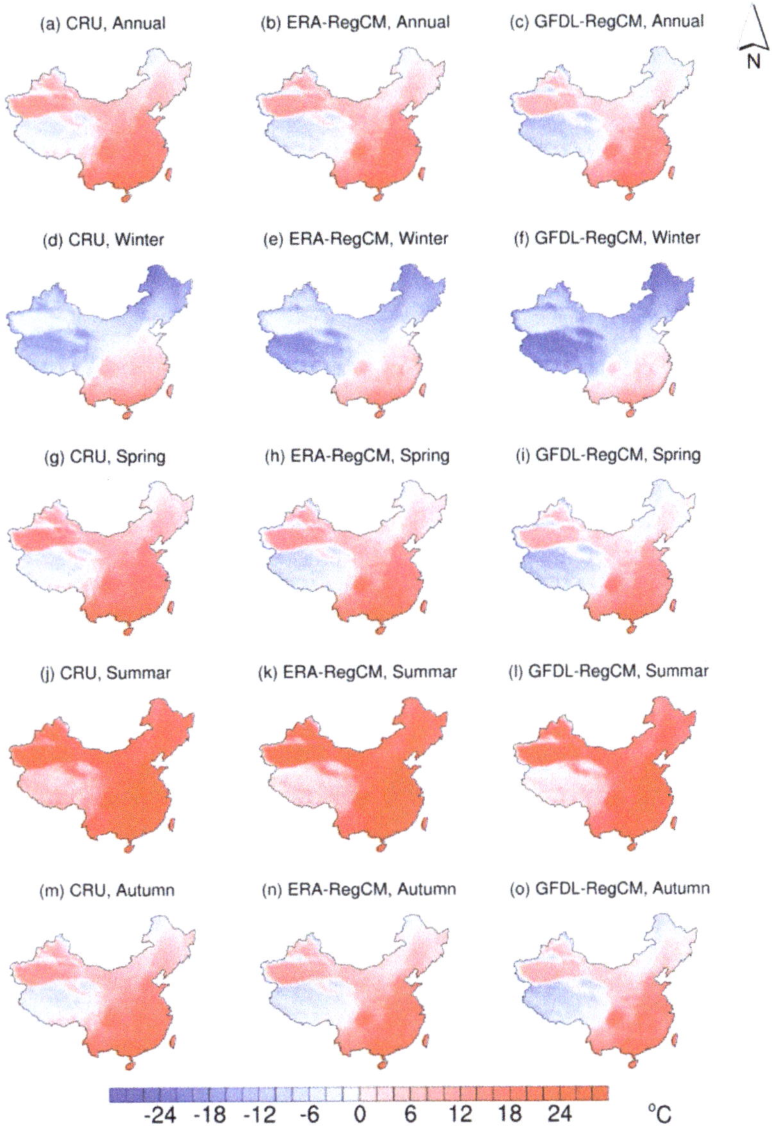

Figure 1. Observed and RegCM-simulated near-surface temperature over China. The annual and seasonal averages are calculated with respect to the baseline period of 1986 to 2005.

The performance of RegCM with respect to precipitation is less satisfactory than for temperature. Although it is able to generate the observed wet-in-the-southeast and dry-in-the-northwest precipitation pattern, underestimations can be found over the southeastern part of the domain and overestimations over the northwestern part (Figure 2a–c). This over- and underestimation pair was shown to be related to the simulation bias in vapor pressure by Lu et al. [22], who further argued that the bias in vapor pressure could be associated with the bias in temperature. There is also an apparent dry bias near the Sichuan Basin and a wet bias near the southeastern edge of the Tibetan Plateau, which could be related to the configuration of the cumulus convective scheme [22]. The spatial correlation (Table S1) is lower for precipitation than for temperature, which is consistent with the above results.

The seasonal precipitation over China shows clear monsoon features (more precipitation in summer and less in winter), which is shared by the two sets of RegCM results.

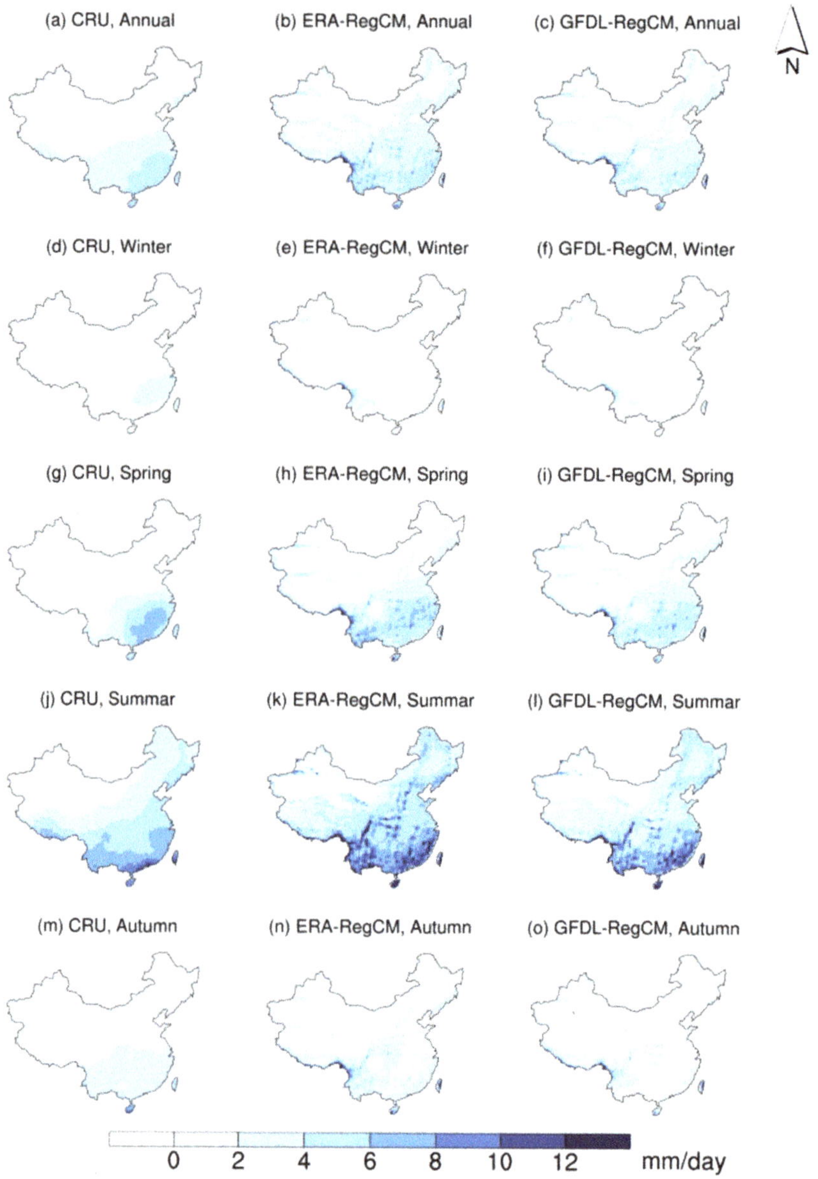

Figure 2. Observed and RegCM-simulated precipitation over China. The annual and seasonal averages are calculated with respect to the baseline period of 1986 to 2005.

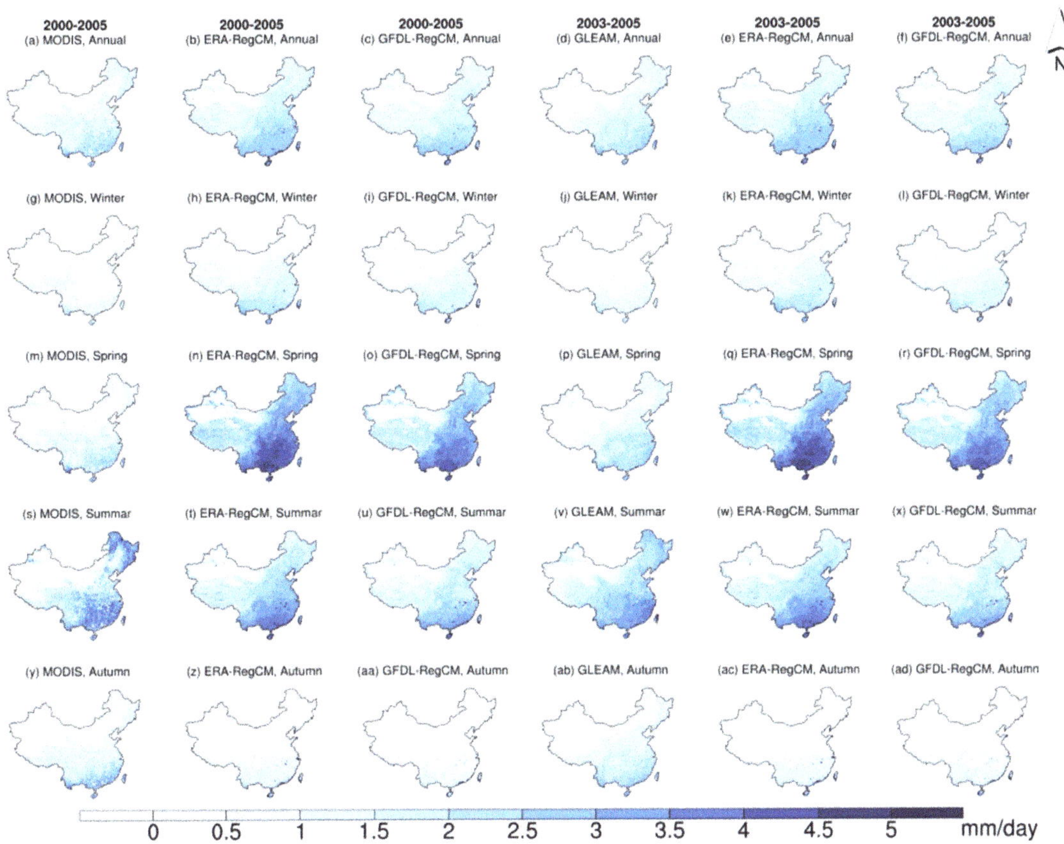

Figure 3. Remotely sensed and RegCM-simulated evapotranspiration over China. The annual and seasonal averages are calculated with respect to the period of 2000 to 2005 for MODIS and 2003 to 2005 for GLEAM.

The spatial pattern of actual evapotranspiration from MODIS and GLEAM (Figure 3a,d) shows more regional details than the observed precipitation pattern from CRU, which is in part due to the higher resolution of remote sensing data than that of the gridded observation, and in part due to evapotranspiration's closer relationship with the geophysical characteristics of the domain than precipitation. The two sets of remote sensing data show similar annual patterns; subtle differences exist sporadically over the domain, which could be explained by the different skills of the two datasets over different land-use types [51]. RegCM shows better skills in simulating evapotranspiration than precipitation (Figure 3b,c,e,f), although minor overestimations can be noticed. As shown in Table S1, for annual average evapotranspiration, the spatial correlations between RegCM results and GLEAM are higher than 0.8, and those for MODIS are higher than 0.6, indicating a reasonable performance for RegCM in evapotranspiration. The spatial correlation between RegCM results and the observed tank evaporation from NIMC is 0.50 and 0.42, respectively, for the two rounds of hindcast simulations. The lower correlation between RegCM and NIMC could be related to the difference between tank evaporation and actual evapotranspiration. In terms of the seasonal variations, RegCM demonstrates overestimations in spring and winter, and underestimations in summer and autumn.

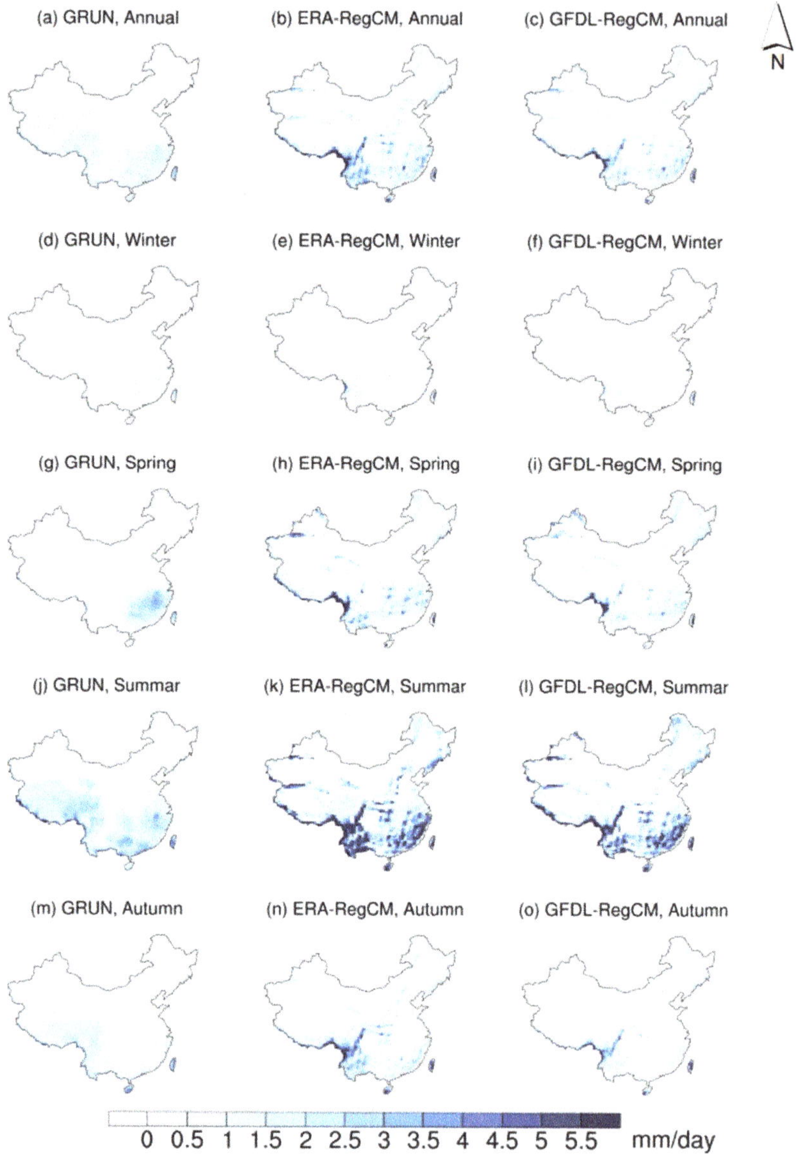

Figure 4. Reconstructed and RegCM-simulated runoff over China. The annual and seasonal averages are calculated with respect to the baseline period of 1986 to 2005.

The RegCM-generated spatial patterns for runoff show considerably larger biases than for other variables. Apparent overestimations can be spotted near the southeastern corner of the Tibetan Plateau, which is likely to be related to the wet bias in precipitation that occurs at the same location. Slight underestimations in runoff can be found over the entire Tibetan Plateau, where overestimations in evapotranspiration can be identified; the latter is likely to be the cause of the former. The spatial correlations between the annual average runoff of the two sets of RegCM results and GRUN are 0.67 and 0.66, respectively (Table S1). From the seasonal perspective, the GRUN reconstructed runoff is high in summer and low in winter. This monsoon feature is well captured by RegCM. As shown by the spatial correlation, for runoff, RegCM shows better skills in summer/autumn (ranges between 0.65 and 0.69) than in winter/spring (between 0.37 and 0.53).

The RegCM-simulated domain average annual cycles for precipitation, evapotranspiration, and runoff are also examined (shown in Figure 5). The two sets of RegCM results present similar features (please refer to Figure S2 in the Supplementary Materials for a direct comparison of the two sets of simulation results). All three variables show a peak in their annual cycles during the monsoon period (May to September), which is consistent with previous results. On domain average, a considerable part of the precipitation is balanced by evapotranspiration, and a smaller portion is attributed to the runoff. The difference between the simulated precipitation and evapotranspiration is also plotted in Figure 5. As indicated by the surface water budget equation, the amount of precipitation that is not balanced by the other two variables contributes to the moisture storage in soil. It can be observed that the difference between precipitation and evapotranspiration is larger than runoff in early spring, and this relationship reverses in autumn, which indicates water storage in spring and dissipation in autumn.

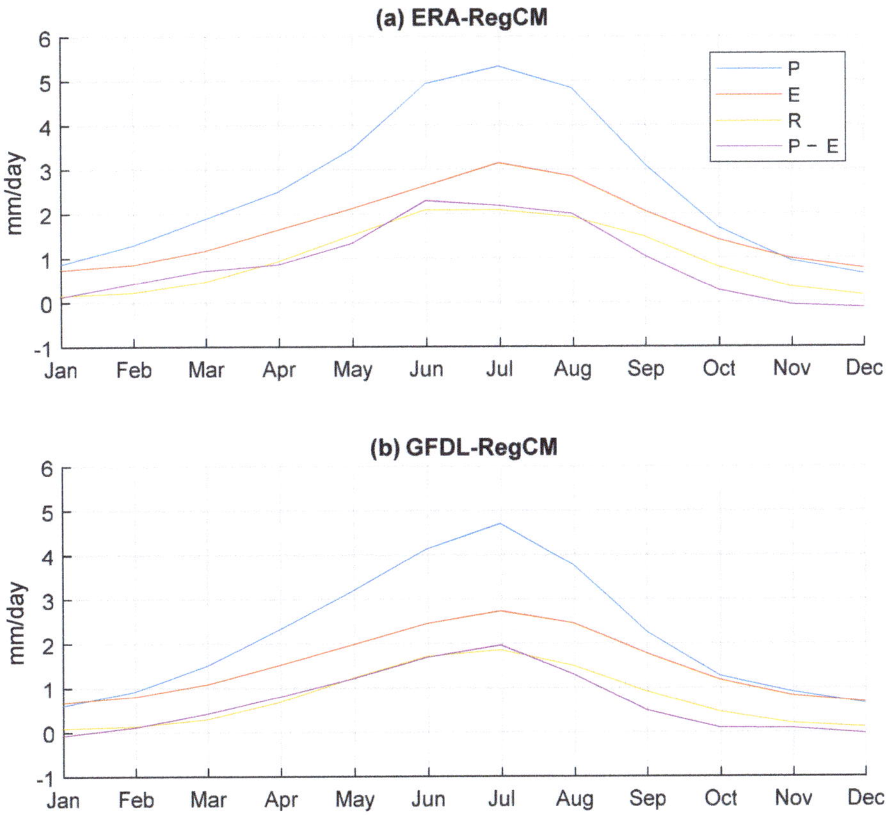

Figure 5. Simulated annual cycles for precipitation, evapotranspiration, runoff, and the difference between precipitation and evapotranspiration in the baseline period.

3.2. *Future Changes in Precipitation, Evapotranspiration, and Runoff*

3.2.1. Precipitation

Having reasonable skills in reproducing the historical climate over China, RegCM is subsequently used to project future changes in water cycle components. The projected changes in precipitation over China in three future periods and under two emission scenarios are shown in Figure 6. The left three columns are for RCP4.5 and the right three for RCP8.5; for each scenario, the three columns, respectively, indicate 2030s, 2050s, and 2080s. The rows in the figure are the annual and seasonal averages. On annual average, increases in precipitation can be found in most parts of the domain. Precipitation changes of the two RCPs show certain similarities. For example, in the 2080s, precipitation increases of larger than 0.3 mm/day are expected in parts of the Tibetan Plateau, Yellow River Basin, Haihe River Basins, Yangtze Plain, and southern coastal hilly regions under both scenarios. In general, the area experiencing increased precipitation is larger under RCP8.5, especially in the Tibetan Plateau, where more pronounced increases (over 0.9 mm/day in the southeastern corner) can be observed as well.

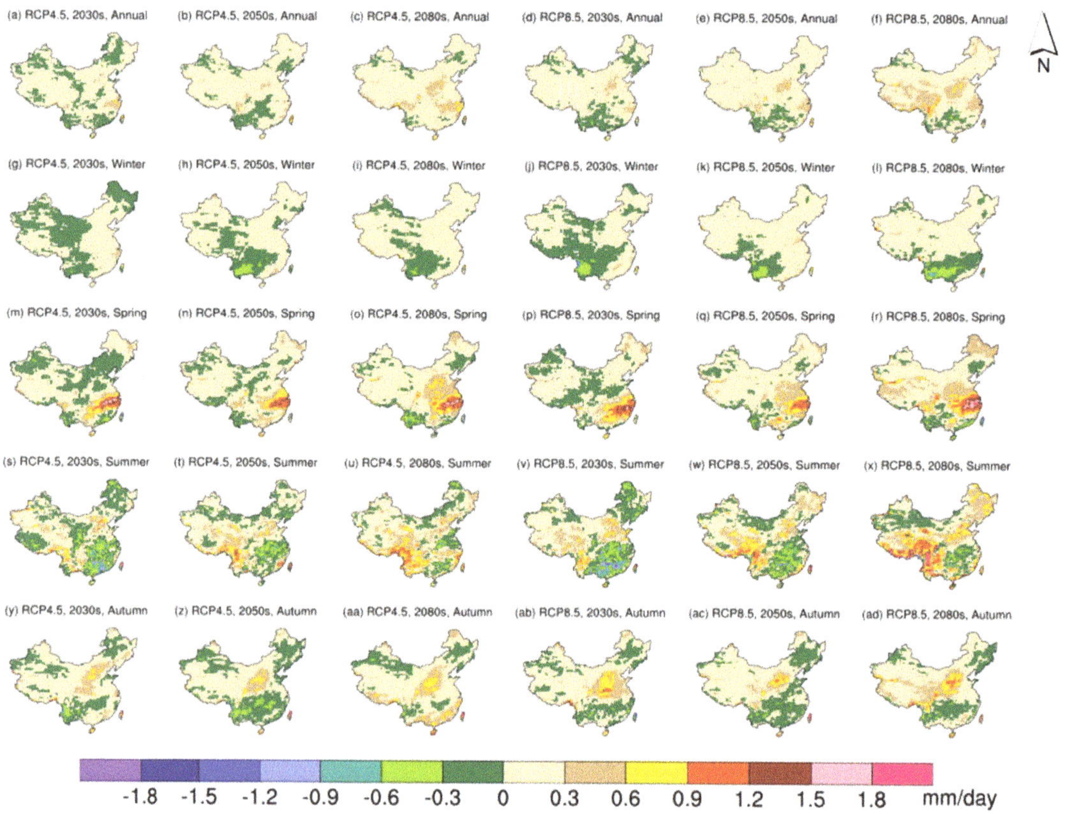

Figure 6. Projected changes in precipitation over China.

Precipitation changes demonstrate apparent seasonal variations. In winter, precipitation decreases of over 0.3 mm/day can be found in parts of the Yunnan–Guizhou Plateau in the 2050s and 2080s under RCP4.5 and in the 2030s and 2050s under RCP8.5. In the 2080s under RCP8.5, more severe changes over larger areas can be noticed; parts of the Pearl River Basin are projected with precipitation decreases of over 0.3 mm/day and parts of the Yunnan–Guizhou Plateau of over 0.6 mm/day. Summer precipitation exhibits similar patterns of changes in the 2050s under the two scenarios, where decreases of over 0.3 mm/day are to be found in the middle and lower reaches of the Yangtze River Basin, and increases of over 0.9 mm/day are expected near the Hengduan Mountains located in the southeastern corner of the Tibetan Plateau. In the 2080s, the spatial distributions of summer precipitation change under the two scenarios are quite different. Under RCP4.5, precipitation decreases of over 0.3 mm/day are likely to occur in the very north of northeastern China, parts of the Yangtze Plain, and parts of the Pearl River Basin, while increases of over 0.9 mm/day are projected in the Hengduan Mountains and the southeastern coastal hilly regions. In comparison, under RCP8.5, decreases in summer precipitation mainly occur in the middle and lower reaches of the Yangtze River Basin (over 0.3 mm/day), while increases of over 0.9 mm/day are to be found in the southern parts of the Tibetan Plateau, Hengduan Mountains, and parts of the Haihe River Basin. The changes in summer precipitation may be related to variations in its major moisture transport branches, which are the transportations by the Indian monsoon, Southeast Asian monsoon, and midlatitude westerlies, as shown by Simmonds [52]. The increases in summer precipitation in southeastern China and decreases in central south and southeastern China under both scenarios could indicate an enhanced Indian summer monsoon and a subsided Southeast Asian summer monsoon. In spring and autumn, precipitation increases are to be seen in the Yangtze Plain (can reach over 1.8 mm/day) and the Yellow River Basin (over 0.9 mm/day), respectively.

The annual series of domain average precipitation under both RCPs are shown in Figure 7. For both scenarios, precipitation demonstrates an evident increasing trend (although a decreasing trend can be observed between 2050 and 2060 under RCP4.5). The Mann–Kendall test confirms the statistical significance of the trends at an α level of 0.05 (when the period of 2010 to 2100 is considered as a whole). The magnitudes of trends, given by Sen's slope estimator, are 0.02 and 0.03 mm/day per decade under RCP4.5 and RCP8.5, respectively (as shown in Table 2). Some seasonal trends are also statistically significant; summer precipitation shows increasing trends of 0.02 and 0.05 mm/day per decade under the two scenarios, and autumn precipitation increases at a rate of 0.02 mm/day per decade under RCP8.5.

Table 2. Future trends (mm/day per decade) of domain average precipitation, evapotranspiration, and runoff (*P*, *E*, and *R* denote precipitation, evapotranspiration, and runoff, respectively).

		Period	Annual	Winter	Spring	Summer	Autumn
P	RCP4.5	2030s	0.02	−0.04	0.02	0.15	−0.05
		2050s	−0.05	0.02	−0.09	−0.06	−0.03
		2080s	0.01	−0.01	0	−0.01	0.02
		2010~2100	**0.02**	0.01	0.02	**0.02**	0.01
	RCP8.5	2030s	0.04	−0.01	0.06	0.05	0.05
		2050s	**0.07**	0	0.01	**0.15**	0.07
		2080s	−0.02	−0.02	0.02	0.03	−0.06
		2010~2100	**0.03**	0.01	0.02	**0.05**	**0.02**
E	RCP4.5	2030s	0.03	0.01	0.02	**0.05**	0.02
		2050s	0.01	0	0.02	0.03	0
		2080s	0	0	0.01	0	−0.01
		2010~2100	**0.01**	**0.01**	**0.02**	**0.01**	**0.01**
	RCP8.5	2030s	**0.03**	0	0.03	**0.05**	0.02
		2050s	**0.03**	0.01	**0.03**	**0.04**	**0.05**
		2080s	0.02	0	0.02	**0.04**	0
		2010~2100	**0.03**	**0.01**	**0.03**	**0.04**	**0.02**
R	RCP4.5	2030s	0	0	−0.02	0.02	−0.01
		2050s	**−0.04**	0	−0.01	**−0.09**	−0.04
		2080s	**0.01**	**0.01**	0.03	−0.02	0.01
		2010~2100	0.01	0.01	0.01	0.01	**0.01**
	RCP8.5	2030s	0.02	0	0.01	0.04	0.05
		2050s	**0.04**	0.01	0.02	**0.07**	**0.04**
		2080s	−0.02	0	0.02	−0.06	−0.01
		2010~2100	0.01	**0.01**	0	0	**0.01**

Notes: Numbers in bold font indicate trends that are statistically significant at α = 0.05.

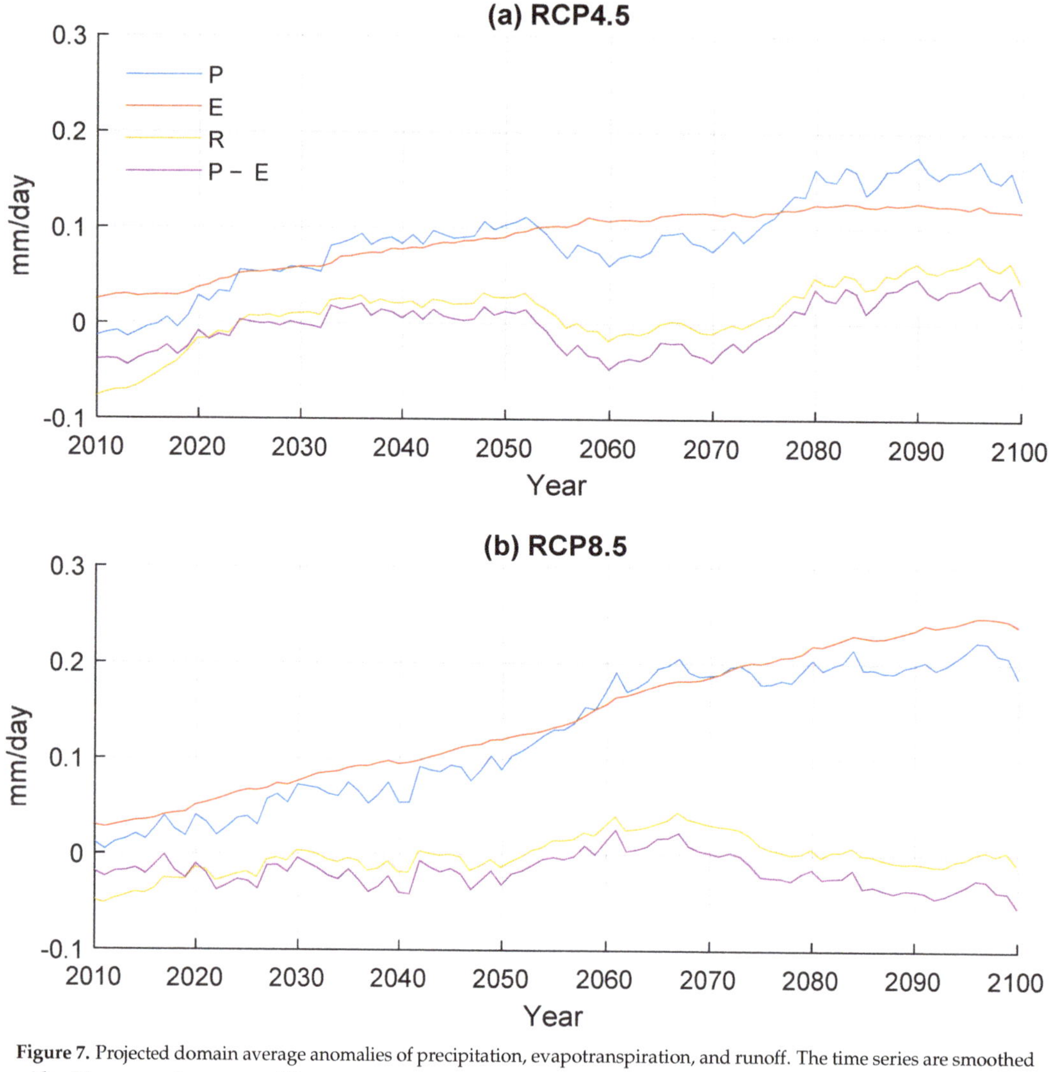

Figure 7. Projected domain average anomalies of precipitation, evapotranspiration, and runoff. The time series are smoothed with a 20-year moving average filter.

Precipitation changes (with respect to the baseline period) in the three future periods under both scenarios are listed in Table 3. Under RCP4.5, the annual average precipitation is projected to increase by 0.06, 0.08, and 0.16 mm/day, respectively, in the three future periods (statistically significant at an α level of 0.05). Under RCP8.5, precipitation change in the 2030s is not statistically significant, and the increases in the 2050s and 2080s are 0.12 and 0.2 mm/day, respectively. Spring and summer precipitation is expected to undergo larger increases than that in winter and autumn. These numbers are consistent with the changes in the annual cycles of precipitation as shown in Figure 8, in which large precipitation increases can be found from April to September in the 2050s and 2080s under both scenarios.

Table 3. Projected changes (mm/day) in precipitation, evapotranspiration, and runoff at different future periods (*P*, *E*, and *R* denote precipitation, evapotranspiration, and runoff, respectively).

		Periods	Annual	Winter	Spring	Summer	Autumn
P	RCP4.5	2030s	**0.06**	0.02	**0.11**	0	**0.1**
		2050s	**0.08**	0.02	**0.15**	**0.11**	0.04
		2080s	**0.16**	0.06	**0.2**	**0.2**	**0.17**
	RCP8.5	2030s	0.07	0	**0.16**	0.01	**0.12**
		2050s	**0.12**	**0.07**	**0.18**	**0.15**	0.08
		2080s	**0.2**	0.03	**0.26**	**0.32**	**0.16**
E	RCP4.5	2030s	**0.06**	0.01	**0.07**	**0.1**	**0.06**
		2050s	**0.1**	**0.04**	**0.11**	**0.16**	**0.08**
		2080s	**0.12**	**0.05**	**0.16**	**0.18**	**0.11**
	RCP8.5	2030s	**0.08**	**0.02**	**0.07**	**0.14**	**0.07**
		2050s	**0.14**	**0.06**	**0.15**	**0.23**	**0.11**
		2080s	**0.22**	**0.08**	**0.25**	**0.38**	**0.18**
R	RCP4.5	2030s	0.01	0.01	**0.05**	**−0.08**	**0.06**
		2050s	0	**0.01**	0.04	**−0.09**	**0.06**
		2080s	**0.05**	**0.03**	**0.06**	0	**0.12**
	RCP8.5	2030s	0	0.01	0.06	**−0.12**	**0.07**
		2050s	0	**0.03**	0.05	**−0.13**	**0.07**
		2080s	0	**0.03**	0.04	**−0.16**	**0.1**

Notes: Numbers in bold font indicate trends that are statistically significant at $\alpha = 0.05$.

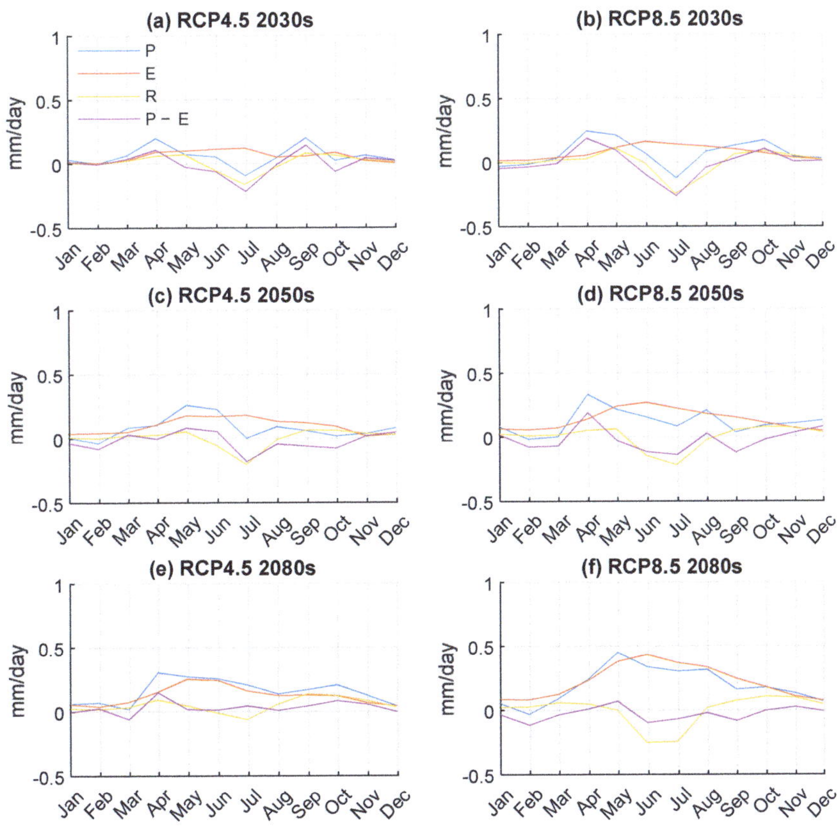

Figure 8. Projected changes in the annual cycles of precipitation, evapotranspiration, and runoff.

Precipitation changes with respect to different warming levels under the two scenarios are shown in Table 4. The magnitudes of changes under the same warming level are similar under different scenarios. For example, with a domain average warming of 2 °C, the annual average precipitation is likely to increase by 0.12 and 0.13 mm/day under RCP4.5 and RCP8.5, respectively. This phenomenon is reasonable because, according to the Clausius–Clapeyron equation, the increase in the water holding capacity of the atmosphere is the same given the same temperature increase. The change in precipitation is not necessarily the same since the actual amount of water vapor available can be different. At a warming level of 2 °C, spring and summer precipitation is also projected to increase by 0.15 and 0.17 mm/day under RCP4.5, and by 0.21 and 0.17 mm/day under RCP8.5. When domain average warming reaches 4 °C under RCP8.5, the annual, spring, and summer precipitation are projected to increase by 0.19, 0.24, and 0.27 mm/day, respectively.

Table 4. Projected changes (mm/day) in precipitation, evapotranspiration, and runoff at different warming levels (*P*, *E*, and *R* denote precipitation, evapotranspiration, and runoff, respectively).

		Warming Level	Annual	Winter	Spring	Summer	Autumn
P	RCP4.5	1 °C	**0.06**	0.02	**0.13**	−0.01	**0.08**
		1.5 °C	**0.08**	0	**0.15**	0.1	**0.08**
		2 °C	**0.12**	0.07	**0.15**	**0.17**	**0.1**
	RCP8.5	1 °C	0.03	−0.02	0.1	0.03	0
		1.5 °C	0.05	0.02	**0.14**	0.01	0.04
		2 °C	**0.13**	0.06	**0.21**	**0.17**	0.08
		3 °C	**0.2**	0.04	**0.22**	**0.34**	**0.19**
		4 °C	**0.19**	0.05	**0.24**	**0.27**	**0.2**
E	RCP4.5	1 °C	**0.06**	0.01	**0.07**	**0.1**	**0.06**
		1.5 °C	**0.07**	**0.02**	**0.08**	**0.13**	**0.06**
		2 °C	**0.12**	**0.05**	**0.16**	**0.19**	**0.09**
	RCP8.5	1 °C	**0.04**	0.02	**0.04**	**0.09**	0.03
		1.5 °C	**0.09**	**0.03**	**0.1**	**0.18**	**0.07**
		2 °C	**0.14**	**0.06**	**0.14**	**0.22**	**0.12**
		3 °C	**0.18**	**0.06**	**0.2**	**0.29**	**0.17**
		4 °C	**0.21**	**0.08**	**0.23**	**0.36**	**0.18**
R	RCP4.5	1 °C	0.01	0.01	**0.06**	**−0.08**	**0.06**
		1.5 °C	0.02	0.01	0.06	−0.05	**0.06**
		2 °C	0.02	**0.02**	0.04	−0.06	**0.08**
	RCP8.5	1 °C	−0.03	−0.01	0.03	**−0.14**	0.01
		1.5 °C	−0.02	0.01	0.03	**−0.16**	**0.05**
		2 °C	0.01	**0.03**	0.06	**−0.11**	**0.08**
		3 °C	**0.04**	**0.03**	0.04	−0.06	**0.13**
		4 °C	0	**0.03**	0.03	**−0.15**	**0.1**

Notes: Numbers in bold font indicate trends that are statistically significant at $\alpha = 0.05$.

3.2.2. Evapotranspiration

The projected changes in evapotranspiration over China are shown in Figure 9. Similar to the spatial pattern of precipitation changes, the annual average evapotranspiration is likely to increase over most parts of the domain. The area experiencing increased evapotranspiration is larger than that for precipitation, but the magnitude of the increase is smaller when compared with precipitation. Annual average evapotranspiration changes are within ±0.3 mm/day for most of the time under both scenarios, except that increases of over 0.3 mm/day are projected in the Hengduan Mountains, Yunnan–Guizhou Plateau, and southeastern coastal hilly regions in the 2080s under RCP8.5. Intra-annual variations can also be observed for evapotranspiration changes. The most pronounced changes are to occur in summer, in which evapotranspiration increases of over 0.3 mm/day can be found over the entire domain except parts in northern and northeastern China in the

2080s under RCP8.5, and increases of over 0.6 mm/day are to be seen in the Hengduan Mountains. In spring, evaporation increases of over 0.3 mm/day are projected in areas between the Yangtze River and Pearl River Basins in the 2080s under RCP4.5, and in most of the southern parts of the domain in the same period under RCP8.5. In autumn, evapotranspiration increases of over 0.3 mm/day can be found in the Hengduan Mountains. Evapotranspiration changes in winter are within ±0.3 mm/day for all future periods under both scenarios.

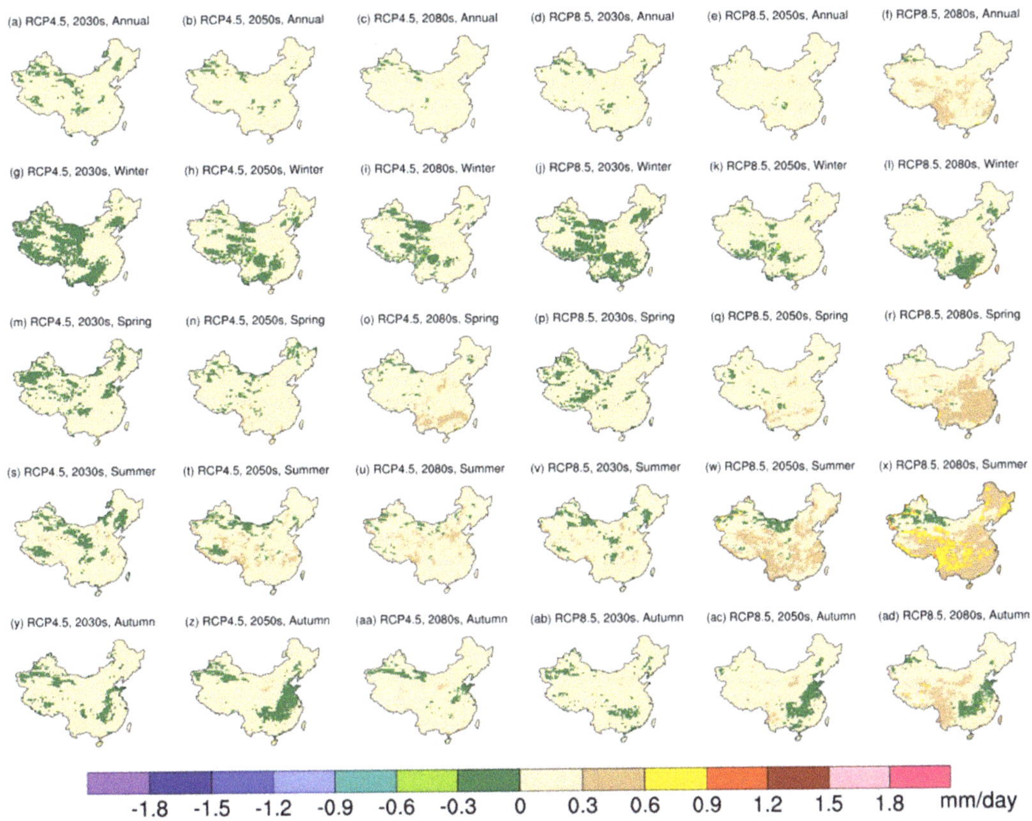

Figure 9. Projected changes in evapotranspiration over China.

The annual average evapotranspiration time series is shown in Figure 7, which appears to be less fluctuating than those of the other two variables. Evident increasing trends can be observed under both scenarios, the magnitudes of which are 0.01 and 0.03 mm/day per decade (statistically significant at an α level of 0.05), respectively (Table 2). All seasonal trends are exclusively statistically significant, which are 0.01, 0.02, 0.01, and 0.01 mm/day per decade, respectively, for winter, spring, summer, and autumn under RCP4.5 and 0.01, 0.03, 0.04, and 0.02 mm/day per decade under RCP8.5. Annual and seasonal evapotranspiration changes in the three future periods under both scenarios are all statistically significant, except for winter in the 2030s under RCP4.5 (Table 3). Under RCP4.5, annual average evapotranspiration is projected to increase by 0.06, 0.1, and 0.12 mm/day, respectively, for the 2030s, 2050s, and 2080s, while, under RCP8.5, the increases are 0.08, 0.14, and 0.22 mm/day. Such increases in evapotranspiration can also be observed in Figure 8, in which evapotranspiration changes are always above zero and larger increases can be found in the monsoon months. At a domain average warming of 2 °C, annual average evapotranspiration is projected to increase by 0.12 and 0.14 mm/day under the two scenarios (Table 4), which are close to the amount of precipitation increases. For the 4 °C warming period under RCP8.5, annual evapotranspiration is likely to increase by 0.21 mm/day, and seasonal evapotranspiration by 0.08, 0.23, 0.36, and 0.18 mm/day for winter, spring, summer, and autumn, respectively.

3.2.3. Runoff

Figure 10 shows the projected changes in future runoff. In general, the annual and seasonal variations in runoff share a certain resemblance with those in precipitation, and only the area experiencing increased runoff is considerably smaller. Runoff changes in most parts of the domain are within ±0.3 mm/day in the three future periods under both scenarios. Under RCP4.5, runoff increases of over 0.3 mm/day can be found in the Yangtze Plain in the 2080s. Under RCP8.5, over 0.3 mm/day decreases in runoff are likely to occur in parts of the Yunnan–Guizhou Plateau in the 2050s and also in parts of the Yangtze River Basin in the 2080s. Seasonal changes in future runoff also demonstrate distinct characteristics. Among all seasons, runoff changes in summer are most severe. For example, under RCP4.5, runoff increases of over 1.2 mm/day are projected in the southeastern coastal hilly regions in the 2080s, which can be related to the increased precipitation in this area. Under RCP8.5, a considerable part of the Yangtze River Basin is projected to experience a runoff reduction of over 0.6 mm/day in the 2080s, which is likely to be caused by the simultaneous decrease in precipitation and increase in evapotranspiration. In winter, changes in runoff are within ±0.3 mm/day in most parts of the domain, with decreases in the central and southwestern parts and increases elsewhere. Spring runoff is projected to increase in the Yangtze Plain by at least over 0.3 mm/day in the future. In autumn, parts of the Yellow River and Haihe River Basins are to receive runoff increases of over 0.3 mm/day. The observed changes in runoff are closely related to the corresponding changes in precipitation.

As shown in Figure 7, the interannual variation in runoff is highly correlated with those in precipitation and precipitation minus evapotranspiration (hereafter as $P - E$). In addition, the changes in $P - E$ are slightly larger than those in runoff, which are consistent with the observations of Zhang et al. [2]. They offered two possible explanations: problems with model spinup or water balance closure, and changes in terrestrial water storage as a result of global warming [2]. The latter case indicates a reduction in future soil moisture, which may bring a negative influence on agriculture. The time series of annual average runoff do not exhibit apparent trends, which is consistent with the lack of a significant trend as shown in Table 2. Previous studies also noticed that the annual runoff series does not demonstrate a trend in the historical period; for example, no statistically significant trend can be detected in the Huaihe River Basin, according to Yu et al. [53]. Significant weak trends of 0.01 mm/day per decade are projected in autumn under RCP4.5 and in winter and autumn under RCP8.5. In comparison with precipitation and evapotranspiration, there is no statistically significant change in annual average runoff except for the 2080s under RCP4.5 (0.05 mm/day). Statistically significant runoff reductions are projected in summer, which are −0.08 and −0.09 mm/day in the 2030s and 2050s under RCP4.5 and −0.12, −0.13, and −0.16 mm/day in the three future periods under RCP8.5. Such a runoff reduction in summer is also evident in Figure 8. These observations are partially consistent with those of Zhang et al. [2], who noticed that changes in precipitation, evapotranspiration, runoff, and $P - E$ mainly occur in the wet season with simultaneous increases in all variables. In this study, the most pronounced changes also occur in the wet season. The projected reduction in runoff and $P - E$ could be related to the difference in the domain selection between this and Zhang et al.'s studies. There is no significant change in annual average runoff under all warming levels and emission scenarios, except that an increase of 0.04 mm/day is projected at a warming level of 3 °C under RCP8.5. At a warming level of 2 °C, runoff changes of 0.02 and 0.08 mm/day are likely to occur in winter and autumn under RCP4.5, and 0.03, −0.11, and 0.08 mm/day in winter, summer, and autumn under RCP8.5. With a domain average warming of 4 °C, runoff in winter, summer, and autumn is to change by 0.03, −0.15, and 0.01 mm/day under RCP8.5.

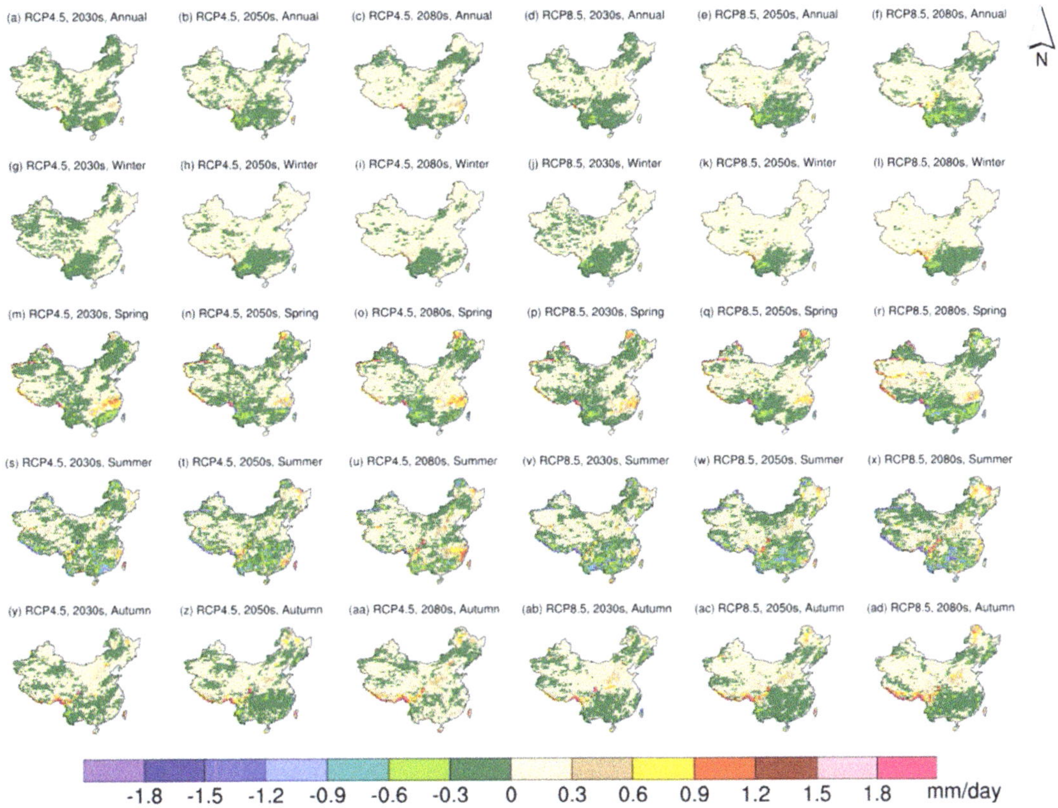

Figure 10. Projected changes in runoff over China.

4. Discussions and Conclusions

In this study, the long-term variations in the water cycle over China are studied using RegCM. The performance of RegCM in terms of temperature, precipitation, evapotranspiration, and runoff is validated through comparisons of the model-generated results with gridded observations, remote sensing data, and reconstructed data. The results show that RegCM can reasonably capture the spatial and seasonal variations in these variables, although certain biases exist, such as a cold bias in the entire domain, a dry and wet bias pair in the southeastern and northwestern parts of the domain, some over- and underestimations of evapotranspiration, respectively, in winter/spring and summer/autumn, and some over- and underestimations of runoff near the Tibetan Plateau.

Long-term projections of precipitation, evapotranspiration, and runoff under two emission scenarios are then developed. The results show that increased annual average precipitation and evapotranspiration can be found in most parts of the domain, while a smaller part of the domain is projected with increased runoff.

For precipitation, the regions most affected by global warming are the Yangtze Plain, Yellow River and Haihe River Basins, and southeastern parts of the Tibetan Plateau, where over 0.3 mm/day increases are expected in the 2080s under both scenarios. The projected increase in precipitation in the Yellow River and Haihe River Basins can also be observed in a CMIP6 GCM ensemble according to Tian et al. [54]. It is worth noting that although the projected precipitation increase in the Tarim Basin is within 0.3 mm/day, the percentage change could be large considering its low annual average total precipitation, which is why several studies (e.g., [55,56]) identified it as an area vulnerable to climate change. Precipitation increase has been shown to be among the driving factors for the increased flood frequency in the Tarim River Basin since the 1980s [57]; the increase in future precipitation as projected by this and the previous studies may indicate increased flood risks in this area, which suggests the need for relevant flood prevention measures.

For evapotranspiration, areas experiencing evident increases are the Hengduan Mountains, Yunnan–Guizhou Plateau, and southeastern coastal hilly regions; the magnitude of change is 0.3 mm/day in the 2080s under RCP8.5. The apparent evapotranspiration increase in southeastern coastal hilly regions is also noted by Su et al. [58]. In terms of seasonal variations, summer and spring are likely to see larger increases in evapotranspiration; this feature is consistent with Ma et al.'s observation [59]. The developed projections in evapotranspiration can be used to evaluate climate change impacts on drought conditions through the calculation of evapotranspiration deficit; this index, compared with those that are based on precipitation and soil moisture, could more effectively reflect moisture deficiency in ecosystems [60].

For runoff, the regions most affected are the Yangtze Delta, Yunnan–Guizhou Plateau, and parts of the Yangtze River Basin, with increases of 0.3 mm/day for the first region in the 2080s under RCP4.5 and decreases of over 0.3 mm/day for the latter two regions in the 2080s under RCP8.5. The projected decrease in runoff in the middle reaches of the Yangtze River Basin is consistent with the results from Xing et al.'s study [61]. In addition, the runoff reduction in the Yunnan–Guizhou Plateau and the upper reaches of the Yangtze River Basin is also noticed by Zhai et al. in their ensemble projection of runoff [62]. Extreme high and low runoff are often related to flood and drought hazards [63], and the runoff projection developed in this study could help identify regions vulnerable to increased flood and drought risks, and thus support flood mitigation and water resource management [64].

In summary, future precipitation and evapotranspiration are likely to increase over China in the wet season, while runoff decreases. The projected changes in precipitation minus evapotranspiration are larger than those in runoff, implying a possible decrease in soil moisture. It is important that future variations in the water cycle components be considered when designing flood and drought mitigation measures.

Extensions of this study can be conducted with respect to the current limitations. For example, more sophisticated bias correction techniques can be applied to the model results. In addition, more GCMs can be used to drive RegCM so that an ensemble can be constructed for more robust projections.

Supplementary Materials: The following are available online at https://www.mdpi.com/article/10.3390/rs13193832/s1, Figure S1: Location of NIMC weather stations. Note: Red dots indicate selected stations for evapotranspiration data. The NIMC tank evaporation data contains a large number of missing data; the stations are selected if the percentage of missing data is less than 50% for each season. Figure S2: Simulated annual cycles for precipitation, evapotranspiration, runoff, and the difference between precipitation and evapotranspiration in the baseline period. The R2's for the two sets of RegCM results are 0.88, 0.92, 0.83, and 0.82, respectively for precipitation, evapotranspiration, runoff, and the difference between precipitation and evapotranspiration. Table S1: Spatial correlation between RegCM-generated results and observations, remote sensing data, and reconstructed data. Table S2: Spatial correlation between raw GFDL data and observations.

Author Contributions: Conceptualization, C.L., G.H. and X.W.; Data curation, C.L. and X.W.; Formal analysis, C.L.; Funding acquisition, G.H.; Investigation, C.L.; Methodology, C.L. and X.W.; Project administration, G.H.; Resources, C.L. and G.H.; Software, C.L. and X.W.; Supervision, G.H.; Validation, C.L., G.H., X.W., G.W., J.Z. and T.S.; Visualization, C.L.; Writing—original draft, C.L.; Writing—review and editing, C.L. and X.W. All authors have read and agreed to the published version of the manuscript.

Funding: This research was supported by the National Key Research and Development Plan (2016YFA0601502), Natural Sciences Foundation (U2040212), Canada Research Chair Program, and Natural Science and Engineering Research Council of Canada.

Data Availability Statement: The gridded observation data are obtained from the Climate Research Unit (https://crudata.uea.ac.uk/cru/data/hrg/; accessed on 18 March 2019). Remote sensing datasets are obtained from the Land Processes Distributed Active Archive Center (https://lpdaac.usgs.gov/; accessed on 9 September 2021) and the Global Land Evaporation Amsterdam Model (https://www.gleam.eu/; accessed on 29 June 2021). The reconstructed runoff data are obtained from Ghiggi et al. [42]. The station-based observations are obtained from the National Meteorological Information Center of China (data.cma.cn; accessed on 9 April 2019). The reanalysis data used to drive RegCM are obtained from the European Centre for Medium-Range Weather Forecasts (https://www.ecmwf.int/en/forecasts/datasets/reanalysis-datasets/era-interim; accessed on 22 November 2017). Global climate model data are obtained from the Geophysical Fluid Dynamics Laboratory and downloaded from the Earth System Grid Federation (https://esgf-node.llnl.gov/search/cmip5/; accessed on 4 April 2017).

Acknowledgments: We are very grateful for the helpful inputs from the editor and anonymous reviewers.

Conflicts of Interest: The authors declare no conflict of interest.

References

1. Sun, Y.; Solomon, S.; Dai, A.; Portmann, R.W. How Often Will It Rain? *J. Clim.* **2007**, *20*, 4801–4818. [CrossRef]
2. Zhang, W.; Zhou, T.; Zhang, L.; Zou, L. Future Intensification of the Water Cycle with an Enhanced Annual Cycle over Global Land Monsoon Regions. *J. Clim.* **2019**, *32*, 5437–5452. [CrossRef]
3. Tan, L.; Schultz, D.M. Damage classification and recovery analysis of the Chongqing, China, floods of August 2020 based on social-media data. *J. Clean. Prod.* **2021**, *313*, 127882. [CrossRef]
4. Trenberth, K.E.; Fasullo, J.T. Regional Energy and Water Cycles: Transports from Ocean to Land. *J. Clim.* **2013**, *26*, 7837–7851. [CrossRef]
5. Hanel, M.; Rakovec, O.; Markonis, Y.; Máca, P.; Samaniego, L.; Kyselý, J.; Kumar, R. Revisiting the recent European droughts from a long-term perspective. *Sci. Rep.* **2018**, *8*, 9499. [CrossRef]
6. Shrestha, N.K.; Wang, J. Water Quality Management of a Cold Climate Region Watershed in Changing Climate. *J. Environ. Inform.* **2019**, *35*, 56–80. [CrossRef]
7. Xie, W.P.; Yang, J.S.; Yao, R.J.; Wang, X.P. Impact Study of Impoundment of the Three Gorges Reservoir on Salt-Water Dynamics and Soil Salinity in the Yangtze River Estuary. *J. Environ. Inform.* **2020**, *36*, 11–23. [CrossRef]
8. Yao, Y.; Huang, G.; An, C.; Chen, X.; Zhang, P.; Xin, X.; Jian, S.; Agnew, J. Anaerobic digestion of livestock manure in cold regions: Technological advancements and global impacts. *Renew. Sustain. Energy Rev.* **2020**, *119*, 109494. [CrossRef]
9. Wang, X.; Wang, P.F.; Wang, C.; Chen, J.; Hou, J.; Miao, L.Z.; Feng, T.; Yuan, Q.S. Taxonomic and Functional Responses of Sediment Bacterial Community to Anthropogenic Disturbances in the Yarlung Tsangpo River on the Tibetan Plateau. *J. Environ. Inform.* **2019**, *35*, 23–33. [CrossRef]
10. Wu, H.P.; Chen, J.; Zeng, G.M.; Xu, J.J.; Sang, L.H.; Liu, Q.; Dai, J.; Xiong, W.P.; Yuan, Z.; Wang, Y.Q.; et al. Effects of Early Dry Season on Habitat Suitability for Migratory Birds in China's Two Largest Freshwater Lake Wetlands after the Impoundment of Three Gorges Dam. *J. Environ. Inform.* **2019**, *36*, 82–92. [CrossRef]
11. Xinhua Net Latest flood on China's Yangtze River Passes Three Gorges Dam. 2020. Available online: www.xinhuanet.com/english/2020-08/22/c_139310599.htm (accessed on 16 July 2021).
12. China News 63 Million People Affected and 219 Killed in This Year's Flood Disaster. 2020. Available online: www.chinanews.com/sh/2020/08-13/9263764.shtml (accessed on 16 July 2021). (In Chinese)
13. Li, M.; Jiang, Z.; Zhou, P.; Le Treut, H.; Li, L. Projection and possible causes of summer precipitation in eastern China using self-organizing map. *Clim. Dyn.* **2020**, *54*, 2815–2830. [CrossRef]
14. Dong, Q.; Wang, W.; Shao, Q.; Xing, W.; Ding, Y.; Fu, J. The response of reference evapotranspiration to climate change in Xinjiang, China: Historical changes, driving forces, and future projections. *Int. J. Climatol.* **2020**, *40*, 235–254. [CrossRef]
15. Yan, Z.; Zhou, Z.; Liu, J.; Han, Z.; Gao, G.; Jiang, X. Ensemble Projection of Runoff in a Large-Scale Basin: Modeling with a Global BMA Approach. *Water Resour. Res.* **2020**, *56*, e2019WR026134. [CrossRef]
16. Gu, L.; Chen, J.; Yin, J.; Xu, C.; Zhou, J. Responses of Precipitation and Runoff to Climate Warming and Implications for Future Drought Changes in China. *Earth's Future* **2020**, *8*, e2020EF001718. [CrossRef]
17. Giorgi, F. Thirty Years of Regional Climate Modeling: Where Are We and Where Are We Going next? *J. Geophys. Res. Atmos.* **2019**, *124*, 5696–5723. [CrossRef]

18. Giorgi, F.; Hewitson, B.; Arritt, R.; Gutowski, W.; Knutson, T.; Landsea, C.; Christensen, J.; Hulme, M.; Von Storch, H.; Whetton, J.P.; et al. Regional Climate Information—Evaluation and Projections. In *Climate Change 2001: The Scientific Bases*; Cambridge University Press: Cambridge, UK, 2001; pp. 583–638.
19. Walton, D.; Berg, N.; Pierce, D.; Maurer, E.; Hall, A.; Lin, Y.; Rahimi, S.; Cayan, D. Understanding Differences in California Climate Projections Produced by Dynamical and Statistical Downscaling. *J. Geophys. Res. Atmos.* **2020**, *125*, e2020JD032812. [CrossRef]
20. Fan, D.; Jiang, Z.; Tian, Z.; Dong, G.; Sun, L. The Effects of Climate Change on Chinese Medicinal Yam Over North China Under the High-Resolution PRECIS Projection. *Earth Space Sci.* **2021**, *8*, e2021EA001804. [CrossRef]
21. Ashfaq, M.; Cavazos, T.; Reboita, M.S.; Torres-Alavez, J.A.; Im, E.-S.; Olusegun, C.F.; Alves, L.; Key, K.; Adeniyi, M.O.; Tall, M.; et al. Robust late twenty-first century shift in the regional monsoons in RegCM-CORDEX simulations. *Clim. Dyn.* **2021**, *57*, 1463–1488. [CrossRef]
22. Lu, C.; Huang, G.; Wang, X. Projected changes in temperature, precipitation, and their extremes over China through the RegCM. *Clim. Dyn.* **2019**, *53*, 5859–5880. [CrossRef]
23. Pan, X.D.; Zhang, L.; Huang, C.L. Future Climate Projection in Northwest China with RegCM4.6. *Earth Space Sci.* **2020**, *7*, e2019EA000819. [CrossRef]
24. Gao, S. Dynamical downscaling of surface air temperature and precipitation using RegCM4 and WRF over China. *Clim. Dyn.* **2020**, *55*, 1283–1302. [CrossRef]
25. Jiang, Z.; Tian, Z.; Dong, G.; Sun, L.; Zhang, P.; Buonomo, E.; Fan, D. High-Resolution Projections of Mean and Extreme Precipitation over China by Two Regional Climate Models. *J. Meteorol. Res.* **2020**, *34*, 965–985. [CrossRef]
26. Grell, G.A.; Dudhia, J.; Stauffer, D.R. A Description of the Fifth-Generation Penn State/NCAR Mesoscale Model (MM5). 1994. Available online: https://opensky.ucar.edu/islandora/object/technotes:170 (accessed on 15 July 2021).
27. Steiner, A.L.; Pal, J.S.; Giorgi, F.; Dickinson, R.E.; Chameides, W.L. The coupling of the Common Land Model (CLM0) to a regional climate model (RegCM). *Theor. Appl. Climatol.* **2005**, *82*, 225–243. [CrossRef]
28. Oleson, K.W.; Lawrence, D.M.; Bonan, G.B.; Drewniak, B.; Huang, M.; Koven, C.D.; Levis, S.; Li, F.; Riley, W.J.; Subin, Z.M.; et al. Technical Description of Version 4.5 of the Community Land Model (CLM). 2013. Available online: https://opensky.ucar.edu/islandora/object/technotes:515 (accessed on 29 August 2021).
29. Niu, G.-Y.; Yang, Z.-L.; Dickinson, R.E.; Gulden, L.E. A simple TOPMODEL-based runoff parameterization (SIMTOP) for use in global climate models. *J. Geophys. Res.* **2005**, *110*, D21106. [CrossRef]
30. Branstetter, M.L. *Development of a Parallel River Transport Algorithm and Applications to Climate Studies*; The University of Texas at Austin: Austin, TX, USA, 2001.
31. Dee, D.P.; Uppala, S.M.; Simmons, A.J.; Berrisford, P.; Poli, P.; Kobayashi, S.; Andrae, U.; Balmaseda, M.A.; Balsamo, G.; Bauer, P.; et al. The ERA-Interim reanalysis: Configuration and performance of the data assimilation system. *Q. J. R. Meteorol. Soc.* **2011**, *137*, 553–597. [CrossRef]
32. Dunne, J.P.; John, J.G.; Adcroft, A.J.; Griffies, S.M.; Hallberg, R.W.; Shevliakova, E.; Stouffer, R.J.; Cooke, W.; Dunne, K.A.; Harrison, M.J.; et al. GFDL's ESM2 Global Coupled Climate–Carbon Earth System Models. Part I: Physical Formulation and Baseline Simulation Characteristics. *J. Clim.* **2012**, *25*, 6646–6665. [CrossRef]
33. Dunne, J.P.; John, J.G.; Shevliakova, E.; Stouffer, R.J.; Krasting, J.P.; Malyshev, S.L.; Milly, P.C.D.; Sentman, L.T.; Adcroft, A.J.; Cooke, W.; et al. GFDL's ESM2 Global Coupled Climate–Carbon Earth System Models. Part II: Carbon System Formulation and Baseline Simulation Characteristics. *J. Clim.* **2013**, *26*, 2247–2267. [CrossRef]
34. Harris, I.; Osborn, T.J.; Jones, P.; Lister, D. Version 4 of the CRU TS monthly high-resolution gridded multivariate climate dataset. *Sci. Data* **2020**, *7*, 109. [CrossRef]
35. Sun, Q.; Miao, C.; Duan, Q.; Ashouri, H.; Sorooshian, S.; Hsu, K. A Review of Global Precipitation Data Sets: Data Sources, Estimation, and Intercomparisons. *Rev. Geophys.* **2018**, *56*, 79–107. [CrossRef]
36. Running, S.; Mu, Q.; Zhao, M.; Moreno, A. MOD16A2GF MODIS/Terra Net Evapotranspiration Gap-Filled 8-Day L4 Global 500 m SIN Grid V006 [Data Set]. 2019. Available online: https://lpdaac.usgs.gov/ (accessed on 9 September 2021).
37. Martens, B.; Miralles, D.G.; Lievens, H.; van der Schalie, R.; de Jeu, R.A.M.; Fernández-Prieto, D.; Beck, H.E.; Dorigo, W.A.; Verhoest, N.E.C. GLEAM v3: Satellite-based land evaporation and root-zone soil moisture. *Geosci. Model Dev.* **2017**, *10*, 1903–1925. [CrossRef]
38. Miralles, D.G.; Holmes, T.R.H.; De Jeu, R.A.M.; Gash, J.H.; Meesters, A.G.C.A.; Dolman, A.J. Global land-surface evaporation estimated from satellite-based observations. *Hydrol. Earth Syst. Sci.* **2011**, *15*, 453–469. [CrossRef]
39. Mu, Q.; Zhao, M.; Running, S.W. MODIS Global Terrestrial Evapotranspiration (ET) Product (NASA MOD16A2/A3) Algorithm Theoretical Basis Document Collection 5; Missoula. 2013. Available online: https://modis-land.gsfc.nasa.gov/pdf/MOD16ATBD.pdf (accessed on 8 September 2021).
40. Liu, Z.; Shao, Q.; Liu, J. The Performances of MODIS-GPP and -ET Products in China and Their Sensitivity to Input Data (FPAR/LAI). *Remote Sens.* **2014**, *7*, 135–152. [CrossRef]
41. Yang, X.; Yong, B.; Ren, L.; Zhang, Y.; Long, D. Multi-scale validation of GLEAM evapotranspiration products over China via ChinaFLUX ET measurements. *Int. J. Remote Sens.* **2017**, *38*, 5688–5709. [CrossRef]
42. Ghiggi, G.; Humphrey, V.; Seneviratne, S.I.; Gudmundsson, L. GRUN: An observation-based global gridded runoff dataset from 1902 to 2014. *Earth Syst. Sci. Data* **2019**, *11*, 1655–1674. [CrossRef]

43. Li, J.; Miao, C.; Wei, W.; Zhang, G.; Hua, L.; Chen, Y.; Wang, X. Evaluation of CMIP6 Global Climate Models for Simulating Land Surface Energy and Water Fluxes During 1979–2014. *J. Adv. Model. Earth Syst.* **2021**, *13*, e2021MS002515. [CrossRef]
44. Ding, Y.; Xu, J.; Wang, X.; Cai, H.; Zhou, Z.; Sun, Y.; Shi, H. Propagation of meteorological to hydrological drought for different climate regions in China. *J. Environ. Manag.* **2021**, *283*, 111980. [CrossRef]
45. Luo, Y.; Yang, Y.; Yang, D.; Zhang, S. Quantifying the impact of vegetation changes on global terrestrial runoff using the Budyko framework. *J. Hydrol.* **2020**, *590*, 125389. [CrossRef]
46. Trenberth, K.E.; Dai, A.; Rasmussen, R.M.; Parsons, D.B. The Changing Character of Precipitation. *Bull. Am. Meteorol. Soc.* **2003**, *84*, 1205–1218. [CrossRef]
47. Mann, H.B. Nonparametric Tests Against Trend. *Econometrica* **1945**, *13*, 245. [CrossRef]
48. Kendall, M.G. *Rank Correlation Methods*, 4th ed.; Charles Griffin: London, UK, 1975.
49. Theil, H. A rank-invariant method of linear and polynomial regression analysis. In *Henri Theil's Contributions to Economics and Econometrics*; Springer: Dordrecht, The Netherlands; Volume 23, pp. 345–381.
50. Sen, P.K. Estimates of the Regression Coefficient Based on Kendall's Tau. *J. Am. Stat. Assoc.* **1968**, *63*, 1379. [CrossRef]
51. Khan, M.S.; Liaqat, U.W.; Baik, J.; Choi, M. Stand-alone uncertainty characterization of GLEAM, GLDAS and MOD16 evapotranspiration products using an extended triple collocation approach. *Agric. For. Meteorol.* **2018**, *252*, 256–268. [CrossRef]
52. Simmonds, I.; Bi, D.; Hope, P. Atmospheric Water Vapor Flux and Its Association with Rainfall overChina in Summer. *J. Clim.* **1999**, *12*, 1353–1367. [CrossRef]
53. Yu, B.Y.; Wu, P.; Sui, J.; Ni, J.; Whitcombe, T. Variation of Runoff and Sediment Transport in the Huai River—A Case Study. *J. Environ. Inform.* **2020**, *35*, 138–147. [CrossRef]
54. Tian, J.; Zhang, Z.; Ahmed, Z.; Zhang, L.; Su, B.; Tao, H.; Jiang, T. Projections of precipitation over China based on CMIP6 models. *Stoch. Environ. Res. Risk Assess.* **2021**, *35*, 831–848. [CrossRef]
55. Xu, H.; Chen, H.; Wang, H. Future changes in precipitation extremes across China based on CMIP6 models. *Int. J. Climatol.* **2021**. [CrossRef]
56. Wu, Y.; Guo, J.; Lin, H.; Bai, J.; Wang, X. Spatiotemporal patterns of future temperature and precipitation over China projected by PRECIS under RCPs. *Atmos. Res.* **2021**, *249*, 105303. [CrossRef]
57. Zhang, Q.; Gu, X.; Singh, V.P.; Sun, P.; Chen, X.; Kong, D. Magnitude, frequency and timing of floods in the Tarim River basin, China: Changes, causes and implications. *Glob. Planet. Chang.* **2016**, *139*, 44–55. [CrossRef]
58. Su, B.; Zhao, J.; Wang, Y.; Tao, H.; Gao, C.; Liu, F.; Li, X.; Jiang, T. Spatial and Temporal Variation of Actual Evapotranspiration in China under the 1.5 °C and 2.0 °C Global Warming Scenarios (In Chinese). *Chin. J. Agrometeorol.* **2018**, *39*, 293–303. [CrossRef]
59. Ma, X.; Zhao, C.; Tao, H.; Zhu, J.; Kundzewicz, Z.W. Projections of actual evapotranspiration under the 1.5 °C and 2.0 °C global warming scenarios in sandy areas in northern China. *Sci. Total Environ.* **2018**, *645*, 1496–1508. [CrossRef]
60. Kim, D.; Ha, K.; Yeo, J. New Drought Projections Over East Asia Using Evapotranspiration Deficits From the CMIP6 Warming Scenarios. *Earth's Future* **2021**, *9*, e2020EF001697. [CrossRef]
61. Xing, W.; Wang, W.; Zou, S.; Deng, C. Projection of future runoff change using climate elasticity method derived from Budyko framework in major basins across China. *Glob. Planet. Chang.* **2018**, *162*, 120–135. [CrossRef]
62. Zhai, R.; Tao, F.; Xu, Z. Spatial–temporal changes in runoff and terrestrial ecosystem water retention under 1.5 and 2 °C warming scenarios across China. *Earth Syst. Dyn.* **2018**, *9*, 717–738. [CrossRef]
63. Zhai, R.; Tao, F.; Lall, U.; Fu, B.; Elliott, J.; Jägermeyr, J. Larger Drought and Flood Hazards and Adverse Impacts on Population and Economic Productivity Under 2.0 than 1.5 °C Warming. *Earth's Future* **2020**, *8*, e2019EF001398. [CrossRef]
64. Kundzewicz, Z.; Su, B.; Wang, Y.; Xia, J.; Huang, J.; Jiang, T. Flood risk and its reduction in China. *Adv. Water Resour.* **2019**, *130*, 37–45. [CrossRef]

remote sensing

MDPI

Communication

Glacier Velocity Changes in the Himalayas in Relation to Ice Mass Balance

Yu Zhou [1], Jianlong Chen [1,2] and Xiao Cheng [3,*]

1 Southern Marine Science and Engineering Guangdong Laboratory (Zhuhai), School of Earth Sciences and Engineering, Sun Yat-sen University, Zhuhai 519000, China; zhouyu36@mail.sysu.edu.cn (Y.Z.); chenjlong25@mail2.sysu.edu.cn (J.C.)
2 Guangdong Geological Survey Institute, Guangzhou 510062, China
3 Southern Marine Science and Engineering Guangdong Laboratory (Zhuhai), School of Geospatial Engineering and Science, Sun Yat-sen University, Zhuhai 519000, China
* Correspondence: chengxiao9@mail.sysu.edu.cn

Abstract: Glacier evolution with time provides important information about climate variability. Here, we investigated glacier velocity changes in the Himalayas and analysed the patterns of glacier flow. We collected 220 scenes of Landsat-7 panchromatic images between 1999 and 2000, and Sentinel-2 panchromatic images between 2017 and 2018, to calculate surface velocities of 36,722 glaciers during these two periods. We then derived velocity changes between 1999 and 2018 for the early winter period, based on which we performed a detailed analysis of motion of each individual glacier, and noted that the changes are spatially heterogeneous. Of all the glaciers, 32% have sped up, 24.5% have slowed down, and the rest 43.5% have remained stable. The amplitude of glacier slowdown, as a result of glacier mass loss, is significantly larger than that of speedup. At regional scales, we found that glacier surface velocity in winter has uniformly decreased in the western part of the Himalayas between 1999 and 2018, while increased in the eastern part; this contrasting difference may be associated with decadal changes in accumulation and/or melting under different climatic regimes. We also found that the overall trend of surface velocity exhibits seasonal variability: summer velocity changes are positively correlated with mass loss, i.e., velocity increases with increasing mass loss, whereas winter velocity changes show a negative correlation. Our study suggests that glacier velocity changes in the Himalayas are spatially and temporally heterogeneous, in agreement with studies that previously highlighted this trend, emphasising complex interactions between glacier dynamics and environmental forcing.

Keywords: glacier velocity changes; Himalayas; Landsat-7; Sentinel-2; ice mass balance

Citation: Zhou, Y.; Chen, J.; Cheng, X. Glacier Velocity Changes in the Himalayas in Relation to Ice Mass Balance. *Remote Sens.* **2021**, *13*, 3825. https://doi.org/10.3390/rs13193825

Academic Editors: Yi Luo and Gareth Rees

Received: 20 August 2021
Accepted: 22 September 2021
Published: 24 September 2021

1. Introduction

Glaciers are sensitive to climate variability and are a major contributor to global sea level rise [1–7]. It is of great importance to understand glacier evolution with time because it provides direct evidence for climate change [2,8,9]. The Himalayas host the largest volume of glaciers outside the polar regions, which also contribute importantly to water resources for the Indus and Ganges basins [10–12]. Due to the difficult accessibility of high mountain areas, remote sensing has been a powerful tool for studying the Himalayan glaciers. Researchers have used satellite altimetry (e.g., [2,13,14]) and optical satellite stereo imagery (e.g., [7,15,16]) to quantify glacier mass balance in the Himalayas. Although the estimates derived from different techniques vary, they consistently show that the Himalayan glaciers are experiencing significant thinning and mass loss, thereby affecting ice fluxes and river discharge. The thinning rate is also suggested to have accelerated in the past 40 years, which is possibly driven by atmospheric warming and associated energy fluxes [15–17]. Recently, Dehecq et al. [9] investigated the response of glacier flow to mass changes at regional scales. They estimated time-series glacier velocities from 2000 to 2016, using Landsat-7 optical satellite images [18], and found that the variability in velocity changes within a large region can be explained solely by changes in ice thickness, i.e., ice mass balance [9]. Their study provides a novel way for estimating ice mass balance in the Himalayas as glacier velocity changes can be easily measured with satellite images.

Glacier surface velocity in summer has been heavily exploited, e.g., [9,18]. Velocity estimates in [9] show that glaciers in the Himalayas have experienced significant slowdown in the past two decades. However, seasonal variability of regional glacier motion remains unclear. The aim of this study is to explore the long-term winter velocity and its changes. We first derived glacier velocities for two periods, early winter in 1999–2000 and 2017–2018, using Landsat-7 and Sentinel-2 image pairs, respectively. High-resolution satellites provide a powerful tool for monitoring ice and snow [19,20]. We chose Sentinel-2 over Landsat-7 for mapping the present glacier motion because it has been tested to have a better geometric and radiometric quality [21]. By differencing the Landsat-7 and Sentinel-2 derived velocities, we map velocity changes over nearly two decades, and, combining the data with glacier mass balance, we show the complex patterns of glacier flow in the Himalayas.

2. Study Area

The Himalaya front (Figure 1) stretches over 3000 km from the west to the east, containing more than 36,000 glaciers of different sizes (Randolph Glacier Inventory, RGI 6.0). The topography increases rapidly across the front, from 200 m in the south to over 5000 m in the north, entering the Tibetan Plateau. Evolution of glaciers in different parts of the Himalayas is affected by different climatic regimes. In the western part, snow accumulation is controlled by westerly atmospheric circulations, so Hindu Kush, Spiti Lahaul and Karakoram receive most accumulation during winter. In the eastern part, the Indian summer monsoon dominates the accumulation in West Nepal, East Nepal, Bhutan and Nyainqentanglha [2,12,22–24]. The extreme topography creates additional complexity; precipitation at high-altitude regions has been suggested to be 2–10 times higher than that at low-altitude regions [25,26]. As a result, glaciers in the Himalayan front exhibit contrasting variabilities in evolution and mass balance [2,7,9]. Shaded relief of the Himalaya region generated from the 3 arc second Shuttle Radar Topography Mission (SRTM) DEM [27] is shown in Figure 1.

3. Data and Methods

In this study, we focused primarily on surface velocity changes of the Himalayan glaciers at decadal scales. Satellite optical images were used to generate glacier velocity maps at different times. We collected 40 pairs of Landsat-7 Level-1T data between 1999 and 2000 to calculate glacier velocity during this period. Each panchromatic Landsat-7 Level-1T image covers an area of 185 km × 170 km, with a spatial resolution of 15 m. A total of 70 panchromatic Sentinel-2A/B Level-1C image pairs were obtained to calculate glacier velocity between 2017 and 2018. Each Sentinel-2 Level-1C image has a footprint of 100 km × 100 km and a spatial resolution of 10 m. Dehecq et al. [9] analysed velocity changes in summer. We are interested to explore whether glacier velocities exhibit seasonal variations, so all the 110 pairs of Landsat-7 and Sentinel-2 images used in this study were acquired during winter, centred around December. We also collected glacier geometry data including length, area, slope and thickness from RGI 6.0 and [28], along with satellite-derived glacier elevation changes from [7], for a comprehensive analysis of glacier velocity changes and the possible driving factors.

We estimated glacier velocities by applying cross-correlation using the COSI-Corr software package. Optical correlation is implemented in the frequency domain with an accuracy ~1/10 of the input pixel size [29,30]. We used a correlation window of 64 pixels × 64 pixels as a first step, followed by 32 pixels × 32 pixels for a second run, with a step of 16 pixels × 16 pixels (160 m) for Sentinel-2A/B data and 10 pixels × 10 pixels (150 m) for Landsat-7 data. The resulting east-west and north-south components of the displacement were filtered using the non-local means algorithm [30]. The purpose of filtering is to use the correlation to excluded snow-covered regions where the correlation is relatively low. We then used RGI 6.0 to mask out the non-glacial areas and generated two annual velocity fields for 1999–2000 and 2017–2018, for all the 36,722 glaciers within the Himalayas, as shown in Figure 2.

The velocity uncertainty was estimated from the overlapping areas between adjacent pairs. As shown in Figure 3, the Landsat-7 and Sentinel-2 derived velocities have a measurement error of -3.4 ± 11.6 m yr^{-1} and 0.4 ± 4.6 m yr^{-1}, respectively. Since all the image pairs are centred around December with little interannual fluctuations, the uncertainty depends largely on radiometric quality and image resolution, indicating that Sentinel-2 imagery has a better radiometric quality than Landsat-7 imagery [21], and higher resolution improves pixel matching [20], hence yielding more consistent results.

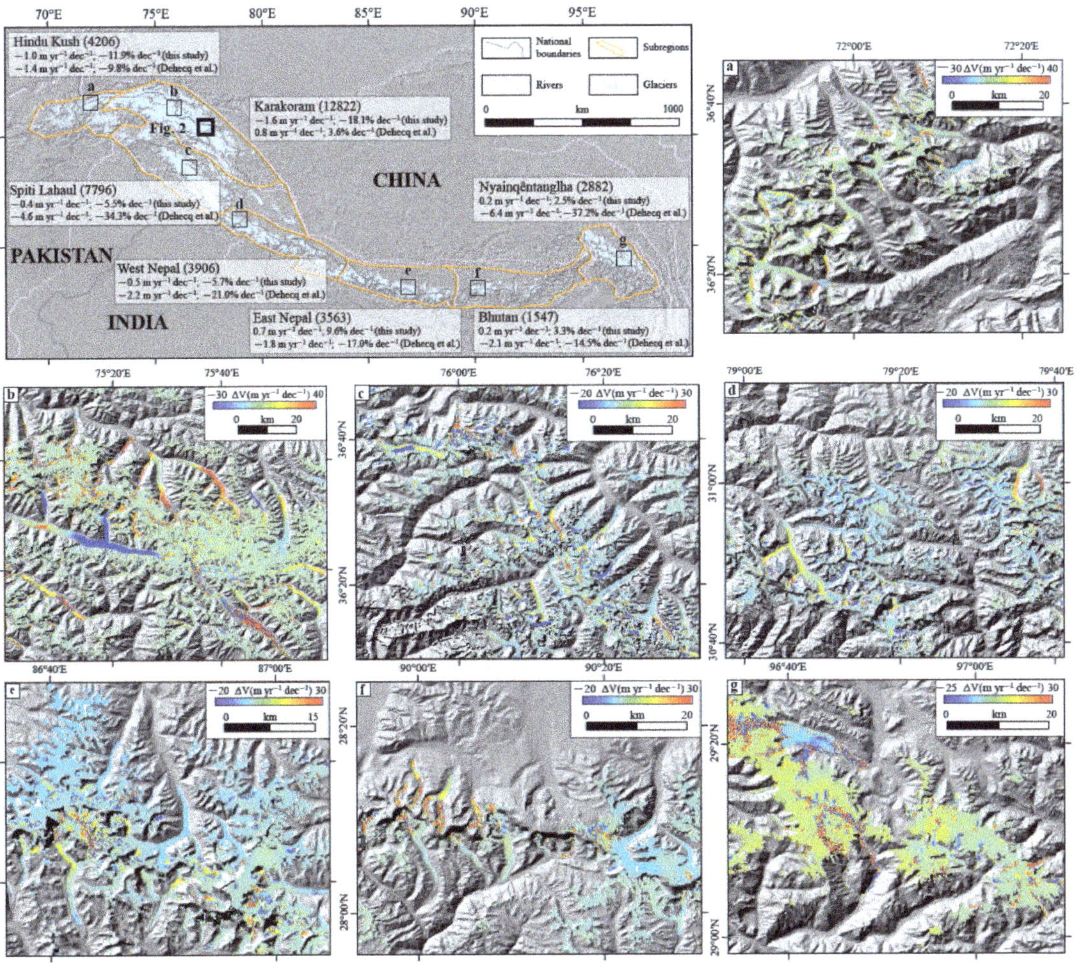

Figure 1. Glaciers are highlighted in light blue. The total number of glaciers in each subregion is labelled. Black box marks the location of Figure 2. We calculated the mean value of velocity changes of each subregion and compared our estimates to [9]. (**a–g**) shows examples of velocity changes in subregions. See Table 1 for details.

To calculate velocity changes, we re-sampled the Landsat-7 and Sentinel-2 derived velocity maps onto the same grid with a spacing of 160 m, and then differenced the two velocities on a pixel basis (Figure 1). Unlike [9], who based their analysis of glacier velocity changes mainly on regions, we conducted our analysis based on individual glacier. For each glacier, we computed its velocity and the associated change by averaging all the pixel values that cover the glacier.

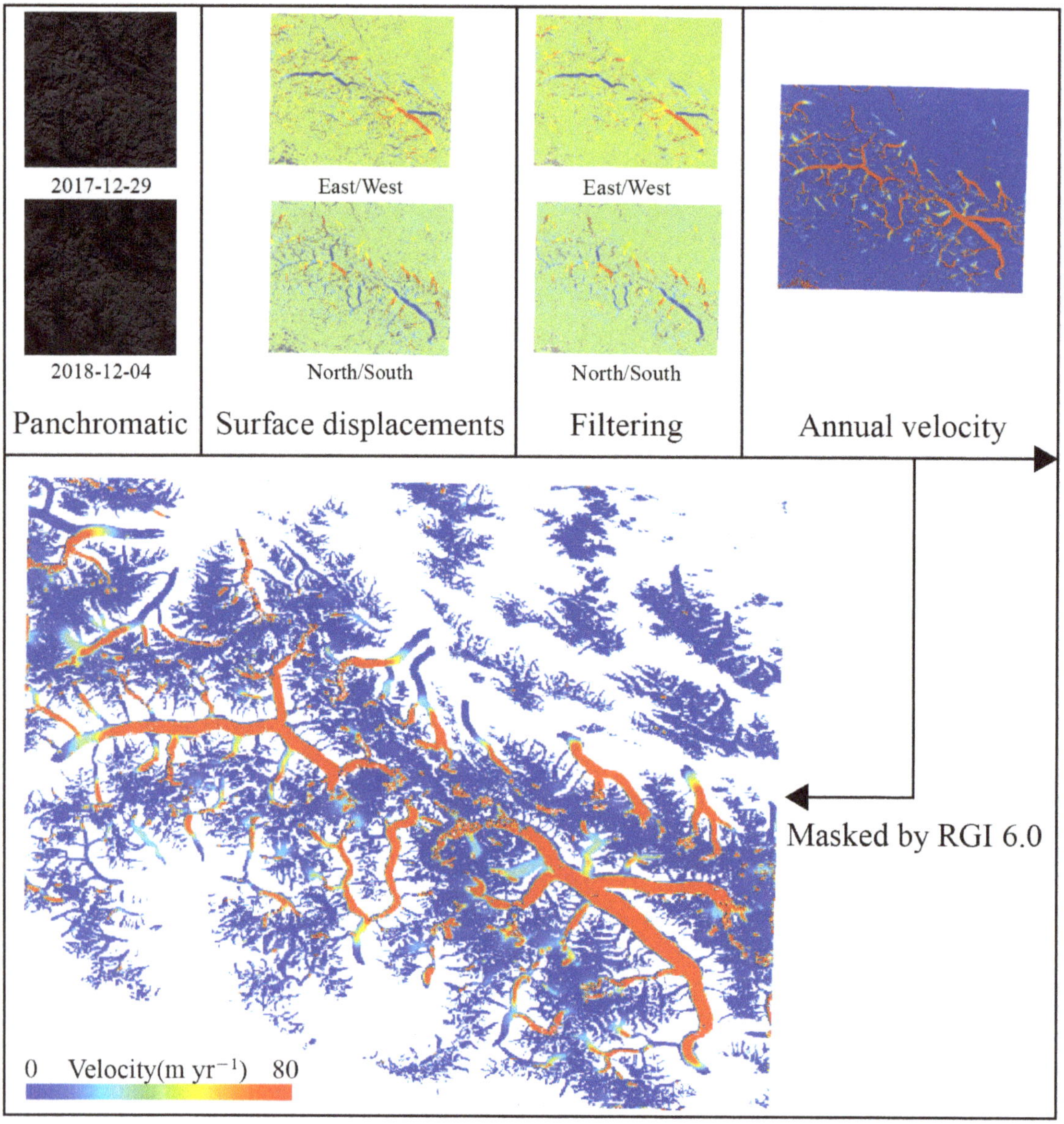

Figure 2. Deriving glacier surface velocity using optical correlation. The Sentinel-2 optical data were provided by the European Space Agency (ESA). See Figure 1 for location.

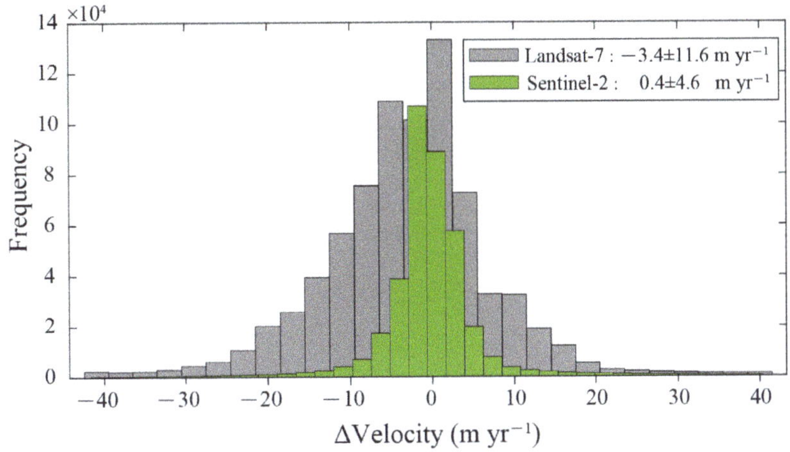

Figure 3. Velocity uncertainties. The Landsat-7 and Sentinel-2 derived velocities have a measurement error of -3.4 ± 11.6 m yr^{-1} and 0.4 ± 4.6 m yr^{-1}, respectively.

4. Results

A total of 36,722 glaciers were included in our study (from the RGI 6.0 database). As shown in Table 1, 43.5% of the glaciers had a stable velocity during 1999–2018 (the difference between Landsat-7 and Sentinel-2 velocities is no more than 3 m yr^{-1}), 32.0% sped up (velocity changes > 3 m yr^{-1}) and 24.5% slowed down (velocity changes < -3 m yr^{-1}). Although speedup glaciers outnumber slowdown glaciers, the average amplitude of glacier speedup (6.3 m yr^{-1}) is much smaller than slowdown (-12.3 m yr^{-1}). At regional scale, more glaciers in the east of the Himalayas, i.e., West Nepal (33.4%), East Nepal (56.2%), Bhutan (34.1%) and Nyainqentanglha (41.3%), have experienced speedup than the west in Hindu Kush (22.6%), Spiti Lahaul (19.6%) and Karakoram (17.3%). Karakoram has seen the largest proportion of glacier slowdown (31.5%), followed by West Nepal (29.1%), Hindu Kush (28.3%), Bhutan (22.1%), Nyainqentanglha (21.9%), East Nepal (19.6%) and Spiti Lahaul (18.7%). The greatest speedup (about 130 m yr^{-1} in RGI60-14.04875) and slowdown (about -250 m yr^{-1} in RGI60-14.0300) also occurred in Karakoram.

We analysed glacier surface velocity in combination with glacier geometry data including area, length, slope and thickness. We used Sentinel-2 derived velocities in the analysis, given the smaller measurement uncertainties. The results (Figure 4) showed that small glaciers (area < 20 km^2), regardless of non-surge and surge types, exhibit complex flow velocities, which are poorly correlated with area ($R = 0.13$ for non-surge type and $R = 0.22$ for surge type), length ($R = 0.12$ and $R = 0.21$), slope ($R = -0.08$ and $R = 0.02$) and thickness ($R = -0.01$ and $R = 0.09$). Flow velocities of large (area ≥ 20 km^2), non-surge-type glaciers show a positive correlation with both area ($R = 0.69$) and length ($R = 0.78$), i.e., the size of glaciers, suggesting faster motion with increasing sizes, as glacier flow laws would predict. Surge-type glaciers do not show an evident correlation with any of the factors. We also noted that glacier velocities appear to be independent of, or at least not linearly correlated with, both slope and thickness. Similarly, we analysed the relationship between glacier velocity changes and glacier size and thickness. As shown in Figure 4, velocity changes of both non-surge- and surge-type glaciers are completely independent of glacier geometry, indicated by the very low correlation ($R \approx 0$).

Table 1. Statistics of glacier velocity changes in the Himalayas. Stable glaciers are defined as the amplitude of differences between Landsat-7 and Sentinel-2 velocities, i.e., velocity changes between 1999 and 2018, $\leq$3 m yr^{-1}. Velocity changes >3 m yr^{-1} are regarded as speedup and <−3 m yr^{-1} as slowdown. The overall changes were calculated by averaging all the glaciers within each subregion, in order to be comparable with the results in [9].

	Number of Glaciers	Speedup	Slowdown	Stable	Speedup (m yr^{-1})		Slowdown (m yr^{-1})		Overall (m yr^{-1} dec^{-1})	
					Average	Maximum	Average	Maximum	This Study	[9]
Hindu Kush	4206	22.6%	28.3%	49.1%	5.2	29.6 (RGI60-14.24088)	−11.5	−186.0 (RGI60-14.22312)	−1.0	−1.4
Karakoram	12,822	17.3%	31.5%	51.2%	6.6	127.7 (RGI60-14.04875)	−13.7	−256.7 (RGI60-14.03000)	−1.6	0.8
Spiti Lahaul	7796	19.6%	18.7%	61.7%	6.4	81.5 (RGI60-14.19261)	−12.1	−179.0 (RGI60-14.17193)	−0.4	−4.6
West Nepal	3906	33.4%	29.1%	37.5%	5.5	27.2 (RGI60-14.12426)	−9.7	−51.4 (RGI60-15.05231)	−0.5	−2.2
East Nepal	3563	56.2%	19.6%	24.2%	6.3	55.8 (RGI60-15.09859)	−12.3	−62.0 (RGI60-15.09861)	0.7	−1.8
Bhutan	1547	34.1%	22.1%	43.8%	6.6	39.5 (RGI60-15.02102)	−8.5	−61.3 (RGI60-15.01907)	0.2	−2.1
Nyainqentanglha	2882	41.3%	21.9%	36.8%	6.9	95.9 (RGI60-15.12714)	−12.4	−135.6 (RGI60-15.12081)	0.2	−6.4
Total	36,722	32.0%	24.5%	43.5%	6.3	127.7 (Karakoram)	−12.3	−256.7 (Karakoram)	−0.3	/

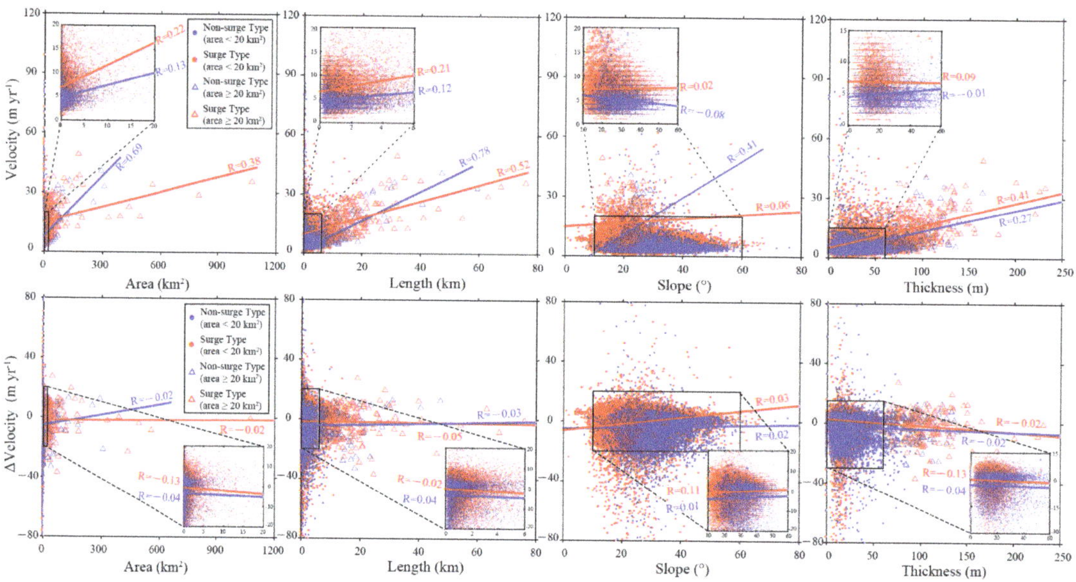

Figure 4. Glacier velocity and velocity changes versus area, length, slope and thickness. Dots and triangles stand for glaciers with an area of smaller than 20 km^2 and over 20 km^2, respectively. Non-surge- and surge-type glaciers are separated by blue and red. Lines show the linear regression analysis of the corresponding group of data.

5. Discussion

5.1. Regional Patterns of Surface Velocity Changes

Dehecq et al. [9] calculated time-series velocity anomalies for 11 subregions in the Himalayas. An analysis of glacier changes at regional scales allows us to explore cryospheric responses to climatic forcing. Here, we followed their line of analysis and focused on the 7 subregions that stretch along the range front. In order to compare our estimates to theirs, we averaged the velocity differences of all glaciers in each subregion and calculated the rate of change (velocity differences divided by 1.8 decades). Velocity changes (Table 1) show that, on average, glaciers in Hindu Kush (−1.0 m yr^{-1} decade^{-1} for 4206 glaciers), Karakoram (−1.6 m yr^{-1} decade^{-1} for 12,822 glaciers), Spiti Lahaul (−0.4 m yr^{-1} decade^{-1} for 7796 glaciers) and West Nepal (−0.5 m yr^{-1} decade^{-1} for 3906 glaciers) have slowed down from 1999 to 2018, whereas a slight acceleration has occurred in the east: East Nepal (0.7 m yr^{-1} decade^{-1} for 3563 glaciers), Bhutan (0.2 m yr^{-1} decade^{-1} for 1547 glaciers) and Nyainqentanglha (0.2 m yr^{-1} decade^{-1} for 2882 glaciers). Our results differ from [9]'s, who found that all subregions have experienced a slowdown between 2000 and 2017, with a changing rate varying from -6.4 m yr^{-1} decade^{-1} in Nyainqentanglha to -1.4 m yr^{-1} decade^{-1} in Hindu Kush, except for Karakoram, which experienced a small speedup (0.8 m yr^{-1} decade^{-1}). The inconsistency is likely to be resulted from seasonal changes in glacier flow. The images used in our study were collected in December, so our estimates are velocities during winter, while [9]'s velocity estimates are centred around June. Past studies (e.g., [31–33]) have shown that, during summer, when melting occurs, glaciers flow much faster with stronger temporal and spatial variations. Velocity changes during winter (0$\sim$1 m yr^{-1} decade^{-1}) are considerably smaller than the changes during summer (1$\sim$6 m yr^{-1} decade^{-1}) (see Figure 1 and Table 1), possibly due to strong spatio-temporal variations in melting in the Himalayas. From the winter velocity changes, we also observed a contrasting difference between the western (slowdown) and eastern (speedup) parts of the Himalayan front, indicating heterogeneous changes in accumulation and/or melting under different climatic regimes.

5.2. Linking Surface Velocity Changes with Glacier Mass Balance

To further investigate the driving factors of glacier velocity changes, we use the empirical power law relation between glacier surface velocity V and driving stress τ [34–37]:

$$V = A\tau^m \tag{1}$$

where A and m are positive constants, related to ice rheology, bed topography and flow mechanisms (ice deformation and basal sliding) [34,35]. m has been estimated to vary from 1 (flow over soft sediments, [38]) to 4 (high subglacial pressure, [9]) under different circumstances.

Taking the derivative of V with respect to τ, we have:

$$dV = Am\tau^{(m-1)}d\tau \tag{2}$$

Combining Equations (1) and (2), we have:

$$\frac{dV}{V} = m\frac{d\tau}{\tau} \tag{3}$$

Assuming that changes in the driving stress ($d\tau$) are induced by mass balance (dM) and ignoring other factors, as proposed by [9], i.e., $d\tau = C_1 dM + C_2$, we have:

$$\frac{dV}{V} = \frac{m}{\tau}C_1 dM + \frac{m}{\tau}C_2 \tag{4}$$

where C_1 and C_2 are assumed constant, relating the driving stress and mass balance. Dehecq et al. [9] analysed velocity changes and mass balance data and found that, at regional scales, summer velocity changes are positively correlated with mass balance: $\frac{dV}{V} \sim 1.25dM$ (see Figure 5).

To test whether velocity changes exhibit any seasonal variability, we applied Equation (4) to analyse the winter velocity estimates. We calculated the average values of velocity V and its change dV for each of the glaciers, based on which we determined the overall dV and V for each subregion. The measurements of glacier mass balance (dM) were taken from [7]. Although estimates of glacier mass balance in Brun et al. [7] did not take seasonal variability into account, an earlier study in [2] has shown that the long-term trend of dM is consistent between seasons (the amplitude differs slightly). Therefore, using the average long-term trend of dM should not affect the linear relationship between dV and dM. As shown in Figure 5, glacier velocity changes $\frac{dV}{V}$ are negatively correlated with mass balance ($\frac{dV}{V} \sim -0.96dM$), suggesting that ice mass loss promotes glacier motion in winter. This is in contrasting difference with summer velocity changes in the study by [9], which states that mass loss drives glacier slowdown. Such seasonal variability indicates different mechanisms of glacier mass loss in the Himalayas. Increased melt conditions in the early winter period, November and December, could enhance velocity, as noted by Bocchiola et al. [39], Pelto et al. [40] in the eastern Himalaya. During summer, mass losses are more likely to be resulted from increased ablation zone melting, given that precipitation has not been significant [41,42]. The result suggests that velocity changes in the Himalayas are temperature driven, rather than accumulation process driven.

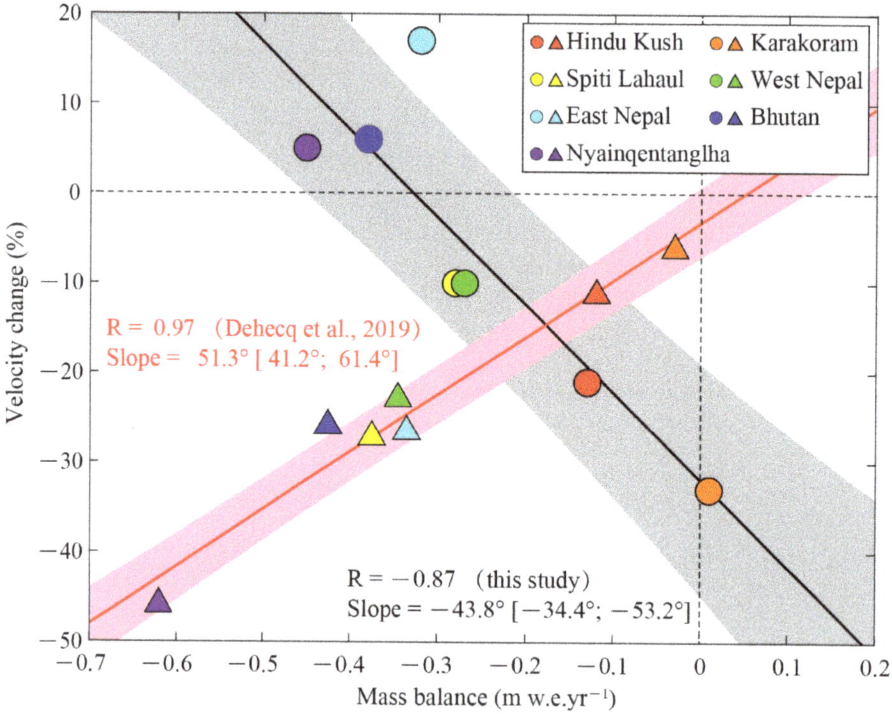

Figure 5. Glacier surface velocity changes versus mass balance. Mass balance data were taken from [7]. Shaded bands indicate 68% confidence interval. Winter velocity changes (filled circles, in this study) are negatively correlated with mass balance, with a correlation of $R = -0.87$. Summer velocity changes (filled triangles, [9]) are positively correlated with mass balance.

6. Conclusions

In this study, we used Landsat-7 and Sentinel-2 optical imagery to investigate glacier surface velocity and the associated changes in the Himalayas. We analysed flow patterns of individual glacier along the Himalayan mountain front and found that glacier velocity changes exhibit an evident heterogeneity at different spatial scales. Of all the 36,722 glaciers, 32% have experienced speedup, 24.5% have slowed down, and the rest, 43.5%, have remained stable. At regional scales, the amplitude of velocity changes is significantly larger in summer [9] than that in winter (this study). The decreasing velocities in winter between 1999 and 2018 in the western part of the Himalayas, in contrast to the increasing velocities in the eastern part, may be caused by changes in accumulation and/or melting under different climatic regimes. Accelerated flow in Eastern Nepal may be impacted by melting conditions, where more melting has been observed. We also observed that glacier velocity changes in winter are controlled by mass balance, as suggested by [9]; however, unlike summer velocity changes that are positively correlated with mass balance, winter velocity changes show a negative correlation. Our study suggests that glacier velocity changes in the Himalayas are more spatially and temporally heterogeneous than what was previously thought, emphasising complex interactions between glacier dynamics and environmental forcing.

Author Contributions: Y.Z. and J.C. performed data processing and interpreted the results. Y.Z. wrote the original draft. J.C. and X.C. reviewed the paper. All authors have read and agreed to the published version of the manuscript.

Funding: The APC was funded by the Second Tibetan Scientific Expedition and Research Program (STEP) (2019QZKK0901).

Data Availability Statement: Landsat-7 images were downloaded from https://earthexplorer.usgs.gov/ (accessed on 23 January 2020); Sentinel-2 images were downloaded from https://scihub.esa.int/ (accessed on 23 January 2020). The RANDOLPH GLACIER INVENTORY (RGI 6.0) is freely available from https://www.gtn-g.ch/data_catalogue_rgi/ (accessed on 23 January 2020).

Acknowledgments: This work was supported by the Second Tibetan Plateau Scientific Expedition and Research Program (STEP) (2019QZKK0901) and Innovatiom Group Project of Southern Marine Science and Engineering Guangdong Laboratory (Zhuhai) (No. 311021008).

Conflicts of Interest: The authors declare that they have no known competing financial interests or personal relationships that could have appeared to influence the work reported in this paper.

References

1. Kaser, G.; Großhauser, M.; Marzeion, B. Contribution potential of glaciers to water availability in different climate regimes. *Proc. Natl. Acad. Sci. USA* **2010**, *107*, 20223–20227. [CrossRef] [PubMed]
2. Kääb, A.; Berthier, E.; Nuth, C.; Gardelle, J.; Arnaud, Y. Contrasting patterns of early twenty-first-century glacier mass change in the Himalayas. *Nature* **2012**, *488*, 495–498. [CrossRef] [PubMed]
3. Bolch, T.; Kulkarni, A.; Kääb, A.; Huggel, C.; Paul, F.; Cogley, J.G.; Frey, H.; Kargel, J.S.; Fujita, K.; Scheel, M.; et al. The state and fate of Himalayan glaciers. *Science* **2012**, *336*, 310–314. [CrossRef] [PubMed]
4. Yao, T.; Thompson, L.; Yang, W.; Yu, W.; Gao, Y.; Guo, X.; Yang, X.; Duan, K.; Zhao, H.; Xu, B.; et al. Different glacier status with atmospheric circulations in Tibetan Plateau and surroundings. *Nat. Clim. Chang.* **2012**, *2*, 663–667. [CrossRef]
5. Kraaijenbrink, P.; Bierkens, M.; Lutz, A.; Immerzeel, W. Impact of a global temperature rise of 1.5 degrees Celsius on Asia's glaciers. *Nature* **2017**, *549*, 257–260. [CrossRef] [PubMed]
6. Ren, J.; Jing, Z.; Pu, J.; Qin, X. Glacier variations and climate change in the central Himalaya over the past few decades. *Ann. Glaciol.* **2006**, *43*, 218–222. [CrossRef]
7. Brun, F.; Berthier, E.; Wagnon, P.; Kääb, A.; Treichler, D. A spatially resolved estimate of High Mountain Asia glacier mass balances from 2000 to 2016. *Nat. Geosci.* **2017**, *10*, 668–673. [CrossRef]
8. Lutz, A.; Immerzeel, W.; Shrestha, A.; Bierkens, M. Consistent increase in High Asia's runoff due to increasing glacier melt and precipitation. *Nat. Clim. Chang.* **2014**, *4*, 587–592. [CrossRef]
9. Dehecq, A.; Gourmelen, N.; Gardner, A.S.; Brun, F.; Goldberg, D.; Nienow, P.W.; Berthier, E.; Vincent, C.; Wagnon, P.; Trouvé, E. Twenty-first century glacier slowdown driven by mass loss in High Mountain Asia. *Nat. Geosci.* **2019**, *12*, 22–27. [CrossRef]
10. Immerzeel, W.W.; Droogers, P.; De Jong, S.; Bierkens, M. Large-scale monitoring of snow cover and runoff simulation in Himalayan river basins using remote sensing. *Remote Sens. Environ.* **2009**, *113*, 40–49. [CrossRef]
11. Guo, W.; Liu, S.; Xu, J.; Wu, L.; Shangguan, D.; Yao, X.; Wei, J.; Bao, W.; Yu, P.; Liu, Q.; et al. The second Chinese glacier inventory: Data, methods and results. *J. Glaciol.* **2015**, *61*, 357–372. [CrossRef]
12. Bookhagen, B.; Burbank, D.W. Toward a complete Himalayan hydrological budget: Spatiotemporal distribution of snowmelt and rainfall and their impact on river discharge. *J. Geophys. Res. Earth Surf.* **2010**, *115*. [CrossRef]
13. Neckel, N.; Kropáček, J.; Bolch, T.; Hochschild, V. Glacier mass changes on the Tibetan Plateau 2003–2009 derived from ICESat laser altimetry measurements. *Environ. Res. Lett.* **2014**, *9*, 014009. [CrossRef]
14. Kääb, A.; Treichler, D.; Nuth, C.; Berthier, E. Brief Communication: Contending estimates of 2003–2008 glacier mass balance over the Pamir–Karakoram–Himalaya. *Cryosphere* **2015**, *9*, 557–564. [CrossRef]
15. Maurer, J.; Schaefer, J.; Rupper, S.; Corley, A. Acceleration of ice loss across the Himalayas over the past 40 years. *Sci. Adv.* **2019**, *5*, eaav7266. [CrossRef] [PubMed]
16. Ragettli, S.; Bolch, T.; Pellicciotti, F. Heterogeneous glacier thinning patterns over the last 40 years in Langtang Himal, Nepal. *Cryosphere* **2016**, *10*, 2075–2097. [CrossRef]
17. Zhou, Y.; Li, Z.; Li, J.; Zhao, R.; Ding, X. Glacier mass balance in the Qinghai–Tibet Plateau and its surroundings from the mid-1970s to 2000 based on Hexagon KH-9 and SRTM DEMs. *Remote Sens. Environ.* **2018**, *210*, 96–112. [CrossRef]
18. Dehecq, A.; Gourmelen, N.; Trouvé, E. Deriving large-scale glacier velocities from a complete satellite archive: Application to the Pamir–Karakoram–Himalaya. *Remote Sens. Environ.* **2015**, *162*, 55–66. [CrossRef]

19. Shugar, D.; Jacquemart, M.; Shean, D.; Bhushan, S.; Upadhyay, K.; Sattar, A.; Schwanghart, W.; McBride, S.; de Vries, M.V.W.; Mergili, M.; et al. A massive rock and ice avalanche caused the 2021 disaster at Chamoli, Indian Himalaya. *Science* **2021**, *373*, 300–306. [CrossRef] [PubMed]
20. Muhuri, A.; Gascoin, S.; Menzel, L.; Kostadinov, T.; Harpold, A.; Sanmiguel-Vallelado, A.; Moreno, J.I.L. Performance Assessment of Optical Satellite Based Operational Snow Cover Monitoring Algorithms in Forested Landscapes. *IEEE J. Sel. Top. Appl. Earth Obs. Remote Sens.* **2021**, *14*, 7159–7178. [CrossRef]
21. Kääb, A.; Winsvold, S.H.; Altena, B.; Nuth, C.; Nagler, T.; Wuite, J. Glacier remote sensing using Sentinel-2. part I: Radiometric and geometric performance, and application to ice velocity. *Remote Sens.* **2016**, *8*, 598. [CrossRef]
22. Azam, M.F.; Ramanathan, A.; Wagnon, P.; Vincent, C.; Linda, A.; Berthier, E.; Sharma, P.; Mandal, A.; Angchuk, T.; Singh, V.; et al. Meteorological conditions, seasonal and annual mass balances of Chhota Shigri Glacier, western Himalaya, India. *Ann. Glaciol.* **2016**, *57*, 328–338. [CrossRef]
23. Sun, Y.; Jiang, L.; Liu, L.; Sun, Y.; Wang, H. Spatial-temporal characteristics of glacier velocity in the Central Karakoram revealed with 1999–2003 Landsat-7 ETM+ Pan Images. *Remote Sens.* **2017**, *9*, 1064. [CrossRef]
24. Fujita, K. Effect of precipitation seasonality on climatic sensitivity of glacier mass balance. *Earth Planet. Sci. Lett.* **2008**, *276*, 14–19. [CrossRef]
25. Immerzeel, W.; Wanders, N.; Lutz, A.; Shea, J.; Bierkens, M. Reconciling high-altitude precipitation in the upper Indus basin with glacier mass balances and runoff. *Hydrol. Earth Syst. Sci.* **2015**, *19*, 4673–4687. [CrossRef]
26. Dumont, M.; Gascoin, S. Optical remote sensing of snow cover. In *Land Surface Remote Sensing in Continental Hydrology*; Elsevier: Amsterdam, The Netherlands, 2016; pp. 115–137.
27. Farr, T.G.; Rosen, P.A.; Caro, E.; Crippen, R.; Duren, R.; Hensley, S.; Kobrick, M.; Paller, M.; Rodriguez, E.; Roth, L.; et al. The shuttle radar topography mission. *Rev. Geophys.* **2007**, *45*. [CrossRef]
28. Farinotti, D.; Huss, M.; Fürst, J.J.; Landmann, J.; Machguth, H.; Maussion, F.; Pandit, A. A consensus estimate for the ice thickness distribution of all glaciers on Earth. *Nat. Geosci.* **2019**, *12*, 168–173. [CrossRef]
29. Leprince, S.; Barbot, S.; Ayoub, F.; Avouac, J.P. Automatic and precise orthorectification, coregistration, and subpixel correlation of satellite images, application to ground deformation measurements. *Geosci. Remote Sens. IEEE Trans.* **2007**, *45*, 1529–1558. [CrossRef]
30. Ayoub, F.; Leprince, S.; Keene, L. *Users Guide to COSI-CORR Co-Registration of Optically Sensed Images and Correlation*; California Institute of Technology: Pasadena, CA, USA, 2009; p. 38.
31. Hooke, R.L.; Calla, P.; Holmlund, P.; Nilsson, M.; Stroeven, A. A 3 year record of seasonal variations in surface velocity, Storglaciären, Sweden. *J. Glaciol.* **1989**, *35*, 235–247. [CrossRef]
32. Lemos, A.; Shepherd, A.; McMillan, M.; Hogg, A.E. Seasonal variations in the flow of land-terminating glaciers in Central-West Greenland using Sentinel-1 imagery. *Remote Sens.* **2018**, *10*, 1878. [CrossRef]
33. Kraaijenbrink, P.; Meijer, S.W.; Shea, J.M.; Pellicciotti, F.; De Jong, S.M.; Immerzeel, W.W. Seasonal surface velocities of a Himalayan glacier derived by automated correlation of unmanned aerial vehicle imagery. *Ann. Glaciol.* **2016**, *57*, 103–113. [CrossRef]
34. Glen, J. Experiments on the deformation of ice. *J. Glaciol.* **1952**, *2*, 111–114. [CrossRef]
35. Glen, J.W. The creep of polycrystalline ice. *Proc. R. Soc. Lond. Ser. A Math. Phys. Sci.* **1955**, *228*, 519–538.
36. Weertman, J. On the sliding of glaciers. *J. Glaciol.* **1957**, *3*, 33–38. [CrossRef]
37. Goldsby, D.; Kohlstedt, D.L. Superplastic deformation of ice: Experimental observations. *J. Geophys. Res. Solid Earth* **2001**, *106*, 11017–11030. [CrossRef]
38. MacAyeal, D.R. Large-scale ice flow over a viscous basal sediment: Theory and application to ice stream B, Antarctica. *J. Geophys. Res. Solid Earth* **1989**, *94*, 4071–4087. [CrossRef]
39. Bocchiola, D.; Bombelli, G.M.; Camin, F.; Ossi, P.M. Field Study of Mass Balance, and Hydrology of the West Khangri Nup Glacier (Khumbu, Everest). *Water* **2020**, *12*, 433. [CrossRef]
40. Pelto, M.; Panday, P.; Matthews, T.; Maurer, J.; Perry, L.B. Observations of Winter Ablation on Glaciers in the Mount Everest Region in 2020–2021. *Remote Sens.* **2021**, *13*, 2692. [CrossRef]
41. Wagnon, P.; Vincent, C.; Arnaud, Y.; Berthier, E.; Vuillermoz, E.; Gruber, S.; Ménégoz, M.; Gilbert, A.; Dumont, M.; Shea, J.; et al. Seasonal and annual mass balances of Mera and Pokalde glaciers (Nepal Himalaya) since 2007. *Cryosphere* **2013**, *7*, 1769–1786. [CrossRef]
42. Wagnon, P.; Brun, F.; Khadka, A.; Berthier, E.; Shrestha, D.; Vincent, C.; Arnaud, Y.; Six, D.; Dehecq, A.; Ménégoz, M.; et al. Reanalysing the 2007–19 glaciological mass-balance series of Mera Glacier, Nepal, Central Himalaya, using geodetic mass balance. *J. Glaciol.* **2021**, *67*, 117–125. [CrossRef]

remote sensing

MDPI

Article

The UVSQ-SAT/INSPIRESat-5 CubeSat Mission: First In-Orbit Measurements of the Earth's Outgoing Radiation

Mustapha Meftah [1,*], Thomas Boutéraon [1], Christophe Dufour [1], Alain Hauchecorne [1], Philippe Keckhut [1], Adrien Finance [1,2], Slimane Bekki [1], Sadok Abbaki [1], Emmanuel Bertran [1], Luc Damé [1], Jean-Luc Engler [1], Patrick Galopeau [1], Pierre Gilbert [1], Laurent Lapauw [1], Alain Sarkissian [1], André-Jean Vieau [1], Patrick Lacroix [1], Nicolas Caignard [1], Xavier Arrateig [1], Odile Hembise Fanton d'Andon [2], Antoine Mangin [2], Jean-Paul Carta [3], Fabrice Boust [4], Michel Mahé [5] and Christophe Mercier [6]

1 Université de Versailles Saint-Quentin-en-Yvelines, Université Paris-Saclay, Sorbonne Université (SU), CNRS, LATMOS, 11 Boulevard d'Alembert, 78280 Guyancourt, France; Thomas.Bouteraon@latmos.ipsl.fr (T.B.); Christophe.Dufour@latmos.ipsl.fr (C.D.); Alain.Hauchecorne@latmos.ipsl.fr (A.H.); Philippe.Keckhut@latmos.ipsl.fr (P.K.); Adrien.Finance@latmos.ipsl.fr (A.F.); Slimane.Bekki@latmos.ipsl.fr (S.B.); sadok.abbaki@latmos.ipsl.fr (S.A.); emmanuel.bertran@latmos.ipsl.fr (E.B.); Luc.Dame@latmos.ipsl.fr (L.D.); Jean-Luc.Engler@latmos.ipsl.fr (J.-L.E.); Patrick.Galopeau@latmos.ipsl.fr (P.G.); Pierre.Gilbert@latmos.ipsl.fr (P.G.); laurent.lapauw@latmos.ipsl.fr (L.L.); Alain.Sarkissian@latmos.ipsl.fr (A.S.); andre-jean.vieau@latmos.ipsl.fr (A.-J.V.); patrick.lacroix@latmos.ipsl.fr (P.L.); nicolas.caignard@latmos.ipsl.fr (N.C.); xavier.arrateig@latmos.ipsl.fr (X.A.)

2 ACRI-ST—CERGA, 10 Avenue Nicolas Copernic, 06130 Grasse, France; oha@acri-st.fr (O.H.F.d.); Antoine.Mangin@acri-st.fr (A.M.)

3 Carta-Rouxel, Campus de l'Innovation des Yvelines, 17 rue Albert Thomas, 78130 Les Mureaux, France; jp.carta@lexat.fr

4 ONERA, 8 Chemin de la Hunière, 91123 Palaiseau, France; fabrice.boust@onera.fr

5 Radio-Club de Saint-Quentin-en-Yvelines F6KRK, 1 bis av. des Frênes, 78180 Montigny le Bretonneux, France; mahe_michel@yahoo.fr

6 AMSAT-Francophone, 14 bis rue des Gourlis, 92500 Rueil Malmaison, France; christophe.mercier@amsat-f.org

* Correspondence: Mustapha.Meftah@latmos.ipsl.fr; Tel.: +33-1-80-28-51-79

Citation: Meftah, M.; Boutéraon, T.; Dufour, C.; Hauchecorne, A.; Keckhut, P.; Finance, A.; Bekki, S.; Abbaki, S.; Bertran, E.; Damé, L.; et al. The UVSQ-SAT/INSPIRESat-5 CubeSat Mission: First In-Orbit Measurements of the Earth's Outgoing Radiation. *Remote Sens.* **2021**, *13*, 1449. https://doi.org/10.3390/rs13081449

Academic Editor: Xander Wang

Received: 24 March 2021
Accepted: 5 April 2021
Published: 8 April 2021

Publisher's Note: MDPI stays neutral with regard to jurisdictional claims in published maps and institutional affiliations.

Abstract: UltraViolet & infrared Sensors at high Quantum efficiency onboard a small SATellite (UVSQ-SAT) is a small satellite at the CubeSat standard, whose development began as one of the missions in the International Satellite Program in Research and Education (INSPIRE) consortium in 2017. UVSQ-SAT is an educational, technological and scientific pathfinder CubeSat mission dedicated to the observation of the Earth and the Sun. It was imagined, designed, produced and tested by LATMOS in collaboration with its academic and industrial partners, and the French-speaking radioamateur community. About the size of a Rubik's Cube and weighing about 2 kg, this satellite was put in orbit in January 2021 by the SpaceX Falcon 9 launch vehicle. After briefly introducing the UVSQ-SAT mission, this paper will present the importance of measuring the Earth's radiation budget and its energy imbalance and the scientific objectives related to its various components. Finally, the first in-orbit observations will be shown (maps of the solar radiation reflected by the Earth and of the outgoing longwave radiation at the top of the atmosphere during February 2021). UVSQ-SAT is one of the few CubeSats worldwide with a scientific goal related to climate studies. It represents a research in remote sensing technologies for Climate observation and monitoring.

Keywords: climate observation and monitoring; earth radiation budget; nanosatellite; IPCC

1. Introduction

UVSQ-SAT is a nanosatellite designed to observe the Sun and the Earth, mainly for observing essential climate variables. It allows, among other things, measurements of the solar radiation reflected by the Earth or outgoing shortwave radiation (OSR) and of the outgoing longwave radiation (OLR) at the top of the atmosphere (TOA). A detailed description of the UVSQ-SAT mission and its satellite is given by Meftah et al. (2020) [1]. The main technical and scientific goal is to implement an agile demonstrator allowing to perform measurements of the Earth's radiation balance at the TOA. The UVSQ-SAT mission will allow to validate the principle of several miniaturized technologies as a first step to accurately measure the Earth's energy imbalance (EEI) with a constellation of small satellites. Regarding the educational objectives, the UVSQ-SAT program seeks to promote the transmission and valorization of scientific and technical knowledge through the innovation. This one feeds the creation of value and allows the implementation of technological breakthroughs and the associated risk-taking, the "NewSpace", and technology transfer.

Developed from 2017, UVSQ-SAT is based on a CubeSat architecture [2,3], which represents a satellite format defined in 1999 by the California Polytechnic State University and Stanford University (USA). UVSQ-SAT was developed and implemented by LATMOS and its INSPIRE international partners. The INSPIRE consortium was formed in 2015, with the objectives of developing small satellites to perform science missions, building a project-based learning curriculum for space science and engineering, and establishing a supporting network of ground stations. UVSQ-SAT/INSPIRESat-5 fully meets these objectives. A representative mockup of the UVSQ-SAT satellite is provided in Figure 1 left.

Figure 1. (**Left**) Representative mockup of the UVSQ-SAT satellite. (**Right**) Transporter 1 mission (SpaceX) with 143 commercial and government satellites on-board, including UVSQ-SAT. Credits: SpaceX.

Before being launched into space, a test campaign of the UVSQ-SAT CubeSat was carried out from August to October 2020. It typically involves several mechanical tests, electromagnetic compatibility (EMC) tests, calibration and performance tests, thermal vacuum and thermal balance tests, magnetic tests, and end to end tests. The satellite has been fully tested with French facilities (Figure 2) at LATMOS, at the Plateforme d'Intégration et de Tests (PIT) of the Observatoire de Versailles Saint-Quentin-en-Yvelines (OVSQ), at French space agency (CNES), and at the Office National d'Etudes et de Recherches Aérospatiales (ONERA). A successful mechanical testing campaign of the UVSQ-SAT CubeSat was mandatory to get a launch clearance. It involved quasi-static testing (test carried out with a static load of 15 g to simulate the acceleration of the launch vehicle), shock testing, as well as random vibration (test levels are shown in Figure 3 left). Random vibration was carried out to simulate the loads of the upper stages of the SpaceX Falcon 9 launcher and is the most severe test. Before and after each random axis test run, a resonance search (Figure 3 right) was done with a frequency sweep of low-level sinusoidal vibrations to characterize main resonant modes of the UVSQ-SAT CubeSat, to determine if the tests loads have changed the mechanical properties of the CubeSat (first eigenmode close to 418 Hz), and to show possible deficiencies. Thermal bakeout was mandatory (+60 °C) for outgassing volatile materials of the CubeSat. A thermal vacuum cycling test (temperatures profile given in Figure 4) was also carried out to simulate the temperature conditions in orbit and to demonstrate that the CubeSat is able to survive, without loss of integrity or functionality. After performing all performances and environmental tests with success and confirmation of frequency coordination, UVSQ-SAT was ready to go in space.

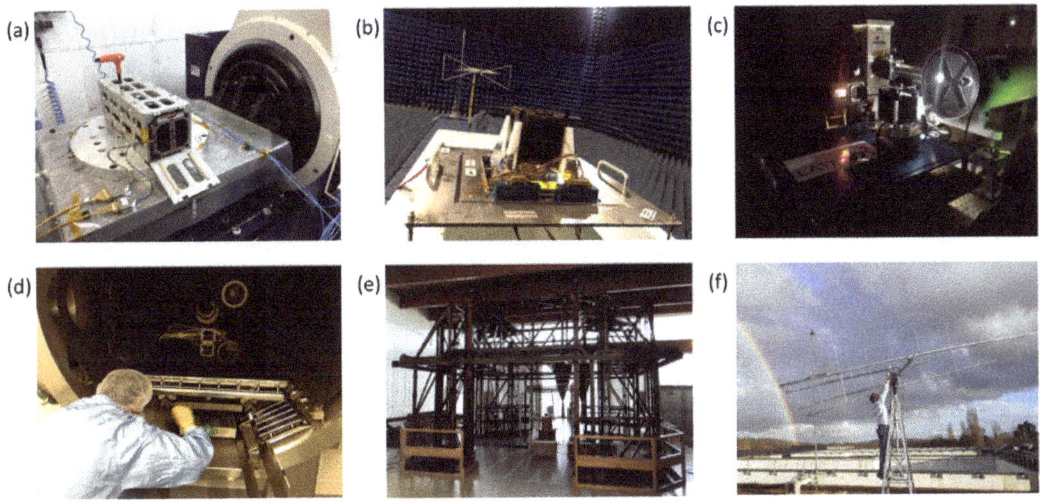

Figure 2. UVSQ-SAT environmental tests and calibrations carried out during the year 2020. (**a**) Vibration. (**b**) EMC tests. (**c**) Calibration with a Xenon lamp and a black body. (**d**) Thermal vacuum and thermal balance tests. (**e**) Magnetic cleanliness verification of the satellite. (**f**) End to end tests with the UHF/VHF LATMOS ground-based station (telemetry at 437.020 MHz, telecommand at 145.905 MHz).

UVSQ-SAT was launched on 24, January 2021 by the Falcon 9 rocket in the frame of the "Transporter 1" mission (Figure 1 right). Since then, the satellite is on a sun-synchronous orbit around the Earth at an altitude of about 533 km (515 km at perigee), an inclination of 97.5010°, an eccentricity of 0.0014455, a right ascension at the ascending node of 128.4244°, and an argument at perigee of 94.1573°. The duration of an orbit is about 95.18 min. The selected orbit presents eclipses, which allow to realize calibrations with observations towards the deep space or only towards the Earth.

This manuscript presents first UVSQ-SAT in-orbit measurements of the Earth's outgoing radiation. Section 2 presents the objectives of the UVSQ-SAT mission and the importance of measuring the Earth's radiation budget. The expected scientific requirements on the measurement of the Earth's energy imbalance will be presented. They are given for the UVSQ-SAT demonstrator and the future constellation (Terra-F) which must be disruptive. Section 3 presents the methods used to reconstruct the maps of the albedo and the OLR at the TOA. They are applicable to the observations by the UVSQ-SAT's instruments and the model used. Section 4 shows the first observations made by the UVSQ-SAT mission and the preliminary results. Section 5 highlights the perspectives related to this study and finally in Section 6, the conclusions are presented.

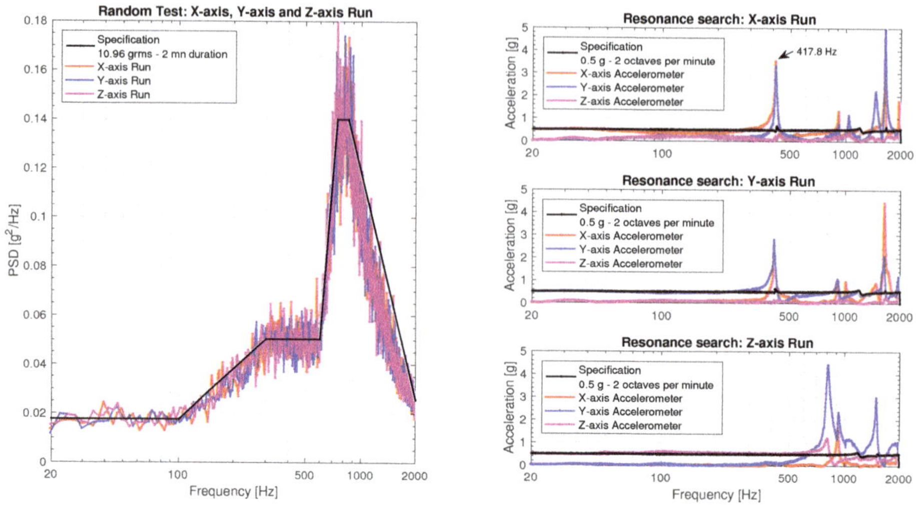

Figure 3. (**Left**) Random vibration profile was run in three orthogonal axes, referred as X, Y, Z. (**Right**) Resonance search carried out on UVSQ-SAT before and after each random run (September 2020).

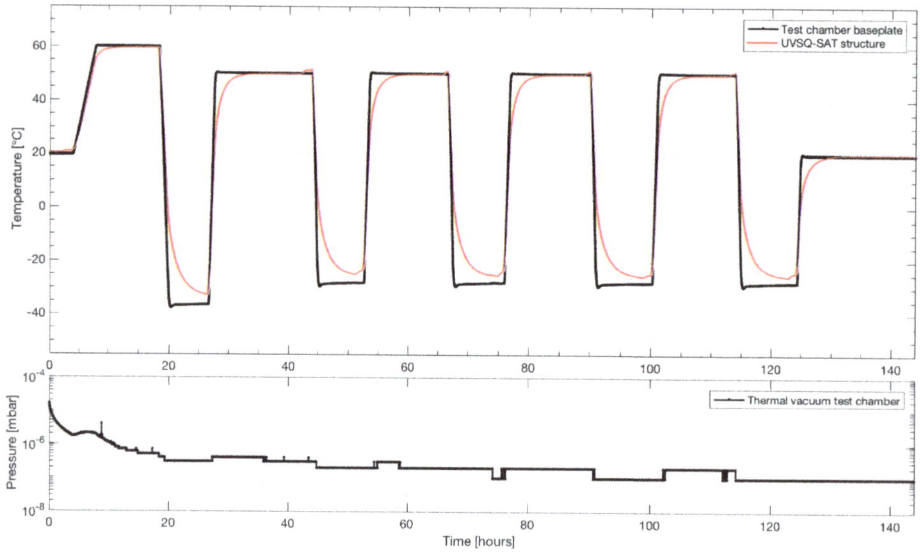

Figure 4. Profile of the UVSQ-SAT thermal vacuum cycling tests carried out in September 2020.

2. Scientific Objectives and Requirements

2.1. Scientific Objectives

One of the objectives of the UVSQ-SAT mission is the study of the Earth's energy budget (ERB) and its energy imbalance [1]. Climate is largely determined by its energy balance at the top of the atmosphere, which regulates the overall energy content of the Earth system, i.e., the atmosphere and ocean (IPCC, 2013). The system is heated by the absorption of incoming solar energy (shortwave) and is cooled by terrestrial infrared emission (longwave) to space. Almost 30% of the incoming solar energy is not absorbed but reflected back to space. Human activities have led to increased levels of heat-trapping greenhouse gases (GHGs) in the atmosphere, with a decrease in the amount of terrestrial radiation that can escape, resulting in an Earth energy imbalance. As a result, the Earth system is heating up. The most obvious sign of this is the long-term increase in global surface temperatures, which increases outgoing terrestrial infrared emission and thus tends to restore the Earth's energy balance. However, as illustrated by the apparent slowdown in global surface warming observed between 1998 and 2012 ("climate hiatus"), global average surface temperature, with its large decadal internal variability, is certainly not the most appropriate indicator of the EEI and associated energy accumulation in the Earth system [4]. Since the EEI drives climate change, its continuous monitoring has been identified as a fundamental diagnostic for analyzing climate variability and anticipating future changes. Closing the Earth's energy balance is considered a key step to further improve our understanding of global climate change [5]. Ideally, the EEI should be measured continuously from space. To date, existing satellite instruments have not been able to measure solar and terrestrial radiative fluxes with sufficient accuracy to directly determine the absolute value of the EEI on their own, although measurements of relative changes in the EEI are much more reliable [6–8]. Note that determining EEI requires very accurate global radiative measurements because EEI results from the difference of several terms (incoming solar energy, reflected incoming solar energy, outgoing terrestrial infrared energy) that are two orders of magnitude higher than EEI. For this reason, satellite radiative flux data are often anchored in some way to ocean heat content data, as most of the excess energy does not accumulate at the surface or in the atmosphere, but is absorbed by the oceans [5,8]. Although ocean heat content is considered a better indicator of EEI, its global assessment remains challenging, mainly due to insufficient and inconsistent data coverage, and uncertainties in global scale extension [9,10]. Overall, this lack of accurate satellite measurements of EEI is a major problem because climate change is primarily a perturbation of the Earth's energy balance. In addition, satellite measurements of solar and terrestrial energy fluxes also represent critical constraints and benchmarks for the evaluation and improvement of climate models, including their representations of the radiative effects of clouds and aerosols (IPCC, 2013). Earth's radiation budget data are therefore a key source of information for distinguishing external from internal climate variability and for assessing the sensitivity of climate to different forcings (including volcanic eruptions, solar variability, and anthropogenic aerosols).

EEI determination represents a complex problem where the first relatively detailed balances of the energy exchanges of our planet and its atmosphere were published in 1997 by Kiehl and Trenberth [11]. An update of the ERB was completed by Stephens et al. [12] in 2012. Figure 5 illustrates the flow of energy through Earth's atmosphere. Table 1 provides ERB main contributions and EEI during the 2000–2010 period. Figure 5 shows how solar radiation warms our planet, and how this energy becomes temporarily trapped as it fluxes away from Earth's surface as longwave infrared radiation. This energy trap produces the greenhouse effect, which represents the main driver of global warming [13].

EEI is the difference between the total solar irradiance (TSI) and the OSR combined with the OLR thermally emitted. At the TOA, the EEI was +0.58 $\pm$ 0.15 Wm^{-2} for the period 2005–2010 according to Hansen et al. (2011) [14]. It was estimated at +0.60 $\pm$ 0.40 Wm^{-2} between 2000 and 2010 by Stephens et al. (2012) [12]. A recently published study by Von Schuckmann et al. (2020) [15] shows that the EEI is currently estimated to be +0.87 $\pm$ 0.12 Wm^{-2} over the period 2010–2018 compared to +0.47 $\pm$ 0.10 Wm^{-2} over the period 1971–2018. This positive EEI describes the excess of heat in the Earth system, mainly due to human activity.

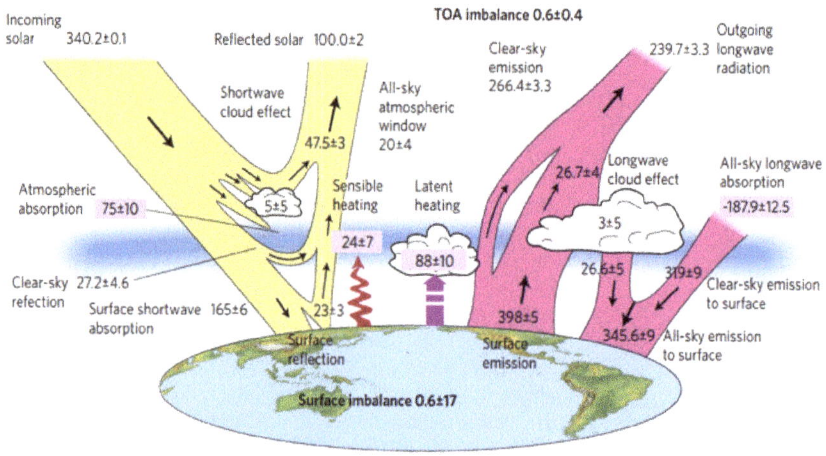

Figure 5. Radiation budget diagram of the Earth's atmosphere from Stephens et al. (2012) [12]. ERB for the period 2000–2010, where all flows are expressed in Wm^{-2}.

Table 1. ERB main contributions and EEI during the period 2000–2010 from Stephens et al. (2012) [12].

Contributor/Imbalance	Acronym	Flux
Incoming solar radiation	TSI/4	$340.2 \pm 0.1\,Wm^{-2}$
Outgoing shortwave radiation	OSR	$100.0 \pm 2.0\,Wm^{-2}$
Outgoing longwave radiation	OLR	$239.7 \pm 3.3\,Wm^{-2}$
TOA Earth energy imbalance	EEI at TOA	$+0.6 \pm 0.4\,Wm^{-2}$
Surface Earth energy imbalance	EEI at surface	$+0.6 \pm 17.0\,Wm^{-2}$

Figure 6 shows the evolution of the OLR and the EEI (12-month rolling averages) over the last four decades. Changes in sea surface temperature (SST) during different phases of the El Niño Southern Oscillation (ENSO) and in cloud properties are often reflected in changes in OLR. The EEI strong variations correspond to volcanic eruptions that deposit aerosols in the stratosphere, thus cooling Earth by reflecting sunlight back to space. In 1982, El Chichón ejected million metric tons of sulfur dioxide into the stratosphere. In 1991, Mount Pinatubo represents the largest stratospheric disturbance since Krakatoa eruption in 1883, dropping global temperatures and increasing ozone depletion. Then, the global energy imbalance of the Earth is not constant over time. It depends on natural variability and the complexity of different climate forcings. The time series of measurements from the Clouds and the Earth's Radiant Energy System (CERES) instrument is compared to the ERA 5 model, which represents the fifth generation of atmospheric reanalyses of global climate from the European Centre for Medium-Range Weather Forecasts (ECMWF). These results highlight the impact of anthropogenic radiative forcing during a period where the TSI has decreased since 1980 and where solar cycle 24 (24th cycle since 1755) is much weaker than the previous two cycles. The various studies [12,14,15] conclusively show the dominant role of the human-created greenhouse effect in driving global climate change. The EEI represents the most critical measure of the state of the Earth's climate. It sets expectations for future climate change (taking into account the Earth's inertia).

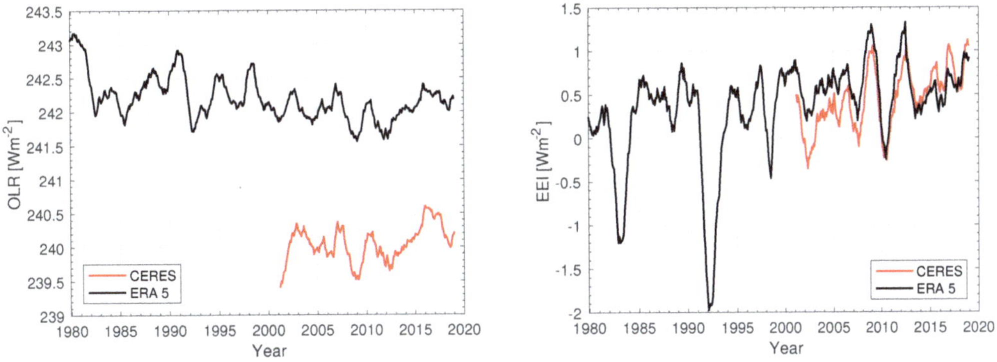

Figure 6. Time evolution of the OLR and the EEI from CERES measurements and the ERA 5 model.

The EEI represents the best approach to characterize global warming compared to other commonly used parameters (CO_2 concentration and Earth surface temperature). This highlights the importance of space-based missions that perform such measurements, including the UVSQ-SAT pathfinder mission associated with its possible small satellites constellation [1] named Terra-F. Further efforts to implement new space-based observing systems are more than necessary, as the EEI represents additional global warming that will occur without additional changes in radiative forcings. Direct measurements of changes in the EEI are of paramount importance in determining the rate of climate change on regional and global scales.

2.2. Scientific Requirements and Uncertainties on EEI Measurements

Accurately measuring the absolute value of the EEI at the TOA and its variability over time is still a scientific and technological challenge. The relevant scientific objective is to be able to detect any long-term global trend with a high accuracy (stability of at least ± 0.2 Wm^{-2} per decade), but also to be able to obtain fine spatial (a few tens of km) and high temporal resolutions (three hours as for weather forecasting) with an accuracy of a few Wm^{-2}. With current technologies, we do not know if all these requirements are achievable with instruments on board satellites, but they are very important to better determine the changes.

To date, the best estimate of the EEI is obtained from the change in heat stored in the ocean, commonly referred as ocean heat content (OHC). However, the absorption of heat by the ocean acts as a "buffer". This causes the rate of surface warming to slow down. The OHC measurements made by the Argo automated floats are excellent, but do not provide answers to the "short term" dynamics. To solve this issue, we need fine spatial and temporal resolutions as previously stated. For this, we need a constellation of satellites with instruments that combine measurements with a narrow field of view (FOV) and a wide FOV. The key to solve this problem is related to the resolution of the diurnal variation, which depends on the physical parameterization of the models. The "long term" is associated with the convolution of quick processes and dynamic phenomena at longer temporal and spatial scales of the atmosphere-ocean system. Thus, the best strategy for measuring the EEI is to track the "long term" temporal evolution of the OHC since most of the excess heat (89% over 1970–2018) is absorbed by the ocean [15]. These measurements must be combined with observations at the TOA using satellite-based instruments to provide "short term" information. Measurements with good spatio-temporal resolutions are crucial for advancing our understanding of climate change, as the radiative balance is driven in part by the radiative impacts of aerosols and clouds, which are highly spatially and temporally variable and are still relatively poorly quantified [16]. Currently, we do not know whether cloud-related changes will amplify, mitigate, or perhaps have only a small effect on the increase in the Earth's global temperature due to anthropogenic radiative forcings. Furthermore, we do not know the impact of radiative forcing related to anthropogenic aerosols that alter cloud properties. Global climate models have limitations with respect to these two issues. This highlights the importance of space-based missions to make accurate measurements with excellent spatio-temporal resolutions.

Observations are important to validate climate models and the magnitude of the energy imbalance at the TOA. Incident solar flux is an input to the model. The solar radiation reflected from the Earth or OSR and the OLR are calculated at the model level and must be compared to those measured. Table 2 summaries the expected science requirements for disruptive measurements of the energy imbalance components. In addition, some objectives of the UVSQ-SAT mission are presented. UVSQ-SAT is a science and technology demonstrator, one of whose objectives is to obtain an accurate short-term measurement of the EEI with a relative uncertainty that must be less than ± 1.0 Wm^{-2} over the course of a year. The expected spatial resolution of the UVSQ-SAT measurements is in the order of 1000 km. Forward, the method of Gristey et al. (2017) [17] will be used to improve the accuracy of the measurements thanks to a spherical harmonic analysis.

Table 2. Scientific requirements for the components of the Earth's radiation budget. High scientific relevance for a satellites constellation (Terra-F) and UVSQ-SAT expected performances.

Requirements	**Scientific Relevance**			
–	Absolute uncertainty	Stability per decade	Spatial resolution	Time resolution
TSI	± 0.54 Wm^{-2} at 1σ	± 0.14 Wm^{-2} at 1σ	–	24 h
OSR	± 1.00 Wm^{-2} at 1σ	± 0.10 Wm^{-2} at 1σ	10–100 km	Diurnal cycle
OLR	± 1.00 Wm^{-2} at 1σ	± 0.10 Wm^{-2} at 1σ	10–100 km	Diurnal cycle
Requirements	**UVSQ-SAT Performances**			
–	Absolute uncertainty	Stability per year	Spatial resolution	Time resolution
TSI	–	–	–	–
OSR	± 10 Wm^{-2} at 1σ	± 5 Wm^{-2} at 1σ	1000 km	15 days
OLR	± 10 Wm^{-2} at 1σ	± 1 Wm^{-2} at 1σ	1000 km	15 days

3. UVSQ-SAT Data Method and Map Reconstruction of the Variables (Observations and Model)

3.1. UVSQ-SAT Data Processing to Obtain Observation Time Series

3.1.1. Instrumental Equations of the Earth's Radiative Sensors

The UVSQ-SAT scientific payload [1] consists of twelve miniaturized Earth's radiative sensors (ERS) based on thermopiles for monitoring the incoming solar radiation and the outgoing terrestrial radiation. Each face of the UVSQ-SAT satellite has two ERS sensors, which have different optical coatings (carbon nanotubes or optical solar reflector). The properties of the coatings were characterized in lab (solar absorption, hemispherical emissivity, bidirectional reflectance distribution function (BRDF)). The principle of these ERS sensors with their associated coatings is to convert thermal energy into electrical energy. The output voltage (V) is passively induced from the thermopile proportional to the heat flux (Wm^{-2}) through the sensor or similarly the temperature gradient across the thin-film substrate and number of thermocouple junction pairs. The output voltage is conditioned by an electronic unit and converted into analog-to-digital units (ADU) and is read by the UVSQ-SAT onboard computer (OBC) thanks to a serial peripheral interface (SPI) bus, and stored by the OBC. Once data are retrieved, the ADU signals S_{ADU} are converted into physical units to express the incident flux measurements Φ_{in} on an ERS sensor using the transfer function (Equation (1)):

$$\Phi_{in} = \left(\frac{S_{ADU}(T_s, T_b) - V_{ADU}^{ref}(T_b)}{N_{samp}} + C_1(T_s, T_b) \right) \times \frac{1}{G(T_b)} \times \frac{1}{\mathrm{Sens}(T_s)} + C_2(T_s) \tag{1}$$

where V_{ADU}^{ref} is a reference voltage, N_{samp} is the number of samples for a measurement, $G(T_b)$ is the gain of the electronic unit, Sens is the sensitivity of the sensor (calibration in lab), C_1 and C_2 are corrective offsets, which depend on the temperature of the sensor. In the case of thermopiles, $C_2 = -\epsilon\sigma T_s^4$. ϵ is the hemispherical emissivity of the thermopile coating and T_s is the ERS sensor temperature. T_b represents the temperature of the sensors electronic board.

The total incident flux (Φ_{in}) on each ERS sensor is the sum of the solar incident flux, the planetary incident flux (Φp_{in}), and the albedo incident flux (Φa_{in}). The incident solar flux is considered to be known (TSI is measured precisely elsewhere). By combining information from two ERS sensors on the same face (sensor with carbon nanotubes (solar absorption close to 1) and sensor with optical solar reflector (solar absorption less that 0.1)), we get the albedo (a) of the planet and the outgoing longwave radiation (OLR) at a measurement point located in latitude and longitude on the world map (Equations (2) and (3) are given as example if a satellite face observes at the nadir):

$$\Phi p_{in} = \epsilon \times \mathrm{OLR} \times \left(\frac{R}{R + z_{sat}} \right)^2 \tag{2}$$

$$\Phi a_{in} = \alpha \times a \times \left(\frac{R}{R + z_{sat}} \right)^2 \times \cos(\xi) \times \mathrm{TSI} \times \left(\frac{1\,au}{d_s} \right)^2 \tag{3}$$

where R is Earth's radius, z_{sat} is the satellite altitude, α is the ERS (coating) solar absorption, ξ is the solar zenith angle, $1\,au$ is one astronomical unit, and d_s is the distance UVSQ-SAT—Sun.

During satellite eclipse periods, two ERS sensors on the same face measure only OLR, which allow to obtain direct measurements and inter-calibration of the sensors.

3.1.2. Instrumental Equations of the Optical Sensors Based on Photodiodes

The UVSQ-SAT scientific payload has also six photodiodes to measure the total solar irradiance and the outgoing shortwave radiation, and four additional photodiodes for observing the Sun (three measure the total solar irradiance and one measures the UV solar spectral irradiance in the Herzberg continuum). Each face of the satellite has at least one photodiode who observes external fluxes. A photodiode is a semiconductor p-n junction device that converts light into an electrical current. The resulting current is converted into ADU and stored by the UVSQ-SAT onboard computer. Once data are retrieved, the ADU signals Sp_{ADU} are converted into physical units to express the measurements Φ_{pho} using the transfer function (Equation (4)):

$$\Phi_{pho} = \left(\frac{Sp_{ADU}(T_{pho}, T_b) - V_{ADU}^{ref}(T_b)}{Np_{samp}} + Cp_1(T_{pho}, T_b) \right) \times \frac{1}{G_{pho}(T_b)} \times \frac{1}{\mathrm{S_{pho}}(\mathrm{T_{pho}})} \tag{4}$$

where Np_{samp} is the number of samples for a measurement, G_{pho} is the gain of the electronic unit, $S_{pho}(T_{pho})$ is the responsivity of the photodiode (calibration done by the manufacturer), Cp_1 is a corrective offset, which depends on the temperature of the photodiode. T_{pho} is the satellite structure temperature close to the photodiode.

For the six photodiodes that measure the total solar irradiance and the outgoing shortwave radiation, the albedo of the Earth can be obtained using an equation similar to that proposed in Equation (3).

3.1.3. Methodology for Obtaining UVSQ-SAT Attitude and Position Time Series

UVSQ-SAT has a new 3-axis accelerometer/gyroscope/compass (Teach' Wear (TW)) for providing satellite attitude estimation [1]. Using the TW observations, it is possible to achieve sufficient attitude estimation accuracy for a satellite using regular Kalman filter algorithm. Another way is to use a new method based on a multilayer perceptron network to determine the UVSQ-SAT satellite attitude [18].

Otherwise, we have a good knowledge of the UVSQ-SAT CubeSat position (latitude, longitude, altitude) according to the Two-Line Element sets (TLE) that are generally used to predict the position of a satellite. For each measurement carried-out by the instruments, we have its location on the world map. Figure 7 shows an example of ground track measurements location of the UVSQ-SAT satellite during February 2021, which represents the path on the surface of the Earth directly below the satellite's trajectory. The knowledge of the satellite position allows to calculate notably the solar zenith angle (from latitude, longitude, solar declination, Universal Time) and to rigorously locate the measurements area.

Figure 7. Location (latitude and longitude) of UVSQ-SAT observations (red colored dots on the world map) during the month of February 2021.

3.2. Map Reconstruction Method from UVSQ-SAT Observation Time Series

Time series of the albedo and the OLR at the TOA are obtained from the observations made by the dedicated sensors on-board UVSQ-SAT.

Physically, each measurement of these time series represents the integral of the signal of interest (OLR, albedo) and depends in practice on a large number of parameters among which the bidirectional reflectance distribution function of the Earth's surface, the opacity of the atmosphere, the spectral and angular sensitivities of the sensors and their FOV. In particular, concerning the opacity, it depends on factors such as aerosol composition, clouds, temperature or pressure.

In a first approach, we consider that each measurement results from the contribution of a Gaussian distribution $G(\theta_{i,j})$ of points located at the surface of the Earth. Let the surface of the Earth modeled by a regular grid whose coordinates are expressed with latitude and longitude. Each pixel (i, j) of the grid is defined by its longitude λ_i and latitude ϕ_j coordinates and by its area S_{ij} (Equation (5)):

$$S_{ij} = R^2 \times \cos(\phi_j) \, \Delta\lambda_i \, \Delta\phi_j \tag{5}$$

where $\Delta\lambda_i$, $\Delta\phi_j$ are size in radians of the pixel at longitude λ_i and latitude ϕ_j.

The satellite has a location $\lambda_{sat}, \phi_{sat}$ and an altitude z_{sat}. The angle $\theta_{i,j}$ at the center of the Earth (nadir) between the satellite and the pixel (i, j) is given by Equation (6). The angle $\alpha_{i,j}$ between the nadir of the satellite and the pixel (i, j) is given by Equation (7). The satellite elevation $\beta_{i,j}$ (complement of zenith angle) seen by the pixel (i, j) is given by Equation (8).

$$\theta_{i,j} = \mathrm{acos}[\cos(\phi_j)\cos(\phi_{sat})\cos(\lambda_i - \lambda_{sat}) + \sin(\phi_j)\sin(\phi_{sat})] \quad (6)$$

$$\alpha_{i,j} = \mathrm{atan}\left(\frac{R\sin(\theta_{i,j})}{z_{sat} + R\,(1 - \mathrm{acos}(\theta_{i,j}))}\right) \quad (7)$$

$$\beta_{i,j} = \frac{\pi}{2} - \alpha_{i,j} - \theta_{i,j} \quad (8)$$

The view angle $\Omega_{i,j}$ under which the pixel (i, j) is seen by the satellite is given by Equations (9) and (10).

$$\Omega_{i,j} = S_{ij}\, z^2 \cos(\alpha_{i,j})\sin(\beta_{i,j}) \qquad \text{if} \qquad \beta_{i,j} \geq 0 \quad (9)$$

$$\Omega_{i,j} = 0 \qquad \text{if} \qquad \beta_{i,j} < 0 \quad (10)$$

Figure 8 represents the different angles between the satellite, the nadir, and the observation location. This latter location is seen by the satellite according to a Gaussian distribution $G(\theta_{i,j})$, which is defined by Equation (11):

$$G(\theta_{i,j}) = \exp\left(-\frac{\alpha^2}{2\sigma^2}\right) \quad (11)$$

with σ, the variance of the Gaussian. σ is related to the FOV via $\sigma = \mathrm{FOV}/2$. For example, for an average altitude of 540 km and a FOV of 135°, which are the characteristics of UVSQ-SAT, the diameter of the ground track is about 2600 km.

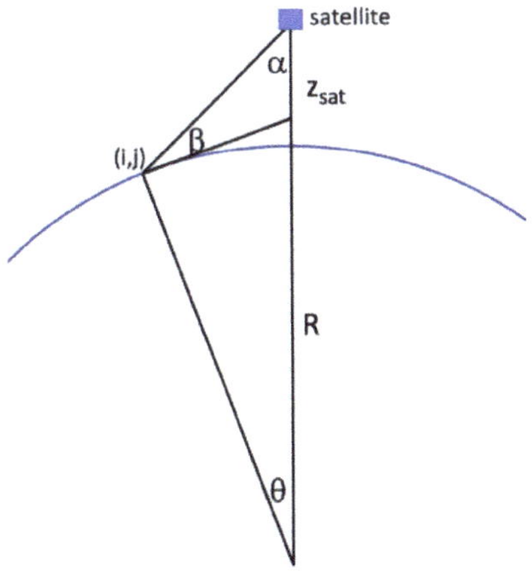

Figure 8. Visualization of the satellite in orbit, characteristic angles and pixel seen by the satellite.

The observed flux by UVSQ-SAT $F_{sat}(\lambda_{sat}, \phi_{sat})$ is given by (Equation (12)):

$$F_{sat}(\lambda_{sat}, \phi_{sat}) = \frac{\sum_{i,j} \Omega_{i,j} G_{i,j} F_{i,j}}{\sum_{i,j} \Omega_{i,j} G_{i,j}} \tag{12}$$

Finally, to reconstruct the map of what was measured by UVSQ-SAT, i.e., $F_{sat}(\lambda_{sat}, \phi_{sat})$, we calculate the functions $\Omega_{i,j}$, $G_{i,j}$ for the satellite track to find the flux $F_{i,j}$ associated with each pixel.

3.3. Reconstruction Method of the ERA 5 Maps to Compare with the UVSQ-SAT Maps

To compare the UVSQ-SAT measurements and a model (ERA 5), it is necessary to have similar inputs that means, i.e., to have the same observations time series (temporal) and similar FOV (spatial) at TOA.

Figure 9 shows the monthly averaged maps of the albedo and the OLR in February 2021 (for the ERA 5 model). All data of the ERA 5 model were used to obtain these maps (temporal resolution of 1 h and a pixel resolution of 721 × 1440).

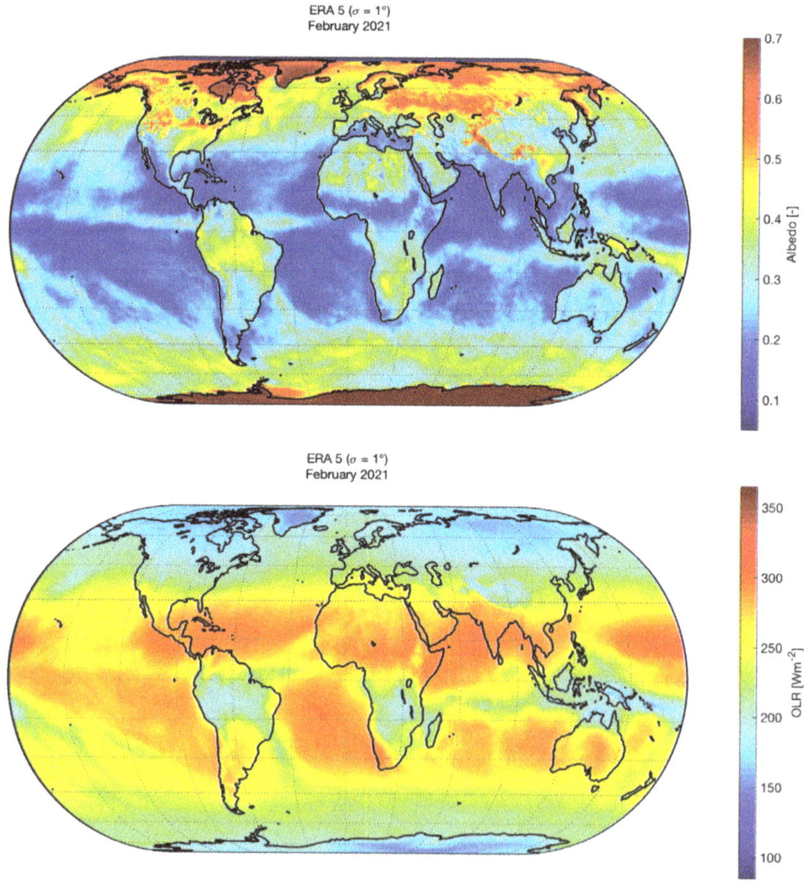

Figure 9. (**Top**) Albedo from ERA 5 during the month of February 2021. (**Bottom**) OLR from ERA 5 during the month of February 2021.

Albedo is computed by the ratio of the difference of the incident shortwave radiation and the net shortwave radiation by the incident shortwave radiation. Surfaces that reflect a lot of the light falling on them have a high albedo. Surfaces that don't reflect much light have a low albedo. Albedo values vary across the globe with latitude. At the poles, it can be greater than 0.7 in some areas. This is a result of the lower solar angle present at the poles but also the higher presence of fresh snow, ice, and smooth open water. In the tropics (23.5°N to 23.5°S), albedo is between 0.1 to 0.4. OLR values vary also across the globe with latitude, and is primarily sensitive to near-surface and atmospheric temperatures, the air humidity, and the presence of clouds, which are related to the intensity of convective activity as well as the latitude and altitude dependence of the variability. Low OLR values (<200 Wm^{-2}) associated with deep atmospheric convection are found over the equatorial land masses, Amazon basin, and in the western equatorial Pacific. These maps highlight the finesse of the observations (notably in equatorial Atlantic ocean). These details will disappear when increasing the FOV, and consequently degrade the spatial resolution. Considering the ground track of UVSQ-SAT in February 2021 (with same spatio-temporal observations), it is possible to reconstruct new albedo and OLR maps from the ERA 5 model. The impact of the FOV (related to σ) also has a significant effect on the map reconstructions. Figure 10 shows the reconstructed map of the albedo for the ERA 5 model for $\sigma = 1°$ and $\sigma = 180°$. Figure 11 shows the reconstructed map of the OLR for the ERA 5 model for $\sigma = 1°$ and $\sigma = 180°$.

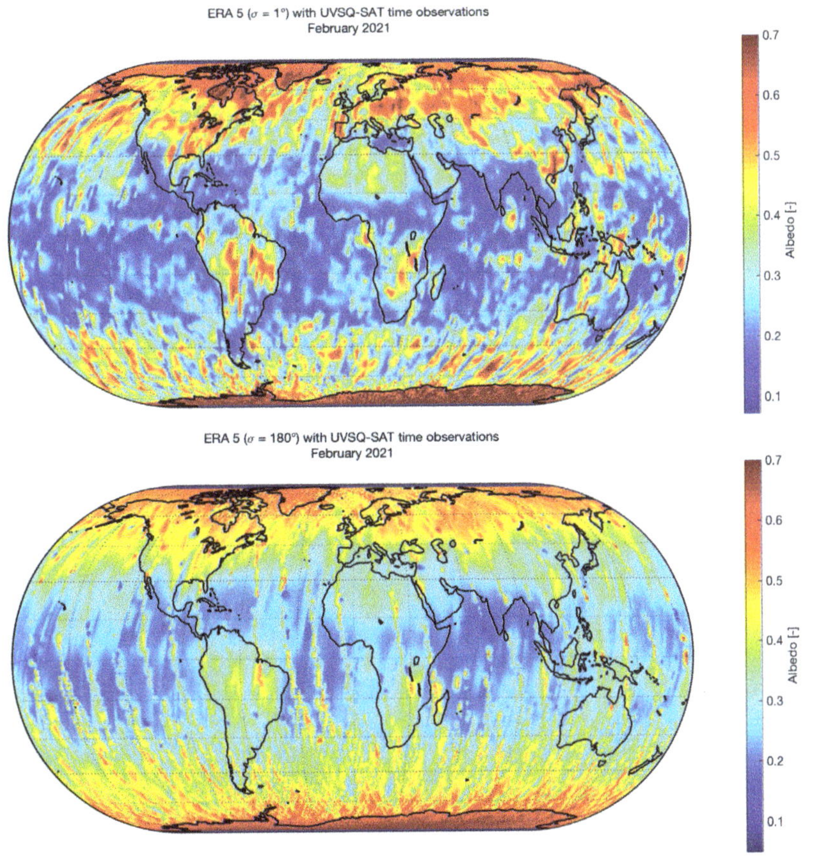

Figure 10. (**Top**) Albedo (ERA 5) during the month of February 2021 from same observation than UVSQ-SAT (σ = 1°). (**Bottom**) Albedo (ERA 5) from same observation than UVSQ-SAT (σ = 180°).

These reconstructed maps are slightly different from the ERA 5 monthly averaged data (Figure 9). The selected data (same spatio-temporal resolution than UVSQ-SAT) related to the ground track of the satellite (Figure 7) have an impact on the reconstructed maps (clearly visible for albedo when increasing σ). Low OLR values in the equatorial Atlantic ocean disappear (Figure 11 Bottom) due to the lack of data and low spatial resolution.

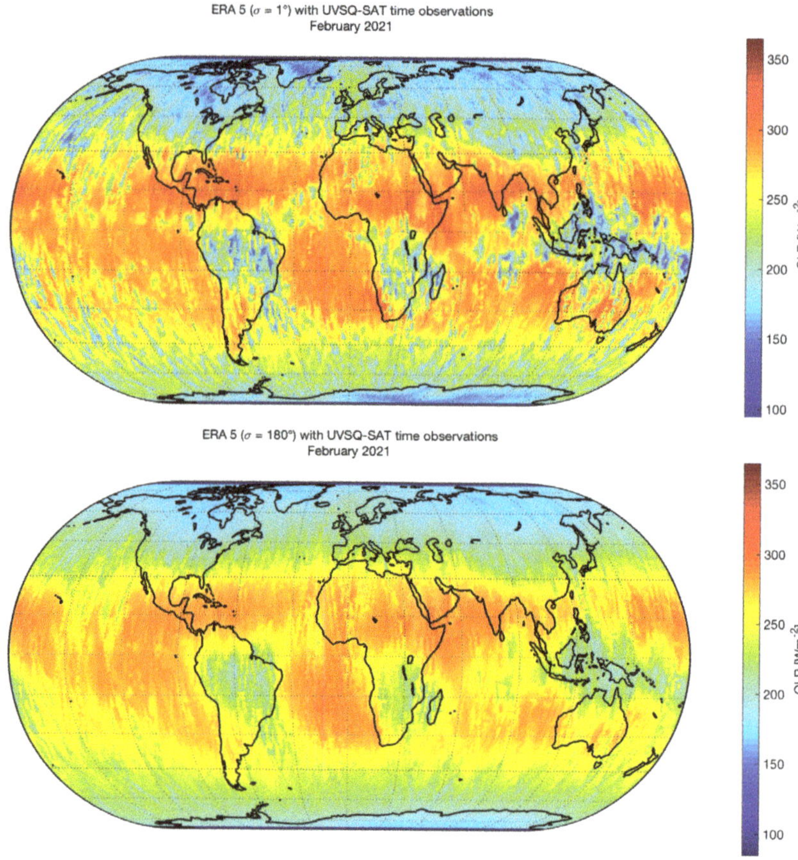

Figure 11. (**Top**) OLR (ERA 5) during the month of February 2021 from same observation than UVSQ-SAT (σ = 1°). (**Bottom**) OLR (ERA 5) from same observation than UVSQ-SAT (σ = 180°).

These simulations show the importance for doing the right comparisons between model and observations (same spatio temporal resolution). That's why, other methods are under investigation to improve the finesse of the details to be restored (albedo and OLR maps). In a second approach, it is planned to use the spherical harmonics method developed by Gristey et al. (2017) [17] to reconstruct the maps of the quantities of interest. The value measured by the satellite F_{sat} is expressed as follows (Equation (13)):

$$F_{\text{sat}}(\lambda, \theta) = \sum_{l=0}^{L} \sum_{m=0}^{l} [C_{lm} \overline{Y}_{lm}^{C}(\theta, \lambda) + S_{lm} \overline{Y}_{lm}^{S}(\theta, \lambda)] + e \tag{13}$$

with θ, λ, the colatitude and longitude of the point considered, $\overline{Y}_{lm}^{C}$ and $\overline{Y}_{lm}^{S}$, the spatially integrated spherical harmonics, C_{lm}, S_{lm}, the coefficients that we seek to determine and e, the error on the measurements. The coefficients C_{lm} and S_{lm} are then estimated by the least squares method.

4. UVSQ-SAT First Observations

Since February 2021, UVSQ-SAT has been measuring, among other things, solar radiation reflected from the Earth and OLR. UVSQ-SAT does not have an active attitude control system. Therefore, it is equipped with several photodiodes and ERS sensors on all sides to perform scientific measurements. Six photodiodes allow to measure both the TSI and the OSR. The albedo is obtained from these measurements considering the TSI as known and measured with other space-based instruments. Twelve ERS sensors allow to measure the TSI, the OSR and the OLR with a wide FOV in a complete hemisphere. The whole absorbed radiations generate temperature gradients on the ERS thermopile which delivers a voltage proportional to the absorbed flux (Equation (1)). Six of the twelve ERS thermopiles are coated with carbon nanotubes in order to absorb all incident radiation from short to long wavelengths (absorption very close to 1). The BRDF of the carbon nanotubes coating was measured with a goniophotometer. The optical coating shows a weak angular dependence. During eclipses, these detectors measure only the OLR. Out of eclipses, the carbon nanotube-based thermopiles measure the TSI, the OSR, and the OLR. There are also six ERS thermopiles, which are coated with optical solar reflectors to mainly measure the OLR.

4.1. UVSQ-SAT Time Series

From the measurements made by the different sensors of UVSQ-SAT and using the instrumental equations described in Sections 3.1.1 and 3.1.2, we obtain the evolution of the albedo and the OLR. Figure 12a shows the evolution of the signal measured by the six photodiodes over time (example given for about 2 observation orbits). The eclipse periods are clearly identifiable by the absence of signal. Figure 12b shows a good agreement between the calculated albedo and thus modeled from the ERA 5 data using different FOV (referred as σ values). On Figure 12c, the satellite temperatures evolve in accordance with the UVSQ-SAT thermal qualification (Figure 4). These temperature and the signal measured by the ERS sensors allow to retrieve the OLR (Figure 12d), which is directly observed during eclipse periods. Once again, the measured OLR fluxes show a good agreement with the modeled ones from the ERA 5 data.

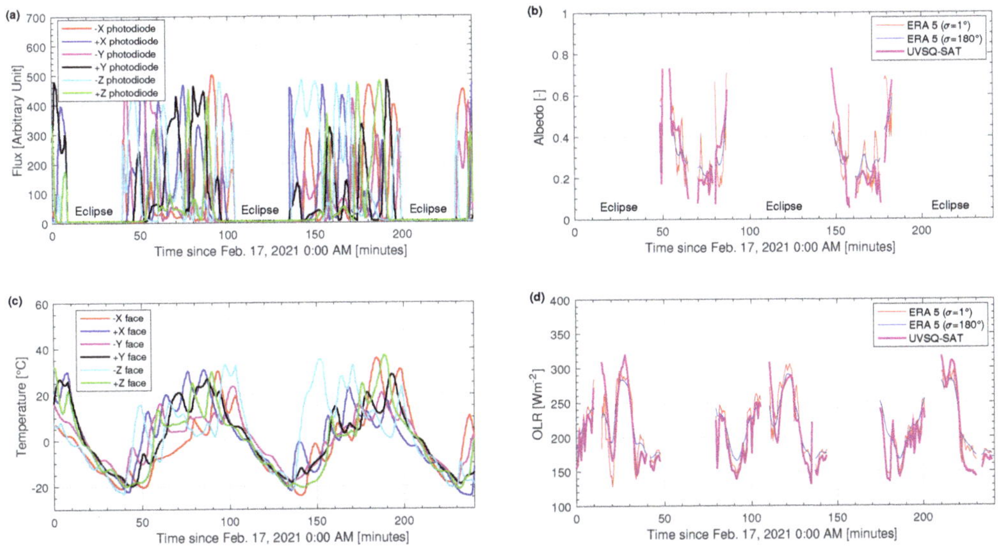

Figure 12. (**a**) Evolution of the measured flux by the six photodiodes during a given period. (**b**) Evolution of the albedo modeled from the ERA 5 data and for UVSQ-SAT observations. (**c**) Evolution of the satellite temperatures. (**d**) Evolution of the OLR.

The accumulation of OLR measurement points over time for the Earth's location points will provide averages to characterize the evolution of the OLR over time. It is important to have new record of the OLR to bring new lights and to continue historical records (Figure 6). A long term continuous record is possible using a new disruptive space observational system based on UVSQ-SAT pathfinder.

4.2. UVSQ-SAT Maps Reconstruction

UVSQ-SAT maps reconstruction (albedo and OLR) are based on UVSQ-SAT time series. Currently, the reconstruction of each map is based on a Gaussian function whose parameters are defined by the satellite altitude and its FOV. In addition, a method based on a deep learning (DL) algorithm has been developed to obtain the satellite attitude in order to better determine the Earth's albedo and OLR [18]. The satellite attitude can also be reconstructed from the measurements of an inertial measurement unit (IMU). UVSQ-SAT is equipped with a gyrometer to measure angular velocity, an accelerometer to measure gravity and linear acceleration, and a magnetometer to measure the Earth's magnetic field. The magnetometer allows to measure the direction and the intensity of the magnetic field. The objective is to gather these inertial and magnetic data in order to reconstruct the attitude of the satellite. Figure 13 shows the intensity of the Earth's magnetic field measured by UVSQ-SAT thanks to its different sensors.

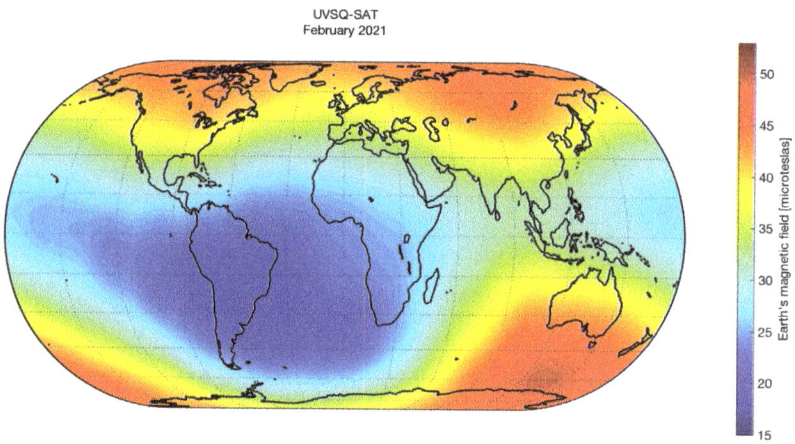

Figure 13. Earth's magnetic field intensity measured by UVSQ-SAT instruments. The calibrations are being validated to consolidate the absolute measurement bias.

Regarding the albedo and OLR, preliminary results have been obtained by applying the methods described in this study and in Meftah et al. (2020) [1]. Figure 14 shows a reconstructed map of the Earth's albedo (February 2021) obtained from UVSQ-SAT instruments. The albedo due to the atmosphere comes from Rayleigh backscattering in the short wavelength range and from clouds, which contribute about two thirds of the total albedo. The effect of the TSI at the TOA is stronger in the tropics. The high albedo at polar latitudes due to the presence of snow and ice on the ground has an important effect. This is of interest to follow this evolution over time. There are also local differences due to the albedo of cloudy regions (intertropical zone of convergence) or the ground (Sahara). The albedo of the ground depends strongly on its nature, it can be very high for fresh snow (about 0.75) and low for vegetation (below 0.2). The albedo of the ocean remains very low and of the order of 0.1. It depends on the distribution of waves.

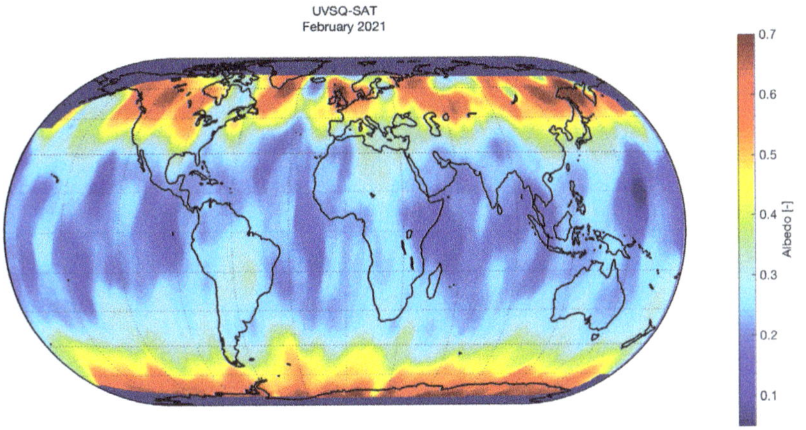

Figure 14. Earth's albedo at TOA measured by UVSQ-SAT instruments. The dark blue areas correspond to regions without data.

The OLR at the TOA (Figure 15) also has a latitude dependence. The high latitudes are cooler and emit less infrared radiation. The humid tropical regions are clearly visible. In tropical and equatorial regions, the weak radiation emerging at the TOA at long wavelengths is due to the presence of clouds at high altitudes. These clouds absorb the radiation emitted by the Earth's surface. Therefore, as they are cold, they emit a weak outgoing radiation in space.

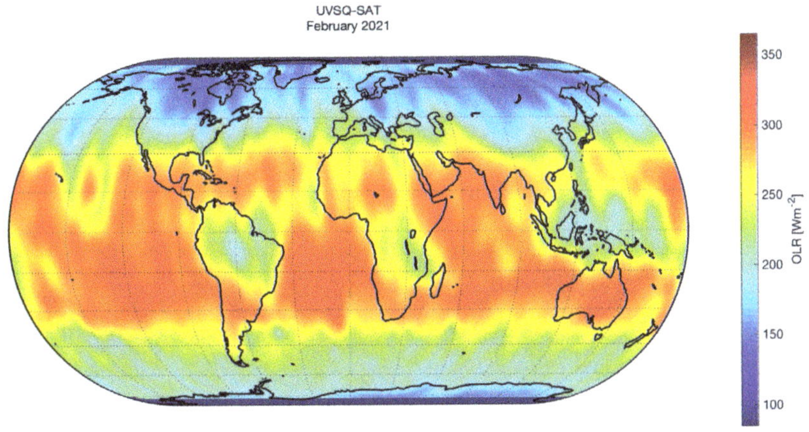

Figure 15. OLR at the TOA measured by UVSQ-SAT instruments.

UVSQ-SAT maps (Figures 14 and 15) look identical to the maps obtained with the ERA 5 model using the same observations time series (Figures 10 and 11). The UVSQ-SAT sensors used have wide fields of view. The reconstructions maps are associated with wide fields of view.

5. Perspectives—Toward a Satellites Constellation for Climate Studies

The objective of the UVSQ-SAT mission is to validate the principle of several miniaturized technologies at the highest level of technological maturity (TRL 9) in order to be able to accurately measure the EEI using a constellation (homogeneous and or heterogeneous) of small satellites. It will be necessary to combine simple instruments (wide FOV and wide wavelength band) with more complex instruments (narrow FOV and using on-board calibration sources to follow the evolution of the degradation of the instruments in orbit). The Radiometer Assessment using Vertically Aligned Nanotubes (RAVAN) 3U CubeSat mission [19] develops a similar approach to demonstrate technologies for the measurement of Earth's radiation budget in order to implement a satellites constellation.

Gristey et al. (2017) [17] showed how a constellation of satellites with broadband radiometers can provide both spatial and temporal information to accurately measure the ERB. Wiscombe and Chiu (2013) [20] proposed a constellation of space-based instruments, capable of overcoming sampling limitations and providing global and diurnal measurements of the Earth's outgoing radiation with accuracy. With advances in technology and miniaturization of instruments, it is possible to envision a constellation of small satellites capable of measuring outgoing and variable terrestrial radiation from the Earth. The "UVSQ-SAT" concept is fully integrated in this approach and shows a way.

With a single low-orbiting satellite (UVSQ-SAT), it is possible to obtain "satisfactory" mappings of the OSR and the OLR at the TOA only every month (with a loss of short time-scale information). Meftah et al. (2020) [1] show that a constellation of at least 50 satellites is necessary to have at least a daily mapping. Indeed, a constellation of at least 50 satellites is a prerequisite to capture spatio-temporal variations (every three hours and ideally a resolution of a few km). This would be fundamental data to validate climate models. These data are used to validate models and the magnitude of energy imbalances at the TOA. The model mesh size is about 100–200 km. The Baseline Surface Radiation Network (BSRN) and Atmospheric Radiation Measurements Program network (ARM) can be used on local and regional scale models. They are mainly used in the calibration and use of spatial data. These data are of primary importance to help validate spatial data and numerical model simulations.

UVSQ-SAT and its future constellation (Figure 16) address the same scientific theme as Far-infrared Outgoing Radiation Understanding and Monitoring (FORUM), but from a completely different angle. FORUM will be put into orbit in 2026. It will measure the Earth's infrared emission spectrum at high spectral resolution. Libera (named for the daughter of agricultural goddess Ceres in ancient Roman mythology) represents another important mission to measure the Earth's outgoing radiative energy. The Libera instrument will fly on National Oceanic and Atmospheric Administration (NOAA) operational Joint Polar Satellite System-3 (JPSS-3) satellite, which is scheduled to launch by December 2027.

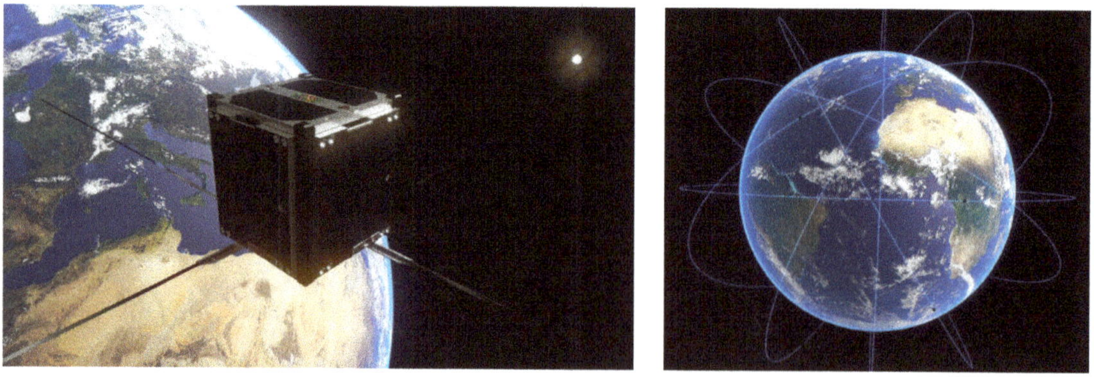

Figure 16. UVSQ-SAT and its potential future satellites constellation named Terra-F.

6. Conclusions

The UVSQ-SAT mission was launched in January 2021. Since February 2021, the satellite has been observing the Earth and the Sun. The first results were obtained using the methods described by Meftah et al. (2020) [1]. These results are very encouraging and clearly show that small satellites represent a fast method to answer key scientific questions. In the era of NewSpace, i.e., an era benefiting from technological advances in miniaturization and reasonable costs to access space, this mission clearly shows that a constellation of small satellites dedicated to the measurement of the EEI is feasible.

At a time when the climate emergency is obvious and the signals of global warming of the Earth are largely above the level of natural variability, one of the greatest challenges remains the difficulty of obtaining adequate funding for research and especially for missions dedicated to the study of climate (EEI at the TOA, spectral solar irradiance, vertical temperature profile, ozone, aerosols, etc.). These missions must rely on the implementation of satellites constellation to ensure sufficient spatial and temporal coverage. Satellites constellation dedicated to the study of climate and science are now a matter of course [21]. The UVSQ-SAT prototype opens the way to a potential Terra-F constellation composed of 50 small satellites with the necessary performance to meet the scientific need.

Other solutions to measure the EEI are possible with two "traditional" satellites placed at the L1 and L2 Lagrange points in order to make measurements of our planet on the illuminated and non-illuminated portions. The main goal is to make measurements of the EEI at the TOA and its variability in time in order to detect any long-term global trend with a stability of at least ± 0.2 Wm^{-2} per decade but also to be able to obtain spatial resolutions of a few tens of km and temporal resolutions of three hours with an accuracy of a few Wm^{-2}. These measurements must be performed on wide wavelength bands.

Today, it is becoming increasingly clear that small satellites are not only a fast way to answer key scientific questions but also contribute qualitatively to scientific research.

Author Contributions: Conceptualization, M.M. (Mustapha Meftah), T.B., C.D., A.H., P.K., A.F., S.B., S.A., E.B., L.D., J.-L.E., P.G. (Patrick Galopeau), P.G. (Pierre Gilbert), L.L., A.S., A.-J.V., P.L., X.A., O.H.F.d., A.M., J.-P.C., F.B., M.M. (Michel Mahé) and C.M.; Data curation, M.M. (Mustapha Meftah), T.B., C.D., A.H., A.F. and N.C.; Formal analysis, M.M. (Mustapha Meftah), T.B., C.D., A.H. and A.F.; Funding acquisition, M.M. (Mustapha Meftah); Investigation, M.M. (Mustapha Meftah), E.B., P.L. and N.C.; Methodology, M.M. (Mustapha Meftah), T.B., C.D., A.H., P.K., A.F., S.B., S.A., L.D., J.-L.E., P.G. (Patrick Galopeau), P.G. (Pierre Gilbert), L.L., A.S., A.-J.V. and X.A.; Project administration, M.M. (Mustapha Meftah); Resources, M.M. (Mustapha Meftah); Software, M.M. (Mustapha Meftah), T.B., C.D., A.H., P.K. and A.F.; Supervision, M.M. (Mustapha Meftah); Validation, M.M. (Mustapha Meftah), T.B. and C.D.; Visualization, M.M. (Mustapha Meftah), T.B., A.H. and A.F.; Writing—original draft, M.M. (Mustapha Meftah), T.B., C.D., A.H., P.K., A.F., S.B., S.A., E.B., L.D., J.-L.E., P.G. (Patrick Galopeau), P.G. (Pierre Gilbert), L.L., A.S., A.-J.V., P.L., N.C., X.A., O.H.F.d., A.M., J.-P.C., F.B., M.M. (Michel Mahé) and C.M.; Writing—review and editing, M.M. (Mustapha Meftah), T.B., C.D. and A.H. All authors have read and agreed to the published version of the manuscript.

Funding: This research was mainly funded by Centre National de la Recherche Scientifique (CNRS, France), Université de Versailles Saint-Quentin-en-Yvelines (UVSQ, France), and Agence Nationale de la Recherche (ANR, France). This work was supported by the Programme National Soleil Terre (PNST, France) of CNRS/INSU (France) co-funded by Centre National d'Études Spatiales (CNES, France) and Commissariat à l'énergie atomique (CEA, France).

Acknowledgments: The UVSQ-SAT team acknowledges support from CNRS, UVSQ, the Sorbonne Université (SU, France), the Université Paris-Saclay (France), the Office National d'Etudes et de Recherches Aérospatiales (ONERA, France), CNES, the Laboratory for Atmospheric and Space Physics (LASP, USA), the National Central University (NCU, Taiwan), and the Nanyang Technological University (NTU, Singapore). The authors acknowledge the multiple and very fruitful discussions about the Earth's energy imbalance which took place during the conception and writing-up of the unsuccessful European Space Agency EArth enerGy imbalance ExploreR (EAGER) proposal. The authors thankfully acknowledge the Ministère de l'Enseignement supérieur, de la Recherche et de l'Innovation (MESRI, France) for their support. Finally, the authors would like to thank the referees for their valuable comments which helped to improve the manuscript.

Conflicts of Interest: The authors declare no conflict of interest.

References

1. Meftah, M.; Damé, L.; Keckhut, P.; Bekki, S.; Sarkissian, A.; Hauchecorne, A.; Bertran, E.; Carta, J.P.; Rogers, D.; Abbaki, S.; et al. UVSQ-SAT, a pathfinder cubesat mission for observing essential climate variables. *Remote Sens.* **2020**, *12*, 92. [CrossRef]
2. Twiggs, R.J. Space system developments at Stanford University: From launch experience of microsatellites to the proposed future use of picosatellites. In Proceedings of the Small Payloads in Space, San Diego, CA, USA, 7 November 2000; Horais, B.J., Twiggs, R.J., Eds.; Volume 4136, pp. 79–86. [CrossRef]
3. Puig-Suari, J.; Schoos, J.; Turner, C.; Wagner, T.; Connolly, R.; Block, R.P. CubeSat developments at Cal Poly: The standard deployer and PolySat. In Proceedings of the Small Payloads in Space, San Diego, CA, USA, 7 November 2000; Horais, B.J., Twiggs, R.J., Eds.; 2000; Volume 4136, pp. 72–78. [CrossRef]
4. Xie, S.P.; Kosaka, Y. What caused the global surface warming hiatus of 1998–2013? *Curr. Clim. Chang. Rep.* **2017**, *3*, 128–140. [CrossRef]
5. von Schuckmann, K.; Palmer, M.D.; Trenberth, K.E.; Cazenave, A.; Chambers, D.; Champollion, N.; Hansen, J.; Josey, S.A.; Loeb, N.; Mathieu, P.P.; et al. An imperative to monitor Earth's energy imbalance. *Nat. Clim. Chang.* **2016**, *6*, 138–144. [CrossRef]
6. Loeb, N.G.; Manalo-Smith, N.; Su, W.; Shankar, M.; Thomas, S. CERES top-of-atmosphere Earth radiation budget climate data record: Accounting for in-orbit changes in instrument calibration. *Remote Sens.* **2016**, *8*, 182. [CrossRef]
7. Allan, R.P.; Liu, C.; Loeb, N.G.; Palmer, M.D.; Roberts, M.; Smith, D.; Vidale, P.L. Changes in global net radiative imbalance 1985–2012. *Geophys. Res. Lett.* **2014**, *41*, 5588–5597. [CrossRef] [PubMed]
8. Trenberth, K.E.; Fasullo, J.T.; Von Schuckmann, K.; Cheng, L. Insights into Earth's energy imbalance from multiple sources. *J. Clim.* **2016**, *29*, 7495–7505. [CrossRef]
9. Boyer, T.; Domingues, C.M.; Good, S.A.; Johnson, G.C.; Lyman, J.M.; Ishii, M.; Gouretski, V.; Willis, J.K.; Antonov, J.; Wijffels, S.; et al. Sensitivity of global upper-ocean heat content estimates to mapping methods, XBT bias corrections, and baseline climatologies. *J. Clim.* **2016**, *29*, 4817–4842. [CrossRef]
10. Cheng, L.; Trenberth, K.E.; Fasullo, J.; Boyer, T.; Abraham, J.; Zhu, J. Improved estimates of ocean heat content from 1960 to 2015. *Sci. Adv.* **2017**, *3*, e1601545. [CrossRef] [PubMed]
11. Kiehl, J.T.; Trenberth, K.E. Earth's annual global mean energy budget. *Bull. Am. Meteorol. Soc.* **1997**, *78*, 197–208. [CrossRef]
12. Stephens, G.L.; Li, J.; Wild, M.; Clayson, C.A.; Loeb, N.; Kato, S.; L'Ecuyer, T.; Stackhouse, P.W.; Lebsock, M.; Andrews, T. An update on Earth's energy balance in light of the latest global observations. *Nat. Geosci.* **2012**, *5*, 691–696. [CrossRef]
13. Trenberth, K.E.; Fasullo, J.T. Tracking Earth's Energy. *Science* **2010**, *328*, 316–317. [CrossRef] [PubMed]
14. Hansen, J.; Sato, M.; Kharecha, P.; von Schuckmann, K. Earth's energy imbalance and implications. *Atmos. Chem. Phys.* **2011**, *11*, 13421–13449. [CrossRef]
15. Von Schuckmann, K.; Cheng, L.; Palmer, M.D.; Hansen, J.; Tassone, C.; Aich, V.; Adusumilli, S.; Beltrami, H.; Boyer, T.; Cuesta-Valero, F.J.; et al. Heat stored in the Earth system: Where does the energy go? *Earth Syst. Sci. Data* **2020**, *12*, 2013–2041. [CrossRef]
16. Pachauri, R.K.; Allen, M.R.; Barros, V.R.; Broome, J.; Cramer, W.; Christ, R.; Church, J.A.; Clarke, L.; Dahe, Q.; Dasgupta, P.; et al. *Climate Change 2014: Synthesis Report. Contribution of Working Groups I, II and III to the Fifth Assessment Report of the INTERGOVERNMENTAL Panel on CLIMATE CHANGE*; IPCC: Geneva, Switzerland, 2014.
17. Gristey, J.J.; Chiu, J.C.; Gurney, R.J.; Han, S.C.; Morcrette, C.J. Determination of global Earth outgoing radiation at high temporal resolution using a theoretical constellation of satellites. *J. Geophys. Res. Atmos.* **2017**, *122*. [CrossRef]
18. Finance, A.; Meftah, M.; Dufour, C.; Boutéraon, T.; Bekki, S.; Hauchecorne, A.; Keckhut, P.; Sarkissian, A.; Damé, L.; Mangin, A. A New Method Based on a Multilayer Perceptron Network to Determine In-Orbit Satellite Attitude for Spacecrafts without Active ADCS Like UVSQ-SAT. *Remote Sens.* **2021**, *13*, 1185. [CrossRef]
19. Swartz, W.; Lorentz, S.; Papadakis, S.; Huang, P.; Smith, A.; Deglau, D.; Yu, Y.; Reilly, S.; Reilly, N.; Anderson, D. RAVAN: CubeSat Demonstration for Multi-Point Earth Radiation Budget Measurements. *Remote Sens.* **2019**, *11*, 796. [CrossRef] [PubMed]
20. Wiscombe, W.; Chiu, C. The next step in Earth radiation budget measurements. In *AIP Conference Proceedings*; American Institute of Physics: College Park, MD, USA, 2013; Volume 1531, pp. 648–651.
21. Baker, D.N.; Chandran, A.; Chang, L.; Macdonald, M.; Meftah, M.; Millan, R.M.; Park, J.H.; Kumar, P.; Price, C.; von Steiger, R.; et al. An International Constellation of Small Spacecraft. *Space Res. Today* **2020**, *208*, 23–28. [CrossRef]

remote sensing

MDPI

Communication

Current Status and Variation since 1964 of the Glaciers around the Ebi Lake Basin in the Warming Climate

Lin Wang [1], Changbin Bai [1] and Jing Ming [2,*]

1 State Key Laboratory of Cryospheric Science, Northwest Institute of Eco-Environment and Resources, Chinese Academy of Sciences, Lanzhou 730000, China; wanglin@lzb.ac.cn (L.W.); baichangbin@nieer.ac.cn (C.B.)
2 Beacon Science & Consulting, Malvern, SA 5061, Australia
* Correspondence: petermingjing@hotmail.com

Abstract: This work analyzed the spatial and temporal variations of the glaciers in the Ebi Lake basin during the period 1964 to 2019, based on the 1st and 2nd Chinese Glacier Inventories (CGI) and remote sensing data; this is believed to be the first long-term comprehensive remote sensing investigation on the glacier change in this area, and it also diagnosed the response of the glaciers to the warming climate by analyzing digital elevation modeling and meteorology. The results show that there are 988 glaciers in total in the basin, with a total area of 560 km^2 and average area of 0.57 km^2 for a single glacier. The area and number of the glaciers oriented north and northeast are 205 km^2 (327 glaciers) and 180 km^2 (265 glaciers), respectively. The glaciers are categorized into eight classes as per their area, which are less than 0.1, 0.1–0.5, 0.5–1.0, 1.0–2.0, 2.0–5.0, 5.0–10.0, 10.0–20.0, and greater than 20.0 km^2, respectively. The smaller glaciers between 0.1 km^2 and 10.0 km^2 account for 509 km^2 or 91% in total area, and, in particular, the glaciers smaller than 0.5 km^2 account for 74% in the total number. The glacial area is concentrated at 3500–4000 m in altitude (512 km^2 or 91.4% in total). The number of glaciers in the basin decreased by 10.5% or 116, and their area decreased by 263.29 km^2 (-4.79 km^2 a^{-1}) or 32% (-0.58% a^{-1}) from 1964 to 2019; the glaciers with an area between 2.0 km^2 and 5.0 km^2 decreased by the largest, -82.60 km^2 or -40.67% in the total area at -1.50 km^2 a^{-1} or -0.74% a^{-1}), and the largest decrease in number (i.e., 126 glaciers) occurs between 0.1 km^2 and 0.5 km^2. The total ice storage in the basin decreased by 97.84–153.22 km^3 from 1964 to 2019, equivalent to 88.06–137.90 km^3 water (taking 0.9 g cm^{-3} as ice mass density). The temperature increase rate in the basin was +0.37 °C $decade^{-1}$, while the precipitation was +13.61 mm $decade^{-1}$ during the last fifty-five years. This analysis shows that the increase in precipitation in the basin was not sufficient to compensate the mass loss of glaciers caused by the warming during the same period. The increase in temperature was the dominant factor exceeding precipitation mass supply for ruling the retreat of the glaciers in the entire basin.

Keywords: Ebi Lake basin; decelerated-melting glaciers; climate change; climate response

Citation: Wang, L.; Bai, C.; Ming, J. Current Status and Variation since 1964 of the Glaciers around the Ebi Lake Basin in the Warming Climate. *Remote Sens.* **2021**, *13*, 497. https://doi.org/10.3390/rs13030497

Academic Editor: Xander Wang
Received: 29 December 2020
Accepted: 28 January 2021
Published: 30 January 2021

1. Introduction

The cryosphere consists of snow, ice, and permafrost on and below the Earth's land and ocean surfaces, and it is one of the major components in the climate system [1]. The accelerated shrinkage of the cryosphere in the context of global warming and the subsequent impacts on the sustainability of the Anthroposphere have attracted unprecedented attention and raised deep concern, because the High-Asian mountain glaciers are crucial for buffering against drought [2] and protecting life from drought [3]. These mountain glaciers are an essential part of the global cryosphere and have shown generally varying degrees of continuous retreat in recent decades [4–6]. As projected by the Intergovernmental Panel on Climate Change (IPCC), even in the mild emission scenario (RCP4.5), the Asian glaciers would disappear by ~50% by the end of the century [7].

China has the most developed mountain glaciers in low and middle latitudes [8], and these glaciers are a vital water resource in arid northwest China and High Asia [9]. As global temperatures increase, China's glacial covers are generally in negative mass balance and showing a retreating and thinning trend [10–12]. The Xinjiang region is an arid and semi-arid area of China rich in mountain glaciers. The Tianshan and Altai Mountains are two main mountains developing concentrated mountain glaciers. The glaciers in the Tianshan Mountains had experienced rapid mass loss (averaged from −24.6 mm w.e. a^{-1} in 1957–1970 to −444.6 mm w.e. a^{-1} in 1971–2009) during the second half of the 20th century [13]; while in Altai, over a quarter of mountain glaciers were projected to shrink in RCP4.5 by 2100 [14].

The Ebi Lake basin is located in the northwestern Xinjiang region, China (Figure 1). The increasingly irrigated area and population in the basin over the last 60 years have led to increased water consumption in the basin, increasing tension between supply and demand, and the ecological degradation [15–17]. This lake had shrunk by 50% from 1955 to 2013, as reported by NASA in 2014 [18]. Another study stated that the water area of the Ebi Lake showed a significantly decreasing trend, although the precipitation was increasing from 2001 to 2016 [19]. Therefore, it is essential to assess the regional glaciers change in a timely manner to monitor glacier water resources and assess their impact on the water resource supply in the basin. This aspect's research has conceptual and practical significances for water security, especially for industrial and agricultural production and economic development in this ecologically vulnerable area.

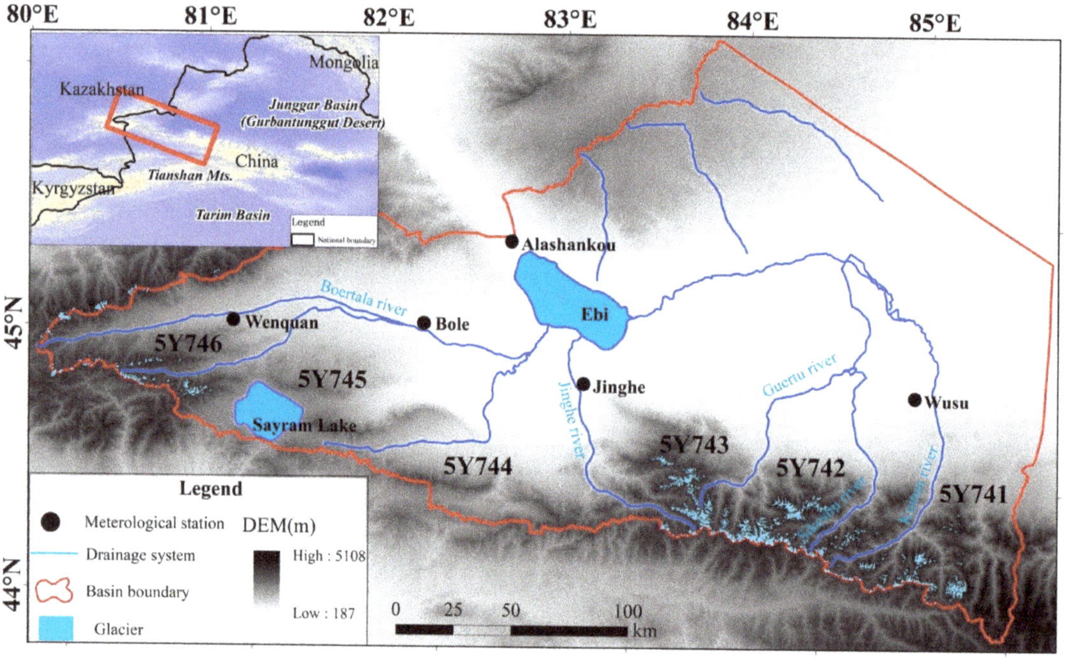

Figure 1. Study area with the Ebi Lake basin (5Y74) and related river and lake basins, where Kuitun River, Sikeshu River, Jinghe River, Daheyanzi River, Sayram Lake, and Bortala River sub-basins are annotated with their CGI codes of 5Y741, 5Y742, 5Y743, 5Y744, 5Y746, and 5Y746, respectively.

In recent years, numerous studies used the topographic maps and remote sensing data of the Tianshan Mountains, regional watersheds, and typical reference glaciers and revealed that the regional glaciers show retreating and thinning [20–26]. So far, very few studies on the glacier change status have been carried out in the Ebi Lake Basin. For example, Wang et al. used remote sensing data to partially reveal that there were around 450 glaciers in the Ebi Lake Basin, Tianshan, undergoing significant mass loss glacier changes from 1964 to 2004 [27]; Zhang et al. estimated the change of the Haxilegen 51 glacier in the basin from 1964 to 2006 and its response to climate [28]. To some extent, these studies explored the changes of an individual glacier or several glaciers in the basin. Still, the holistic and more detailed picture of these glaciers' change is not complete and update.

Dramatic change has had been with these glaciers in number, area, and ice volumes from 1964 to 2009. According to the 1st CGI, there were 1104 glaciers with the total glacial area of 823 km^2 and the ice storage of 47.54 km^3 by 1964; while to the 2nd CGI, the number of glaciers here decreased to 1000, the glacier area to 598 km^2, and the ice storage to 32.45 km^3 by 2009 [8]. However, the update status of the varying glaciers and their association with the regional climate change during the most recent decade (2009–2019) is little known, which are crucial for policy-makers to take proper measures to adapt climate change and mitigate the impact of the change. Therefore, based on the first and second Chinese Glacier Inventories (CGIs) and the most updated glacier vector data released in 2019, this work assesses the glaciers' change in the Ebi Lake Basin over the last 55 years (1964–2019), more systematically and accurately. Furthermore, we will discuss the response of glaciers in the basin to the warming climate and suggest the measures of how to use the regional water resource for sustainability rationally.

2. Materials and Methods

2.1. Study Area

The Ebi Lake Basin (43°38′~45°52′N, 79°53′~85°02′E) is this study area and located at the hinterland of Asia and Europe (Figure 1). The basin lies to the northern foothills of the western Tianshan Mountains and in the southwest of the Junggar Basin. It is surrounded by mountains from three sides—the north, west, and south—and in the east is China's second largest desert, Gurbantunggut Desert. The basin is concurrently in the northern hemispheric temperate zone with the significantly continental arid climate, mostly windy weather. The annual average temperature in the basin moderates between 6.6 and 7.8 °C, the annual precipitation between 116.0 and 169.2 mm, and the potential evapotranspiration from 1500 to 2000 mm [29,30]. The Ebi Lake is the largest lake in the basin and is the largest saltwater lake with about 500 km^2 in Xinjiang. The plains in the basin are accompanied with little precipitation and runoff, and mountain precipitation and snowmelt water from the peripheral alpine glaciers are the main sources of river runoff. The Ebinur Lake Basin (code 5Y74 in the CGIs) consists of six sub-basins, the Quitun River Basin (5Y741), the Sikeshu River Basin (5Y742), the Jing River Basin (5Y743), the Daheyanzi River Basin (5Y744), the Sailimu Lake Basin (5Y745), and the Bortala River Basin (5Y746) (Figure 1). The glaciers around the basin are typically continental glaciers developing along the valleys.

2.2. Data

2.2.1. Remote Sensing Images

At present, the publicly accessible high spatial resolution remote sensing images mainly include

- Landsat-8 OLI remote sensing images (30 m),
- Aster remote sensing images (15 m), and
- Sentinel-2 MSI remote sensing images (10 m).

Here, we use the Sentinel-2 MSI remote sensing images to retrieve the most accurate interpretation of the glacier boundaries. The image data have a revisit spanning time of five days and were acquired from June to September of 2019, which was the melting season of the glaciers. We use the Google Earth Engine (GEE) platform, a cloud-based geospatial processing platform, to quickly and efficiently filter high-quality, cloud-free Sentinel-2 MSI remote sensing images (referring to the work in [31]) and avoid laborious data collection, storage, organization, and preprocessing work.

2.2.2. The 1st and 2nd CGIs Data

The first CGI data, completed in 2002, are mainly derived from the 1960s' aerial-photographed topographic maps, and the watershed of the Ebi Lake basin includes 13 sheets of 1:50,000 topographic maps. Moreover, the second CGI data, released in 2014, integrate remote-sensing data, topographic maps, and digital elevation model data. Both times of CGI data are available on the National Tibetan Plateau Third Pole Environment Data Center (http://westdc.westgis.ac.cn/).

2.2.3. Digital Elevation Data

The digital elevation model is derived from the SRTM (Shuttle Radar Topography Mission), measured jointly by the NASA and the National Mapping Agency (NIMA), USA. This work uses the revised version 4.1 data with a spatial resolution of 90 m [32]. This version of SRTM data is provided by the CIAT (International Center for Tropical Agriculture) with a new interpolation algorithm. The nominal absolute elevation and planimetric accuracy of the data are ±16 m and ±20 m [33], respectively.

2.2.4. Meteorology

The meteorological data were obtained from the China Meteorological Data Service. (CMDS). The service provides the daily meteorological dataset collected from nearly 700 ground baseline stations in nationwide China. It can be accessed freely by registered educational and academic users, and here is the direct web link http://data.cma.cn/data/cdcdetail/dataCode/SURF_CLI_CHN_MUL_DAY_V3.0.html. In this work, the monthly temperature and precipitation data were retrieved from the CMDS dataset recorded by the five meteorological stations in the lower Ebi Lake Basin during 1964–2017 (Table 1).

Table 1. The locations of the five meteorological stations in the Ebi Lake Basin.

Code by CMDS	Name	Latitude (N)	Longitude (E)	Elevation (m)
51232	Alashankou	45°11′	82°35′	284.8
51238	Bole	44°54′	82°04′	531.9
51330	Wenquan	44°58′	81°01′	1354.6
51334	Jinghe	44°37′	82°54′	320.1
51346	Wusu	44°26′	84°40′	478.8

2.3. Methods

2.3.1. Confining Glacial Boundaries

Automatic interpretation methods applied to remotely sensed images for retrieving glacier boundaries are popular at present for simplifying and expediting the process [34]; however, the presence of snow, shadows, moraines, and water bodies makes it difficult to guarantee the accuracy of the automatic-interpretation methods in obtaining glacier boundaries. Visual interpretation is to obtain glacier-boundary information with remote sensing images based on existing glaciological knowledge, and this method is the most credible method to retrieve glacier boundaries [35]. Taking into account that the relatively small study area and accuracy, this study adopts visual interpretation for the latest remote sensing images to manually delineate glacier boundaries. With the ArcGIS version 10.5 software [36] and the first and second CGI cataloguing data, the glacier boundaries around the Ebi Lake Basin were vectorized and corrected by artificially visual interpretation in the Google Earth images, and the derived glacier boundaries were further amended with expert opinions.

2.3.2. Calculating Glacier Area and Volume

Changes in glacier area can be reflected by the difference in glacier area between the two periods, indicated by the rate of change in the glacier area and the relative rate of change in glacier area, using the following equation,

$$AC = (A_1 - A_0)/(t_1 - t_0), \tag{1}$$

$$\text{and } AAC = (A_1 - A_0)/[A_0 \times (t_1 - t_0)] \times 100\%, \tag{2}$$

where AC is the varying rate of glacier area (km^2/a), AAC is the relative varying rate of glacier area (%/a), A is glacier area (km^2), t indicates the year, and the subscript 0 and 1 indicate the starting and ending years for the specific glacial areas, respectively.

The glacier ice volume is not a directly approachable measurement here, because there were no direct measurements of ice thickness in most glaciers. This research used the volume–area empirical relationship by Gärtner-Roer et al. [37]:

$$V = c \times A^e, \tag{3}$$

where V is glacier ice storage (km^3), A is glacier area (km^2), and c and e are the empirical coefficients by Gärtner-Roer et al. [37], respectively.

2.3.3. Assessing the Uncertainty

The accuracy evaluation of glacier boundaries obtained from remote sensing images is important but difficult to determine. In processing remote sensing images, errors mainly come from glacier boundary extraction uncertainties and image resolution and alignment errors. Glacier boundary extraction uncertainties can be reduced through field validation and glaciological experience, and the uncertainty in remotely sensed images can be calculated through the following formulas (4 [38] and 5 [20]) to calculate their error, based on the formula for calculating the uncertainty of glacier length and area variation [39–41].

$$U_T = \sqrt{\Sigma\lambda^2} + \sqrt{\Sigma\varepsilon^2} \tag{4}$$

$$\text{and } U_A = 2U_T \sqrt{\Sigma\lambda^2} + \sqrt{\Sigma\varepsilon^2} \tag{5}$$

where U_T is the uncertainty of glacier length, λ is the impact resolution, ε is the alignment error of satellite images and the boundary layers, and U_A is the uncertainty of glacier area. In this study, λ equals 10 m for the Sentinel-2 images, and ε equals $\frac{1}{4}$ pixel. The calculated length and area uncertainties in this study are ±12.5 m and ±0.00025 km^2, respectively.

3. Results

3.1. Updated Status of Glaciers in the Ebi Lake Basin in 2019

Overall, the number of glaciers in the Ebi Lake Basin investigated in 2019 was 988, with an area of ~560 km^2, and the averaged area of a single glacier is 0.57 km^2. The glaciers with the area between 1.0 and 5.0 km^2 account for the largest share (40%) of the total glacier area in the study area, followed by those with areas of <0.5 km^2 (21%), 5.0–10.0 km^2 (17%), 0.5–1.0 km^2 (16%), and >10.0 km^2 (6%). The Sikeshu River Basin (5Y742) has the most extensively developed glaciers among the six sub-basins on all scales, while the Sayram Lake and Daheyanzi River basins have barely visible glaciers in Figure 1. The only two glaciers larger than 10 km^2 are both located in the Sikeshu River Basin (Figure 2a). 74% of the glaciers in number have areas smaller than 0.5 km^2 and are majorly located in the Kuitun, Sikeshu, and Bortala River sub-basins. The larger glaciers are much fewer distributed (Figure 2b). The glacial resources in the Ebi Lake basin are concentrated in four sub-basins (5Y741, 5Y742, 5Y743, and 5Y746), with the number of 974 (98.6%) and the area of ~557 km^2 (99.6%), respectively.

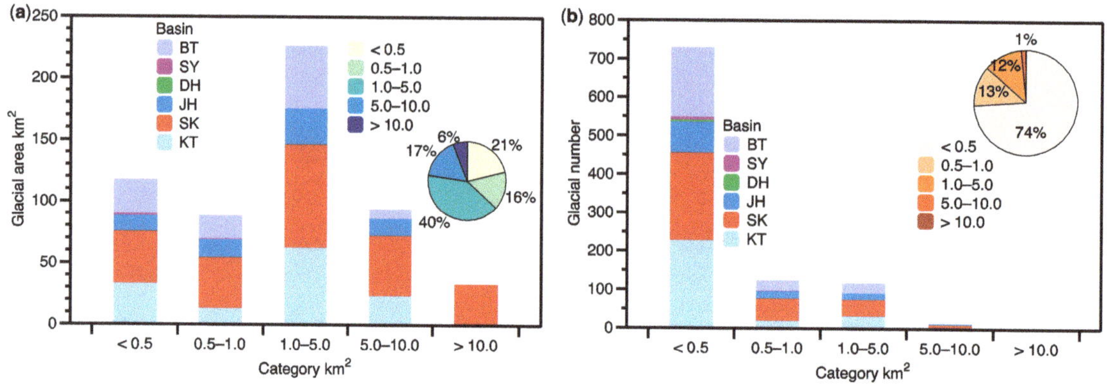

Figure 2. The glacial areas (**a**) and numbers of glaciers (**b**) in each sub-basin categorized into <0.5, 0.5–1.0, 1.0–5.0, 5.0–10.0, and >10.0 km^2, respectively, as of 2019. KT, SK, JH, DH, BT, and SY denote the Kuitun (5Y741), Sikeshu (5Y742), Jinghe (5Y743), Daheyanzi (5Y744), Bortala (5Y746) Rivers, and the Sayram Lake (5Y745) sub-basin, respectively. The percent shares of the glaciers in area and number are shown in the respective pie charts.

In the Ebi Lake Basin, there is around 400 km^2 of glaciers, accounting for 72% in a total spread from 3500 to 4000 m in elevation; 11% and 17% in area developed under 3500 m and between 4000 and 4500 m, respectively; and only 1% glacial area developed above 4500 m as of 2019 (Figure 3a). Most glaciers developed in the north (75% in area and 65% in number) and the east (15% in area and 22% in number) orientations (Figure 3b), implicated by the developing conditions of mountain glaciers if the north orientation indicates colder air masses, and the east implies richer water vapor sources in the Tianshan Mountains.

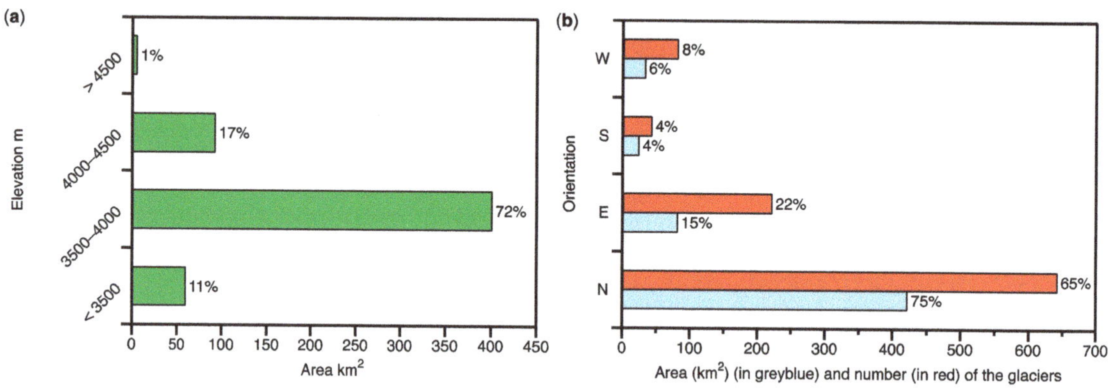

Figure 3. (**a**) The distribution of glacier areas in elevation, and (**b**) that of glacier areas and numbers in orientation in the Ebi Lake basin investigated in 2019.

3.2. Changes of the Glaciers Relative to the 1st and 2nd CGIs

Table 2 present the glaciers in area and number investigated by the first and second CGIs (1964 and 2009, respectively) and this work up to 2019. The glaciers in the Ebi Lake Basin have changed a lot since the first and second CGIs either in area or in number. Almost all measures had decreased from the first CGI investigation to this work in 2019, except that the glacier number increased from 281 by the second CGI to 285 by this work. The sole exception was probably because of some separated glacier branches resulting from the shrinking area.

Table 2. Glaciers in area and number in the six drainage sub-basins as per three inventories.

Drainage Basin		1st CGI (by 1964)		2nd CGI (by 2009)		This Work (in 2019)	
Name	Code	Area (km^2)	Number	Area (km^2)	Number	Area (km^2)	Number
Kuitun River	5Y741	201.12	309	147.67	281	133.11	285
Sikeshu River	5Y742	336.25	364	259.66	342	249.54	337
Jinghe River	5Y743	96.20	129	74.81	118	69.70	118
Daheyanzi River	5Y744	4.17	34	0.38	7	0.15	5
Sayram Lake	5Y745	4.28	13	2.17	9	2.25	9
Bortala River	5Y746	181.04	255	113.75	243	105.02	234
Ebi Lake	5Y74	823.06	1104	598.44	1000	559.77	988

Note: The red underlined number indicates the increasing count.

In more detail, the number of the Ebi Lake Basin's glaciers decreased from 1104 to 998 (-2 a^{-1} or -2% a^{-1}), and the area decreased by 263 km^2 (-4.79 km^2 a^{-1} or -0.58% a^{-1}) during the period 1964–2019. The overall rate of glacier area decreasing during 1964–2019 was -4.79 km^2 a^{-1}, and the rate during 2009–2019 (-3.87 km^2 a^{-1}) was slower than that during 1964–2009 (-4.99 km^2 a^{-1}) (Figure 4). The shrinking rate of the glaciers in the Ebi Lake Basin was slowed by 22% or 1.12 km^2 a^{-1} in the last decade, comparing with the period 1964–2009. Figure 5 shows some typical glacier boundaries in the basin defined by the second CGI and 2019 investigations, respectively. Compared with the boundary by 2009, the glacial area showed a general shrinkage in 2019.

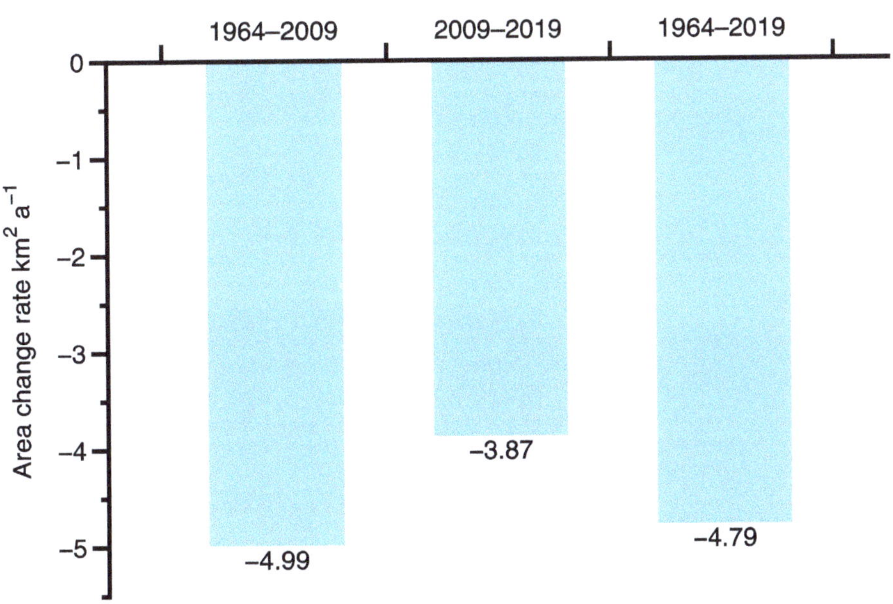

Figure 4. The area changing rates of the glaciers in the Ebi Lake basin during 1964–2009, 2009–2019, and 1964–2019, respectively.

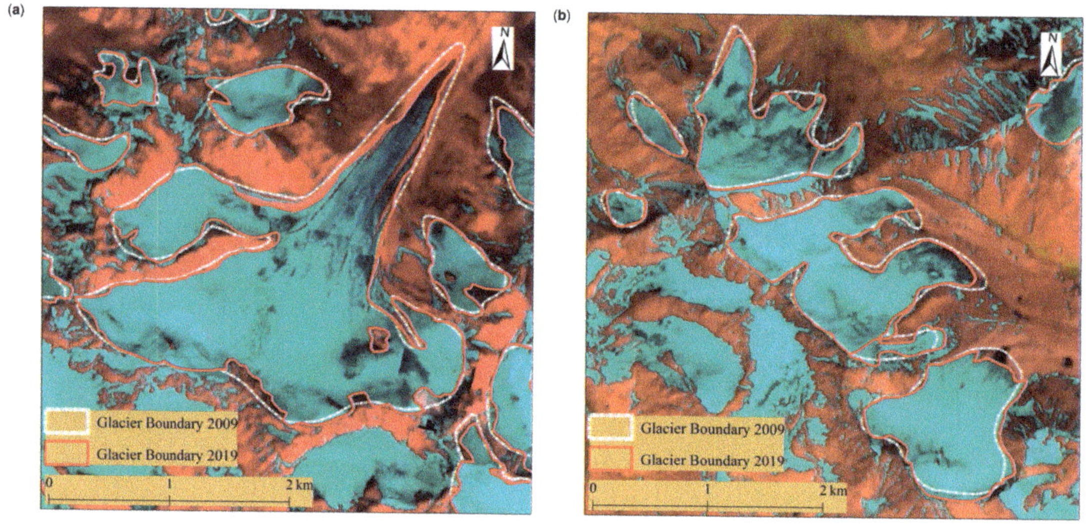

Figure 5. The land cover images with two typical ice bodies of the Ebi Lake Basin, defined by (**a**) the second CGI and (**b**) 2019 investigations (retrieved from the Sentinel-2 satellite with the Google Earth Engine in 2019), where the greenish colors indicate the ice.

4. Discussion

4.1. Spatial and Volume Variations of Glaciers in the Ebi Lake Basin since the 1st and 2nd CGIs

To understand the spatial variations of the Ebi Lake Basin's glaciers, we show their varying rates in a geographic map (Figure 6). The area and number of glaciers in the individual sub-basin of the Ebi Lake basin showed a decreasing trend from 1964 to 2019. The Bortala River Basin (5Y746) had the largest decrease in glacier area (111.80 km^2) with a rate of change of -2.03 km^2 a^{-1}, followed by the Sikeshu River Basin (5Y742) (-101.90 km^2, or -1.85 km^2 a^{-1}) and the Quitun River Basin (5Y741) (-94.62 km^2, -1.72 km^2 a^{-1}); the largest number of glaciers disappeared in the Sikeshu River Basin (5Y742) and the Daheyanzi River Basin (5Y744), losing 27 and 28 glaciers, respectively. From the analysis of the relative rate of glacier area change, the glaciers in the Daheyanzi River Basin (5Y744) were retreating fastest (-1.81 % a^{-1}), followed by the Selimu Lake Basin (5Y745) with a retreating rate of -1.23 % a^{-1}, and the glacier resources in the two sub-basins are on the verge of extinction. During 1964–2009, the vanished glaciers spread from west to east, while they concentrated in the west during 2009–2019.

This work used various methods to calculate the ice storages derived from different investigations and their variations in the basin (Table 3). The results show that the basin glacier ice reservation decreased by 97.84–153.22 km^3 between 1964 and 2019, the equivalent water equivalent was 88.06–137.90 km^3, taking 0.9 g cm^{-3} as ice density. The ice reservation decreased by 39.18%–41.23%, or -1.78~-2.79 km^3 a^{-1}. Compared with the glacier area reduction rate, the reduction rate of ice is greater, indicating that ice storage is more sensitive to regional warming. The glaciers in the basin are experiencing dramatic retreating and thinning.

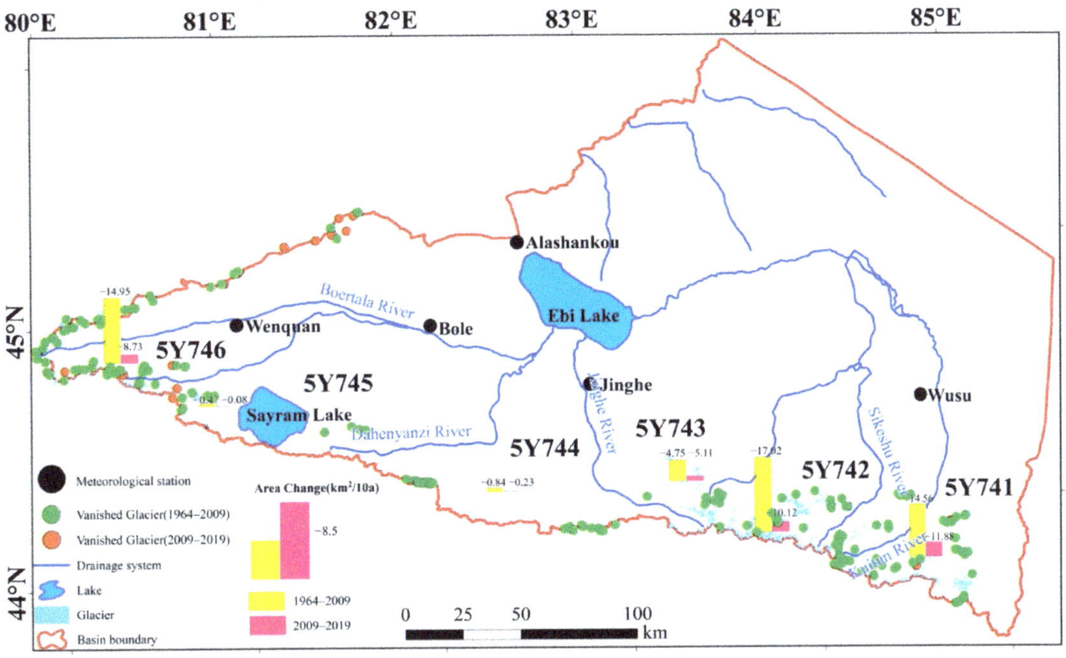

Figure 6. Area changes of glaciers at drainage scale in the Ebinur Lake Basin.

Table 3. The glacier volumes in the Ebi Lake Basin estimated with different empirical approaches.

Method	Ice Volume of Glaciers (km^3)			Glacier Volume Change (2019s—FCGI)			Source
	FCGI $A = 823.06$ km^2	SCGI $A = 598.44$ km^2	2019s $A = 559.77$ km^2	km^3	km^3 a^{-1}	% a^{-1}	
$V = 0.0285 * A^{1.357}$	257.68	167.21	152.72	−104.96	−1.91	−0.74	[42]
$V = 0.0298 * A^{1.379}$	312.32	201.25	183.54	−128.78	−2.34	−0.75	[43]
$V = 0.0365 * A^{1.375}$	372.40	240.29	219.18	−153.22	−2.79	−0.75	[44]
$V = 0.037 * A^{1.314}$	250.66	164.89	151.04	−99.62	−1.81	−0.72	[38]
$V = 0.04 * A^{1.35}$	345.06	224.41	205.06	−140.00	−2.55	−0.74	[45]
$V = 0.0433 * A^{1.29}$	249.69	165.52	151.85	−97.84	−1.78	−0.71	[46]
Average	297.97	193.93	177.23	−120.74	−2.20	−0.74	This work

4.2. Retreat of Glaciers in the Ebi Lake Basin Compared with the Greater Tianshan Region

In recent years, high-resolution satellite image data have been widely used in the dynamic monitoring of glaciers, and it has become possible to study glacier changes in large regions. To study the change of the glaciers in the Ebi Lake Basin in the greater Tianshan, this study compares the glacier changes here with those in the other typical mountains and watersheds of the Tianshan (Table 4). It is found that the trend of glaciers change in the Ebi Lake is consistent with the trends in other regions, that is, smaller-size glaciers have less reduction. In general, the glacial retreat rates in the greater Tianshan region were varying. The glaciers in the Ebi Lake basin retreated at a rate of 0.58% a^{-1}, making the Ebi Lake Basin one of the fastest retreating areas in the Tianshan region. The glacier retreat rate in the western section of the Tianshan is larger, followed by the middle section and the smallest in the eastern section, and the glacier retreat rate on the northern slope is higher than on the southern slope. Regional climate variation (temperature and precipitation) and the individual glaciers' sizes are the main influencing factors on glacier area in the Ebi Lake Basin. The number of glaciers in the region is dominated by those with an area of smaller than 0.5 km^2 (730 glaciers, or 73.9% of the total number of glaciers).

Table 4. Comparison of the area change of the glaciers in the greater Tianshan range.

Study Area	Study Period	Area Change (km^2)	Rate of Change in Area	Data	Source
Tianshan	1960–2010	—	$-0.22\%\ a^{-1}$ (-11.50%)	—	[24]
Tailan River Basin	1972–2011	−50.06	$-0.22\%\ a^{-1}$ (-11.50%)	Topographic map, ETM+	[47]
Gaizi River Basin	1960–1999	−188.1	$-0.26\%\ a^{-1}$ (-10.00%)	Topographic map, TM, ETM+	[21]
Harlik mountains	1959–2001	−13.4	$-0.27\%\ a^{-1}$ (-11.40%)	Topographic map, TM, ETM+	[48]
Kaidu River Basin	1963–2000	−38.5	$-0.31\%\ a^{-1}$ (-11.60%)	Topographic map, TM, ETM+	[21]
Middle Tianshan Mountains	1963–2000	−7	$-0.35\%\ a^{-1}$ (-13.00%)	MSS, SPOT, ETM+	[49]
Jinghe River Basin	1964–2004	−13.9	$-0.38\%\ a^{-1}$ (-15.20%)	Topographic map, ASTER	[50]
Urumqi River Basin	1962–1992	−6.65	$-0.45\%\ a^{-1}$ (-13.80%)	Topographic map, Aerial photograph	[51]
Kukesu River Basin	1963–2004	−50.06	$-0.46\%\ a^{-1}$ (-18.99%)	Topographic map, ASTER	[52]
Bogda peak area	1972–2005	−31.2	$-0.49\%\ a^{-1}$ (-21.60%)	Topographic map, ASTER, SPOT	[26]
Manas River Basin	1972–2013	−159.02	$-0.60\%\ a^{-1}$ (-24.61%)	TM, ETM+	[53]
Alatau Mountains	1990–2011	−137.77	$-0.92\%\ a^{-1}$ (-20.24%)	TM, ETM+	[54]
Ebinur Lake Basin	1964–2019	−263.29	$-0.58\%\ a^{-1}$ (-31.90%)	Topographic map, Sentinel2 MSI	This study

4.3. Factors and Their Roles in the Change of Glaciers in the Ebi Lake Basin

Precipitation and temperature are the main factors affecting glacier development. The interannual changes of these two factors together determine the nature, development and evolution of glaciers. Temperature determines ablation, and precipitation affects accumulation. Therefore, it is prerequisite to study the climate in the watershed to project the glaciers' evolution. As per the geography of the study area, this work selected five meteorological stations, i.e., Alashankou (51232), Bole (51238), Wenquan (51330), Jinghe (51334), and Wusu (51346), as the climatic reference for the study area (Figure 1).

The yearly average temperature and precipitation, and their anomalies at the five meteorological stations in winter and summer from 1964–2017, are shown in Figure 7. The temperature averaged from the five stations shows an increasing trend of 0.37 °C $(10a)^{-1}$ (0.001 confidence level) from 1964 through 2017, 2.6 times as high as the global mean rate, i.e., 0.14 °C $(10a)^{-1}$ from 1951 to 2012 [4]. The temperatures of the watershed in summer and winter both show increasing trends (0.51 °C $(10a)^{-1}$ in winter and 0.21 °C $(10a)^{-1}$ in summer, respectively), and the temperature increase in winter is 2.5 times faster than the summer. Similar to the temperature, precipitation records at all five weather stations show an average increasing trend (13.61 mm $(10a)^{-1}$ at 0.001 confidence level); both summer and winter precipitation in the watershed show increasing, greater in summer (4.78 mm $(10a)^{-1}$) than in winter (2.63 mm $(10a)^{-1}$). The increasing temperature and precipitation trends in the Ebi Lake Basin are consistent with the climate shift theory from warm dry to warm wet in northwest China [55]. The mass loss of global mountain glaciers induced by a one degree increase in temperature needs 25–35% increase in precipitation to compensate for the loss [56]. In the context of this combined hydrothermal climate change in the Ebi Lake Basin, precipitation has increased, but temperature also increases to melt the glaciers, and the increased precipitation does not add enough mass to the glaciers and compensate for their mass loss. With increasing mass deficiency, glacier mass loss accelerates, and mass balance lines rise, leading to the widespread glacier retreat.

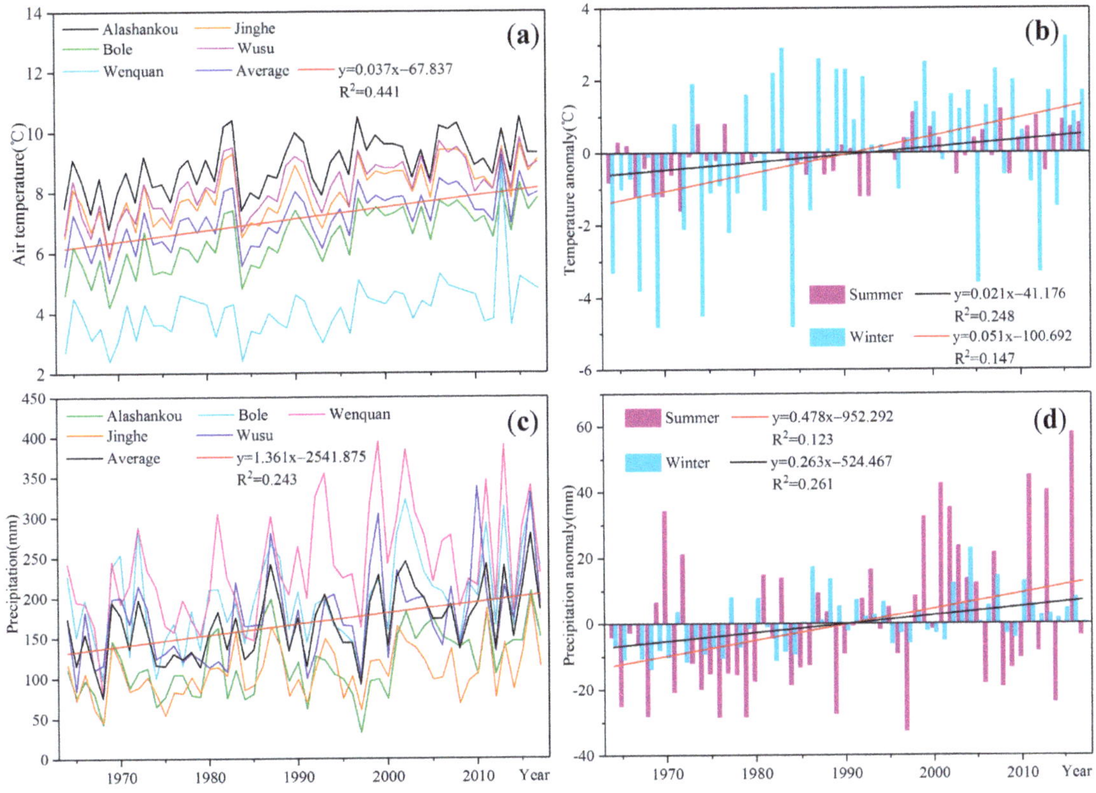

Figure 7. The interannual variations (**a**,**c**) and winter–summer anomalies (**b**,**d**) of temperature and precipitation in the Ebi Lake basin.

5. Conclusions

The number of glaciers in the Ebi Lake Basin in the 2019 ablation season was 988 with the total area of 559.77 km^2. The glaciers in the basin are mainly characterized by individuals with the area between 0.1 and 10.0 km^2 summed 509.05 km^2, ~91% of the total glacial area in the basin. There are 730 glaciers smaller than 0.5 km^2, accounting for 74% of the total number of glaciers in the basin. The glaciers are concentrated between 3500 m and 4000 m in elevation, with a total area of 512 km^2, ~91% of the total.

During 1964–2019, the number and area of the Ebi Lake Basin's glaciers had decreased by 116 (10.5%) and 263 km^2 (32%) at the rate of -4.79 km^2 a^{-1} or -0.58% a^{-1}, respectively. Glaciers with an area between 2.0 and 5.0 km^2 had the largest reduction of -82.60 km^2 (~41%) with the rate of -1.5 km^2 a^{-1} or -0.74% a^{-1}. The glaciers in the basin were all in retreat from 1964 through 2019, and the glaciers lower than 3200 m disappeared in 2019, and those between 3500 m and 4000 m dominated the total glacial area. Those north- and northeast-oriented glaciers in the basin had the largest area and number. The glaciers in each sub-basin of the Ebi Lake basin showed a decreasing trend from 1964 to 2019; noticeably, these glaciers in the recent decade (2009–2019) showed a slower retreating trend, compared with the investigations by the first and second CGIs. The Bortala River Basin (5Y746) has the largest decrease in glacier area (111.80 km^2) with a rate of -2.03 km^2 a^{-1}, followed by the Sikeshu River Basin (5Y742) (-101.90 km^2 or -1.85 km^2 a^{-1}) and the Quitun River Basin (5Y741) (-94.62 km^2 or -1.72 km^2 a^{-1}). The ice storage in the basin during the last 55 years had decreased by 97.84–153.22 km^3. The equivalent water equivalent was 88.06–137.90 km^3 with −1.78~−2.79 km^3 a^{-1} or −0.71~−0.75 % a^{-1}.

The temperature in the basin had increased by 0.36 °C (10a)$^{-1}$ during 1964–2017, much faster than the global mean, and the annual precipitation in the basin also showed an increasing trend of 12.06 mm (10a)$^{-1}$. The temperature and precipitation trends in the basin are consistent with the climate shift from warm-dry to warm-wet in northwest China. Although the precipitation in the basin has increased, the increase in precipitation was not sufficient to compensate for the mass loss of the glaciers by the increased temperatures, leading widespread retreating of the glaciers in the basin. This work may be the first report on the status of the glaciers in the Ebi Lake Basin and revealed their most recent change since the second China Glacier Investigation. The main points can be helpful for policymakers to take measures to mitigate the impact of climate change on the region.

Author Contributions: Conceptualization, L.W. and J.M.; methodology, L.W.; software, C.B. and J.M.; validation, L.W., C.B., and J.M.; formal analysis, C.B.; investigation, L.W.; resources, J.M.; data curation, L.W. and C.B.; writing—original draft preparation, L.W. and C.B.; writing—review and editing, J.M.; visualization, C.B.; supervision, J.M.; project administration, L.W.; funding acquisition, L.W. All authors have read and agreed to the published version of the manuscript.

Funding: This research is funded by the National Key Research and Development Program of China (grant number 2020YFF0304400), the Second Tibetan Plateau Scientific Expedition and Research (STEP) program (grant number 2019QZKK0201), the State Key Laboratory of Cryospheric Sciences (grant number SKLCS-ZZ-2020), and the Key Research Program of Frontier Sciences of Chinese Academy of Sciences (grant number QYZDB-SSW-SYS024).

Institutional Review Board Statement: Not applicable.

Informed Consent Statement: Not applicable.

Data Availability Statement: The data presented in this study are available on request from the corresponding author.

Acknowledgments: We thank the TPDC for providing the first and second CGI data; NASA, NIMA, and CIAT for providing the version 4.1 SRTM data; and the CMDS for providing the meteorology data. We also thank the two anonymous reviewers for their suggestions during the revising procedure.

Conflicts of Interest: The authors declare no conflict of interest.

References

1. WGII. Climate Change 2014: Impacts, Adaptation, and Vulnerability-AR5 Summary for Policymakers. *WGII AR5 Phase I Rep. Launch* **2014**, *1*, 31.
2. Pritchard, H.D. Asia's glaciers are a regionally important buffer against drought. *Nature* **2017**, *545*, 169–174. [CrossRef] [PubMed]
3. Pritchard, H.D. Asia's shrinking glaciers protect large populations from drought stress. *Nature* **2019**, *569*, 649–654. [CrossRef] [PubMed]
4. Intergovernmental Panel on Climate Change. *Climate Change 2013—The Physical Science Basis: Working Group I Contribution to the Fifth Assessment Report of the Intergovernmental Panel on Climate Change*; Cambridge University Press: Cambridge, UK, 2014. [CrossRef]
5. Immerzeel, W.W.; Lutz, A.F.; Andrade, M.; Bahl, A.; Biemans, H.; Bolch, T.; Hyde, S.; Brumby, S.; Davies, B.J.; Elmore, A.C.; et al. Importance and vulnerability of the world's water towers. *Nature* **2019**, *577*, 364–369. [CrossRef] [PubMed]
6. Zemp, M.; Huss, M.; Thibert, E.; Eckert, N.; McNabb, R.; Huber, J.; Barandun, M.; Machguth, H.; Nussbaumer, S.U.; Gartner-Roer, I.; et al. Global glacier mass changes and their contributions to sea-level rise from 1961 to 2016. *Nature* **2019**, *568*, 382–386. [CrossRef]
7. Kraaijenbrink, P.D.A.; Bierkens, M.F.P.; Lutz, A.F.; Immerzeel, W.W. Impact of a global temperature rise of 1.5 degrees Celsius on Asia's glaciers. *Nature* **2017**, *549*, 257–260. [CrossRef]
8. Liu, S.Y.; Yao, X.J.; Guo, W.Q.; Xu, J.L.; Shangguan, D.; Wei, J.; Bao, W.; Wu, L. The contemporary glaciers in China based on the Second Chinese Glacier Inventory. *Acta Geogr. Sin.* **2015**, *70*, 3–16. [CrossRef]
9. Shi, Y.; Liu, S. Pre-estimation for the response of China glaciers to global warming in the 21st century. *Chin. Sci. Bull* **2000**, *45*, 434–438. [CrossRef]
10. Sun, M.; Liu, S.; Yao, X.; Guo, W.; Xu, J. Glacier changes in the Qilian Mountains in the past half-century: Based on the revised First and Second Chinese Glacier Inventory. *J. Geogr. Sci.* **2018**, *28*, 206–220. [CrossRef]

11. Xing, W.; Li, Z.; Zhang, H.; Zhang, M.; Liang, P.; Mu, J. Spatial-temporal variation of glacier resources in Chinese Tianshan Mountains since 1959. *Acta Geogr. Sin.* **2017**, *72*, 1594–1605.
12. Yao, T.; Thompson, L.; Yang, W.; Yu, W.; Gao, Y.; Guo, X.; Yang, X.; Duan, K.; Zhao, H.; Xu, B. Different glacier status with atmospheric circulations in Tibetan Plateau and surroundings. *Nat. Clim. Chang.* **2012**, *2*, 663–667. [CrossRef]
13. Liu, Q.; Liu, S. Response of glacier mass balance to climate change in the Tianshan Mountains during the second half of the twentieth century. *Clim. Dyn.* **2015**, *46*, 303–316. [CrossRef]
14. Zhang, Y.; Enomoto, H.; Ohata, T.; Kitabata, H.; Kadota, T.; Hirabayashi, Y. Projections of glacier change in the Altai Mountains under twenty-first century climate scenarios. *Clim. Dyn.* **2016**, *47*, 2935–2953. [CrossRef]
15. Ma, L.; Wu, J.; Abuduwaili, J. Variation in aeolian environments recorded by the particle size distribution of lacustrine sediments in Ebinur Lake, northwest China. *SpringerPlus* **2016**, *5*, 481. [CrossRef] [PubMed]
16. Wu, J.; Yu, Z.; Wang, N. Possible solar forcing of 400-year wet–dry climate cycles in northwestern China. *Clim. Chang.* **2009**, *96*, 473–482. [CrossRef]
17. Tao, S.; Fang, J.; Ma, S.; Cai, Q.; Xiong, X.; Tian, D.; Zhao, X.; Fang, L.; Zhang, H.; Zhu, J.; et al. Changes in China's lakes: Climate and human impacts. *Natl. Sci. Rev.* **2019**, *7*, 132–140. [CrossRef]
18. NASA. Shrinking Lake Ebinur. Available online: https://earthobservatory.nasa.gov/images/84785/shrinking-lake-ebinur (accessed on 27 January 2021).
19. Liu, H.; Chen, Y.; Ye, Z.; Li, Y.; Zhang, Q. Recent Lake Area Changes in Central Asia. *Sci. Rep.* **2019**, *9*, 16277. [CrossRef]
20. Farinotti, D.; Longuevergne, L.; Moholdt, G.; Duethmann, D.; Mölg, T.; Bolch, T.; Vorogushyn, S.; Güntner, A. Substantial glacier mass loss in the Tien Shan over the past 50 years. *Nat. Geosci.* **2015**, *8*, 716–722. [CrossRef]
21. Liu, S.; Ding, Y.; Zhang, Y.; Shangguan, D.; Li, J.; Han, H.; Wang, J.; Xie, C. Impact of the glacial change on water resources in the Tarim River Basin. *Acta Geogr. Sin.* **2006**, *61*, 482–490.
22. Wang, P.; Li, Z.; Cao, M.; LI, H.-l. Variation of Qingbingtan glacier No. 72 in Mt. Tuomuer region during past 45 years. *Sci. Geogr. Sin.* **2010**, *30*, 962–967.
23. Wang, P.; Li, Z.; Li, H.; Wu, L.; Jin, S.; Zhou, P. Changes of ice-thickness and volume for representative glaciers in Tianshan Mountains in the past 50 years. *Acta Geogr. Sin.* **2012**, *67*, 929–940.
24. Wang, S.; Zhang, M.; Li, Z.; Wang, F.; Li, H.; Li, Y.; Huang, X. Response of glacier area variation to climate change in Chinese Tianshan Mountains in the past 50 years. *Acta Geogr. Sin.* **2011**, *66*, 38–46.
25. Xu, J.; Liu, S.; Guo, W.; Zhang, Z.; Wei, J.; Feng, T. Glacial area changes in the Ili River catchment (Northeastern Tian Shan) in Xinjiang, China, from the 1960s to 2009. *Adv. Meteorol.* **2015**, *2015*. [CrossRef]
26. Li, K.; Li, Z.; Gao, W.; Wang, L. Recent glacial retreat and its effect on water resources in eastern Xinjiang. *Chin. Sci. Bull* **2011**, *56*, 3596–3604. [CrossRef]
27. Wang, L.; Li, Z.; Wang, F.; Edwards, R. Glacier shrinkage in the Ebinur lake basin, Tien Shan, China, during the past 40 years. *J. Glaciol.* **2014**, *60*, 245–254. [CrossRef]
28. Zhang, H.; Li, Z.; Wang, P.; Huai, B. Variation of Haxilegen No.51 Glacier at the Headwater of Kuitun River in Tianshan Mountains and its response to climate change. *Arid Zone Res.* **2015**, *32*, 88–93.
29. Abuduwailil, J.; Zhaoyong, Z.; Fengqing, J. Evaluation of the pollution and human health risks posed by heavy metals in the atmospheric dust in Ebinur Basin in Northwest China. *Environ. Sci. Pollut. Res.* **2015**, *22*, 14018–14031. [CrossRef]
30. Zhang, A.; Ding, J.; Zhang, J.; Dong, Y. The variation characteristics of actual evapotranspiration in the Ebinur Lake oasis and its relationship with landscape pattern. *J. Glaciol. Geocryol.* **2020**, *42*, 1–9.
31. Gorelick, N.; Hancher, M.; Dixon, M.; Ilyushchenko, S.; Thau, D.; Moore, R. Google Earth Engine: Planetary-scale geospatial analysis for everyone. *Remote Sens. Environ.* **2017**, *202*, 18–27. [CrossRef]
32. Reuter, H.I.; Nelson, A.; Jarvis, A. An evaluation of void-filling interpolation methods for SRTM data. *Int. J. Geogr. Inf. Sci.* **2007**, *21*, 983–1008. [CrossRef]
33. Jarvis, A.; Reuter, H.; Nelson, A.; Guevara, E. Hole-Filled Seamless SRTM Data V4: International Centre for Tropical Agriculture. *Srtm. Csi. Cgiar. Org Accessed* **2008**, *31*.
34. Tsagkatakis, G.; Aidini, A.; Fotiadou, K.; Giannopoulos, M.; Pentari, A.; Tsakalides, P. Survey of Deep-Learning Approaches for Remote Sensing Observation Enhancement. *Sensors (Basel)* **2019**, *19*, 3929. [CrossRef] [PubMed]
35. Raup, B.; Racoviteanu, A.; Khalsa, S.J.S.; Helm, C.; Armstrong, R.; Arnaud, Y. The GLIMS geospatial glacier database: A new tool for studying glacier change. *Glob. Planet. Chang.* **2007**, *56*, 101–110. [CrossRef]
36. Environmental Systems Research Institute. *ArcGIS Desktop: Release 10*; Environmental Systems Research Institute: Redlands, CA, USA, 2011.
37. Gärtner-Roer, I.; Naegeli, K.; Huss, M.; Knecht, T.; Machguth, H.; Zemp, M. A database of worldwide glacier thickness observations. *Glob. Planet. Chang.* **2014**, *122*, 330–344. [CrossRef]
38. Moore, J.C.; Grinsted, A.; Zwinger, T.; Jevrejeva, S. Semiempirical and process-based global sea level projections. *Rev. Geophys.* **2013**, *51*, 484–522. [CrossRef]
39. Ye, Q.; Kang, S.; Chen, F.; Wang, J. Monitoring glacier variations on Geladandong mountain, central Tibetan Plateau, from 1969 to 2002 using remote-sensing and GIS technologies. *J. Glaciol.* **2006**, *52*, 537–545. [CrossRef]
40. Silverio, W.; Jaquet, J.-M. Glacial cover mapping (1987–1996) of the Cordillera Blanca (Peru) using satellite imagery. *Remote Sens. Environ.* **2005**, *95*, 342–350. [CrossRef]

41. Hall, D.K.; Bayr, K.J.; Schöner, W.; Bindschadler, R.A.; Chien, J.Y. Consideration of the errors inherent in mapping historical glacier positions in Austria from the ground and space (1893–2001). *Remote Sens. Environ.* **2003**, *86*, 566–577. [CrossRef]
42. Chen, J.; Ohmura, A. Estimation of Alpine glacier water resources and their change since the 1870s. *IAHS Publ.* **1990**, *193*, 127–135.
43. Macheret, Y.; Cherkasov, P.; Bobrova, L. Tolschina i ob'em lednikov Djungarskogo Alatau po danniy aeroradiozondirovaniya. *Mater. Glyatsiologicheskikh Issled. Khronika Obs.* **1988**, *62*, 59–71.
44. Radić, V.; Hock, R. Regional and global volumes of glaciers derived from statistical upscaling of glacier inventory data. *J. Geophys. Res. Earth Surf.* **2010**, *115*, F01010. [CrossRef]
45. Shiyin, L.; Wenxin, S.; Yongping, S.; Gang, L. Glacier changes since the Little Ice Age maximum in the western Qilian Shan, northwest China, and consequences of glacier runoff for water supply. *J. Glaciol.* **2003**, *49*, 117–124. [CrossRef]
46. Grinsted, A. An estimate of global glacier volume. *Cryosphere* **2013**, *7*, 141. [CrossRef]
47. Huai, B.; Li, Z.; Sun, M.; Zhou, P. Glaciers change in the Tailan River watershed in the last 40 years. *Sci. Geogr. Sin.* **2014**, *34*, 229–236.
48. Wang, Y.; Hou, S.; Lu, A.; LIU, Y.-p. Response of glacier variations in the eastern Tianshan Mountains to climate change, during the last 40 years. *Arid Land Geogr.* **2008**, *31*, 813–821.
49. Li, B.; Zhu, A.-X.; Zhang, Y.; Pei, T.; Qin, C.; Zhou, C. Glacier change over the past four decades in the middle Chinese Tien Shan. *J. Glaciol.* **2006**, *52*, 425–432. [CrossRef]
50. Wang, L.; Li, Z.; Wang, F.; Li, H.; Wang, P. Glacier changes from 1964 to 2004 in the Jinghe River basin, Tien Shan. *Cold Reg. Sci. Technol.* **2014**, *102*, 78–83. [CrossRef]
51. Chen, J.; Liu, C.; Jin, M. Application of the repeated aerial photogrammetry to monitoring glacier variation in the drainage area of the Urumqi River. *J. Glaciolgy Geocryol.* **1996**, *18*, 331–336.
52. Gao, W.; Li, Z.; Li, K.; Zhang, M.-j. Glacier variation in the Kukesu River Basin during 1963–2004 based on remote sensing data and GIS techniques. *Arid Land Geogr.* **2011**, *34*, 252–261.
53. Xu, C.; Wang, F.; Li, Z.; Wang, L.; Wang, P. Glacier variation in the Manas River Basin during the period from 1972 to 2013. *Arid Zone Res.* **2016**, *33*, 628–635.
54. He, Y.; Yang, T.-b.; Du, J.; TIAN, H.-z. Glacier variation of past 22 years in Alatau Mountains of Central Asia based on GIS and RS. *Res. Soil Water Conserv.* **2013**, *20*, 130–134.
55. Shi, Y.S.; Shen, Y.P.; Li, D.L.; Zhang, G.W.; Ding, Y.J.; Hu, R.J.; Kang, E.S. Discussion on the present climate change from warm-dry to warm-wet in northwest China. *Quaternay Sci.* **2003**, *23*, 152–164.
56. Oerlemans, J. Extracting a Climate Signal from 169 Glacier Records. *Science* **2005**, *308*, 675. [CrossRef] [PubMed]

MDPI
St. Alban-Anlage 66
4052 Basel
Switzerland
Tel. +41 61 683 77 34
Fax +41 61 302 89 18
www.mdpi.com

Remote Sensing Editorial Office
E-mail: remotesensing@mdpi.com
www.mdpi.com/journal/remotesensing

www.ingramcontent.com/pod-product-compliance
Lightning Source LLC
Chambersburg PA
CBHW061811210926
51850CB00006B/671
9783036566283